Logic, Algebra, and Computation

NATO ASI Series

Advanced Science Institutes Series

A series presenting the results of activities sponsored by the NATO Science Committee, which aims at the dissemination of advanced scientific and technological knowledge, with a view to strengthening links between scientific communities.

The Series is published by an international board of publishers in conjunction with the NATO Scientific Affairs Division

A Life Sciences B Physics	Plenum Publishing Corporation London and New York
C Mathematical and Physical Sciences D Behavioural and Social Sciences E Applied Sciences	Kluwer Academic Publishers Dordrecht, Boston and London
F Computer and Systems Sciences G Ecological Sciences H Cell Biology I Global Environmental Change	Springer-Verlag Berlin Heidelberg New York London Paris Tokyo Hong Kong Barcelona Budapest

NATO-PCO DATABASE

The electronic index to the NATO ASI Series provides full bibliographical references (with keywords and/or abstracts) to more than 30000 contributions from international scientists published in all sections of the NATO ASI Series. Access to the NATO-PCO DATABASE is possible in two ways:

- via online FILE 128 (NATO-PCO DATABASE) hosted by ESRIN, Via Galileo Galilei, I-00044 Frascati, Italy.
- via **CD-ROM** "NATO-PCO DATABASE" with user-friendly retrieval software in English, French and German (© WTV GmbH and DATAWARE Technologies Inc. 1989).

The CD-ROM can be ordered through any member of the Board of Publishers or through NATO-PCO, Overijse, Belgium.

Series F: Computer and Systems Sciences Vol. 79

Logic, Algebra, and Computation

International Summer School directed by
F. L. Bauer, W. Brauer, G. Huet, J. A. Robinson, H. Schwichtenberg

Edited by

Friedrich L. Bauer

Professor Emeritus
Technische Universität München
Postfach 202 420, W-8000 München, Federal Republic of Germany

Springer-Verlag
Berlin Heidelberg New York London Paris Tokyo
Hong Kong Barcelona Budapest
Published in cooperation with NATO Scientific Affairs Division

Proceedings of the NATO Advanced Study Institute on Logic, Algebra, and Computation held at Marktoberdorf, Federal Republic of Germany, July 25–August 6, 1989.

CR Subject Classification (1991): F.3–4, D.1.3, D.1.6, D.2.1, F.1, I.2.3

Library of Congress Cataloging-in-Publication Data
Logic, algebra, and computation : international summer school directed by F. L. Bauer ... [et al.]/edited by Friedrich L. Bauer. p.cm.—(NATO ASI series. Series F., Computer and system sciences ; no. 79) "Published in cooperation with NATO Scientific Affairs Division." "Proceedings of the NATO Advanced Study Institute on Logic, Algebra, and Computation held at Marktoberdorf, Federal Republic of Germany, July 25–August 6, 1989"—T. p. verso.

ISBN: 978-3-642-76801-9 e-ISBN-13: 978-3-642-76799-9
DOI: 10.1007/978-3-642-76799-9

1. Logic programming—Congresses. 2. Algebra—Congresses. 3. Computational complexity—Congresses. I. Bauer, Friedrich Ludwig, 1924– . II. NATO Advanced Study Institute on Logic, Algebra, and Computation (1989 : Marktoberdorf, Germany) III. Series. QA76.63.L618 1991 005.1—dc20 91-26502

Softcover reprint of the hardcover 1st edition 1991

Typesetting: Camera ready by author

45/3140-543210 – Printed on acid-free-paper

Preface

The Marktoberdorf Summer Schools on Informatics were started in 1970, with the intention to convene every second or third year a group of top researchers in computing, devoted to preach their most recent results to an elite of advanced students — young and most promising people — and prepared to stand their questions, criticism and suggestions. The themes of these Advanced Study Institutes under the sponsorship of the NATO Scientific Affairs Division varied slightly over the years, oscillating more or less around Programming Methodology, as the following list shows:

1970	Data Structures and Computer Systems
1971	Program Structures and Fundamental Concepts of Programming
1973	Structured Programming and Programmed Structures
1975	Language Hierarchies and Interfaces
1978	Program Construction
1981	Theoretical Foundations of Programming Methodology
1984	Control Flow and Data Flow: Concepts of Distributed Programming
1986	Logic of Programming and Calculi of Discrete Design
1988	Constructive Methods in Computing Science
1989	Logic, Algebra, and Computation

Logic, Algebra, and Computation is the theme of the summer school to which this volume is devoted. It is the tenth in succession, but it is also the first in a new series (the "blue" series) that is intended to alternate in future with the traditional (the "red" series) arrangement; in fact the tenth summer school in the "red" series with the title "Programming and Mathematical Method", held in 1990, was the subject of celebrating both its serial number and the twenty years of Marktoberdorf Summer Schools altogether.

The reason for the splitting into two series was the following: With growing reputation of the school, there was increasing demand for participation. For organizational as well as for pedagogic reasons, the number of participants was always limited to roughly 90, and high rejection rates, although unadvisable, happened to be unavoidable. The Summer School steering group, that evolved over the years, felt that repeating the same summer school every other year would take away a lot of the spontaneity that made the school so famous, and organising it every year with a completely new program was too much of a burden for the lecturers. On the other hand, the themes that were focused upon in the summer schools were becoming wider and wider.

Thus, it seemed to be advisable to build up a new series with a new steering group, dealing with themes that the other group most recently did not cover centrally, but allowing overlapping of themes and to some extent even of lecturers. The first summer school under the new regime thus obtained an orientation somewhat closer to logic. In the meantime, preparations for the 1991 school in the "blue" series are well under way. With the title "Logic and Algebra of Specification", it will stay within this line which has turned out to be both useful and attractive, and still it will keep close philosophical ties to the "red" series. Applied logic is programming, programming is applied logic — this is the message the present volume can be reduced to. Computation is a common concern of logic and mathematics, as Alan Turing and John von Neumann demonstrated half a century ago.

It is my pleasure to state that the 1989 Marktoberdorf Summer School, although being the first one in the "blue" series, was already a highlight in the scientific presentation and in the exchange of ideas both by discussions and by numerous gatherings in smaller groups and in private conversation. The ambience for such successful days was provided by the experienced scientific staff from Munich and by the local Marktoberdorf crew. My thanks go to them, in particular to Dr. Werner Meixner who was in charge of the organization of the summer school and who helped me in the preparation of this volume.

Munich, March 1991 F. L. Bauer

Table of Contents

Theory of Computation and Specification over Abstract Data Types, and its Applications

J V Tucker

Department of Mathematics and Computer Science
University College of Swansea
Singleton Park
Swansea, SA2 8PP
Wales

Abstract

The theory of computable functions on abstract data types is outlined. The problems of extending the theory to establish the scope and limits of the specification and verification of functions are described. Applications of this general mathematical theory to hardware design, program verification and logic programming are discussed.

Keywords: abstract data types; many sorted algebras; computable functions; specifiable functions; Church-Turing Thesis; deterministic parallel computation; synchronous concurrent algorithms; theory of hardware.

INTRODUCTION

In these lecture notes I will sketch a theory of functions and relations that are computable, specifiable, and verifiable, uniformly over a class of many sorted algebras. Such a class K of algebras, closed under isomorphism, is a mathematical model of the semantics an abstract data type. Because algebra provides a general theory of data, the theory of computation, specification and verification for functions and relations over K is general, and has applications

NATO ASI Series, Vol. F 79
Logic, Algebra, and Computation
Edited by F. L. Bauer

in many areas of computer science. Among the application areas discussed in these notes are the following:

- *synchronous concurrent algorithms and architectures, and their use in hardware design;*
- *program specification and verification by Floyd-Hoare logics;*
- *logic programming modules and abstract data types.*

My research on this theory, and on these apparently disparate applications, is part of a programme that aims at the precise formulation and analysis of the many interesting notions of computability, specifiability, and verifiability that may be found in different areas of computer science and mathematics. The material presented in these lecture notes is based mainly on the results of longstanding collaborative work with J I Zucker (McMaster University) and B C Thompson (University College of Swansea).

Origins

The mathematical theory presented in these notes is a generalisation of the "classical" computability theory for functions and relations on the natural numbers. It concerns models of computation, and methods of specification and verification, for functions and relations on *any* abstract algebra, or class of algebras. This theory of computability for algebras originates with work by E Engeler in 1965. Subsequent well-publicised work in mathematical logic by H Friedman and Y N Moschovakis, and in computer science by, for example, D Luckham, D Park, M Paterson, R Constable, D Gries, Z Manna, A Chandra, and A J Kfoury has also been significant: for further information see the surveys in Greibach [75], Shepherdson [85] and Tucker and Zucker [88]. There are many interesting and useful models of computation for the definition of the class of computable functions on an algebra or class of algebras; the models of computation used in these lectures are those developed in Tucker and Zucker [88].

Summary

In Section 1 some important concepts concerning abstract data types will be explained, including the rôle of classes of algebras, and of constructions with algebras.

Section 2 describes the model of computation for functions on algebras and classes of algebras defined in Tucker and Zucker [88]. It includes a *Generalised Church-Turing Thesis* for *deterministic parallel computation* on abstract data types. The model of computation is that of *recursively defined functions on algebras and classes of algebras.* An important point is the distinction between *primitive recursion* and *course-of-values recursion:* these constructs are equivalent on the natural numbers, but are not equivalent on abstract algebras in general. This simple model of computation will be applied to hardware design in Section 4. It also allows us to discuss an important duality that exists between the methods of defining the operations of algebras by recursion and by equational specification.

Section 3 is concerned with the specification and verification of the computable functions on an algebra or class of algebras. It begins with the distinction between the specification of a task, by means of a family

$$R = \{R_A : A \in K\}$$

of relations R_A on each algebra A in a class K of algebras, and the specification of the algorithms and programs that accomplish the task, by means of a family

$$f = \{f_A : A \in K\}$$

of functions f_A on each A in a class K. The essential point is that a program α on K *correctly implements a task* specified by a family R of relations on K if it correctly computes a family f of *selection functions* for R on K, i.e. a family f of functions f_A such that for each $A \in K$

$$\forall x[\exists y R_A(x,y) \Rightarrow f_A(x)\downarrow \text{ and } R_A(x, f_A(x))].$$

Thus, in this approach, the study of the specification and verification of programs becomes the study of relations, their selection functions, and the programs that compute the selection functions.

The relations of the form

$$R_A(x,y) \text{ and } \exists y R_A(x,y)$$

for $A \in K$ characterise the computational tasks, and are discussed in terms of the ideas of

effective testability of specifications;
definability of specifications in a logical language; and
nondeterministic computation.

Important classes of sets are the class of *semicomputable sets*, which generalise recursively enumerable sets of natural numbers, and the class of *projections of semicomputable sets*, which are sets of the form

$$\exists y R(x,y)$$

where $R(x,y)$ is *semicomputable.* A *Church-Turing thesis for specification* on abstract data types is formulated in terms of the selection functions for projections of semicomputable sets. Applications of the specification theory outlined earlier can then made to *program specification and verification by means of Floyd-Hoare logics*, and to *logic programming with abstract data types*. The verification of computable selection functions against their specifications is also discussed.

The work in Sections 2 and 3, on the scope and limits of methods for computation, specification, and verification on abstract data types, is in collaboration with J I Zucker and is a continuation of work on subjects that we met in the preparation of our book: see Tucker and Zucker [88,89] and Tucker, Wainer and Zucker [90]. For example, the model of computation in Section 2 was developed to prove that important sets of program states, such as the *weakest preconditions* and *strongest post conditions* of programs and conditions, were expressible in a weak second order many sorted logical language. At the heart of this exercise is a theorem that characterises the projectively semicomputable sets (see 3.3 and 3.4).

The main application of the theory of computable functions in these notes is explained in Section 4. The objective of the application is to provide a general mathematical theory of computation for the design and analysis of hardware.

A *synchronous concurrent algorithm* (sca) is a network of processors and channels that compute and communicate in parallel, and are synchronised by a global clock. Such algorithms compute on infinite streams of data, and are deterministic because of the clock. Most hardware is constructed from components that are examples of scas. Specific examples of families of scas include: *systolic algorithms; neural nets; cellular automata;* and *coupled map lattice dynamical systems.*

In Section 4, synchronous concurrent computation is formalised using *the course-of-values recursive functions over classes of stream algebras.* The study of this and other mathematical models of scas, and their application to hardware design, is a substantial task that involves research on mathematical theories, case studies in hardware design, practically oriented formal methods, and software tools. The subject of synchronous concurrent algorithms and architectures has been motivated and developed independently of the theory of computation on abstract data types. Indeed the subject must be sufficiently specialised to retain its independence for it to be useful in the world's work. However, the discussion here will make the point that important parts of the mathematical theory of scas can be obtained from an application of the theory of computation on abstract data types.

The theory of scas originates in my work with B C Thompson on the important special case of an sca with unit time delays throughout the network; this case involves the use of primitive recursive functions over stream algebras and has been studied in depth. Significant contributions to the mathematical theory of scas and to the development of case studies have been made by: K Meinke (Swansea), N A Harman (Swansea), S M Eker (Royal Holloway and Bedford New College, London) and K Hobley (Oxford). The application of the ideas to the construction of software for simulation has begun in work by A R Martin (IBM, Austin); and to verification, using existing theorem provers, in work with *Nuprl* by J Derrick and G Lajos (Leeds), and with *OBJ3* by V Stavridou and S M Eker (Royal Holloway and Bedford New College, London). The application of the theory to VLSI design has been considered by A Williams (Manchester). A study of the connection between scas and discrete space, discrete time dynamical systems is in progress in my work with B C Thompson and A V Holden (Leeds). Some of these contributions will be surveyed in Section 4.

Prerequisities

The objective of these notes is to introduce the reader to the theory of computation on abstract data types, and to encourage him or her to study it and its applications in detail. For this the reader should have some knowledge of universal algebra, and computability theory on the natural numbers. Interest in applications, particularly those to hardware, is also necessary. Appropriate references are made in the notes; however, it has not been possible for me to provide a survey of all relevant literature on this occasion.

Acknowledgements
I would like to thank the Directors of the Marktoberdorf Summer School on *Logic, algebra and computation, 1989*, for the opportunity of delivering lectures on my work. The idea of making a survey, or rather a progress report, of this kind originated in an invitation to address the meeting *J W de Bakker: 25 Jaar semantiek* in Amsterdam, April 1989.

I am indebited to J I Zucker and B C Thompson for many years of stimulating and enjoyable work on the collaborative projects described here. In addition, I would like to thank P Abdulla, J A Bergstra, J Derrick, S Eker, K Hobley, A V Holden, G Lajos, K Meinke, T E Rihll, H Simmons, V Stavridou, V Stoltenberg-Hansen, B C Thompson, S S Wainer, and J I Zucker for valuable conversations in the course of preparing these lecture notes.

Parts of this work was supported by SERC Research Grant GR/F/59070.

1. ALGEBRAS, MODULES AND ABSTRACT DATA TYPES

We summarise some basic ideas about algebras and abstract data types that we will need later. For further basic material on algebra see Ehrig and Mahr [85] and Meinke and Tucker [90]; for advanced material see Malcev [73].

1.1 Modules and algebras

In the theory of abstract data types, computation is characterised by a programming construct called a *module*, the semantics of which is a class K of *many sorted algebraic structures, algebras* or *models*.

A *many sorted algebraic structure* A consists of a family of non-empty sets of data, called *carriers*, and families of operations and relations on the data. A *signature* for a many sorted algebra is a set Σ of notations for (i) the names of the carriers, called *sorts*, and (ii) the names and types of the functions and relations.

Terms and closed terms are defined in the usual way. A signature is *non-void* if there is a closed term of every sort. A Σ algebra with non-void signature must have non-empty carriers and, in particular, *some* element in each carrier can be named by a closed term. We will assume that signatures are non-void. A Σ algebra is called *minimal* if *each* element in each carrier can be named by a closed term.

In algebra, two Σ algebras are identified if, and only if, they are *isomorphic*.

A many sorted algebraic structure A, possibly satisfying some further properties, can be used to model semantically a *concrete implementation* of a module. A class K of structures, again possibly satisfying some further properties, can be used to model semantically the module more abstractly as a *class of implementations*. Two implementations are identified if, and only if, they are isomorphic. These ideas generalise to any data the following ideas about the natural numbers published by R Dedekind in 1888.

Consider the abstract data type of the natural numbers. We imagine an algebra N with infinite carrier $\{0,1,2,...\}$, constant 0, and operations of successor $succ(n)=n+1$, addition $add(n,m)=n+m$ and multiplication $mult(n,m)=n.m$. Each specific representation of the natural numbers, in binary or decimal notation systems for example, forms an algebra that models semantically a concrete implementation of the natural numbers; and the class of all isomorphic copies of notation systems with the same signature models semantically the abstract data type of the natural numbers.

One common definition of an abstract data type is that it is an isomorphism type i.e. a class of all isomorphic copies of an algebra: see Goguen, Thatcher, Wagner and Wright [78]. However we use the following:

1.1.1 Abstract data type The semantics of an *abstract data type* is a class K of many sorted algebras, of the same signature, that is closed under isomorphism i.e. if $A \in K$ and $A \cong B$ then $B \in K$.

Among the properties of algebras that are used to characterise meaningful or useful implementations, and classes of implementations, are: *minimality, initiality, and computability* (see 1.3).

1.2 Modularisation

Consider the general ideas that all programs are modules and that programming is the creation of new modules from old modules:

A programmer creates a program module P from a component module C in order to obtain a task module T for a user.

Consider a programming task in the simple case when the semantics of modules are defined by particular implementation algebras.

Suppose we want to compute a function F and a relation R on a set A using operations f_i and relations r_j for $i \in I$ and $j \in J$. Then we can define a *task algebra*

$$(A:F,R)$$

and a *component algebra*

$$(A: f_i, r_j \text{ for } i \in I \text{ and } j \in J).$$

The program may introduce a new data set B and functions g_p and relations s_q, for $p \in P$ and $q \in Q$, and we may define a *program algebra*

$$(A, B: f_i, g_p, F, r_j, s_q, R \text{ for } i \in I, j \in J, p \in P \text{ and } q \in Q).$$

To a user the finished product is the task algebra, and only the function F and the relation R are visible, for the f_i, g_p and r_j ,s_q are hidden.

Of course this simplified description of the semantics of modules must be generalised to take account of the case of many sets, the role of isomorphisms, and the case when the modules define classes of implementation algebras.

The point is that algebras arise as mathematical models of components, programs and tasks. This discussion also motivates the following definition.

1.2.1 Reducts Let Σ and Σ' be signatures with $\Sigma \subseteq \Sigma'$. If A is a Σ' algebra then the Σ *reduct* of the Σ' algebra A is the algebra

$$A/_{\Sigma}$$

of signature Σ, consisting of the carriers of A named by the sorts of Σ, and equipped with the functions and relations of A named by the function and relation symbols of Σ.

Clearly, in the discussion above, the component algebra and the task algebra are reducts of the program algebra.

1.3 Classes of models

For the purposes of specification and verification, we expect K to be a subclass of the class $Mod(T)$ of all Σ models of some axiomatic theory T over Σ. Among the classes of interest are:

$K=Mod(T)$	$K=\{A\in Mod(T)\text{: } A \text{ is finite}\}$
$K=\{A\in Mod(T)\text{: } A \text{ is computable}\}$	$K=\{A\in Mod(T)\text{: } A \text{ is semicomputable}\}$
$K=\{A\in Mod(T)\text{: } A \text{ is initial}\}$	$K=\{A\in Mod(T)\text{: } A \text{ is final}\}$.

Each of these subclasses formalises an interesting aspect of the semantics of T. For example, consider *initiality* and *computability*. If model A is initial then it is (isomorphic to) a certain form of *standard implementation* by computer, (namely that defined by a *term rewriting system* derived from T). To study the computation of F and R with respect to this subclass K is to study the properties of a program module for F and R uniformly across all standard implementations.

That a model A is computable means that it is *implementable in some way* by computer, according to the "classical" Church-Turing Thesis. To study the properties of F and R with respect to this subclass K is to study the properties of a program module for F and R uniformly across all possible computer implementations. See Bergstra and Tucker [80,83,87] and Meseguer and Goguen [85].

1.4 Constructions and higher-order computation and logic

In computing or defining functions and relations on an algebra A we may need to add carriers, operations and relations to make a new algebra B such that A is a reduct of B. We may then study those functions and relations on A that can be computed or defined on the augmented algebra B.

Many sorted algebras provide a unified framework for representing computation and logic over augmented structures. For example, two useful forms of second order computation and logic over a structure A can be represented by applying first order computation and logic to certain extended structures, $\underline{A}$ and A^*, being A with streams, and arrays adjoined, respectively.

1.4.1 Augmentation of booleans and natural numbers Let A be an algebra and add to the carriers of A a set $B=\{tt,ff\}$ of booleans. To the operations of A, add the constants $tt, ff \in B$ and the operations of conjunction and negation on B. With the booleans we may represent relations on A as boolean-valued functions.

Let A be an algebra and add to the carriers of A a set $N = \{0,1,2,...\}$ of natural numbers. To the operations of A, add the constant $0 \in N$ and the operations of successor, addition and multiplication on N.

Clearly, A is a reduct of each of these algebras. Structures augmented with the booleans and natural numbers we will call *standard algebras*.

1.4.2 Augmentation of time cycles and streams Let A be an algebra and add to the carriers of A a set $T=\{0,1,2,...\}$ of natural numbers and, for each carrier A_i of A, the set $[T \to A_i]$ of all functions $T \to A_i$, called *streams* over A_i. To the operations of A we add the constant 0 and operation of successor $t+1$ on T, and $eval_i : T \times [T \to A_i] \to A_i$ defined by

$$eval_i(t,a) = a(t).$$

Let this new algebra be $\underline{A}$; clearly, A is a reduct of $\underline{A}$.

1.4.3 Augmentation of arrays Let A be an algebra and add to the carriers of A a set of natural numbers T and the constant 0 and operation of successor $t+1$ on T. Next add an undefined symbol u_i to each carrier A_i of A, and extend the operations by strictness. This forms an algebra A_u with carriers

$$A_{i,u} = A_i \cup \{u_i\}.$$

We model a *finite array* over A_i by a pair (α, l) where

$$\alpha : T \to A_{i,u} \text{ and } l \in T \text{ such that } \alpha(t) = u \text{ for all } t > l.$$

Thus the pair models an infinite array, uninitialised almost everywhere. Let A_i^* be the set of all such pairs. We add these sets to the algebra A_u to create carriers of the array algebra A^*. The new constants and operations of A^* are: the everywhere undefined array; functions that evaluate an array at a number address; update an array by an element at a number address; evaluate the length and update the length of an array. If Σ is the signature of A then let Σ^* be the signature of A^*. Clearly A, is a reduct of A^*.

We will use both extensions $\underline{A}$ and of A^* in the next sections. The second augmentation is an enhancement of the array algebras that we made in Tucker and Zucker [88]. It has many interesting extensions and applications.

2. COMPUTABILITY THEORY

We begin by summarising some basic ideas about recursive functions on standard algebras. A useful reference for computable functions on the natural numbers is Cutland [80]. For further basic material on generalisations of computability theory see Shepherdson [85], Tucker and Zucker [88] and Fitting [81]; for some advanced material see Fenstad [80].

2.1 Computable functions on abstract data types

In Tucker and Zucker [88] we have examined some classes of functions over an algebra that are generated from basic functions, including

operations of the algebra;
projection functions; and
conditional functions;

by means of

sequential composition;
parallel composition;
simultaneous primitive recursion;
simultaneous course-of-values (cov) recursion; and
search operators.

These function building operations are defined for many sorted algebras whose domains include the booleans and the natural numbers - the *standard algebras* of 1.4.1. The concepts are straightforward generalisations of the corresponding well-known concepts associated with the natural numbers; we will define them shortly. Relations we treat as boolean-valued functions.

Let A be a standard algebra. We consider the important set

$CIND(A)$

of *course-of-values (cov) inductively definable partial functions;* these are the functions on A that are formed from the basic functions by combining sequential and parallel composition, course-of-values recursion, and least number search.

If course-of-values recursion is replaced by primitive recursion in this definition then we obtain the set

$IND(A)$

of *inductively definable partial functions.*

In either case, the functions are defined by a *parallel deterministic model* of computation. The vectorisation and simultaneous recursions are responsible for the parallelism. The numerical control of the recursions is responsible for the particularly simple form of determinism. There are many useful kinds of recursion more general in form than course-of-values recursion, but the aim is to work with the simplest form equivalent in power to these others.

2.2 Basic functions

Let A be a standard algebra which for simplicity of notation we assume has only one carrier in addition to the booleans B and natural numbers N. We are interested in all functions on the carriers of A of the form

$$f: D_1 \times D_2 \times ... \times D_n \rightarrow R_1 \times R_2 \times ... \times R_m$$

where D_i and R_j are A, N or B. For simplicity, we will often abuse notation and write such functions $f: A^n \rightarrow A^m$ and hence ignore the details of the mixture of carriers in the domains and ranges of functions; this should not cause confusion.

The basic functions are as follows:

The operations of the algebra including the operations on B and N as in 1.4.1.

The projection functions $\pi_{i,n}: A^n \rightarrow A$, for any n and $i=1,...,n$, defined by

$$\pi_{i,n}(a_1,...,a_n) = a_i$$

for $a_1,...,a_n \in A$.

The conditional functions of the form $f: B \times A^2 \rightarrow A$ defined by

$$f(b,x,y) = x \quad \text{if } b = tt$$
$$\qquad\quad = y \quad \text{if } b = ff.$$

2.3 Parallel and sequential composition

Let $f_1,...,f_m : A^n \rightarrow A$. Then the function $f: A^n \rightarrow A^m$ defined by

$$f(a) = (f_1(a),...,f_m(a))$$

for $a \in A$ is defined by *parallel composition* of the $f_1,...,f_m$.

Let $g: A^n \rightarrow A^k$ and $h: A^k \rightarrow A^m$. Then the function $f: A^n \rightarrow A^m$ defined by

$$f(a) = h(g(a))$$

for $a \in A$ is defined by *sequential composition* of the g and h.

We now define the main constructs of *course-of-values* and *primitive recursion.*

2.4 Course-of-values recursion

Let $g: A^n \rightarrow A^m$ and $h: T \times A^n \times A^{mp} \rightarrow A^m$.

Let $\delta_i: T \times A^n \rightarrow T$ for $i = 1,...,p$ be functions which are subject to the condition that

$$\delta_i(t,a) < t$$

for all $t > 0$ in T and $a \in A^n$.

Then the function $f: T \times A^n \rightarrow A^m$ defined by

$$f(0,a) = g(a)$$
$$f(t,a) = h(t, a, f(\delta_1(t,a),a),...., f(\delta_p(t,a),a))$$

for $t > 0$ is defined by *(simultaneous) course-of-values recursion* of g, h and δ_i on A.

Let us consider this definition in more detail. Each of the functions g and h has m coordinate functions

$$g_i: A^n \rightarrow A$$
$$h_i: T \times A^k \times A^{mp} \rightarrow A.$$

The functions δ_i for $i = 1,\dots,p$,

$$\delta_i: T \times A^n \rightarrow T$$

give times at which p previous values are used and hence the functions are subject to the condition that

$$\delta_i(t,a) < t$$

for all $t > 0$ and all a.

The maps f_i for $i=1,\dots,m$ are defined by the following:

$$f_1(0,a) = g_1(a)$$

$$\dots$$

$$f_m(0,a) = g_m(a)$$

$$f_1(t,a) = h_1(t,a,f_1(\delta_1(t,a),a),\dots,f_m(\delta_1(t,a),a),\dots f_1(\delta_p(t,a),a),\dots,f_m(\delta_p(t,a),a))$$

$$\dots$$

$$f_m(t,a) = h_m(t,a,f_1(\delta_1(t,a),a),\dots f_m(\delta_1(t,a),a),\dots,f_1(\delta_p(t,a),a),\dots f_m(\delta_p(t,a),a)).$$

Of course, the functions g and h need not depend on all the arguments. But in the case displayed above, each f_i depends on t, a, and p previous values of each of $f_1,\dots, f_m$.

Notice the following: let d_i: $T \times A^k \rightarrow T$ be *any* function. Then we can define for $t > 0$

$$\delta_i(t,a) = min(d_i(t,a),t-1)$$

such that $\delta_i(t,a) < t$.

There are several simple conditions we may impose on the δ_i. For instance, we may assume that a fixed constant delay d_i is assigned so that

$$\delta_i(t,a) = min(t - d_i,\ t-1).$$

2.5 Primitive recursion

If we take $\delta_i(t,a) = t-1$ for $t > 0$ then we can rewrite the equations for the map f as follows.

Let $g: A^n \rightarrow A^m$ and $h: T \times A^n \times A^m \rightarrow A^m$.

Then the function $f: T \times A^n \rightarrow A^m$ defined by

$$f(0,a) = g(a)$$
$$f(t+1,a) = h(t, a, f(t,a))$$

is defined by *(simultaneous) primitive recursion* of g and h on A.

Let us again unfold this definition to see the simultaneity. Each of the functions g and h has m coordinate functions

$g_i: A^n \to A$

$h_i: T \times A^k \times A^m \to A.$

The maps f_i for $i=1,...,m$ are defined by the following:

$$f_1(0,a) = g_1(a)$$

$$\ldots$$

$$f_m(0,a) = g_m(a)$$

$$f_1(t+1,a) = h_1(t, a, f_1(t,a),..., f_m(t,a))$$

$$\ldots$$

$$f_m(t+1,a) = h_m(t, a, f_1(t,a),..., f_m(t,a)).$$

2.6 More on primitive recursion

Let $g: A^n \to A^n$. Then the function $f: T \times A^n \to A^m$ defined by

$$f(t,a) = g(g(g(...g(a)))) \quad (t \text{ times})$$

or

$$f(t,a) = g^t(a)$$

is defined by *primitive iteration* of g on A.

Proposition *Let F be a set of functions on A containing the projection functions and closed under parallel and sequential composition. Then f is definable by primitive recursion over elements of F if, and only if, f is definable by primitive iteration over elements of F.*

2.7 Search

Let $g: T \times A^n \to B$. Then the function $f: A^n \to T$ defined by

$$f(a) = (\text{least } t \in T)[\ g(t,a) = tt \text{ and for } i = 0,...,t\text{-}1,\ g(i,a) = ff]$$

is defined by *least number search operator* on g.

2.8 Formal definitions

To define formally the computable functions of interest we must provide a syntax for the recursive defintions, and a semantics in terms of the functions they compute. This is done by inductively defining appropriate recursion *schemes* over Σ. A scheme α describes the structure of the recursive definition. The scheme computes a function α_A on an algebra A and computes a family

$$\alpha_K = \{\alpha_A : A \in K\}$$

of functions uniformly for $A \in K$.

Exercise Reconstuct the above definitions in the many sorted case, and define a syntax and semantics for schemes over Σ.

2.9 Computation on A^*

For many purposes it is convenient to present the computable functions on A using primitive recursion and the least number search operator on the array algebra A^* of 1.4.3. This is possible because the models of computation apply to any standard algebra and we have the following fact:

Theorem *Let f be a function on A. Then*
$$f \in CIND(A) \text{ if, and only if, } f \in IND(A^*).$$

We will often speak of the computable functions on A and K and mean either the cov inductively definable functions on A and K, or the inductively definable functions on A^* and K^*.

Much more work on translating recursions is necessary as the subject is not well understood even in the case of the natural numbers: see Simmons [88].

2.10 Other characterisations and a thesis

There are *many* models of computation that may be used to define the same classes of computable functions on abstract data types. These models are summarised in Shepherdson [85] and Tucker and Zucker [88]. The basic theory of the model used here is taken from Tucker and Zucker [88] in which it is argued that whilst the informal idea of "effective calculability" is ill-defined when generalised to an abstract setting, the ideas of "deterministic computation" and "operational semantics" are meaningful and, furthermore, the following is true:

A Church-Turing thesis for computation on abstract data types
Consider a deterministic programming language over an abstract data type D. The set of functions and relations on a structure A, representing an implementation of the abstract data type D, that can be programmed in the language, is contained in the set of cov inductively definable functions and relations on A. The class of functions and relations over a class K of structures, representing a class of implementations of the abstract data type D, that can be programmed in the language, uniformly over all implementations in K, is contained in the class of cov inductively definable functions and relations over K.

We will use this thesis in Section 4 and the form restated in terms of inductive definability on A^* and K^* in Section 3.

2.11 A duality between recursion theory and algebra

The definitions of the recursive functions given above are closely related to equations and conditional equations. In the case of computability theory on the natural numbers, there is a pleasing and very useful duality theory that connects recursive definitions (and their associated operational and denotational semantics) with equations (and their associated logic, term rewriting systems, and initial algebra semantics). This duality between recursive definitions and algebraic specifications on the natural numbers is developed and exploited in my research with J A Bergstra about the power of algebraic specification techniques to define computable algebras of various kinds: see Bergstra and Tucker [79, 80, 83, 87], for example.

With a general theory of computability on arbitrary abstract algebras the essentials of *some* of the ideas and techniques used in the proofs of theorems about computable algebras is revealed more clearly, and are found to have practical applications. Consider the following generalisation of a basic lemma in Bergstra and Tucker [80] (which was published as Bergstra and Tucker [87]).

Let A be an algebra and f a primitive recursive function on A. Let algebra A_f be A augmented by all the subfunctions involved in the definition of f, obtained from the parse tree of a primitive recursive scheme for the function f in a certain straight forward way.

In the terminology of 1.2, A is the component algebra, (A,f) is the task algebra, and A_f is the program algebra.

Let (Σ_f, E_f) be the signature and equations obtained by adding the corresponding names for these functions and equations obtained from the definition of the primitive recursive function f.

Theorem *Let (Σ, E) be an algebraic specification of the component algebra A. Let f be a primitive recursive function over A. Then the program algebra*

$$A_f \cong I(\Sigma_f, E_f)$$

and hence (Σ_f, E_f) is an algebraic specification with hidden functions for the task algebra (A,f).

This generalisation was proved in Thompson and Tucker [90] as part of our work on synchronous concurrent algorithms, to be discussed in Section 4. Using a very detailed proof of this fact, the functional definition of f over Σ can be compiled into an algebraic specification (Σ_f, E_f) in preparation for machine processing; B C Thompson has written such a compiler. The theorem can be extended to course-of-values recursive functions by the fact that A^* is definable by adding conditional equations to an algebraic specification for A, and using the theorem in 2.9. This theme has been further developed for *total functions* defined by arbitrary recursive definitions with fixed point semantics in Meinke [90].

The point of the duality or equivalence is to unite ideas and techniques for
(i) defining computable functions on sets, and packaging them in algebras,
(ii) specifying and reasoning about algebras.

The duality relates models of computation (operational semantics, term rewriting), models of specification (fixed points, initial algebras), and logics for reasoning (equational logic and its model theory). This duality can be extended from computable functions to relate semicomputable relations of various kinds with the theory of Horn clauses: see Tucker and Zucker [89].

3. SPECIFICATION THEORY

We now consider ideas about specification that complement our theory of computation.

3.1 Specification and computation by means of relations and functions

Let us assume that there are three basic components of a computation:

(1) a specification of a task to be performed or problem to be solved;

(2) specifications for algorithms whose input-output behaviour accomplish the task or solve the problem; and

(3) algorithms with appropriate input-output behaviour.

In the language of logic, we consider input-output behaviour both intensionally and extensionally. To progress with a mathematical analysis we assume further that

(1') a specification of the task or problem is defined by a *relation* R on algebras, given uniformly over a class K;

(2') specifications of algorithms for the task or problem are defined by *functions* on algebras, given uniformly over a class K; and

(3') algorithms are defined by *programs that compute functions* on algebras, uniformly over a class K.

Thus we use a relational paradigm for problem specification and a functional paradigm for *both* solution specification and solution computation.

How are these notions connected?

Consider the specification of a task on a set A by means of a relation R.

Informal Idea of Tasks *Given input-output condition $R \subseteq A^n \times A^m$, for all $x \in A^n$, calculate some $y \in A^m$ such that $R(x,y)$, if such exists. The set*

$$S = \{ x \in A^n : \exists y R(x,y) \},$$

which defines the limits of the task, may be called the domain of the task.

3.1.1 Formal Specifications of Tasks Given input-output condition R, compute *any* partial function $f: A^n \to A$ that satisfies for all $x \in A^n$ the property that f is *totally correct* with respect to R in the sense that

$$\exists y R(x,y) \text{ implies } f(x)\downarrow \text{ and } R(x,f(x)).$$

There are variants of this formal idea of a task such as those requiring several, or all, functions with the above property to be computed; or those concerning functions f which are *partially correct* in the sense that

$$\exists y R(x,y) \text{ and } f(x)\downarrow \text{ implies } R(x,f(x)).$$

Example Consider the set $\mathbf{R}$ of real numbers and the input-output relation R

$$R(a,b,c,x) \equiv ax^2 + bx + c = 0$$

defined on $\mathbf{R}^4$. A task is: given a,b,c find x such that $R(a,b,c,x)$, i.e. find *some* root of the quadratic equation with coefficients a,b,c. The domain of the task is

$$S = \{(a,b,c) \in \mathbf{R}^3 : \exists x R(a,b,c,x)\},$$
$$= \{(a,b,c) \in \mathbf{R}^3 : \exists x(ax^2 + bx + c = 0)\}.$$

In this case we know that there are two roots for each $(a,b,c) \in \mathbf{R}^3$, and that they are given by the functions

$$f_1(a,b,c) = [-b + \sqrt{(b^2 - 4ac)}] / 2a$$

$$f_2(a,b,c) = [-b - \sqrt{(b^2 - 4ac)}] / 2a.$$

These functions are totally correct for the task defined by R. Furthermore, the domain of the task is

$$S = \{ (a,b,c) \in \mathbf{R}^3 : b^2 - 4ac \geq 0\}.$$

In general, the solution of equations in mathematics fits our mould for the specification and computation of tasks. Let us define precisely the situation we wish to consider.

3.1.2 Selection functions over A and K Let A be an algebra which, for simplicity of notation, we assume has one carrier, and let R a subset of A^{n+m}. A *selection function* for R is a partial map $f: A^n \to A^m$ such that

$$\forall x[\ \exists y R(x,y) \Rightarrow f(x)\downarrow \text{ and } R(x,f(x))].$$

Such functions are also called *realising functions* or *witnessing functions.* Notice that we assume that f is total on the domain of the task. Let $Sel_A(R)$ be the set of all selection functions on A for R.

This idea may be extended to uniform computation on K in the usual way by using families of relations and selection functions:

Let K be a class and

$$R = \{R_A : A \in K\}$$

a family of relations on K, where R_A is a subset of A^{n+m} for all $A \in K$.

A *selection function* for R on K is a family

$$f = \{f_A : A \in K\}$$

of maps $f_A: A^n \to A^m$ such that

$$\forall x[\exists y R_A(x,y) \Rightarrow f_A(x)\downarrow \text{ and } R_A(x,f_A(x))].$$

for all $A \in K$. We will also write:

$$K \models \forall x[\exists y R_A(x,y) \Rightarrow f(x)\downarrow \text{ and } R_A(x,f(x))].$$

The precise connection between program, program specification and task specification is this:

Computation and Specification Hypotheses *Algorithms and programs compute functions over a class K. The functions to be computed are the selection functions for relations over a class K. Thus a program α on K correctly implements a task specified by R on K if it correctly computes a selection function f for R on K, i.e. for all $A \in K$*

$$\forall x[\exists y R_A(x,y) \Rightarrow \alpha_A(x)\downarrow \text{ and } R_A(x,\alpha_A(x))].$$

This formulation raises a number of questions about the specification of tasks:

What properties of specifications are characteristic of their role in testing?

What properties of specifications are necessary for them to be expressible in a formal logical language?

How can we reason about specifications and the functions and programs that implement them?

By attempting to answer some of these questions we will refine the ideas given above. In Section 2, we gave a theory of computation for functions and relations (seen as boolean valued functions) on *K*, but we must now consider relations on *K* in their role as specifications of problems and tasks. The first mathematical problem is to classify the relations that may be allowed in the specifications of tasks, using the computability theory. The specification of a task in 3.1.1 involves the input-output relation $R(x,y)$ and the relation $\exists y R(x,y)$ which is the domain of the task. What can be said about the relations?

3.2 Effective testing and semicomputability

For a specification *R* to be *effectively testable* we must require that for any given data *x* and *y* we can effectively test whether or not $R(x,y)$ is true - that is, whether or not a given *y* is a valid output for a given input *x*.

Effective Testing Hypothesis *Given an input-output relation R, there is a program such that, for any given data x and y, the program terminates, and if R(x,y) is true then we discover that y is a correct output for x; but if R(x,y) is false then we discover that y is not a correct output for input x.*

However, we may take the view that in order to explore the limits of this specification method a weaker testing condition to place on a specification is more useful:

Positive Testing Hypothesis *Given an input-output relation R, there is a program such that, for any given data x and y, if R(x,y) is true then the program terminates and we discover that y is a correct output for x.*

We may also formulate *negative testing* in the obvious way. In order to place conditions on *R* that formalise testing we must study the notions of computable and semicomputable sets on algebras and classes.

Recall from 2.9 and 2.10 that the computable functions are the course-of-values inductively definable functions on A and K; or, equivalently, the inductively definable functions on A^* and K^*.

We define a *computable* relation R on A to be a relation with computable characteristic function on A: there is a computable function $g: A^n \to B$ such that for all x,

$R(x) \Leftrightarrow g(x) = tt$ and $-R(x) \Leftrightarrow g(x) = ff$.

We define a *semicomputable* relation R on A to be a relation that is the domain of a partial computable function on A: there is a computable function g: $A^n \to A$ such that for all x,

$$R(x) \Leftrightarrow g(x)\downarrow.$$

A relation R is *cosemicomputable* if its complement $-R$ is semicomputable.

Not every set that is semicomputable over A is computable over A.

Lemma *R is computable on A if, and only if, R and its complement -R is semicomputable on A.*

To define an algorithmically testable input-output specification $R(x,y)$ we might require that it be computable, semicomputable, or cosemicomputable. The obvious case to choose first is that R is computable.

Now we must consider the domain $\exists y R(x,y)$ of the task.

A relation R is *projectively computable* if it is a projection of a computable relation: there is a computable total function $g: A^n \times A^m \to B$ such that for all x,

$R(x) \Leftrightarrow (\exists y \in A^m)[g(x,y) = tt]$.

A relation R is *projectively semicomputable* if it is a projection of a semicomputable relation: there is a computable partial function $g: A^n \times A^m \to A$ such that for all x,

$R(x) \Leftrightarrow (\exists y \in A^m)[g(x,y)\downarrow]$.

Here we must be careful to distinguish between projections on A and on A^*. The ideas of cov inductively computable, semicomputable and cosemicomputable on A correspond with those of inductively computable, semicomputable and cosemicomputable on A^*. However, when using projections there is a difference between projecting over the elements A, in the case of $CIND(A)$, and over the finite sequences of A^*, in the case of $IND(A^*)$.

According to the weakest of the above hypotheses on testing,

the specifiable tasks are those whose domains are the projectively inductively semicomputable sets over A^*.

Not every set that is projectively semicomputable over A^* is semicomputable over A^*. Indeed, the equivalence of the classes of sets is a strong property on A and K. However, in the case of classes of *minimal structures*, it can be proved that the notions are equivalent.

It can be shown, by means of later considerations in 3.3, that the notions of *projectively semicomputable* and *projectively computable* are equivalent on A^*.

Hence:

Effectively testable specifications *The effectively specifiable tasks are those whose domains are the projectively inductively computable sets over A^*.*

3.3 Definability of specifications
We will now consider the relations suitable for task specifications in terms of their expressibility or definability in a formally defined specification language. From this point of view, some important conditions on the class of input-output relations R are the following:

Formal Specification Language Hypothesis *There is a logical language L in which input-output relations and task domain relations may be defined and that, in principle, L facilitates machine processing of specifications.*

Any language L chosen to express the relations may be required to satisfy this condition:

Adequacy Condition *The language L is sufficiently expressive to specify each input-output relation and task domain relation of computational interest.*

What is of computational interest can be determined by the model of computation, or, more precisely, by the class of functions its defines. For instance, the adequacy condition leads to the following mathematical question:

For each computable partial function f does there exist a formula in L to define a relation R that specifies f as a selection function, and, in particular, is f uniquely defined as its selection function?

The existence of *some* relation is a triviality; let us examine uniqueness. Consider a computable partial function $f\colon A^n \to A$. Let P_f be the *graph* of f, so that for all $x \in A^n$, $y \in A$

$$P_f(x,y) \text{ if, and only if, } f(x){\downarrow}y.$$

Then f is defined uniquely as the only selection function for P_f.

Clearly many properties of computations can be defined in terms of P_f. For example, *termination* for f is this: for $x \in A^n$,

$$\exists y P_f(x,y) \text{ if, and only if, } f(x){\downarrow}.$$

Thus, to answer the adequacy question positively it is sufficient to answer the following:

For each computable function f does there exist a formula in L to define P_f?

A converse idea for adequacy is this:

Soundness Condition *The language L cannot specify tasks that cannot be implemented.*

This leads to the following soundness question:

If R is defined by a formula in L, does it have a computable selection function?

Consider the language $L(\Sigma^*)$ of first order formulae over the signature Σ^* of A^*. This many sorted language is certainly machine processable. Let us test it against the adequacy and soundness conditions.

It is easy to show that not every $L(\Sigma^*)$ formula has a computable selection function. Hence there are tasks that may be defined but not performed. We will now show that what may be computed may be specified.

Theorem *For any computable partial function f on K^* the graph P_f of f is unformly definable over K^* by a Σ_1 formula of $L(\Sigma^*)$.*

This leads us to consider a sublanguage $\Sigma_1{}^*$ of $L(\Sigma^*)$ over K^* that is adequate for specifying tasks we can actually perform according to our computability theory.

From Tucker and Zucker [88], we can characterise the relations by their definability in a logical language suitable for formal specifications.

Theorem *A relation R is projectively semicomputable over K^* if, and only if, R is uniformly definable over K^* by a Σ_1 formula of $L(\Sigma^*)$.*

From this analysis of definability it is possible to show:

Lemma *R is projectively computable on K^* if, and only if, R is projectively semicomputable on K^*.*

There are different views of the nature of specification to be found in, for example, the theories of program verification by Floyd-Hoare logic; program design by stepwise program development; program synthesis from specifications; and logic programming. Does the mathematical analyses of these and other views lead us to consider the same class of sets?

3.4 Program specification and verification by Floyd-Hoare logic

Floyd-Hoare verification is an area of application for some of the material in 3.3, where we again meet the projectively semicomputable sets on A^* and K^*. It was this application that interested us in specification theory originally, when working on Tucker and Zucker [88]. In this monograph, the semantic and correctness theories of **while** programs, **while array** programs, and recursive programs established in de Bakker [80] are generalised to allow programs that compute not simply on the natural numbers but on any abstract data type. In addition, error or exceptional states are introduced into the semantics, which arise when a variable is called in a computation without first being initialised. These enhancements involved mathematical work on: many sorted algebras A

and classes K of many sorted algebras; error and exceptions in semantics and proof rules; weak second order logic assertion languages; and the theory of computable functions on A and K. Let us summarise the connection with 3.3.

Consider an imperative programming language $\boldsymbol{P}$ over K, with a deterministic implementation. Let $State(A)$ be the set of all states for program variables on A and let $M_A(S)$: $State(A) \rightarrow State(A)$ be the input-output behaviour of a program $S \in \boldsymbol{P}$.

Let S be specified by input and output conditions p,q written in an assertion language L over K. Let $\{p\}S\{q\}$ be a specification of S and suppose that

$K \models \{p\}S\{q\}$

means that the specification is valid under *partial correctness* uniformly over K. That is, for each $\sigma \in State(A)$, if $A \models p(\sigma)$ and $M_A(S)(\sigma) \downarrow \sigma'$ then $A \models q(\sigma')$, uniformly for all $A \in K$.

A general question of importance is:

How expressive is the assertion language L with respect to the programming language ***P****?*

More specifically: Does the assertion language allow us to express important sets of states such as the *weakest precondition* and *strongest postcondition?* The definability in L of these sets of states is necessary for practical reasoning and for the proofs of the *completeness theorems* for Floyd-Hoare program correctness logics over L. This question can be formulated precisely as follows:

Expressiveness Question *Is the weakest precondition*

$wp_A(S,q) = \{\sigma \in State(A)\text{: if } M_A(S)(\sigma) \downarrow \sigma' \text{ then } A \models q(\sigma')\}$

definable by L uniformly for all $A \in K$*?*

Suppose we have formulated some deterministic operational semantics that describes computation in terms of sequences of states. Then we have a *computation relation* defined by

$COMP_A(S,\sigma,\tau,\sigma')$ if, and only, if $M_A(S)(\sigma)\downarrow\sigma'$ via some sequence $\tau \equiv \sigma{=}\sigma_1, \ldots ,\sigma_k{=}\sigma'$.

The graph of the state transformer is definable in terms of the relation:

$M_A(S)(\sigma) \downarrow \sigma'$ if, and only if, $(\exists\tau)COMP_A(S,\sigma,\tau,\sigma')$;

and hence so is the weakest precondition.

In Tucker and Zucker [88] the representation is carried out in detail using the first order language $L(\Sigma^*)$ of A^* which is, of course, a weak second order language over A. The method is first to represent the computation relation for $\boldsymbol{P}$ over A by inductively computable functions and relations over A^*. Then to use the fact that inductively computable functions and relations are definable in $L(\Sigma^*)$. The formulae of $L(\Sigma^*)$ used to define the computation relations are Σ_1 since they are represented by projective semicomputable sets (by the theorem in 3.3).

This representation of the programs of $\boldsymbol{P}$ by inductive functions is in fact compilation. In general, proofs of expressiveness involve the following principle: if $\boldsymbol{P_1}$ and $\boldsymbol{P_2}$ are programming languages and $c{:}\boldsymbol{P_1} \rightarrow \boldsymbol{P_2}$ is a compiler then if $\boldsymbol{P_2}$ is expressive then $\boldsymbol{P_1}$ is expressive.

A related question is this: Does the assertion language L allow us to capture the meaning of a program completely?

Determinacy Question *Let* $S, S' \in \boldsymbol{P}$. *If for any* $p, q \in L$

$K \models \{p\}S\{q\}$ *if, and only if,* $K \models \{p\}S'\{q\}$

then is it the case that $S \equiv S'$ *on* K?

Clearly the converse holds for any reasonable assertion language. For information on the determinacy question see Bergstra, Tiuryn and Tucker [82].

3.5 Logic programming

In the case of logic programming, it is also possible to identify the projective semicomputable sets.

A *logic programming language* is a language for specification *and* computation in which the means of computation is deduction in a logical system. More precisely, we think of a program as a module that uses axiomatic theories expressed in formally defined logical languages, such as many sorted first order logic, to define classes of implementation algebras (recall 1.1). This introduces proof theory as the basis for a semantics of computation; and, on the other hand, the model theory of logical systems as a basis for the semantics of specification. Not surprisingly, the subject of proof dominates research on logic programming, mainly in the form of work on practical deduction for implementations, and the subject of the model-theoretic semantics of specification seems hardly to have been examined.

A connection made in logic programming between specification and computation is of the following kind:

Let T *be a theory which axiomatises a class* K_T *of components for programming. Let* R *be a relation which represents a task to be defined uniformly over* K_T. *Let* P *be a logic program with goal relation* R. *Then semantically we interpret* P *in some class* K_P *of models that extends the class* K_T *in order to specify relation* R *and compute selection functions* f *for* R *uniformly over* K_T.

Thus we want to design P to be valid over a class K_P of algebras, i.e. $K_P \subseteq Mod(T \cup P)$, and to compute one or more f such that:

$$K_P \models \forall x[\exists y R(x,y) \Rightarrow f(x)\downarrow \text{ and } R(x,f(x))].$$

Note that the problem of computing selection functions by logical deduction is a motivating problem of proof theory, and of the programs as proofs paradigm.

In Tucker and Zucker [89] we take the basic step of generalising Horn clause computability from Herbrand universes, as explained in Lloyd [84], to *any* structure A or class K of structures, and we show the following:

Theorem *A relation R is projectively semicomputable over K^* if, and only if, R is definable by Horn clauses over K^*.*

This work on logic programming is relevant to the development of the concept of a *logic programming module* which generalises the concept of an algebraic specification of an abstract data type module. There is a close connection between logic programming modules and algebraic specification modules: see Goguen and Meseguer [84] and Tucker and Zucker [89]. An important aspect of this connection is (the generalisation of) the duality theory between computable functions and initial models mentioned in Section 2. Finally, let us note that this idea of a module is, of course, many sorted. Many sorted logic programming has been studied in depth by Walther [87] and Cohn [87], motivated by theorem prover efficiencies made possible by typing.

3.6 Other characterisations and a thesis

The projective semicomputable sets have several other characterisations by means of nondeterministic models for computation including

while-array with initialisation;

while-array with random assignments;

and older notions such as *search computability* in the sense of Y N Moschovakis, and notions of M Fitting, for example.

The research of Section 3, by J I Zucker and myself, is focussed on ideas about *algorithmic models for specifications;* it also includes ideas about *nondeterministic programming*. Our assumptions in 3.1, and subsequent results, lead us to formulate the following general kind of Church-Turing Thesis for specification to complement that for computation in 2.9:

A Church-Turing thesis for specification on abstract data types

Consider an algorthmic specification language over an abstract data type D. The set of relations, and functions, on a structure A, representing an implementation of the abstract data type D, that can be expressed or specified in the language, is contained in the set of projectively inductively definable relations on A^*, and their selection functions, respectively. The class of relations and functions over a class K of structures, representing a class of implementations of the abstract data type D, that can be expressed in the language, uniformly over all implementations of K, is contained in the class of projectively inductively definable relations over K^*, and their selection functions, respectively.

To establish a viable proposal for specification is a more complex task than that for computation: for more details see Tucker and Zucker [89, 90]; further work is in progress.

3.7 Verification

Suppose we have our specifications of tasks defined in the first order specification language $L(\Sigma^*)$, and more specifically in the Σ_1 fragment that is necessary and sufficient for our task specifications by the theorem in 3.3. Then we can use logics to reason about our specifications, and the selection functions and programs that implement them. Some basic mathematical questions that arise as soon as one has a theoretical account of specification and computation are those to do with verification:

Given an axiomatic specification T for the algebras of K, and a proof system Proof(T) for the specification language L, is it possible to prove that all programs meet their specifications written in L?

Is there an L and a Proof(T) on which it is possible to base a Generalised Church-Turing Thesis that characterises the computable functions on K that are formally verifiable.

In Tucker, Wainer and Zucker [90], we consider the language of Σ_1 formulae of $L(\Sigma^*)$ and give a suitable candidate *Proof(T)*. This proof system is based on *T* and the induction principle restricted to the Σ_1 formulae of $L(\Sigma^*)$, i.e. the specifiable relations according to the thesis of the last section. The proof system is denoted Σ_1^**-Ind(T)*.

Recall that the computable functions on *A* and *K* are the cov inductively definable functions on *A* and *K*. Let us call the subclass of functions defined without the use of the least number search operator the *cov recursive functions*; these are total functions on *A* and *K*.

Theorem *Let T be a universal theory true of a class K. Suppose the following specification is provable:*

$$\Sigma_1^*\text{-}Ind(T) \vdash \forall x \exists y R(x,y).$$

where R is Σ_1^*. *Then we can effectively construct a cov recursive scheme* α *that computes a selection function f for R, uniformly on all A in K.*

Conversely, every cov recursive scheme α *gives the unique selection function to some such specification.*

Since termination is Σ_1^* definable, we can deduce the following:

Corolary *Let* α *be a cov inductive scheme on K. Then* α *is provably total in* Σ_1^**-Ind(T) if, and only if,* α *is equivalent to a cov recursive scheme on K.*

4. SYNCHRONOUS CONCURRENT ALGORITHMS AND HARDWARE

A *synchronous concurrent algorithm* (sca) is an algorithm based on a network of modules and channels, computing and communicating data in parallel, and synchronised by a global clock. Synchronous algorithms process infinite streams of input data and return infinite streams of output data. Most importantly an sca is a *parallel deterministic algorithm.* The deterministic nature of these algorithms is established by the clock; nondeterministic concepts arise from the interaction of two or more clocks.

Many types of algorithms and architectures are examples of scas. First and foremost, conventional digital hardware, including all forms of serial and parallel computers and digital controllers, is made from components that are scas. Further, many new specialised models of computation possess the essential features of scas, including: *systolic arrays; neural nets; cellular automata; deterministic data flow;* and *coupled map lattice dynamical systems.* The many types of parallel algorithms and architectures that comprise the class of scas have applications throughout science and engineering. For example, they can be used for the real-time computations found in communication and signal processing, graphics, and instrumentation for process control; for the design of computers with novel parallel architectures and languages; and for computational models of natural phenomena. In many applications, the scas are complicated and require extensive simulation to understand their operation and, in the case of safety-critical applications, mathematical analysis for their verification.

There is a mathematical theory of scas that unifies the study of these apparently disparate types of algorithms and architectures, and provides general methods for their design, analysis and verification. This theory was begun in 1984 by B C Thompson and myself, and has developed gradually in our hands and those of the colleagues mentioned in the Introduction.

This section will sketch some basic ideas about the theory and its applications. It will show how the mathematical theory of scas can be based on the general models of deterministic parallel computing described in Section 2.

For a general survey of the theoretical foundations of hardware see McEvoy and Tucker [90].

4.1 A general model of synchronous concurrent computation

Consider a network of modules that composes a sca. Suppose that the data processed by the network belongs to a single set A. To represent the algorithm, we first collect together the set A of data involved, and the functions f_i specifying the basic modules of the network, to form a many sorted algebra A. To this algebra we adjoin a clock $T=\{0,1,...\}$ and the set $[T\rightarrow A]$ of *streams*, together with simple operations, to form a stream algebra $\underline{A}$ as in 1.4.2. This stream algebra defines the components from which the sca is built; it is a component algebra in the sense of 1.2.

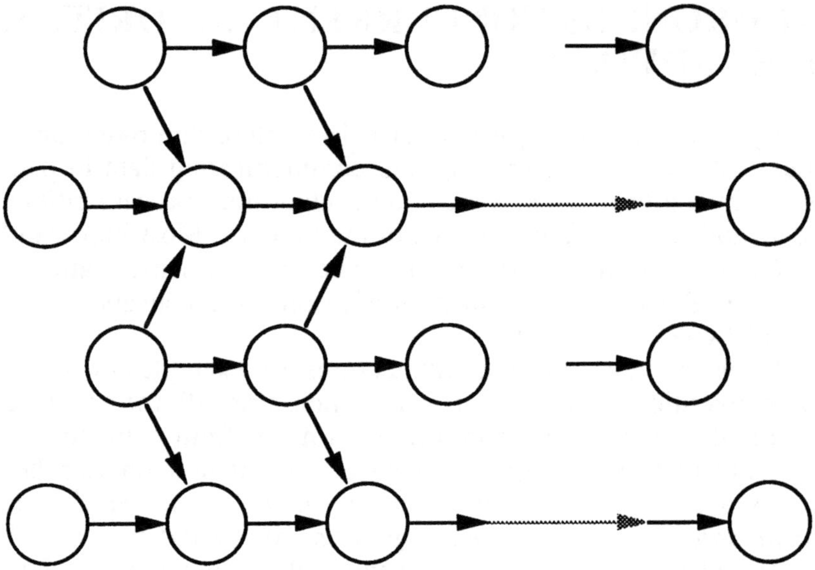

Figure 1: Architecture of synchronous concurrent algorithm.

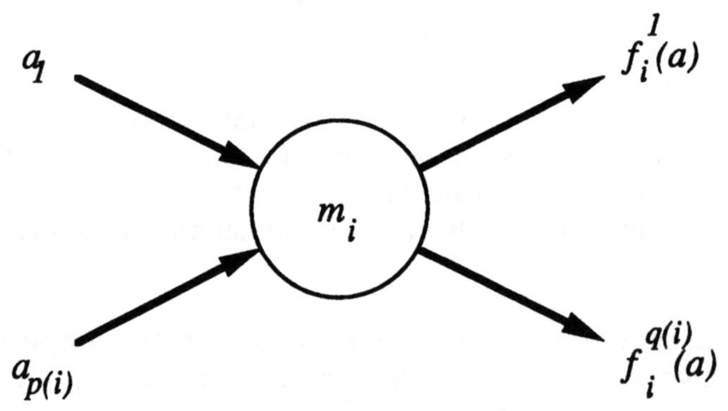

Figure 2: A module

A sca implements a specification that may be a mapping of the form

$$F: [T \to A^n] \to [T \to A^m],$$

called a *stream transformer*. However, we consider this function in the form

$$F: T \times [T \to A^n] \to A^m.$$

The algebra consisting of A, T, $[T \to A]$ and the coordinate functions of F forms a task algebra in the sense of 2.1.

The network and algorithm is then represented by means of the following method.

Suppose the algorithm consists of k modules $m_1, \ldots, m_k$ and, for simplicity, that each module has several input channels, but only *one* output channel. Suppose that the input channels of each module are either all connected to other modules, or are all connected to the input streams. The modules satisfying the latter property are called *input modules*; we suppose that there are n such input modules, and that m modules have been designated *output modules*.

Assumption T (for Termination) *Let us also suppose that each module produces an output at each clock cycle.*

To each module m_i we associate a total function

$$V_i: T \times [T \to A^n] \times A^k \to A$$

which defines the value $V_i(t,a,x) \in A$ that is output from module m_i at time t, if the algorithm is processing input streams $a=(a_1,\ldots,a_n) \in [T \to A^n]$ from the initial state $x=(x_1,\ldots,x_k) \in A^k$ of the network. The algorithm is represented by the parallel composition of functions $V_1,\ldots,V_k$.

More precisely, suppose that each module m_i has $p(i)$ input channels, 1 output channel, and is specified by a function

$$f_i: A^{p(i)} \to A.$$

Suppose that the module m_i is *either* connected to the modules

$$m_{\beta(i,1)}, \ldots, m_{\beta(i,p(i))},$$

or is an input module that is connected to input streams

$$a_{\beta(i,1)}, \ldots, a_{\beta(i,p(i))}.$$

We will assume that for each pair of modules there is a *delay* along the channel to module m_i from module m_j that is specified by the function

$$\delta_{i,j}: T \times [T \to A^n] \times A^k \to T$$

which is subject to the condition that

$$\delta_{i,j}(t,a,x) < t$$

for all $t > 0$ and all a,x.

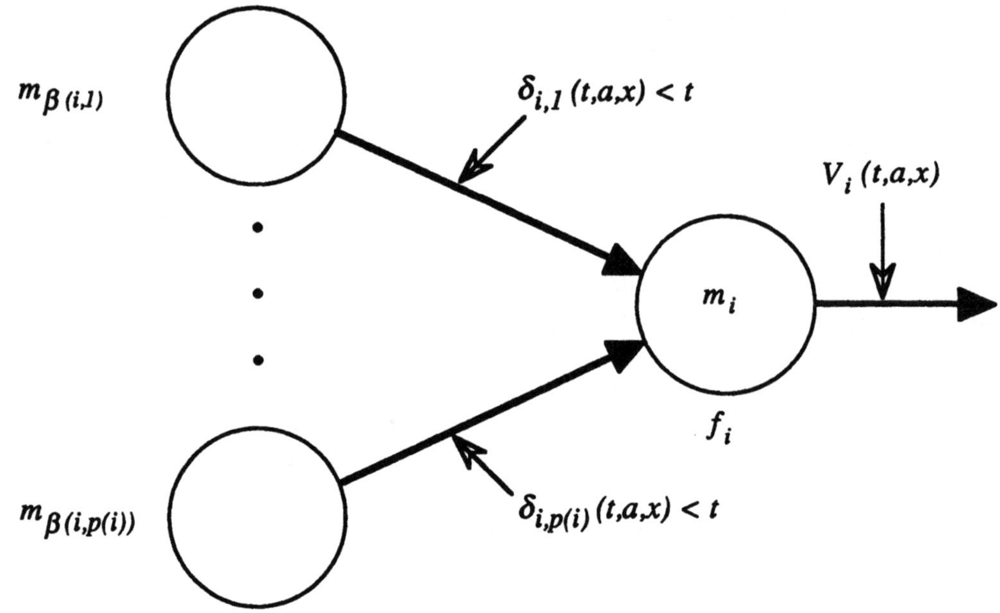

Figure 3: Value functions

The maps V_i for $i=1,...,k$ are defined by the following:

For any m_i,

$V_i(0,a,x) = x_i$.

For each m_i an input module,

$V_i(t,a,x) = f(a_{\beta(i,1)}(\delta_{i,1}(t,a,x)),..., a_{\beta(i,p(i))}(\delta_{i,p(i)}(t,a,x)))$.

For each m_i another module,

$V_i(t,a,x) = f(V_{\beta(i,1)}(\delta_{i,1}(t,a,x),a,x),..., V_{\beta(i,p(i))}(\delta_{i,p(i)}(t,a,x),a,x))$.

Clearly, this definition of the $V_1,...,V_k$ is a *simultaneous course-of-values recursion* on the stream algebra $\underline{A}$.

The algebra consisting of A, T, $[T \rightarrow A]$ together with all the component algebras' operations, and the $V_1,...,V_k$ is a program algebra in the sense of 2.2.

Let us note that we have considered computation over a single algebra A and its stream algebra $\underline{A}$. *In practice*, the above discussion invariably applies to a class K of algebras. For example, often in the case of systolic algorithms, we design for the class of all initial (= standard) models of an axiomatisation of the integers or characters; or for some subclass of the class of all commutative rings.

4.2 A unit delay model

There are several simple conditions we may impose on the $\delta_{i,j}$ that directly reflect operational properties of the modules or channels. For instance, we may assume that a fixed constant delay $d_{i,j}$ is assigned to each channel so that

$\delta_{i,j}(t,a,x) = min(t - d_{i,j}, t-1)$.

If we take $d_{i,j} = 1$ then $\delta_{i,j}(t,a,x) = t - 1$ and we can rewrite the equations for the maps V_i for $i=1,...k$ as follows:

For any m_i,

$V_i(0,a,x) = x_i$.

For each m_i an input module,

$V_i(t+1,a,x) = f(a_{\beta(i,1)}(t),..., a_{\beta(i,p(i))}(t))$

For each m_i another module,

$V_i(t+1,a,x) = f(V_{\beta(i,1)}(t,a,x),..., V_{\beta(i,p(i))}(t,a,x))$

Clearly, this is a *simultaneous primitive recursion* over $\underline{A}$.

4.3 Applying the general theory of computation to scas

Let us consider the relationship between the model for scas and the model for computability theory in Section 2. It seems that the former is a special case of the latter. What can we deduce from this fact?

4.3.1 Scope and limits of synchronous concurrent computation

For example, by the theorem in 2.9, there is a sense in which the general model in 4.1, and the unit delay model in 4.2, are equivalent, thanks to finite sequences. What is the precise meaning of this result in terms of scas?

More intriguing is the observation that the *Generalised Church-Turing Thesis* for deterministic computation in 2.10 can be applied to delimit the class of computable functions over *any* algebra *A*: the class is identified as *CIND(A)* or *IND(A*)*. Thus we have a tool to speed us toward the goal of establishing the scope and limits of computation in hardware by *deriving* an appropriate Church-Turing Thesis for synchronous concurrent computation. This model of synchronous computation will be identified with the notion of *parallel deterministic computation over abstract data types with streams.* A stream transformation

$$F: [T \to A^n] \to [T \to A^m]$$

is identified with its cartesian form

$$F: T \times [T \to A^n] \to A^m$$

in $CIND(\underline{A})$ or $IND(\underline{A}^*)$.

We are working with computable functions over the stream algebra $\underline{A}$. Thus, according to the general principles, we may construct $(\underline{A})^*$ and consider the set of functions $IND(\underline{A}^*)$. However, recalling 1.4.3, we can simplify:

Lemma $IND_A(\underline{A}^*) = IND_A(\underline{A}^u)$.

From this reasoning we can offer the following formulation as a guide:

Thesis for synchronous concurrent algorithms
The functions and stream transformations over a set A that are implementable by synchronous concurrent algorithms made from processors with functional specifications $f_1, \ldots f_k$ *on A are those in* $IND(\underline{\langle A, f_1, \ldots, f_k \rangle}^u)$.

It is important to note that a considerable number of details, concerning the interpretation of technical notions about computable functions in the framework of scas, must be examined. A major technical point is that *Assumption T* (*T* for total) must be replaced by an appropriate *Assumption P* (*P* for partial).

4.3.2 Algebraic specifications

Let us also consider the duality between recursive functions and algebraic specifications. We may apply (the proof of) the result in 2.11 which will take scas, represented by primitive recursive functions over stream algebras, and map them into algebraic specifications, in preparation for machine processing:

Corollary *Let* (Σ, E) *be an algebraic specification of the component stream algebra* $\underline{A}$. *Let V be primitive recursive over* $\underline{A}$, *representing a sca. Let sca program algebra* $\underline{A}_V$ *be* $\underline{A}$ *augmented by all subfunctions involved in the* *definition of V. Let* (Σ_V, E_V) *be the signature and equations corresponding with the primitive recursive function V.*

Then the program algebra $\underline{A}_V \cong I(\Sigma_V, E_V)$ *and hence* (Σ_V, E_V) *is an algebraic specification for the sca task algebra (A, V).*

Some programs based on the algorithmic nature of the proof of the theorem have been constructed and applied to scas by B C Thompson.

This material will be included in a definitive paper on the unit delay model: Thompson and Tucker [90]. From the point of view of the theory of algebraic specifications, notice that this machinery provides a huge number of algebraic specifications that are *not* related to the stack: clocked hardware; systolic arrays; cellular automata; neural nets, for instance!

The nature of the stream algebras, and other second order algebras, has inspired a special theory of higher order algebraic specifications: see Meinke [90].

4.4 The theory of scas as a theory of hardware

We aim to create a comprehensive mathematical theory of hardware design, that is based on the concept of a sca, and on the methods and software tools associated with algebra and logic. In our research to date we have emphasised the analysis of case studies in hardware design and used them to shape our theory, rather than using general mathematical questions. This has enabled us

(i) to establish the theory of scas independently of the theory of computable functions;

(ii) to develop, and to evaluate, practically applicable formal methods and software tools; and

(iii) to use the theory in teaching mathematically well-founded design methods to a range of students of computer science and electronic engineering.

Here is a short survey of the research programme which can be divided into the following categories:

4.4.1 Models We have concentrated on the simple case of scas with unit delays, which is general enough to treat a huge number of interesting examples, as will be described shortly. The first announcement of the methods was made in Thompson and Tucker [85]; and a comprehensive theoretical account of this case was made in Thompson [87]. In Thompson and Tucker [90] we describe in detail the essential ideas of the theory. Most of work to date has concerned scas with unit delay. However, in Hobley [90], case studies of signal processing algorithms with non-unit delays are studied using the course-of-values recursive functions.

In addition to the functional model based on simultaneous recursive functions on $\underline{A}$, which is suited to work on *specification* and *verification*, other models have been considered in the unit delay case:

(i) An imperative model based on concurrent assignments and function procedures on A; this is suited to work on *programming, simulation* and *testing*.
(ii) A directed graph model based on A; this is suited to work on *architecture* and *layout*.

Some models from each of these families have been defined formally by means of small languages, and, in particular, proved to be computationally equivalent. A meticulous study of the equivalence of the concurrent assignment model and the unit delay model is made in Thompson [87]; here there is an emphasis on compiler correctness, and performance is included in the analysis. A summary of the imperative model has been published in Thompson and Tucker [88].

A comprehensive study of the graph model, including its equivalence with other models, is made in Meinke [88]; see also Meinke and Tucker [88].

In formulating and classifying models of synchronous concurrent computation, we are following the pattern of work associated with "classical" and generalised computability theory which aims at formulating a Church-Turing Thesis to establish the scope and limits of computation. The idea of a multi-representational environment, in which it is possible to intercompile between representations depending on one's work in the design of the algorithm, also motivates work on models.

The theory includes a method for the careful analysis of the performance of scas. This has been applied to problems in the hierarchical design of integrated circuits in Williams [89].

The theoretical development of models of scas is strongly influenced by research on case studies. In addition to computer hardware, we have studied the connection between scas and *discrete space, discrete time, continuous or discrete state dynamical systems;* see Holden, Tucker and Thompson [90a,90b]. Thus, for example, infinite sca algorithms and architectures are being studied to analyse the infinite parallelism exhibited in certain dynamical systems, and as a tool for the design of finite scas.

4.4.2 Specification of scas and hardware A substantial study of the specification of scas and their rôle in the process of designing hardware is underway. An important contribution is a very simple mathematical theory of clocks, and retimings between clocks, based on the following notions:

A *clock* is an algebra $T=(\{0,1,2,...\}|\ 0,\ t+1)$. A *retiming* of clock T to clock T' is a function $r: T \rightarrow T'$ such that (i) $r(0)=0$; (ii) r is monotonic; and (iii) r is surjective.

Some theoretical results have been obtained on linear and nonlinear retimings, hierarchical design, and synchronising clocks but the main interest remains the application of the theoretical concepts in detailed case studies of the design of correlators and convolvers; counters; uarts; computers (including RISCII and VIPER).

The emphasis in this area is on the rigorous analysis of methodological models and practical formal methods. Very general methodological frameworks, based on formally defined notions of specifications as stream transformers over many sorted algebras, and their consistent refinement, have been developed.

See: Harman and Tucker [87, 88a, 88b], Harman [89], and Hobley and Tucker [90], Hobley [90].

4.4.3 Derivation of algorithms The systematic and formal derivation or development of scas have been studied in connection with rasterisation algorithms: see Eker and Tucker [87, 88, 89] and Eker [90]. However, derivation is an area that requires further work. There is a large literature on developing systolic algorithms of various kinds, but much of it is *ad hoc*, informal, and application specific. Nevertheless research by H T Kung, P Quinton, and C Lengauer provide us with a platform on which to build an analysis of the algorithm design process that complements our analysis of the specification design process mentioned in 4.4.2. Studies in the transformation of scas have been initiated in connection with a theoretical analysis of compilation of functional to graph descriptions. Using equational specifications for data types and term rewriting techniques, optimising transformations for scas have been defined as preprocessors to simple verified compilers: see Meinke [88].

4.4.4 Verification of algorithms The functional model, based on algebras and recursive functions, is beautifully suited to the specification and verification-by-hand of scas. In addition to the examples mentioned in 4.4.3, large number of case studies of hardware and systolic algorithms for linear algebra and string processing, have been verified: see Thompson and Tucker [85, 90]; Hobley, Thompson and Tucker [88]; and, in particular, Thompson [87], and Hobley [90].

Furthermore, these recursive functions possess methods for representing the computable functions in *many* formal logics and hence, in principle, allow *many* types of machine supported formal verification. Of interest to us are examples of verification systems based on *equational logic, many sorted resolution logic, first order logic, higher order logic, constructive type theory*, including those of K Hanna, N Daeche and M Gordon (*HOL*, based on Church's type theory); R Constable (*Nuprl,* based on Martin Lof's type theory); and J Goguen (*OBJ*, based on term rewriting). Thus work with a number of existing theorem provers would be more useful for demonstrating the usefulness of our tools. Work has begun with *Nuprl* - see Derrick, Lajos and Tucker [90]; and with *OBJ3* - see Eker, Stavridou, and Tucker [90].

However, it is important to note that *the mathematical theory is about formal design and verification techniques at a level of abstraction above these particular logics and their verification systems; hence, the theory is intended to be independent of and applicable to all these logics and their tools.*

4.4.5 Software tools A detailed design of a programming language and programming environment based on the imperative model (i) in 4.4.1 has been undertaken, and a prototype system constructed, by A R Martin. This system includes: the language, which is called *CARESS* (for Concurrent Assignment REpresentation of Synchronous Systems); a C compiler; a preprocessor; and an interactive tracing/debugging tool. A multilingual shell for animating, editing and debugging specifications of scas has been built. The prototype is robust and convenient enough to have been used in undergraduate teaching; an interface and some new tools have been added by M Poole. With this prototype, it is possible to test specifications against their scas automatically. In addition, a test compiler from a functional notation for the recursive functions to *CARESS* has been made.

For the basic principles behind the system and its design, see Martin and Tucker [87] and, especially, Martin [89].

4.5 Some concluding remarks on the theory of scas

Our interest in making a general theory of hardware requires us to maintain an independent theoretical foundation for the study of scas. The connection between sca theory and general computability theory should be viewed as an invaluable tool that helps in the development of the sca theory. We have seen a few illustrations of this point, in applications of the material in Section 2. Further interesting connections are possible, using the material of Section 3. A few interesting and easily motivated topics for research are as follows.

A large number of generalisations of the existing computational models is needed to allow for more complex processing elements, timing characteristics, and architectures; these models must be compared with one another and classified by constructing compilers.

A central problem is to understand the semantics of *partiality* in terms of scas, which affects all aspects of the theory.

It is possible to model asynchronous nondeterministic computation in terms of synchronous deterministic computation in several ways using multiple clocks. The study of the nondeterminism and asynchrony as abstractions of determinism and synchrony is a basic theoretical task that has applications in the practical modelling of hardware.

Research on the functional model and its connections with equational specifications, and with logic programming techniques, is necessary in order to support work on verification and software tools.

6. REFERENCES

J W de Bakker, *Mathematical theory of program correctness*, Prentice Hall, 1980.

J A Bergstra, J Tiuryn and J V Tucker, Floyd's principle, correctness theories and program equivalents, *Theoretical Computer Science*, 17 (1982) 451-476.

J A Bergstra and J V Tucker, A characterisation of computable data types by means of a finite equational specification method, in J W de Bakker and J van Leeuwen (eds.) *Automata, Languages and Programming, Seventh Colloquium, Noordwijkerhout, 1980*, Springer Lecture Notes in Computer Science 81, Springer-Verlag, Berlin, 1980, pp. 76-90.

J A Bergstra and J V Tucker, Algebraic specifications of computable and semi-computable data structures, Research Report IW 121, Mathematisch Centrum, 1980.

J A Bergstra and J V Tucker, Algebraic specifications of computable and semi-computable data types, *Theoretical Computer Science*, 50 (1987) 137-181.

J A Bergstra and J V Tucker, Initial and final algebra semantics for data type specifications: two characterisation theorems, *Society for Industrial and Applied Mathematics (SIAM) J on Computing*, 12 (1983) 366-387.

A G Cohn, A more expressive formulation of many sorted logic, *J. Automated Reasoning* 3 (1987) 113-200.

J Derrick, G Lajos and J V Tucker, Specification and verification of synchronous concurrent algorithms using the *Nuprl* proof development system, Centre for Theoretical Computer Science Report, University of Leeds, in preparation.

J Derrick and J V Tucker, Logic programming and abstract data types, in *Proceedings of 1988 UK IT Conference*, held under the auspices of the Information Engineering Directorate of the Department of Trade and Industry (DTI), Institute of Electrical Engineers (IEE), 1988, pp. 217-219.

H Ehrig and B Mahr, *Fundamentals of algebraic specifications 1 - Equations and initial semantics*, Springer-Verlag, 1985.

S M Eker, Foundations for the design of rasterisation algorithms and architectures, PhD thesis, University of Leeds, in preparation.

S M Eker, V Stavridou and J V Tucker, Verification of synchronous concurrent algorithms using OBJ3. A case study of the Pixel Planes architecture, in Proceedings of Workshop on Designing Correct Circuits, Oxford, 1990, in preparation.

S M Eker and J V Tucker, Specification, derivation and verification of concurrent line drawing algorithms and architectures, in R A Earnshaw (ed.), *Theoretical foundations of computer graphics and CAD,* Springer-Verlag, 1988, pp. 449-516.

S M Eker and J V Tucker, Specification and verification of synchronous concurrent algorithms: a case study of the Pixel Planes architecture, in P M Dew, R A Earnshaw and T R Heywood (eds), *Parallel processing for computer vision and display,* Addison Wesley, 1989, pp.16-49.

J E Fenstad, *General recursion theory: An axiomatic approach,* Springer-Verlag, Berlin, 1980.

M Fitting, *Fundamentals of generalised recursion theory,* North-Holland, Amsterdam, 1981.

F Fogelman Soulie, Y Robert, M Tchuente (eds.), *Automata networks in computer science,* Manchester University Press, 1986.

J A Goguen, J W Thatcher, E G Wagner, and J B Wright, An initial algebra approach to the specification, correctness and implementation of abstract data types, in R T Yeh (ed.), *Current trends in programming methodology: IV Data structuring*, Prentice Hall, 1978, pp. 80-149.

J A Goguen and J Meseguer, Equality, types, modules, and (why not?) generics for logic programming, *J Logic Programming,* 2 (1984) 179-210.

J A Goguen and J Meseguer, Unifying functional, object-oriented and relational programming with logical semantics, Report SRI-CSL-87-7, SRI International, 1987.

S Greibach,*Theory of program structures: schemes, semantics, verification,* Springer Lecture Notes in Computer Science 36, Berlin, 1975.

N A Harman and J V Tucker, Clocks, retimings, and the formal specification of a UART, in G Milne (ed.) *The fusion of hardware design and verification* (Proceedings of IFIP Working Group 10.2 Working Conference), North-Holland, 1988, pp. 375-396.

N A Harman and J V Tucker, The formal specification of a digital correlator I: User specification process, Centre for Theoretical Computer Science, Report 9.87, University of Leeds, 1987. Also in K McEvoy and J V Tucker [90].

N A Harman and J V Tucker, Formal specifications and the design of verifiable computers, in *Proceedings of 1988 UK IT Conference,* held under the auspices of the Information Engineering Directorate of the Department of Trade and Industry (DTI), Institute of Electrical Engineers (IEE), 1988, pp. 500-503.

L A Harrington *et al.* (eds.) *Harvey Friedman's research on the foundations of Mathematics*, North-Holland, 1985.

K M Hobley, The specification and verification of synchronous concurrent algorithms, PhD thesis, University of Leeds, in preparation.

K M Hobley, B C Thompson, and J V Tucker, Specification and verification of synchronous concurrent algorithms: a case study of a convolution algorithm, in G Milne (ed.) *The fusion of hardware design and verification* (Proceedings of IFIP Working Group 10.2 Working Conference), North-Holland, 1988, pp. 347-374.

K Hobley and J V Tucker, Clocks and retimings, in preparation.

A V Holden, J V Tucker and B C Thompson, The computational structure of neural systems, in A V Holden and V I Kryukov (eds.) *Neurocomputers and attention. I: Neurobiology, synchronisation and chaos*, Manchester University Press, 1990.

A V Holden, J V Tucker and B C Thompson, Can excitable media be considered as computational systems? in A V Holden, M Markus, H G Othmer (eds.) *Nonlinear wave processes in excitable media,* Plenum, New York, 1990, pp. 509-516.

A J Kfoury, The pebble game and logics of programs, in Harrington *et al.* [1985], pp. 317-329.

J W Lloyd, *Foundations of logic programming,* Springer-Verlag, 1984.

A I Mal'cev, *Algebraic systems*, Springer-Verlag, 1973.

A I Mal'cev, *The metamathematics of algebraic systems: Collected Papers 1936-1967*, North-Holland, 1971.

A R Martin and J V Tucker, The concurrent assignment representation of synchronous systems, in J W de Bakker, A J Nijman and P C Treleaven (eds.),

PARLE : Parallel Architectures and Languages Europe, Vol II Parallel languages, Springer Lecture Notes in Computer Science 259, Springer-Verlag, 1987, pp.369-386. A revised and expanded edition appears in *Parallel Computing* 9 (1988/89) 227-256.

K McEvoy and J V Tucker (eds.), *Theoretical foundations of VLSI design,* Cambridge University Press, 1990

K Meinke, A graph theoretic model of synchronous concurrent algorithms, PhD Thesis, School of Computer Studies, University of Leeds, Leeds, 1988.

K Meinke, Universal algrebra in higher types, Theoretical Computer Science, to appear.

K Meinke, Fixed point and initial algebra semantics of recursion equations, Computer Science Division Research Report, University College of Swansea, in preparation.

K Meinke and J V Tucker, Specification and representation of synchronous concurrent algorithms, in F H Vogt (ed.) *Concurrency '88*, Springer Lecture Notes in Computer Science 335, Springer-Verlag, 1988, pp.163-180.

K Meinke and J V Tucker, *Universal algebra* in S Abramsky, D Gabbay, T S E Maibaum (eds.) *Handbook of logic in computer science*, OUP, to appear.

J C Shepherdson, Algorithmic procedures, generalised Turing algorithms, and elementary recursion theory, in Harrington *et al.* [1985], pp. 285-308.

H Simmons, The realm of primitive recursion, *Archive Math. Logic*, 27 (1988) 117-188.

B C Thompson, A mathematical theory of synchronous concurrent algorithms. PhD Thesis, School of Computer Studies, University of Leeds, 1987.

B C Thompson and J V Tucker, Theoretical considerations in algorithm design, in R A Earnshaw (ed.), *Fundamental algorithms for computer graphics*, Springer-Verlag, 1985, pp. 855-878.

B C Thompson and J V Tucker, A parallel deterministic language and its application to synchronous concurrent algorithms, in *Proceedings of 1988 UK IT Conference*, held under the auspices of the Information Engineering Directorate of the Department of Trade and Industry (DTI), Institute of Electrical Engineers (IEE), 1988, pp. 228-231.

B C Thompson and J V Tucker, Synchronous concurrent algorithms, Computer Science Division Research Report, University College of Swansea, in preparation.

J V Tucker and J I Zucker, *Program correctness over abstract data types, with error state semantics*, North Holland, 1988.

J V Tucker and J I Zucker, Horn programs and semicomputable relations on abstract structures, *Automata, Languages and Programming 1989, Proceedings of the Sixteenth Colloquium, Stresa,* Springer Lecture Notes in Computer Science 372, Springer-Verlag, 1989, pp.745-760

J V Tucker, S S Wainer and J I Zucker, Provably computable functions on abstract data types, *Automata, Languages and Programming 1990, Proceedings of the Seventeenth Colloquium, Coventry,* Springer Lecture Notes in Computer Science, Springer-Verlag, 1990.

C Walther, *A many sorted calculus based on resolution and paramodulation*, Pitman, 1987.

A Williams, Theoretical and empirical studies in VLSI complexity theory, PhD Thesis, School of Computer Studies, University of Leeds, Leeds, 1989.

FUNDAMENTALS OF DEDUCTIVE PROGRAM SYNTHESIS*

Zohar Manna
Computer Science Department
Stanford University
and
Computer Science Department
Weizmann Institute of Science

Richard Waldinger
Artificial Intelligence Center
SRI International
and
Computer Science Department
Stanford University

INTRODUCTION

This is an introduction to program synthesis, the derivation of a program to meet a given specification. It focuses on the deductive approach, in which the derivation task is regarded as a problem of proving a mathematical theorem.

Let us outline this approach in very general terms. We here construct only applicative (functional) programs. We are given a specification that describes a relation between the input and output of the desired program. The specification does not necessarily suggest any method for computing the output. To construct a program that meets the specification, we prove the existence, for any input object, of an output object that satisfies the specified conditions. The proof is conducted in a background theory that expresses the known properties of the subject domain and describes the primitives of the programming language. The proof is restricted to be sufficiently constructive so that, in establishing the existence of a satisfactory output object, it is forced to indicate a computational method for finding one. That method becomes the basis for a program that can be extracted from the proof.

In principle, many theorem-proving methods can be adapted for program synthesis. We have developed a *deductive-tableau* proof system that is specially intended for this purpose.

In this paper, we shall begin by motivating work on program synthesis and relating it to other software development technology. We shall then introduce the deductive-tableau proof system and show how programs are extracted from tableau proofs.

* This research was supported in part by the National Science Foundation grants CCR-89-04809, CCR-89-11512, and CCR-89-13641, by the Defense Advanced Research Projects Agency under contract N00039-84-C-0211, and by the United States Air Force Office of Scientific Research under contract AFOSR-90-0057.

NATO ASI Series, Vol. F 79
Logic, Algebra, and Computation
Edited by F. L. Bauer

Specifications

Program synthesis begins with a specification; in our case, this is a representation of the relationship between the input and output. A specification should be a description of the purpose or expected behavior of the desired program. Ideally, it is close to the intentions of the users of the system. A good specification is clear and readable; we do not care if it describes an efficient computation, or indeed any computation at all. A program, on the other hand, is primarily a description of a computation, preferably an efficient one.

While many languages have been proposed for specification, we have settled on logic in our own work, because it is quite general and appropriate for deductive methods. If other languages are more appropriate for particular subject domains, it is plausible that they be translated into logic.

Let us give logical specifications for a couple of familiar programs.

Example (sorting specification)

Suppose we would like our programs to sort a list of numbers. Then we may be given the specification

$$sort(\ell) \Leftarrow \begin{cases} \textit{find } z \textit{ such that} \\ \qquad perm(\ell, z) \land ord(z). \end{cases}$$

This specification is presented in a background theory of lists of numbers. For a given input object, the list ℓ, the program must return an output object, the list z, satisfying the condition that $perm(\ell, z)$, i.e., that z is a permutation of ℓ, and that $ord(z)$, i.e., that z is in nondecreasing order. The background theory provides the meaning for the constructs *perm* and *ord*. ∎

Note that the specification provides a clear statement of the purpose of a sorting program, but does not describe how we want the list to be sorted. A sorting program itself (such as *quicksort* or *mergesort*) does describe how the computation is to be performed, but does not state the purpose of the program.

Example (square-root specification)

Suppose we want a program to find a rational approximation to the square root of a nonnegative rational. Then we may be given the specification

$$sqrt(r, \epsilon) \Leftarrow \begin{cases} \textit{find } z \textit{ such that} \\ \qquad \textit{if } \epsilon > 0 \\ \qquad \textit{then } z^2 \leq r \ \land \ r < (z+\epsilon)^2. \end{cases}$$

Here we are given the nonnegative rational r and positive rational error tolerance ϵ as inputs. Our desired output z is less than or equal to $\sqrt{r}$, that is, $z^2 \leq r$, but $z + \epsilon$ is strictly greater than $\sqrt{r}$, that is, $r < (z + \epsilon)^2$. In other words, $\sqrt{r}$ lies in the half-open interval $[z, z + \epsilon)$:

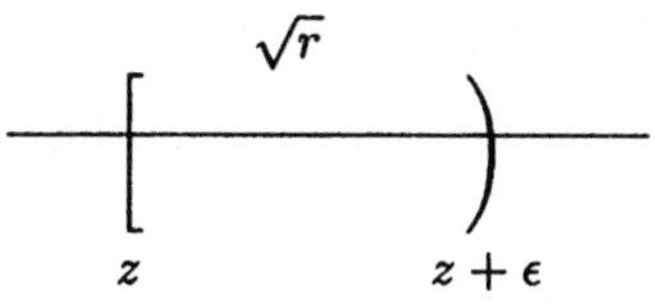

Our background theory is that of the nonnegative rationals. ∎

In general, we shall be dealing with specifications of the form

$$f(a) \Leftarrow \textit{find } z \textit{ such that } \mathcal{Q}[a, z],$$

where $\mathcal{Q}[a, z]$ is a sentence of the background theory.

Deductive Software Technologies

Program synthesis is one of several methods to assist in software development that can be aided by deductive techniques. Here we shall mention some of the other deductive software development methods, with representative references.

- *Program Verification.* Proving that a given program meets a given specification. The method requires that both the specification and the program be given. (Boyer and Moore [BM79]). This is the oldest of the deductive methods.
- *Program Transformation.* Transforming a given program into a more efficient, perhaps less understandable equivalent (Burstall and Darlington [BD77]).
- *Rapid Prototyping.* Assuring a potential user that a specification actually does agree with his expectations (Guttag and Horning [GH80]).
- *Logic Programming.* Executing a program expressed in logic (Kowalski [Kow74]).
- *Debugging.* Exhibiting inputs that cause a program to fail to meet its specification (Shapiro [Sp83]).
- *Modification.* Altering a given program to reflect changes in its specification or environment (Dershowitz [De83]).

In a somewhat different category, we may consider a variety of knowledge-based software development methods (e.g., Smith [Sm85]) which rely on imitating the techniques of the experienced programmer. Automated deduction is exploited here in an auxiliary role; the programming process is not regarded as a task of proving a theorem, but rather as one with many deductive subtasks.

Many researchers in formal methods for software development (e.g., Dijkstra [Dij76]) do regard programming as primarily a deductive process, but are not at all concerned with automating the task. Rather, they intend to provide intellectual tools for the programmer.

All of these methods rely on deductive techniques, and several of them are less ambitious than full program synthesis. By developing more powerful theorem-proving techniques that are specialized to software-engineering applications, we can make progress in several of these areas at once.

OUTLINE OF DEDUCTIVE PROGRAM SYNTHESIS

In this section we shall give a more detailed outline of program synthesis and its relation to mathematical proofs.

In general, we are given a specification

$$f(a) \Leftarrow \textit{find } z \textit{ such that } \mathcal{Q}[a, z].$$

The theorem corresponding to this specification is

$$(\forall a)(\exists z)\mathcal{Q}[a, z].$$

In other words, for every input a, there exists an output z that satisfies the input-output relation $\mathcal{Q}[a, z]$. The proof is restricted to be sufficiently constructive to indicate a method for finding z in terms of a. That method is expressed by a single term $t[a]$, which can be extracted from the proof. The term indicates what substitutions were made for z to allow the proof to go through. The program we produce is then

$$f(a) \Leftarrow t[a].$$

We describe the method as if there were only one input and output. In fact, we can have several of each. If there is more than one output, we define a separate function for each. In the following example, there are two outputs.

Example (front/last derivation outline)

In the theory of finite strings, we would like to construct a program to find, for a given nonempty string s, two outputs: the last character $last(s)$ of s and the string $front(s)$ of all but the last character of s. For example, if s is the string BADA, $front(s)$ is the string BAD and $last(s)$ is the character A.

The program may be specified as follows:

$$\langle front(s),\ last(s)\rangle \Leftarrow \textit{find } \langle z_1, z_2\rangle \textit{ such that} \left[\begin{array}{l} \textit{if } \neg(s = \Lambda) \\ \textit{then } char(z_2) \ \wedge\ s = z_1 * z_2 \end{array}\right].$$

In other words, s is to be decomposed into the concatenation $z_1 * z_2$ of two strings, z_1 and z_2, where z_2 consists of a single character. Here Λ is the empty string. Note that characters are regarded as strings.

The theorem corresponding to the specification is

$$(\forall s)(\exists z_1, z_2)\left[\begin{array}{l} \textit{if } \neg(s = \Lambda) \\ \textit{then } char(z_2) \ \wedge\ s = z_1 * z_2 \end{array}\right].$$

The proof is restricted to be sufficiently constructive to indicate a method for finding z_1 and z_2. In this case, the program we shall extract from the proof is

$$front(s) \Leftarrow \begin{cases} \textit{if } char(s) \\ \textit{then } \Lambda \\ \textit{else } head(s) \cdot front(tail(s)) \end{cases}$$

$$last(s) \Leftarrow \begin{cases} \textit{if } char(s) \\ \textit{then } s \\ \textit{else } last(tail(s)) . \end{cases}$$

Here $head(s)$ and $tail(s)$ are, respectively, the first character and the string of all but the first character of the nonempty string s. Also $char(s)$ is true if s consists of a single character. If c is a character and s is a string, the prefix function $c \cdot s$ yields the result of prefixing c to s. Thus, $c \cdot s$ is the same as $c * s$, but $c \cdot s$ is a basic function defined only for a character and a string. The concatenation function $s_1 * s_2$ is defined in terms of the prefix function, for any two strings s_1 and s_2. ∎

The structure of the proof of the theorem determines the structure of the program we extract. In particular, a case analysis in the proof corresponds to the formation of a conditional or test in the program. The use of the principle of mathematical induction in the proof coincides with the appearance of recursion or other repetitive constructs in the program. If the proof requires some lemmas, the program will invoke some auxiliary subprograms. Of course, different proofs of the theorem may lead to different programs, some of which may be preferable to others.

The phrasing of a specification as a theorem is quite straightforward. If a proof is sufficiently constructive, the extraction of the program is purely mechanical. Thus the main problem of deductive program synthesis is finding a sufficiently constructive proof of the theorem. We now turn our attention to the field of theorem proving, or automated deduction.

Theorem Proving

We may distinguish between automatic and interactive systems, but this distinction is not sharp. Implementers of interactive systems introduce automatic features, to reduce the burden on the user. At the same time, implementers of automatic systems introduce interactive controls, so the user can assist the system to discover proofs that are too difficult for it to find automatically.

Although interactive systems are susceptible to gradual automation, most of them are intended to help the user to check and flesh out a proof he has already outlined by hand, rather than to discover a new proof. The logical framework imbedded in the automatic systems are more conducive to proof discovery.

We may classify automatic theorem provers according to logical theories they focus on.

- *Predicate Logic with Equality.* Much work has exploited the resolution (J. A. Robinson [Ro65]) and paramodulation (Wos and G. Robinson [WR69]) inference rules for these theories. Theorem provers based on these ideas, such as those developed at the Argonne National Laboratory (McCune [Mc89]), regularly settle open questions in mathematics and logic [WW83], admittedly in areas (such as combinatory logic and equivalential calculus) in which human intuition is weak. Recent theorem proving systems in this theory have employed term-rewriting systems (Kapur and Narendran [KN84]) and connection methods (Andrews [An81], Bibel [Bi83]), rather than resolution and paramodulation, as the primary inference rule.
- *Theories with Induction.* A separate body of work focuses on proofs requiring the principal of mathematical induction. The Boyer-Moore system (Boyer and Moore [BM79]) has been motivated by and applied to large problems in program verification, but has also been applied to the interactive reconstruction of large proofs in mathematics and logic, such as the Godel Incompleteness theorem (Shankar [Sn85]).

All of this work is relevant to program synthesis. Yet it is difficult to find an existing system with all the features we need. We require the ability to prove theorems involving the quantifiers and connectives of first-order logic and the mathematical induction principle. The Argonne systems, for example, do well with pure predicate logic but have no facilities for inductive proofs. The Boyer-Moore system, which specializes in proof by induction, does not prove theorems with existential quantifiers.

Many of the interactive systems have grown out of LCF (Gordon et al [GMW79]), which was based on Scott's Logic of Computable Functions. Although these systems are under user control, there is the capability to encode commonly repeated patterns of inference as tactics. The system Isabelle (Paulson [Pa89]) arises from LCF but is generic; that is, it allows us to describe a new logic and then prove theorems in that logic (cf. Felty and Miller [FM88]).

Of particular relevance to program synthesis is the development of interactive systems to prove theorems in constructive logics. The Nuprl system (Constable et al [Con86], cf. Coquand and Huet [CH88], Sato [Sa79], Hayashi and Nakano [HN88]) is based on Martin-Löf's constructive logic (Martin-Löf [M-L82], Nordstrom et al [NPS90]) and has been applied to problems in program derivation as well as mathematics.

Although a derivation proof must be sufficiently constructive to allow us to extract a program, it does not need to be carried out in a constructive logic. Typically, most of a derivation proof has no bearing on what program we extract; it deals with showing that a program fragment extracted from some other part of the proof satisfies some additional conditions. Since many intuitively natural steps are not constructive, it is too constraining to carry out the entire derivation proof in a constructive logic. In our treatment, we adopt a classical logic, restricting it to be constructive only when necessary.

Most theorem-proving systems can be adapted to program synthesis and other software-engineering applications. The deductive framework we employ in this discussion is a hybrid. It incorporates ideas from resolution and inductive theorem proving and it is intended for both interactive and automatic implementation.

PRELIMINARIES

In this section we introduce some formal preliminaries. We are a bit brisk here; the section may be skimmed by those already familiar with these notions. Those wanting a more detailed explanation may refer to Manna and Waldinger ([MW85], [MW90]).

Language

We first define the *expressions* of our language, which consist of the terms and the sentences.

The *terms* include the constants $a, b, c, \ldots$, the variables $u, v, w, \ldots$, and may be constructed by the repeated application of function symbols $f, g, h, \ldots$ to other terms. For example, $f\big(a,\ g(a,x)\big)$ is a term. Also, if $\mathcal{F}$ is a sentence and s and t are terms, the conditional (*if* $\mathcal{F}$ *then* s *else* t) is a term; we call the *if-then-else* operator a *term constructor*.

Atomic sentences (or *atoms*) are constructed by applying predicate symbols $p, q, r, \ldots$ to terms. For example, $p\big(u,\ f\big(a,\ g(a,x)\big)\big)$ is an atomic sentence. We allow both prefix and infix notations for function and predicate symbols. We include the equality symbol $=$ as a predicate symbol.

Sentences include the truth symbols *true* and *false* and the atomic sentences, and may be constructed by the repeated application of the connectives $\wedge$, $\vee$, $\neg$, ... and the quantifiers $(\forall x)$ and $(\exists x)$ to other sentences. We use the notation *if-then* for implication (in place of the conventional arrow or horseshoe). We include a conditional connective *if-then-else*; in other words, if $\mathcal{F}$, $\mathcal{G}$, and $\mathcal{H}$ are sentences then (*if* $\mathcal{F}$ *then* $\mathcal{G}$ *else* $\mathcal{H}$) is also a sentence. We rely on context to distinguish between the conditional connective and the conditional term constructor.

A *closed* expression contains no free (unquantified) variables. A *ground* expression contains no variables at all. A *herbrand* expression is ground and contains no connectives, term constructors, or equality symbols; thus $g(a)$ is a herbrand term and $p\big(a,\ f(a,b)\big)$ is a herbrand atom.

Interpretation and Truth

The truth of a sentence is defined only with respect to a particular interpretation. Intuitively speaking, we may think of an interpretation as a situation or case. We adopt the herbrand notion and define an interpretation as a finite or infinite set of herbrand atoms. Informally, we think of the elements of the interpretation as a complete list of the herbrand atoms that are true in the corresponding situation. The truth-value of any closed sentence with respect to the interpretation is determined by the recursive application of the following semantic rules.

- A herbrand atom $\mathcal{P}$ is true under an interpretation $\mathcal{I}$ if $\mathcal{P} \in \mathcal{I}$.
- If a sentence is not closed, we do not define its truth-value. Thus we do not say whether $p(x)$ is true under $\{p(a)\}$. (Henceforth in this section we speak only of closed sentences.)
- A (closed) sentence $(\mathcal{F} \;\wedge\; \mathcal{G})$ is true under $\mathcal{I}$ if $\mathcal{F}$ and $\mathcal{G}$ are both true under $\mathcal{I}$; similarly for the other connectives.
- A (closed) sentence $(\exists x)\mathcal{F}[x]$ is true under $\mathcal{I}$ if there is a herbrand term t such that $\mathcal{F}[t]$ is true under $\mathcal{I}$; here $\mathcal{F}[t]$ is the result of replacing all free occurrences of x in $\mathcal{F}[x]$ with t. For example, the sentence $(\exists x)p(x)$ is true under the interpretation $\{p(a)\}$ because a is a herbrand term and $p(a)$ is true under $\{p(a)\}$. Similarly for the universal quantifier.
- If (*if* $\mathcal{P}$ *then* s *else* t) is a closed term, a (closed) sentence $\mathcal{F}[$*if* $\mathcal{P}$ *then* s *else* $t]$ is true under $\mathcal{I}$ if the sentence (*if* $\mathcal{P}$ *then* $\mathcal{F}[s]$ *else* $\mathcal{F}[t]$) is true under $\mathcal{I}$.
- For herbrand terms s and t, $s = t$ is true under $\mathcal{I}$ if, for each herbrand atom $\mathcal{P}\langle s\rangle$, $\mathcal{P}\langle s\rangle \in \mathcal{I}$ if and only if $\mathcal{P}\langle t\rangle \in \mathcal{I}$. Here $\mathcal{P}\langle t\rangle$ is obtained from $\mathcal{P}\langle s\rangle$ by replacing exactly one free occurrence of s with t. This holds only when s and t are indistinguishable under $\mathcal{I}$. For example, $a = b$ is true under the interpretation $\{p(a), p(b)\}$, but false under the interpretation $\{q(a,b),\ q(a,a),\ q(b,b)\}$; $q(a,a)$ belongs to the latter interpretation, but $q(b,a)$ does not. In general, if a closed sentence $s = t$ is true under $\mathcal{I}$, we shall also say that $s = t$ under $\mathcal{I}$ or that s and t are *equal* under $\mathcal{I}$.

Models and Theories

An interpretation $\mathcal{I}$ is a *model* for a finite or infinite set of (closed) sentences $\mathcal{S}$ if every sentence in $\mathcal{S}$ is true under $\mathcal{I}$. Thus the interpretations $\{p(a)\}$ and $\{p(b)\}$ are models for the set of sentences $\{(\exists x)p(x),\ p(a) \vee p(b)\}$ but the interpretation $\{p(b)\}$ is not a model for the set of sentences $\{p(a)\}$.

A set of sentences $\mathcal{S}$ implies a sentence $\mathcal{F}$ if $\mathcal{F}$ is true under every model for $\mathcal{S}$. For example, the set $\{p(a)\}$ implies the sentence $(\exists x)p(x)$. The theory TH defined by a set of sentences $\mathcal{S}$ is the set of all closed sentences implied by $\mathcal{S}$. (This is also called the *deductive closure* of $\mathcal{S}$.) We say that the sentences that belong to TH are *valid* in the theory. We call $\mathcal{S}$ the set of *axioms* for the theory TH.

The valid sentences of a theory are true under every model for the theory. The *contradictory* sentences of the theory are defined to be those that are false under every model for the theory. A sentence $\mathcal{F}$ is contradictory in the theory if and only if its negation $\neg\mathcal{F}$ is valid in the theory.

Examples (theories)

The theory defined by the empty set $\{\,\}$ of axioms is ***predicate logic*** PL. For example, $(\exists x)p(x) \vee (\forall x)\neg p(x)$ is a valid sentence of predicate logic. Any interpretation is a model for predicate logic.

The ***total reflexive*** theory TR is defined by the following two axioms.

$$(\forall u)[u \succeq u] \qquad (\textit{reflexivity})$$

$$(\forall u)(\forall v)[u \succeq v \;\vee\; v \succeq u] \quad (\textit{totality})$$

By convention, we shall omit outermost universal quantifiers from axioms. Thus, we may write the axioms for the total reflexive theory TR as

$$u \succeq u \qquad (\textit{reflexivity})$$

$$u \succeq v \;\vee\; v \succeq u \quad (\textit{totality})$$

The sentence

$$(\forall x)(\forall y)(\exists z)[z \succeq x \;\wedge\; z \succeq y]$$

is valid in this theory. ∎

When we say that a (closed) sentence is valid, without specifying a theory, we shall mean that it is valid in predicate logic. If a sentence is valid (in predicate logic), it is valid in any theory.

The models for a theory are the same as the models for its axioms. Intuitively speaking, a model for a theory corresponds to a situation that could possibly happen. For example, an interpretation that contains neither $a \succeq b$ nor $b \succeq a$ is not a model for the total reflexive theory TR, because it violates the ***totality*** axiom.

Substitutions

A ***substitution*** is a set $\{x_1 \leftarrow t_1, \ldots, x_n \leftarrow t_n\}$ of replacement pairs $x_i \leftarrow t_i$, where the x_i are distinct variables, the t_i are terms, and each x_i is distinct from its corresponding t_i. Thus, $\{x \leftarrow y,\ y \leftarrow g(x)\}$ is a substitution but $\{x \leftarrow a,\ x \leftarrow b\}$ and $\{x \leftarrow x\}$ are not. The ***empty*** substitution $\{\,\}$ contains no replacement pairs.

If e is an expression and $\theta : \{x_1 \leftarrow t_1, \ldots, x_n \leftarrow t_n\}$ is a substitution, then $e\theta$, the ***result of applying θ to e***, is obtained by "safely" replacing each free occurrence of x_i in e with the corresponding term t_i. (The "safety" condition requires that certain quantified variables y in e be given a new name y' if some of the terms t_i also contain occurrences of y. For the precise details, see Manna and Waldinger [MW85].) Applying the empty substitution leaves an expression unchanged; that is, $e\{\ \} = e$ for all expressions e. We shall say that any expression $e\theta$ is an *instance* of e.

The *composition* $\theta\lambda$ of two substitutions θ and λ is a substitution with the property that $e(\theta\lambda) = (e\theta)\lambda$ for all expressions e. For example,

$$\begin{aligned} \{x \leftarrow y\}\{y \leftarrow a\} &= \{x \leftarrow a, y \leftarrow a\} \\ \{x \leftarrow y\}\{y \leftarrow x\} &= \{y \leftarrow x\} \\ \{x \leftarrow y\}\{x \leftarrow a\} &= \{x \leftarrow y\}. \end{aligned}$$

Composition is associative but not commutative. The empty substitution is an identity under composition.

A substitution is a permutation if the terms t_i are the same as the variables x_i, in some order. Thus $\{x \leftarrow y,\ y \leftarrow z,\ z \leftarrow x\}$ is a permutation; $\{x \leftarrow y\}$ is not. Permutations are the substitutions that have inverses; that is, π is a permutation if and only if there is some substitution π^{-1} such that $\pi\pi^{-1} = \{\ \}$.

A substitution θ is *more general* than a substitution ϕ if there exists a substitution λ such that

$$\theta\lambda = \phi\,.$$

For example, $\{x \leftarrow y\}$ is more general than $\{x \leftarrow a,\ y \leftarrow a\}$, because $\{x \leftarrow a,\ y \leftarrow a\} = \{x \leftarrow y\}\{y \leftarrow a\}$. It follows that any substitution θ is more general than itself and the empty substitution $\{\ \}$ is more general than any substitution θ.

A substitution θ is a *unifier* of two expressions d and e if

$$d\theta = e\theta\,.$$

For example, $\{x \leftarrow a,\ y \leftarrow b\}$ is a unifier of the two expressions $p(x, b)$ and $p(a, y)$. If two expressions have a unifier, they are said to be *unifiable*.

A unifier of d and e is *most-general* if it is more general than any unifier of d and e. For example, $\{x \leftarrow y\}$ and $\{y \leftarrow x\}$ are most-general unifiers of x and y. The substitution $\theta : \{x \leftarrow a,\ y \leftarrow a\}$ is a unifier of x and y, and both $\{x \leftarrow y\}$ and $\{y \leftarrow x\}$ are more general than θ.

A *unification algorithm* is a procedure for testing whether two expressions are unifiable. If so, it returns a most-general unifier; otherwise, it returns a special object *nil*, say, which is distinct from any substitution.

THE DEDUCTIVE TABLEAU

Our proofs are represented by a two-dimensional structure, the *deductive tableau*. Each row in a tableau contains a sentence, either an *assertion* or a *goal*, and an optional term, the output entry. (In general, in a given row, there may be one output entry for each output of the desired program.) Thus, typical rows in a tableau have the following forms.

$$\begin{array}{|c|c||c|c|c|}\hline \text{assertions} & \text{goals} & f_1(a) & \ldots & f_n(a) \\ \hline\hline \mathcal{A}_1 & & s_1 & & s_n \\ \hline & \mathcal{G}_1 & t_1 & & t_n \\ \hline \multicolumn{2}{c}{} & \multicolumn{3}{c}{\underbrace{\hspace{10em}}_{\text{output columns}}} \end{array} \begin{array}{l} \left.\begin{array}{c} \\ \\ \\ \end{array}\right\} \text{rows} \\ \\ \\ \end{array}$$

The proof itself is represented by the assertions and goals of the tableau; the output entries serve for extracting a program from the proof. Usually we shall speak as if our tableaux have only a single output column, but in fact the results apply when there are several output columns too.

Before we describe the meaning of a tableau, let us look at an example.

Example (deductive tableau)

assertions	goals	$ub(a_1, a_2)$
	$z \succeq a_1 \ \wedge \ z \succeq a_2$	z
$u \succeq u$		
	$a_1 \succeq a_2$	a_1
	true	*if* $a_1 \succeq a_2$ *then* a_1 *else* a_2

As it turns out, this tableau is part of the derivation of a program to find an upper bound for two objects a_1 and a_2 in the total reflexive theory TR. ∎

Suiting a Tableau

We have said that a tableau may represent a proof and a derivation; it may also be regarded as a specification. Specifications describe sets of permissible output objects, which are identified with ground terms. In this section, we gradually define what it means for a ground term to satisfy a tableau.

We first restrict our attention to a particular interpretation and a single row of a tableau.

Definition (suiting a row)

A closed term t *suits* a row $\boxed{\mathcal{A}}\boxed{}\|\boxed{s}$ (or, respectively, $\boxed{}\boxed{\mathcal{G}}\|\boxed{s}$) under an interpretation $\mathcal{I}$ if, for some substitution λ, the following two conditions are satisfied:

- ***truth condition.*** The sentence $\mathcal{A}\lambda$ is closed and false under $\mathcal{I}$ (or, respectively, the sentence $\mathcal{G}\lambda$ is closed and true under $\mathcal{I}$).
- ***output condition.*** If there is an output entry s, the term $s\lambda$ is closed and $s\lambda$ equals t under $\mathcal{I}$.

In case the output entry s is absent, the ***output*** condition holds vacuously. We call λ a *suiting* substitution. ∎

Example (suiting a row)

If $a_1 \succeq a_2$ is true under an interpretation $\mathcal{I}$, the term a_1 suits the row

	$z \succeq a_2$	z

under $\mathcal{I}$. To see this, we take the suiting substitution λ to be $\{z \leftarrow a_1\}$. The ***truth*** condition holds because $(z \succeq a_2)\lambda$, that is, $a_1 \succeq a_2$, is closed and true under $\mathcal{I}$. The ***output*** condition holds because $z\lambda$, that is, a_1, is closed and equal to a_1 under $\mathcal{I}$.

In that example, the term a_1 was actually identical to the instance $z\lambda$ of the output entry z. The conditional term (*if* $a_1 \succeq a_2$ ***then*** a_1 *else* a_2) is also equal to this instance of z under $\mathcal{I}$, because $a_1 \succeq a_2$ is true, but the two terms are not identical. In fact, the conditional term (*if* $a_1 \succeq a_2$ ***then*** a_1 *else* a_2) also suits this row under $\mathcal{I}$. ∎

If a row has no output entry, the ***output*** condition for suiting a row always holds. This means that, under an interpretation, if some closed term suits the row, then any closed term suits the row, since the ***truth*** condition does not depend on the term. In a sense, a missing output entry may be thought of as a "don't care" condition.

We have defined what it means to suit a single row; now we say what it means to suit an entire tableau.

Definition (suiting a tableau)

Under an interpretation, a closed term *suits* a tableau if it suits some row of the tableau. ∎

If we think of the tableau as a specification, and the interpretation as a situation or case, the closed terms that suit the tableau coincide with the outputs that will meet the specification in that case.

Example (suiting a tableau)

Let $\mathcal{T}$ be the following tableau:

	$a_1 \succeq a_2$	a_1
	$\neg(a_1 \succeq a_2)$	a_2

If $a_1 \succeq a_2$ is true under $\mathcal{I}$, then a_1 suits $\mathcal{T}$ under $\mathcal{I}$, with the empty suiting substitution $\{\,\}$. If, on the other hand, $\neg(a_1 \succeq a_2)$ is true under $\mathcal{I}$, then a_2 suits $\mathcal{T}$ under $\mathcal{I}$. In either case, the conditional term (*if* $a_1 \succeq a_2$ *then* a_1 *else* a_2) suits $\mathcal{T}$ under $\mathcal{I}$. ∎

Satisfying a Tableau

The notion of suiting a tableau depends on the interpretation; a term may suit a tableau under one interpretation and not under another. In that sense suiting is analogous to truth for a sentence. We now introduce a notion of "satisfying" a tableau, which is independent of the particular interpretation. That notion is analogous to validity for a sentence.

Definition (satisfying a tableau)

In a theory TH, a closed term t *satisfies* a tableau $\mathcal{T}$ if t suits $\mathcal{T}$ under every model of TH. ∎

If we think of the tableau as a specification, t corresponds to a program that satisfies the specification.

Example (satisfying a tableau)

Suppose $\mathcal{T}$ is the following tableau:

assertions	goals	$f(a_1, a_2)$
	$z \succeq a_2$	z
	$a_2 \succeq a_1$	a_2

Let our background theory be the total reflexive theory TR. Then the closed term

$$t:\ \textit{if}\ a_1 \succeq a_2\ \textit{then}\ a_1\ \textit{else}\ a_2$$

satisfies $\mathcal{T}$ in TR.

To see this, consider an arbitrary model $\mathcal{I}$ for TR. We distinguish between two cases:

Case: $a_1 \succeq a_2$ is true under $\mathcal{I}$

In this case, t equals a_1 under $\mathcal{I}$. Then, t suits the first row

	$z \succeq a_2$	z

under $\mathcal{I}$, as we have seen. Therefore, t suits $\mathcal{T}$ under $\mathcal{I}$.

Case: $a_1 \succeq a_2$ is false under $\mathcal{I}$

In this case, t equals a_2 under $\mathcal{I}$. Also (by the ***totality*** axiom, since $\mathcal{I}$ is a model for the total reflexive theory TR), $a_2 \succeq a_1$ is true under $\mathcal{I}$. Thus, t suits the second row

	$a_2 \succeq a_1$	a_2

under $\mathcal{I}$. Therefore, t suits $\mathcal{T}$ under $\mathcal{I}$.

Thus, for any model $\mathcal{I}$ for the theory TR, t suits $\mathcal{T}$ under $\mathcal{I}$. Hence, t satisfies the tableau in TR. ▮

Equivalence Between Tableaux

We shall introduce two distinct relations of similarity between tableaux. The stronger relation, equivalence, requires that the two tableaux always have the same suiting terms.

Definition (equivalence of tableaux)

Two tableaux $\mathcal{T}_1$ and $\mathcal{T}_2$ are ***equivalent*** in the theory TH, written $\mathcal{T}_1 \leftrightarrow \mathcal{T}_2$, if and only if for every closed term t and every model $\mathcal{I}$ for TH,

t suits $\mathcal{T}_1$ under $\mathcal{I}$
if and only if
t suits $\mathcal{T}_2$ under $\mathcal{I}$. ▮

That is, for $\mathcal{T}_1$ and $\mathcal{T}_2$ to be equivalent in TH, they must have the same suiting terms under each model for the theory. When we say that two tableaux are equivalent, without specifying a theory, we shall mean that they are equivalent in predicate logic. If two tableaux are equivalent (in predicate logic), they are equivalent in any theory.

Examples of equivalent tableaux will be provided by the following basic properties. The proof of one of these properties is provided; the others are similar.

Property (duality)

For any sentences $\mathcal{A}$ and $\mathcal{G}$ and optional term s, we have

$\mathcal{A}$		s

$\longleftrightarrow$

	$\neg\mathcal{A}$	s

	$\mathcal{G}$	s	$\longleftrightarrow$	$\neg\mathcal{G}$		s

■

In other words, any assertion $\mathcal{A}$ is equivalent to a goal $(\neg\mathcal{A})$, with the same output entry s, if any. Any goal $\mathcal{G}$ is equivalent to an assertion $(\neg\mathcal{G})$, also with the same output entry.

It will follow that, for any tableau, we can push all the assertions into the goal column by negating them, or vice versa, obtaining an equivalent tableau. Nevertheless, the distinction between assertions and goals has intuitive appeal and possible strategic power, so we retain it.

The equivalence relation between tableaux has the *substitutivity* property that if we replace any subtableau of a given tableau with an equivalent tableau, we obtain an equivalent tableau. Hence, the *duality* property tells us that we can push any assertion of a tableau into the goal column by negating it, obtaining an equivalent tableau.

Property (renaming)

For any sentences $\mathcal{A}$ and $\mathcal{G}$, optional term s, and permutation π, we have

$\mathcal{A}$		s	$\longleftrightarrow$	$\mathcal{A}\pi$		$s\pi$

	$\mathcal{G}$	s	$\longleftrightarrow$		$\mathcal{G}\pi$	$s\pi$

■

Applying a permutation to a row has the effect of systematically renaming its free variables. For example, by applying the permutation $\pi : \{x \leftarrow y,\ y \leftarrow z,\ z \leftarrow x\}$ to the assertion

$p(x, y)$		$f(x)$

we obtain the assertion

$p(y, z)$		$f(y)$

The property tells us that these two rows are equivalent.

The *renaming* property tells us that we can systematically rename the free variables of any row, obtaining an equivalent tableau.

We prove the *renaming* property for a goal row.

Proof (renaming property)

Suppose the closed term t suits the row

	$\mathcal{G}$	s

under interpretation $\mathcal{I}$, with suiting substitution λ. Then, by the *truth* condition,

(∗) $\mathcal{G}\lambda$ is closed and true under $\mathcal{I}$

and, by the *output* condition,

(†) $s\lambda$ is closed and equal to t under $\mathcal{I}$.

We show that then t also suits the row

	$\mathcal{G}\pi$	$s\pi$

with suiting substitution $\pi^{-1}\lambda$, where π^{-1} is the inverse of the permutation π.

To show this, we show the *truth* condition,

$(\mathcal{G}\pi)(\pi^{-1}\lambda)$ is closed and true under $\mathcal{I}$,

and the *output* condition

$(s\pi)(\pi^{-1}\lambda)$ is closed and equal to t under $\mathcal{I}$.

But these follow from the original conditions (∗) and (†), because, by properties of substitutions, $(\mathcal{G}\pi)(\pi^{-1}\lambda) = \mathcal{G}(\pi\pi^{-1})\lambda = \mathcal{G}\{\,\}\lambda = \mathcal{G}\lambda$, and similarly for s.

In the other direction, we assume that t suits the row

	$\mathcal{G}\pi$	$s\pi$

with suiting substitution λ, and can show that t also suits the original row

	$\mathcal{G}$	s

with suiting substitution $\pi\lambda$. ∎

Property (instance)

For any sentences $\mathcal{A}$ and $\mathcal{G}$, optional term s, and substitution θ, we have

$\mathcal{A}$		s

$\longleftrightarrow$

$\mathcal{A}$		s
$\mathcal{A}\theta$		$s\theta$

	$\mathcal{G}$	s

$\longleftrightarrow$

	$\mathcal{G}$	s
	$\mathcal{G}\theta$	$s\theta$

∎

It follows that we may add to a tableau any instance of any of its rows, obtaining an equivalent tableau. Note that, while the *duality* and *renaming* properties allow us to replace one row with another, the *instance* property requires that we retain the old row while adding the new one. If we replaced the row, we would not necessarily retain equivalence.

The following property allows us to add or remove from a tableau any valid assertion, retaining the tableau's equivalence. We restrict our attention to a fixed theory TH.

Property (valid assertion and contradictory goal)

Suppose $\mathcal{A}$ is a sentence where every ground instance $\mathcal{A}\theta$ is valid in theory TH; suppose $\mathcal{G}$ is a sentence whose every ground instance $\mathcal{G}\theta$ is contradictory in TH. Then for any tableau $\mathcal{T}$ and term s,

$\mathcal{T} \longleftrightarrow$

$\mathcal{T}$		
$\mathcal{A}$		s

$\mathcal{T} \longleftrightarrow$

$\mathcal{T}$		
	$\mathcal{G}$	s

in theory TH. In other words, $\mathcal{A}$ may be added as an assertion, or $\mathcal{G}$ as a goal, to any tableau, yielding an equivalent tableau. ∎

It follows from the ***valid assertions*** property that any row [*true* | | s] or [| *false* | s] can be dropped from any tableau. These are sometimes called ***trivial*** rows.

We have only defined validity in a theory for closed sentences. However, if $\mathcal{A}$ is an assertion in a tableau that is not closed, we shall often say that $\mathcal{A}$ is a valid sentence when we really mean that every closed instance of $\mathcal{A}$ is valid. The ***valid assertion*** property can then be paraphrased to say that a valid assertion can be added to any tableau, preserving its equivalence.

The following property tells us more about what it means for a row to lack an output entry.

Property (no output)

A row (assertion or goal) with no output entry is equivalent to one whose output entry is a new variable, that is, one that does not occur free in the row.

$\mathcal{T}$: [|] $\longleftrightarrow$ $\mathcal{T}_u$: [| u] ∎

The rationale here is that if some closed term suits either of these rows, then any closed term will. More precisely, a closed term t suits $\mathcal{T}$ with suiting substitution $\{x_1 \leftarrow t_1, \ldots, x_n \leftarrow t_n\}$ if and only if t suits $\mathcal{T}_u$ with suiting substitution $\{u \leftarrow t,\ x_1 \leftarrow t_1, \ldots, x_n \leftarrow t_n\}$.

Primitive Expressions

For some purposes, the notion of equivalence is too strong. We may not care if two tableaux are suited by the same set of closed terms, for each model for the theory, so long as they are satisfied by the same set of closed terms. And we may not care if they are satisfied by precisely the same closed terms, so long as they are satisfied by the same closed terms that correspond to computer programs, that is, those that we know how to compute. This latter idea is captured in the notion of primitive terms.

Definition (primitive expression)

Assume we are given a finite set of constant, function, and predicate symbols, called the *primitive set.* An expression is said to be *primitive* if

- It is quantifier-free.
- All of its constant, function, and predicate symbols belong to the primitive set. ▮

Note that a primitive expression may contain any variables.

Intuitively speaking, the primitive expressions are those that we know how to compute, in terms of the variables and the elements of the primitive set. Typically, the primitive set will include the basic operators of the theory plus those for which we have already derived programs. For example, in deriving a program to compute the multiplication function in the theory of the nonnegative integers, we shall typically include the constant symbol 0, the addition function symbol +, and the equality predicate symbol = in the primitive set.

We can now define a relation of similarity, weaker than equivalence, between tableaux.

Definition (primitively similar)

Two tableaux are *primitively similar* in theory TH if they have the same primitive satisfying terms, that is, for every closed primitive term t,

t satisfies $\mathcal{T}_1$ in TH
if and only if
t satisfies $\mathcal{T}_2$ in TH. ▮

Evidently, if two tableaux are equivalent they are primitively similar. Let us give an example to show that primitive similarity is a strictly weaker notion than equivalence.

Example (equivalence versus primitive similarity)

Consider the two tableaux

$\mathcal{T}_p$:

	$p(a)$	a

$\mathcal{T}_q$:

	$q(a)$	a

These tableau are not equivalent. If $\mathcal{I}_p$ is the interpretation $\{p(a)\}$, a suits $\mathcal{T}_p$ under $\mathcal{I}_p$ but a does not suit $\mathcal{T}_q$ under $\mathcal{I}_p$.

On the other hand, in the theory of predicate logic, no closed term satisfies $\mathcal{T}_p$; in particular, no term suits $\mathcal{T}_p$ under the empty interpretation $\{\,\}$, because $p(a)$ is false under $\{\,\}$. Similarly, no closed term suits $\mathcal{T}_q$ in predicate logic either. Hence, the two tableaux are primitively similar, because they are satisfied by exactly the same primitive satisfying terms, namely, none. ▮

If two tableaux are primitively similar, they specify the same class of programs.

PROPERTIES OF DEDUCTION RULES

Deduction rules add new rows to a tableau. They do not necessarily preserve equivalence, but they do preserve primitive similarity, that is, they maintain the set of primitive closed terms that satisfy the tableau. Thus the program specified by the tableau is unchanged by the application of deduction rules.

Definition (soundness)

A rule for adding new rows to a tableau is *sound* in theory TH if the same primitive closed terms satisfy the tableau in TH before and after applying the rule. ▮

We shall guarantee that each of our deduction rules is sound in the background theory.

Let us introduce some terminology for speaking about deduction rules. We use the following notation to describe a rule:

$\mathcal{T}_r$	$\mathcal{A}_r$		
		$\mathcal{G}_r$	
$\mathcal{T}_g$	$\mathcal{A}_g$		
		$\mathcal{G}_g$	

Here the assertions $\mathcal{A}_r$ and the goals $\mathcal{G}_r$ are the *required* rows $\mathcal{T}_r$, which must be present in the tableau if the rule is to be applied. The assertions $\mathcal{A}_g$ and the goals $\mathcal{G}_g$ are the *generated* rows $\mathcal{T}_g$, which may be added to the tableau by the rule.

The *old* tableau is the tableau before the application of deduction rules; if the rule is applicable, the required rows form a subtableau $\mathcal{T}_r$ of the old tableau. The *new* tableau is the tableau after application of the rule; it is the union of the old tableau and the generated tableau $\mathcal{T}_g$.

Although we shall not yet introduce the deduction rules of our system, we mention one of them as an illustration.

Example (if-split rule)

In tableau notation, the ***if-split*** rule is written

	if $\mathcal{A}$ *then* $\mathcal{G}$	s
$\mathcal{A}$		s
	$\mathcal{G}$	s

In other words, if a row of form (*if* $\mathcal{A}$ *then* $\mathcal{G}$) is present in the tableau, then we may add the new assertion $\mathcal{A}$ and the new goal $\mathcal{G}$. The output entry s for the required goal (*if* $\mathcal{A}$ *then* $\mathcal{G}$), if any, is inherited by the generated assertion $\mathcal{A}$ and the generated goal $\mathcal{G}$. ∎

Description of the Derivation Process

At this point we describe the derivation process and relate it to the deductive-tableau notation.

We are given a specification

$$f(a) \Leftarrow \textit{find } z \textit{ such that } \mathcal{Q}[a, z]$$

in theory TH. We assume that z is the only free variable in $\mathcal{Q}[a, z]$. We are also given a set of primitive symbols. (To allow the formation of recursive programs, we shall include f in the primitive set.)

We form the ***initial*** tableau

assertions	goals	$f(a)$
	$\mathcal{Q}[a, z]$	z
$\mathcal{A}_1$		
$\vdots$		
$\mathcal{A}_n$		

The input a is taken to be a constant; the output z is a variable. The assertions $\mathcal{A}_1, \ldots, \mathcal{A}_n$ in the initial tableaux are known to be valid in TH.

The deductive process proceeds by the application of sound deduction rules to the tableau, which add new rows while maintaining the primitive similarity.

The process continues until we obtain a *final* row, either the assertion

false		t

or the goal

	true	t

where t is a ground primitive term. At this point we may stop the derivation process. The program we obtain is

$$f(a) \Leftarrow t.$$

Example (derivation process)

In the theory of finite strings, we would like to derive a program that, for a given nonempty string s, returns the last character of s and the string of all but the last character of s. Our specification is

$$\langle front(s), last(s)\rangle \;\Leftarrow\; \begin{cases} \textit{find}\ \ \langle z_1, z_2\rangle\ \textit{such that} \\ \qquad \textit{if}\ \neg(s = \Lambda) \\ \qquad \textit{then}\ char(z_2)\ \wedge\ s\ =\ z_1 * z_2\,. \end{cases}$$

In other words, we would like to decompose s into the concatenation $z_1 * z_2$ of a string z_1 and a character z_2; then z_2 is the last character of s and z_1 is the string of all but z_2. We assume s is not equal to the empty string Λ. Note that for this program there are two outputs, z_1 and z_2. That is, we need to compute two functions, *front* and *last*. The primitive set includes all the basic constant, function, and predicate symbols of the theory of strings, as well as the function symbols *front* and *last*.

The corresponding initial tableau then contains the goal

assertions	goals	$front(s)$	$last(s)$
	if $\neg(s = \Lambda)$ *then* $char(z_2)\ \wedge\ s = z_1 * z_2$	z_1	z_2

Here the input s is a constant and the outputs z_1 and z_2 are variables. Properties of the theory of strings are also included in the initial tableau as assertions. For instance, the axioms for the concatenation function are represented as the assertions

$\Lambda * y = y$			
if $char(u)$ *then* $(u \cdot y_1) * y_2\ =\ u \cdot (y_1 * y_2)$			

By the application of deduction rules, new rows are added to the tableau, obtaining a primitively similar tableau. The process continues until ultimately we obtain the final goal:

	true	*if* $char(s)$ *then* Λ *else* $head(s) \cdot front(tail(s))$	*if* $char(s)$ *then* s *else* $last(tail(s))$

The program we extract from the proof is then

$$front(s) \Leftarrow \begin{cases} \textit{if } char(s) \\ \textit{then } \Lambda \\ \textit{else } head(s) \cdot front(tail(s)) \end{cases}$$

$$last(s) \Leftarrow \begin{cases} \textit{if } char(s) \\ \textit{then } s \\ \textit{else } last(tail(s)) . \quad \blacksquare \end{cases}$$

The correctness of the derivation process depends on two properties of tableaux. We begin with a definition.

Definition (correctness)

A program $f(a) \Leftarrow t[a]$ is *correct with respect to* the specification

$$f(a) \Leftarrow \textit{find } z \textit{ such that } \mathcal{Q}[a, z]$$

if the sentence

$$(\forall x)\, \mathcal{Q}[x, f(x)]$$

is valid in the theory TH augmented with the additional axiom

$$(\forall x)\big[f(x) = t[x]\big] . \quad \blacksquare$$

The additional axiom we add to the theory is the definition of the new program f.

Property (initial tableau)

If any closed term $t[a]$ satisfies the initial tableau

assertions	goals	$f(a)$
	$\mathcal{Q}[a, z]$	z
$\mathcal{A}_1$		
$\vdots$		
$\mathcal{A}_n$		

in theory TH, the program $f(a) \Leftarrow t[a]$ is correct with respect to the specification

$$f(a) \;\Leftarrow\; \textit{find } z \textit{ such that } \mathcal{Q}[a, z]$$

in TH. ▮

The proof is omitted. A separate proof, also omitted, shows that the program terminates.

Property (final tableau)

A closed term t satisfies any tableau containing the final assertion

false		t

or the final goal

	true	t

in any theory TH. ▮

Let us prove this.

Proof (final tableau property)

Suppose the tableau contains the final goal

	true	t

Then for any model $\mathcal{I}$ of the theory, the ***truth*** condition holds because ***true*** is true under $\mathcal{I}$, and the ***output*** condition holds because t equals t under $\mathcal{I}$. ▮

Justification of a Deduction Rule

Our deductive system will have several deduction rules. Furthermore, if we wish to apply the system to a particular theory, it may be convenient to introduce new rules peculiar to that theory. To establish the soundness of these rules, we introduce a general method for justifying deduction rules.

For each rule, we formulate a justification condition. If the justification condition holds, then the rule is sound. This is the content of the following result.

Property (justification)

A deduction rule is sound in theory TH if the following *justification condition* holds:

for any model $\mathcal{I}_r$ for theory TH,
there exists a model $\mathcal{I}_g$ for TH such that
for any closed primitive term t,
if t suits the generated tableau $\mathcal{T}_g$ under $\mathcal{I}_g$
then t suits the required tableau $\mathcal{T}_r$ under $\mathcal{I}_r$
and
t suits the old tableau $\mathcal{T}_o$ under $\mathcal{I}_g$
if and only if
t suits $\mathcal{T}_o$ under $\mathcal{I}_r$. ∎

The *justification* condition is exactly what we need to establish that when we add the generated rows to the tableau we are not altering the set of primitive satisfying terms.

Proof (justification property)

Suppose that the *justification* condition holds for a deduction rule. We would like to show that the rule is sound. In other words, we must show that the new tableau and the old tableau specify the same class of primitive closed terms. Because we are adding new rows but not deleting any, we cannot lose any primitive closed terms in applying the rule; we merely must ensure that we do not gain any. In other words, we must guarantee that for any primitive closed term t,

if t satisfies $\mathcal{T}_n$ in theory TH
then t satisfies $\mathcal{T}_o$ in TH.

Suppose t does satisfy $\mathcal{T}_n$ in TH; we must show that t also satisfies $\mathcal{T}_o$. Consider an arbitrary model $\mathcal{I}_r$ for TH; we would like to show that

t suits $\mathcal{T}_o$ under $\mathcal{I}_r$

We have supposed that the *justification* condition holds for this deduction rule. Let $\mathcal{I}_g$ be the model corresponding to $\mathcal{I}_r$ whose existence is guaranteed by the *justification* condition. Because we have supposed that t satisfies $\mathcal{T}_n$ in theory TH, we know that

t suits $\mathcal{T}_n$ under $\mathcal{I}_g$.

The new tableau $\mathcal{T}_n$ consists of the original rows $\mathcal{T}_o$ plus the generated rows $\mathcal{T}_g$. To suit the entire tableau $\mathcal{T}_n$, the term must suit one of these two subtableaux. We distinguish between two cases.

Case: t suits $\mathcal{T}_o$ under $\mathcal{I}_g$.

Then, by the *justification* condition,

t suits $\mathcal{T}_o$ under $\mathcal{I}_r$,

as we wanted to show.

Case: t suits $\mathcal{T}_g$ under $\mathcal{I}_g$.

Then, by the *justification* condition again,

t suits $\mathcal{T}_r$ under $\mathcal{I}_r$.

But since $\mathcal{T}_r$ is a subtableau of the old tableau $\mathcal{T}_o$, we have

t suits $\mathcal{T}_o$ under $\mathcal{I}_r$,

as we wanted to show. ∎

The *justification* property can be used to show soundness of rules that do not preserve the equivalence of the tableau. If a rule does preserve equivalence, it is automatically sound, and there is a simpler way to show that it preserves equivalence.

Property (justification for equivalence)

A deduction rules preserves equivalence in theory TH if the following *justification condition for equivalence* holds:

for any model $\mathcal{I}$ for theory TH,
for any closed term t,
if t suits the generated tableau $\mathcal{T}_g$ under $\mathcal{I}$
then t suits the required tableau $\mathcal{T}_r$ under $\mathcal{I}$. ∎

Proof (justification for equivalence property)

Suppose that the *justification condition for equivalence* holds for a deduction rule. We would like to show that the rule preserves equivalence. In other words, for each model $\mathcal{I}$ for TH, we must show that the sets of terms that suit the two tableaux are the same. Because the rule adds rows but never deletes any, we cannot lose any suiting terms in applying the rule, but we must ensure that we do not gain any. In other words, we must show that, for any closed term t,

if t suits $\mathcal{T}_n$ under $\mathcal{I}$
then t suits $\mathcal{T}_0$ under $\mathcal{I}$.

Suppose t does suit $\mathcal{T}_n$ under $\mathcal{I}$; we must show that t also suits $\mathcal{T}_0$. Because t suits $\mathcal{T}_n$, it must suit either the original subtableau $\mathcal{T}_0$ (as we wanted to show) or the generated subtableau $\mathcal{T}_g$ under $\mathcal{I}$.

If t suits $\mathcal{T}_g$ under $\mathcal{I}$, the *justification condition for equivalence* tells us that it also suits $\mathcal{T}_r$, and therefore $\mathcal{T}_0$, under $\mathcal{I}$, as we wanted to show. ∎

Let us use the *justification condition for equivalence* to show the soundness of the *if-split* rule.

Property (soundness of if-split)

The *if-split* rule preserves equivalence of tableaux, and hence is sound, in any theory TH. ∎

The proof of the soundness of the *if-split* rule requires a technical notion that will also be useful later.

Definition (closing substitution)

Let e be any expression, $y_1, \ldots, y_n$ be a complete list of all the free variables in e, and a be a constant. Then the substitution $\lambda_a = \{y_1 \leftarrow a, \ldots, y_n \leftarrow a\}$ is a ***closing substitution*** for e.

In the case in which there are no free variables in e, that is, if e is closed, we take the closing substitution $\lambda_a = \{\ \}$. ∎

Note that, if λ_a is a closing substitution for e, then $e\lambda_a$ is closed.

Proof (soundness of if-split)

We show that the *justification condition for equivalence* holds for the *if-split* rule.

Let $\mathcal{I}$ be a model for the theory TH and t be any closed term. We suppose that t suits the generated tableau $\mathcal{T}_g$ under $\mathcal{I}$ and show that then t suits the required tableau $\mathcal{T}_r$ under $\mathcal{I}$.

If t suits the generated tableau, it must suit at least one of the two rows

$\mathcal{A}$		s
	$\mathcal{G}$	s

We suppose it suits the assertion. Then, for some suiting substitution λ, we have, by the *truth* condition,

$\mathcal{A}\lambda$ is closed and false under $\mathcal{I}$

and, by the *output* condition,

$s\lambda$ is closed and equal to t under $\mathcal{I}$.

Let λ_a be a closing substitution for $\mathcal{G}\lambda$. We are going to show that t suits the required tableau $\mathcal{T}_r$ under $\mathcal{I}$, with suiting substitution $\lambda\lambda_a$.

Because $\mathcal{A}\lambda$ is closed, $\mathcal{A}\lambda\lambda_a$ is identical to $\mathcal{A}\lambda$, and hence is closed and false under $\mathcal{I}$. Therefore, by the semantic rule for if-then, (*if* $\mathcal{A}\lambda\lambda_a$ *then* $\mathcal{G}\lambda\lambda_a$) is closed and true under $\mathcal{I}$; that is, the *truth* condition

(*if* $\mathcal{A}$ *then* $\mathcal{G}$)$\lambda\lambda_a$ is closed and true under $\mathcal{I}$

holds.

Because $s\lambda$ is closed, $s\lambda\lambda_a$ is identical to $s\lambda$, and hence we have the *output* condition

$s\lambda\lambda_a$ is closed and equal to t under $\mathcal{I}$.

This establishes that t suits the required tableau $\mathcal{T}_r$

	if $\mathcal{A}$ *then* $\mathcal{G}$	s

under $\mathcal{I}$, as we wanted to show.

The proof for the case in which t suits the generated goal is the same. ∎

Simplification

Before we introduce the rules of our system, we would like to describe the simplification process. This is a process in which subexpressions of the tableau are replaced by simpler expressions. Simplification can be applied to the assertions, goals, or output entries of the tableau. Subsentences are replaced by equivalent sentences and subterms are replaced by equal terms. The set of simplifications to be applied depends on the background theory, though some simplifications can be applied in any theory. Because the result of a simplification is always simpler than it was before, termination of the simplification process is guaranteed.

Simplification is not regarded as a deduction rule. While a rule adds new rows to a tableau without changing those that are already present, simplification replaces an old row with a new one. Also, while the application of a deduction rule is at the discretion of a user or control strategy, the simplification process is mandatory and automatic. That is, we shall fully simplify all the rows of our tableau before applying any deduction rule.

Example (simplification)

The ***and-two*** simplification

$$\mathcal{F} \wedge \mathcal{F} \Rightarrow \mathcal{F}$$

allows any subsentence of form $(\mathcal{F} \wedge \mathcal{F})$ to be replaced by the corresponding sentence of form $\mathcal{F}$. Applying that simplification, we replace the row

$p(x) \vee (q(a) \wedge q(a))$		$g(x)$

with the corresponding row

$p(x) \vee q(a)$		$g(x)$

∎

We arbitrarily divide our simplifications into categories. The ***true-false*** simplifications replace subsentences containing instances of the truth symbols ***true*** or ***false***. For example, the ***and-true*** simplification

$$\mathcal{F} \wedge \textit{true} \Rightarrow \mathcal{F}$$

and the ***not-false*** simplification

$$\neg\, \textit{false} \Rightarrow \textit{true}$$

are *true-false* simplifications. We provide a full set of these, so that, after simplification, a sentence will contain no proper suboccurrences of the truth symbols *true* or *false*.

There are other *logical* simplifications that are not *true-false* simplifications, such as the *or-two* simplification,

$$\mathcal{F} \lor \mathcal{F} \Rightarrow \mathcal{F},$$

the *cond-term-two* simplification,

$$\textit{if } \mathcal{F} \textit{ then } s \textit{ else } s \Rightarrow s,$$

and the *all-redundant-quantifier* simplification,

$$(\forall x)\mathcal{F} \Rightarrow \mathcal{F}, \quad \text{where } x \text{ does not occur free in } \mathcal{F}.$$

Finally, there are *theory* simplifications, whose application is limited to a particular theory. For example, if our background theory is the nonnegative integers, we include the *plus-zero-right* simplification for addition,

$$u + 0 \Rightarrow u.$$

In the theory of strings, we have the *left-empty* simplification for concatenation,

$$\Lambda * v \Rightarrow v.$$

THE DEDUCTION RULES

We are now ready to introduce the deduction rules of our system. We divide them into several categories.

- *splitting rules*: break down a row into its logical components
- *resolution rule*: performs a case analysis on the truth of a subsentence of two rows
- *equivalence rule*: replaces a subsentence with an equivalent sentence
- *skolemization rules*: remove quantifiers
- *equality rule*: replaces a subterm with an equal term
- *mathematical induction rule*: assumes that the desired program behaves correctly on inputs smaller than the given one.

We shall describe the splitting rules, the resolution rule, the equality rule, and the mathematical induction rule here.

The Splitting Rules

These rules are logically redundant: Any theorem that can be proved with the help of the splitting rules can also be proved without them. Nevertheless, splitting rules often clarify the presentation of a proof.

We include three splitting rules in our system.

Rule (and-split)

In tableau notation:

$\mathcal{A}_1 \wedge \mathcal{A}_2$		s
$\mathcal{A}_1$		s
$\mathcal{A}_2$		s

In other words, as assertion that is a conjunction can be decomposed into its two conjuncts. The output entries of the required assertion, if any, are inherited by the two generated assertions. If the required row has no output entry, neither do the generated rows. ▮

The *or-split* rule is similar.

Rule (or-split)

	$\mathcal{G}_1 \vee \mathcal{G}_2$	s
	$\mathcal{G}_1$	s
	$\mathcal{G}_2$	s

In other words, a goal that is a disjunction can be decomposed into its two disjuncts. ▮

The *and-split* and *or-split* rules reflect the meaning of the tableau: There is an implicit conjunction between the assertions of our tableau and an implicit disjunction between the goals. Note that there is no *or-split* rule for assertions and no *and-split* rules for goals.

We have seen the *if-split* rule.

Rule (if-split)

	if $\mathcal{A}$ *then* $\mathcal{G}$	s
	$\mathcal{G}$	s
$\mathcal{A}$		s

In other words, an implication can be split into an assertion and a goal, its antecedent and consequent, respectively. ▮

The *if-split* rule reflects the intuitive proof method that, to prove a sentence (*if* $\mathcal{A}$ *then* $\mathcal{G}$), assume the antecedent $\mathcal{A}$ and attempt to prove the consequent $\mathcal{G}$. The justification for the *if-split* rule was used to illustrate the *justification property for equivalence.* The justification for the other splitting rules is similar.

Example (if-split rule)

Suppose our tableau contains the goal

	if $\epsilon > 0$ *then* $z^2 \le r \ \wedge \ r < (z+\epsilon)^2$	z

Then we may add its antecedent as an assertion

$\epsilon > 0$		z

and its consequent as the goal

	$z^2 \le r \ \wedge \ r < (z+\epsilon)^2$	z

∎

The Resolution Rule

The *resolution* rule is a nonclausal version of the classical Robinson [Ro65] resolution principle introduced for program synthesis (Manna and Waldinger [MW80]); a similar nonclausal resolution rule was developed independently by Murray [Mu82]. The rule corresponds to a case analysis in an informal argument, and it accounts for the introduction of conditional terms in program derivation. We present it first as it applies to two goals.

Rule (GG-resolution)

$\mathcal{G}_1[\mathcal{P}]$	s
$\mathcal{G}_2[\mathcal{P}']$	t
$\mathcal{G}_1\theta[\mathit{false}]$ $\wedge$ $\mathcal{G}_2\theta[\mathit{true}]$	*if* $\mathcal{P}\theta$ *then* $t\theta$ *else* $s\theta$

More precisely, the rule allows the following inference:

- We take $\mathcal{G}_1$ and $\mathcal{G}_2$ to be goal rows with no free variables in common; we rename the variables of these rows to achieve this if necessary.
- We require that $\mathcal{P}$ and $\mathcal{P}'$ be free, quantifier-free subsentences of $\mathcal{G}_1[\mathcal{P}]$ and $\mathcal{G}_2[\mathcal{P}]$, respectively, that are unifiable. We let θ be a most-general unifier of these sentences; thus $\mathcal{P}\theta$ and $\mathcal{P}'\theta$ are identical. In general, there can be more than one subsentence $\mathcal{P}$ in $\mathcal{G}_1[\mathcal{P}]$ and more than of subsentence $\mathcal{P}'$ in $\mathcal{G}_2[\mathcal{P}]$; we take θ to be a most-general unifier of all these subsentences.
- We replace all occurrences of $\mathcal{P}\theta$ in $\mathcal{G}_1\theta$ with *false*, obtaining $\mathcal{G}_1\theta[\mathit{false}]$; we replace all occurrences of $\mathcal{P}'\theta$ (that is, $\mathcal{P}\theta$) in $\mathcal{G}_2\theta$ with *true*, obtaining $\mathcal{G}_2\theta[\mathit{true}]$.

- We take the conjunction of the results, obtaining $(\mathcal{G}_1\theta[\mathit{false}] \wedge \mathcal{G}_2\theta[\mathit{true}])$. After simplification, this is added to the tableau as a new goal.
- The output entry associated with the new goal is the conditional term (*if* $\mathcal{P}\theta$ *then* $t\theta$ *else* $s\theta$). The test of this conditional is the unified subsentence $\mathcal{P}\theta$. Then *then*-term and the *else*-term are the appropriate instances $t\theta$ and $s\theta$ of the required goals. ∎

Before discussing the ramifications of this rule, we illustrate it with an example.

Example (resolution rule)

We apply the rule to a goal and a copy of itself. Assume our tableau contains the row

	$z^2 \le r \;\wedge\; \neg \boxed{(z+\epsilon)^2 \le r}^{\,-}$	z

(We shall explain the box and minus sign annotations soon.) This row has the variable z in common with itself; therefore, in the copy we rename z to $\widehat{z}$:

	$\boxed{\widehat{z}^2 \le r}^{\,+} \;\wedge \neg \left[(\widehat{z}+\epsilon)^2 \le r\right]$	$\widehat{z}$

The boxed subsentences $\mathcal{P}$: $(z+\epsilon)^2 \le r$ and $\mathcal{P}'$: $\widehat{z}^2 \le r$ are unifiable: A most-general unifier is $\widehat{z} \leftarrow z+\epsilon$. The unified subsentence $\mathcal{P}\theta$ is then $(z+\epsilon)^2 \le r$.

We apply θ to the two rows; the original row is unchanged, but the renamed copy becomes

	$(z+\epsilon)^2 \le r \;\wedge\; \neg \left[\left((z+\epsilon)+\epsilon\right)^2 \le r\right]$	$z+\epsilon$

We replace all copies of $\mathcal{P}\theta$ in the instantiated original row with *false* and all occurrences of $\mathcal{P}\theta$ in the instantiated copy with *true*. The conjunction of the resulting goals is added to the tableau as a new goal

	$z^2 \le r \;\wedge\; (\neg\, \mathit{false})$ $\wedge$ $\mathit{true} \wedge \neg \left[\left((z+\epsilon)+\epsilon\right)^2 \le r\right]$	*if* $(z+\epsilon)^2 \le r$ *then* $z+\epsilon$ *else* z

The output entry of the new goal is a conditional term whose test is the unified subsentence and whose *then*-term and *else*-term are the appropriate instances of the output entries of the two required rows.

The derived row is simplified to

	$z^2 \leq r \ \wedge\ \neg\left[(z+2\epsilon)^2 \leq r\right]$	*if* $(z+\epsilon)^2 \leq r$ *then* $z+\epsilon$ *else* z

The simplifications that were applied to the goal are the ***true-false*** simplifications

$\neg\ \mathit{false} \ \rightarrow\ \mathit{true}$

$\mathcal{F}\ \wedge\ \mathit{true} \ \rightarrow\ \mathcal{F}$

$\mathit{true}\ \wedge\ \mathcal{F} \ \rightarrow\ \mathcal{F}$

and the numerical simplification

$(u+v)+v \ \rightarrow\ u+2v\,.$ ∎

Digression (binary search)

Let us interrupt the exposition a moment and discuss the intuition behind the step in the preceding example.

The given goal

	$z^2 \leq r \ \wedge\ \neg\left[(z+\epsilon)^2 \leq r\right]$	z

is a consequence of the initial goal from the derivation of the rational square-root program. It expresses the fact that we would like to find a nonnegative rational number z that is an approximation within ϵ less than the exact square root of r. That is, $\sqrt{r}$ should lie in the half-open interval $[z, z+\epsilon)$. If we succeed, z will be a suitable output for the program.

The derived row

	$z^2 \leq r \ \wedge\ \neg\left[(z+2\epsilon)^2 \leq r\right]$	*if* $(z+\epsilon)^2 \leq r$ *then* $z+\epsilon$ *else* z

expresses the fact that it suffices to find a nonnegative rational z that is a cruder approximation, within 2ϵ less than the exact square root of r. That is, $\sqrt{r}$ should lie in the interval $[z, z+2\epsilon)$. If so, the conditional term

if $(z+\epsilon)^2 \leq r$
then $z+\epsilon$
else z

will be a suitable output for the program.

Why is this? Note that $z+\epsilon$ is the midpoint of the interval $[z, z+2\epsilon)$. In the case in which $(z+\epsilon)^2 \leq r$, that is, $z+\epsilon \leq \sqrt{r}$, we know that $\sqrt{r}$ lies in the right half of the interval. But then the *then*-term $z+\epsilon$ is within ϵ less than the exact square root of r.

In the alternative case, in which $r < (z+\epsilon)^2$, that is, $\sqrt{r} < z + \epsilon$, we know that $\sqrt{r}$ lies in the left half of the interval $[z, z+2\epsilon)$. But then the *else*-term z is already within ϵ less than the exact square root of r.

In either case, the value of the conditional term is within ϵ less than the exact square root, and hence is a suitable output for our program.

The derived row contains the basis for the idea of binary search, while the given row did not. This discovery was obtained by a mechanical step, a single application of the resolution rule. ∎

No-Conditional Cases

In applying the resolution rule, we normally introduce a conditional term as the output entry for the derived row. There are some cases, however, in which we apply the rule without introducing a conditional.

Suppose that the output entries s and t of the required rows happen to be unified by the substitution θ; that is, $s\theta$ and $t\theta$ are identical. In this case, the conditional output entry

if $\mathcal{P}\theta$
then $t\theta$
else $s\theta$

is simplified by the *cond-term-two* simplification

if $\mathcal{F}$ *then* u *else* u $\rightarrow$ u

to yield $s\theta$. Thus, in this case the rule introduces no conditional term at all. The resulting program is of course more efficient than if the conditional had been introduced. Moreover, if the test $\mathcal{P}\theta$ is not primitive, we may not know how to compute the conditional at all.

Suppose now that one of the two required goals, say $\mathcal{G}_2$, has no output entry. Then, instead of the conditional, the output entry for the generated goal will be simply $s\theta$, where s is the output entry for $\mathcal{G}_1$.

Why is this? By the *no-output* property, the goal $\mathcal{G}_2$ with no output entry is equivalent to one whose output entry is the new variable u, which does not occur free in the row and which is unaffected by θ. The output entry generated by the standard, conditional case of the rule is (*if* $\mathcal{P}\theta$ *then* $u\theta$ *else* $s\theta$), that is, because u is unaffected by θ, (*if* $\mathcal{P}\theta$ *then* u *else* $s\theta$). By the *instance* property, we may add to our tableau the instance of that row whose goal is the same but whose output entry is (*if* $\mathcal{P}\theta$ *then* $s\theta$ *else* $s\theta$), which it again simplifies to $s\theta$. We shall call this the *one-output* case.

Suppose finally that both goals have no output entry; then the derived goal has no output entry either. Why? By the *no-output* property, again, the first goal is equivalent to one with output entry v, where v is a new variable. By the *one-output* case of the resolution rule, we may associate with the goal the output entry $v\theta$, that is, v. But then, by the *no-output* property again, that output entry can be dropped altogether.

The no-conditional cases of the resolution rule will be illustrated after we have introduced the dual versions.

Dual Versions of the Resolution Rule

We have presented the resolution rule as it applies to two goals. With the ***duality*** property, we can justify dual versions of the rule, that apply to two assertions, or to an assertion and a goal. These may be expressed as follows.

Rule (AA-resolution)

$\mathcal{A}_1[\mathcal{P}]$		s
$\mathcal{A}_2[\mathcal{P}']$		t
$\mathcal{A}_1\theta[\mathit{false}]$ $\vee$ $\mathcal{A}_2\theta[\mathit{true}]$		*if* $\mathcal{P}\theta$ *then* $t\theta$ *else* $s\theta$

■

Rule (AG-resolution)

$\mathcal{A}_1[\mathcal{P}]$		s
	$\mathcal{G}_2[\mathcal{P}']$	t
	$\neg(\mathcal{A}_1\theta[\mathit{false}])$ $\wedge$ $\mathcal{G}_2\theta[\mathit{true}]$	*if* $\mathcal{P}\theta$ *then* $t\theta$ *else* $s\theta$

■

Rule (GA-resolution)

	$\mathcal{G}_1[\mathcal{P}]$	s
$\mathcal{A}_2[\mathcal{P}']$		t
	$\mathcal{G}_1\theta[\mathit{false}]$ $\wedge$ $\neg(\mathcal{A}_2\theta[\mathit{true}])$	*if* $\mathcal{P}\theta$ *then* $t\theta$ *else* $s\theta$

■

The notation restrictions and no-conditional cases for these dual versions of the rule are the same as for the original (GG) version.

The justification for these dual versions lies in pushing the assertions into the goal column by negating them and then applying the GG version of the rule. For the AA version, the resulting goal is then pushed back to the assertion column, negating it once more. The resulting assertion

$$\neg \begin{pmatrix} \neg\, \mathcal{A}_1\theta[\mathit{false}] \\ \wedge \\ \neg\, \mathcal{A}_2\theta[\mathit{true}] \end{pmatrix}$$

is then simplified, with the simplification

$$\neg(\neg\mathcal{F} \wedge \neg\mathcal{G}) \quad \rightarrow \quad \mathcal{F} \vee \mathcal{G},$$

to yield

$$\begin{matrix} \mathcal{A}_1\theta[\mathit{false}] \\ \vee \\ \mathcal{A}_2\theta[\mathit{true}]. \end{matrix}$$

The following application of the *resolution* rule illustrates both the AG version and the *one-output* case of the rule.

Example (dual version, no-conditional)

Suppose our tableau includes the assertion

$\boxed{u \succeq u}^{-}$		

and the goal

	$\boxed{z \succeq a_1}^{+} \;\wedge\; z \succeq a_2$	z

We would like to apply the AG version of the resolution rule to these rows. The two rows have no variables in common. The boxed subsentences are unifiable; a most-general unifier is θ: $\{u \leftarrow a_1,\, z \leftarrow a_1\}$. The result of applying the AG version of the resolution rule is then

	$\neg\, \mathit{false}$ $\wedge$ $\mathit{true} \wedge\; a_1 \succeq a_2$	a_1

which simplifies to

	$a_1 \succeq a_2$	a_1

Because the assertion has no output entry, the derived goal has no conditional; this is a *one-output* case of the rule.

The step illustrated is part of the derivation of a program to find an upper-bound for two objects a_1 and a_2 in the total reflexive theory TR. The intuitive content of the derived row is that, in the case in which $a_1 \succeq a_2$, the term a_1 will be a suitable output for the program. ∎

Polarity

The resolution rule is symmetric and could be applied with the roles of the two rows reversed. For instance, in the preceding section, we applied the AG version of the resolution rule to an assertion and a goal. We could also have applied the GA version to the same goal and assertion, obtaining

	$\mathit{false} \wedge a_1 \succeq a_2$ $\wedge$ $\neg\ \mathit{true}$	a_1

which simplifies to the trivial goal

	false	a_1

It is typical that, of the two symmetric ways of applying the rule, one will not advance the proof. In this section, we introduce a syntactic condition that will allow us to avoid many of these fruitless applications of the resolution rule.

Roughly speaking, a subsentence of a tableau is of negative $(-)$ or positive $(+)$ polarity if it is within the scope of an odd or even number, respectively, of negation $(\neg)$ connectives. Thus, in the goal

	$\neg\left([p(x)]^- \ \wedge\ \neg\left([q(y)]^+\right)\right)$	

$p(x)$ is of negative polarity, because it is within the scope of a single negation, but $q(y)$ is of positive polarity, because it is within the scope of two negations. We have annotated each of these subsentences with its polarity symbol.

We regard the antecedent $\mathcal{F}$ of an implication (*if* $\mathcal{F}$ *then* $\mathcal{G}$) as being within the scope of an additonal implicit negation, because (*if* $\mathcal{F}$ *then* $\mathcal{G}$) is equivalent to $((\neg\mathcal{F})$ *or* $\mathcal{G})$. Also, while each goal has positive polarity, we regard each assertion $\mathcal{A}$ as having negative polarity, because we could push it into the goal column by negating it, obtaining $(\neg\mathcal{A})$. We regard both sides $\mathcal{F}$ and $\mathcal{G}$ of an equivalence $(\mathcal{F} \equiv \mathcal{G})$ as having both polarities $(\pm)$ because $(\mathcal{F} \equiv \mathcal{G})$ is equivalent to (*if* $\mathcal{F}$ *then* $\mathcal{G}$) $\wedge$ (*if* $\mathcal{G}$ *then* $\mathcal{F}$); the first occurrence of $\mathcal{F}$ is within the scope of an additional implicit negation, but the second is not; similarly for $\mathcal{G}$. The *if*-part $\mathcal{F}$ of a conditional sentence (*if* $\mathcal{F}$ *then* $\mathcal{G}$ *else* $\mathcal{H}$) or a conditional term (*if* $\mathcal{F}$ *then* s *else* t) also has both polarities.

Example (polarity)

The following sentence is annotated according to its polarities

$\left[\begin{array}{l} \textit{if } (\textit{if } [p(a)]^- \textit{ then } [q(a)]^+)^+ \\ \textit{then } ([p(a)]^\pm \equiv [q(a)]^\pm)^- \end{array}\right]^-$		

Because the sentence is an assertion rather than a goal, its polarity, and that of all its subsentences, are reversed. ▮

Now that we have defined polarity of a subsentence of a tableau, we can use the notion to describe a strategy for restricting the resolution rule.

Definition (polarity strategy, for resolution)

An application of the resolution rule is ***in accordance with the polarity strategy*** if at least one negative occurrence of the unified subsentences $\mathcal{P}$ is replaced by *false* and at least one positive occurrence of the unified subsentences $\mathcal{P}'$ is replaced by *true*. (The positive and negative occurrences we refer to may actually have both polarities.) ▮

We illustrate the polarity strategy with an example.

Example (polarity strategy)

Suppose our tableau contains the two goals

	$\left[\boxed{p(x)}^{\pm} \equiv q(x)\right] \wedge \left(\boxed{p(y)}^{+} \vee \neg q(a)\right)$	$t(x, y)$
	$\boxed{p(a)}^{+}$	a

The boxed subsentences are unifiable, with most-general unifier $\{x \leftarrow a, y \leftarrow a\}$. Therefore, we may apply the resolution rule, obtaining

	$[\textit{false} \equiv q(a)] \wedge [\textit{false} \vee \neg q(a)]$ $\wedge$ *true*	*if* $p(a)$ *then* a *else* $t(a, a)$

which simplifies to

	$\neg q(a)$	*if* $p(a)$ *then* a *else* $t(a, a)$

This application of the rule is in accordance with the polarity strategy: The subsentence $p(x)$, which has negative (in fact, both) polarities, is replaced by *false*; also the subsentence $p(a)$, which has positive polarity, is replaced by *true*.

We can also reverse the roles of the two goals in applying the resolution rule, obtaining

	false $\wedge$ $[true \equiv q(a)] \wedge (true \vee \neg q(a))$	*if* $p(a)$ *then* $t(a,a)$ *else* a

which simplfies to the trivial goal

	false	*if* $p(a)$ *then* $t(a,a)$ *else* a

This application of the rule is in violation of the polarity strategy, because no negative occurrence of the unified subsentences is replaced by *false*. ∎

We have illustrated the polarity strategy with the GG version of the resolution rule. The strategy is precisely the same for the other versions. We must remember, however, that polarities are reversed in assertions.

Violating the polarity strategy does not always cause us to derive a trivial row; furthermore, observing the strategy does not always prevent us from deriving a trivial row. Nevertheless, it can be shown that observing the polarity strategy can never prevent us from completing a proof. In fact, the strategy never even lengthens the proof. Because observing the strategy greatly reduces the number of applications of the rule we must consider, there is little reason to ever apply the resolution rule in violation of the polarity strategy.

Relation with Classical Resolution

The question arises as to how the nonclausal resolution rule presented here relates to the classical clausal resolution principle introduced by Robinson [Ro65]. The clausal version of the rule is only applied to assertions that are in clausal form; that is, they are disjunctions of literals, where each literal is either an atom or the negation of an atom. If we apply the clausal resolution principle to the two clauses

$$\mathcal{P} \vee \mathcal{Q}$$

$$\neg \mathcal{P}' \vee \mathcal{R}$$

(where $\mathcal{P}$ is an atom and $\mathcal{Q}$ and $\mathcal{R}$ are themselves clauses) we obtain

$$\mathcal{Q}\theta \vee \mathcal{R}\theta$$

where θ is a most-general unifier of $\mathcal{P}$ and $\mathcal{P}'$.

On the other hand, if we apply the AA version of the resolution rule to the corresponding two assertions

$\mathcal{P} \vee \mathcal{Q}$		
$\neg \mathcal{P}' \vee \mathcal{R}$		

we obtain the new assertion

$\mathit{false} \vee \mathcal{Q}\theta$ $\vee$ $\neg\, \mathit{true} \vee \mathcal{R}\theta$		

which simplifies to

$\mathcal{Q}\theta \vee \mathcal{R}\theta$		

This assertion corresponds to the same clause produced by the classical resolution rule.

Justification of the Resolution Rule

Let us now justify the resolution rule.

Property (soundness of resolution)

The resolution rule preserves equivalence of tableaux, and hence is sound, in any theory TH. ∎

Proof (soundness of resolution)

Let us reproduce the resolution rule here for convenience:

$\mathcal{T}_r$		$\mathcal{G}_1[\mathcal{P}]$	s
		$\mathcal{G}_2[\mathcal{P}']$	t
$\mathcal{T}_g$		$\mathcal{G}_1\theta[\mathit{false}]$ $\wedge$ $\mathcal{G}_2\theta[\mathit{true}]$	*if* $\mathcal{P}\theta$ *then* $t\theta$ *else* $s\theta$

We show that the rule satisfies the *justification condition for equivalence.* Let $\mathcal{I}$ be a model for TH and r be any closed term. We suppose that r suits the generated tableau $\mathcal{T}_g$ under $\mathcal{I}$ and show that r then suits the required tableau $\mathcal{T}_r$ under $\mathcal{I}$.

If r suits $\mathcal{T}_g$ under $\mathcal{I}$, there must be a suiting substitution λ. In other words, by the *truth* condition,

$$\begin{pmatrix} \mathcal{G}_1\theta[\mathit{false}] \\ \wedge \\ \mathcal{G}_2\theta[\mathit{true}] \end{pmatrix} \lambda, \quad \text{that is,} \quad \begin{pmatrix} \mathcal{G}_1\theta[\mathit{false}]\,\lambda \\ \wedge \\ \mathcal{G}_2\theta[\mathit{true}]\,\lambda \end{pmatrix}$$

is closed and true under $\mathcal{I}$, and, by the *output* condition,

$$\begin{pmatrix} \mathit{if}\ \mathcal{P}\theta \\ \mathit{then}\ t\theta \\ \mathit{else}\ s\theta \end{pmatrix} \lambda, \quad \text{that is,} \quad \begin{matrix} \mathit{if}\ \mathcal{P}\theta\lambda \\ \mathit{then}\ t\theta\lambda \\ \mathit{else}\ s\theta\lambda \end{matrix}$$

is closed and equal to r under $\mathcal{I}$.

It follows that

$\mathcal{G}_1\theta[\mathit{false}]\,\lambda$ is closed and true under $\mathcal{I}$,

$\mathcal{G}_2\theta[\mathit{true}]\,\lambda$ is closed and true under $\mathcal{I}$,

and $\mathcal{P}\theta\lambda$, $t\theta\lambda$, and $s\theta\lambda$ are all closed.

The proof distinguishes between two cases.

Case: $\mathcal{P}\theta\lambda$ is false under $\mathcal{I}$.

In this case, we shall show that r suits the first row

	$\mathcal{G}_1[\mathcal{P}]$	s

of $\mathcal{T}_r$ with suiting substitution $\theta\lambda$.

We must show the ***truth*** condition, that

$\mathcal{G}_1[\mathcal{P}]\theta\lambda$ is closed and true under $\mathcal{I}$.

But $\mathcal{G}_1\theta[\mathit{false}]\lambda$ may be obtained from $\mathcal{G}_1[\mathcal{P}]\theta\lambda$ by replacing some occurrences of the closed subsentence $\mathcal{P}\theta\lambda$ with the sentence *false*, which has the same truth-value in this case. Also $\mathcal{G}_1\theta[\mathit{false}]\lambda$ is itself closed and true under $\mathcal{I}$. This implies the desired ***truth*** condition.

We must also show the ***output*** condition, that

$s\theta\lambda$ is closed and equal to r under $\mathcal{I}$.

But the conditional term (*if* $\mathcal{P}\theta\lambda$ *then* $t\theta\lambda$ *else* $s\theta\lambda$) is, in this case, equal to $s\theta\lambda$ under $\mathcal{I}$. Also the conditional term is closed and equal to r under $\mathcal{I}$. This implies the desired ***output*** condition.

Hence, in this case, r suits the first row of $\mathcal{T}_r$ under $\mathcal{I}$. In the alternative case, in which $\mathcal{P}\theta\lambda$ is true under $\mathcal{I}$, we show that r suits the second row of $\mathcal{T}_r$ under $\mathcal{I}$, again with suiting substitution $\theta\lambda$. Hence, in either case, r suits the required tableau $\mathcal{T}_r$ under $\mathcal{I}$. This shows that the rule satisfies the ***justification condition for equivalence.*** ∎

The Equality Rule

Normally we describe the properties of the functions and relations of our theory by introducing assertions into the tableau. For example, we may describe the $\succeq$ relation of the total reflexive theory TR by introducing axioms into our tableau as assertions:

$u \succeq u$		
$u \succeq v \lor v \succeq u$		

Proven properties may also be introduced into the tableau as additional assertions, such as the following property of the upper-bound function ub:

$ub(u,v) \succeq u \land$ $ub(u,v) \succeq v$		

This approach is not adequate for describing the equality relation, for which we require a large number of so-called ***functional-*** and ***predicate-substitutivity*** axioms, such as

$$\begin{array}{lcl} \textit{if } u = v & \quad\text{and}\quad & \textit{if } u = v \\ \textit{then } f(u, w) = f(v, w) & & \textit{then } p(w, u, x) \equiv p(w, v, x) \end{array}$$

Several such axioms may be required for each function and predicate symbol used in our proof. If we add all the required instances, the strategic ramifications are disastrous. These axioms spawn numerous consequences irrelevant to the theorem at hand.

Most theorem provers that have been successful at working with the equality relation have used special equality rules, rather than representing its properties axiomatically. The equality rule we use here is a nonclausal version of the paramodulation rule (Wos and G. Robinson [WR69]).

We present the rule first as it applies to two assertions.

Rule (AA-equality)

$\mathcal{A}_1[\ell = r]$		s
$\mathcal{A}_2\langle \ell' \rangle$		t
$\mathcal{A}_1\theta[\textit{false}]$ $\vee$ $\mathcal{A}_2\theta\langle r\theta \rangle$		*if* $(\ell = r)\theta$ *then* $t\theta$ *else* $s\theta$

More precisely, the rule allows the following inference:

- We take $\mathcal{A}_1$ and $\mathcal{A}_2$ to be assertion rows with no free variables in common; we rename the variables of these rows to achieve this if necessary.
- We require that $\ell = r$ be a subsentence of $\mathcal{A}_1[\ell = r]$ and ℓ' be a subterm of $\mathcal{A}_2\langle \ell' \rangle$ such that ℓ and ℓ' are unifiable, with most-general unifier θ. Here $\ell = r$ and ℓ' are free and quantifier-free subexpressions. As in the resolution rule, there may be many distinct subsentences $\ell = r$ in $\mathcal{A}_1[\ell = r]$ and many subterms ℓ' in $\mathcal{A}_2\langle \ell' \rangle$; the substitution θ must unify all the appropriate expressions.
- We replace all occurrences of $(\ell = r)\theta$ in $\mathcal{A}_1\theta$ with *false*, obtaining $\mathcal{A}_1\theta[\textit{false}]$; we replace one or more occurrences of $\ell'\theta$ (that is, $\ell\theta$) in $\mathcal{A}_2\theta$ with $r\theta$, obtaining $\mathcal{A}_2\theta\langle r\theta \rangle$. (Because we replace some but not necessarily all occurrences, we use the angle brackets $\langle\ \rangle$ rather than the square brackets $[\ ]$ to denote replacement.)
- We take the disjunction of the results, obtaining $(\mathcal{A}_1\theta[\textit{false}] \vee \mathcal{A}_2\theta\langle r\theta \rangle)$. After simplification, this is added to the tableau as a new assertion.
- The output entry associated with the new assertion is the conditional term $(\textit{if } (\ell = r)\theta \textit{ then } t\theta \textit{ else } s\theta)$. ∎

We have presented the equality rule as it applies to two assertions. As with the resolution rule, we can apply dual versions of the equality rule to an assertion and a goal, or to two goals; the justification of these versions of the rule appeals to the duality property.

Also as with the resolution rule, we only introduce a conditional term into the output entry if both given rows have output entries that fail to be unified by the substitution θ. If only one of the rows has an output entry s, we take $s\theta$ as the new output entry. If both rows have output entries s and t that are unified by θ, we take the unified term $s\theta$ as the new output entry. If both rows have no output entry, neither does the new row.

An application of the rule is in accordance with the polarity strategy if at least one negative occurrence of an equality $\ell = r$ is replaced by *false*; no restriction is imposed on the occurrences of the subterms ℓ'.

The equality rule allows us to replace instances of the left term ℓ with corresponding instances of the right term r. By the symmetry of the equality relation, we can justify a right-to-left version of the rule, which allows us to replace instances of the right term r with corresponding instances of the left term ℓ.

We illustrate the equality rule with an example.

Example (equality rule)

This example is taken from the transformation of a program to reverse a string. We are in the process of deriving an auxiliary subprogram $rev2(s, t)$ to reverse the string s and concatenate it onto the string t.

Our tableau contains the two goals

	$\neg\left(\boxed{s} = \Lambda\right)^{-}$	$rev2\big(tail(s),\ head(s) \cdot t\big)$
	$z = rev\left(\boxed{s}\right) * t$	z

These rows have no variables in common. The boxed subterms are identical and hence unifiable with most-general unifier { }. The result of applying a dual version of the rule, the GG-equality rule, is then

	$\neg$ *false* $\wedge$ $z = rev(\Lambda) * t$	*if* $s = \Lambda$ *then* z *else* $rev2\big(tail(s),\ head(s) \cdot t\big)$

which reduces under simplification to

	$z = t$	*if* $s = \Lambda$ *then* z *else* $rev2\big(tail(s),\ head(s) \cdot t\big)$

Because both terms have output entries, a conditional term is introduced as the new output entry. The application is in accordance with the polarity strategy, because the occurrence of the equality $(s = \Lambda)$ is negative in the tableau. ▮

Example (equality rule)

This example is taken from the derivation of a square-root program in the theory of nonnegative rationals. We assume our tableau contains the assertion

$(\boxed{0 \cdot v} = 0)^-$		

which is an axiom for multiplication, and the goal

	$\boxed{z \cdot z} \leq r \ \wedge$ $r < (z + \epsilon)^2$	z

The two rows have no variables in common. The boxed subterms are unifiable; a most-general unifier is $\{z \leftarrow 0,\ v \leftarrow 0\}$. The result of applying a dual version of the equality rule is then

	$\neg$ *false* $\wedge$ $0 \leq r \ \wedge$ $r < (0 + \epsilon)^2$	0

which reduces under simplification to

	$r < \epsilon^2$	0

(The condition $0 \leq r$ is simplified to *true* in the theory of nonnegative rationals.) Because the given assertion has no output entry, no conditional construct is introduced in applying the rule. The application is in accordance with the polarity strategy, because the occurrence of the equality $(0 \cdot v = 0)$ is negative in the tableau.

The intuitive content of the derived goal is that, in the case in which $r < \epsilon^2$, that is, if $\sqrt{r}$ is in the half-open interval $[0, \epsilon)$, then 0 is a suitable output for the desired square-root program. ▮

The equality rule allows us to discard all the equality axioms, except for the *reflexivity* axiom $u = u$, from our initial tableau, without sacrificing the possibility of completing any derivation.

The Well-Founded Induction Rule

The well-founded induction principle is valuable for program synthesis and other applications because of its generality: The induction principles of all theories turn out to be instances of the well-founded induction rule. In derivation proofs, use of the rule corresponds to the introduction of recursion, or other repetitive constructs, into the derived program. Before we describe the rule, we introduce the notion of a well-founded relation.

Definition

A relation $\prec$ is *well-founded* (in a theory TH) if there are no infinite decreasing sequences in TH, i.e., no sequences $x_1, x_2, x_3, \ldots$ such that

$$x_1 \succ x_2 \quad \text{and} \quad x_2 \succ x_3 \quad \text{and} \quad \ldots \qquad \blacksquare$$

For example, the less-than relation $<$ and the proper-substring relation $\prec_{string}$ are well-founded in the theories of nonnegative integers and strings, respectively. (A string s is a proper substring of a string t, written $s \prec_{string} t$, if s and t are distinct and if the elements of s occur contiguously in t.) On the other hand, the less-than relation $<$ is not well-founded in the theory of nonnegative rationals, because $1, 1/2, 1/4, 1/8, \ldots$ constitutes an infinite decreasing sequence under $<$.

Well-founded relations are of interest to us because of the following property.

Property (well-founded induction principle)

For any well-founded relation $\prec$ in theory TH and any sentence $\mathcal{P}[x]$, any closed instance of the following sentence is valid in TH:

$$\begin{array}{l} \textit{if } (\forall x') \left[\begin{array}{l} \textit{if } (\forall x) \left[\begin{array}{l} \textit{if } x \prec x' \\ \textit{then } \mathcal{P}[x] \end{array}\right] \\ \textit{then } \mathcal{P}[x'] \end{array}\right] \\ \textit{then } (\forall x)\mathcal{P}[x] \end{array}$$

where x' does not occur free in $\mathcal{P}[x]$. ▮

In other words, suppose we are trying to prove that $\mathcal{P}[x]$ is true for every object x. For this purpose, it suffices to consider an arbitrary object x' and show that $\mathcal{P}[x']$ holds under the induction hypothesis that $\mathcal{P}[x]$ is true for every x such that $x \prec x'$. The well-founded induction principle is called *complete* induction or *course-of-values* induction when $\prec$ is taken to be the less-than relation $<$ over the nonnegative integers. It is also called *noetherian* induction.

In the deductive-tableau framework, this principle is represented as a rule.

Rule (well-founded induction)

assertions	goals	$f(a)$
	$\mathcal{Q}[a, z]$	z
if $x \prec_w a$ *then* $\mathcal{Q}[x, f(x)]$		

Here $\mathcal{Q}[a, z]$ is the initial goal of the tableau; we require that z be the only free variable in the row. (If there are several output entries $z_1, \ldots, z_n$, all of them may occur free in the goal.) The relation $\prec_w$ is required to be well-founded in TH. The function symbol f stands for the function we are trying to compute. ∎

The rationale for the induction rule is as follows. We are trying to construct a program to compute a function f that, for a given input a, will yield an output z that will satisfy the input-output relation $\mathcal{Q}[a, z]$. It suffices to conduct the derivation under the induction hypothesis that the function f will behave properly on each input x that is less than a under $\prec_w$. More precisely, we may assume inductively that, for each input x such that $x \prec_w a$, the output $f(x)$ will satisfy the input-output relation $\mathcal{Q}[x, f(x)]$.

Example (well-founded induction rule)

Recall that the initial goal for the *front-last* derivation is

assertions	goals	*front*(s)	*last*(s)
	if $\neg(s = \Lambda)$ *then* $char(z_2) \wedge$ $s = z_1 * z_2$	z_1	z_2

This now says that we would like our program to decompose a nonempty string s into the concatenation of a string z_1 and a character z_2, so that *front*(s) and *last*(s) can be taken to be z_1 and z_2, respectively.

According to the induction rule, we may add to our tableau the new assertion

if $x \prec_w s$ *then if* $\neg(x = \Lambda)$ *then* $char(last(x)) \wedge$ $x = front(x) * last(x)$			

This row corresponds to the induction hypothesis that, for any nonempty input x less than s under $\prec_w$, the functions *front* and *last* will indeed decompose x into the concatenation of string *front*(x) and a character *last*(x). The relation $\prec_w$ can be any well-founded relation. ∎

When the program being derived has more than one input, the well-founded relation $\prec_w$ applies to two tuples of inputs, rather than two individual inputs.

Example (well-founded induction rule)

The initial goal for the rational square-root derivation is

assertions	goals	$sqrt(r, \epsilon)$
	$\mathit{if}\ \epsilon > 0$ $\mathit{then}\ z^2 \leq r \wedge$ $r < (z + \epsilon)^2$	z

According to the well-founded induction rule, we may add to our tableau as an induction hypothesis the assertion

$\mathit{if}\ \langle x, v\rangle \prec_w \langle r, \epsilon\rangle$ $\mathit{then}\ \mathit{if}\ v > 0$ $\mathit{then}\ \big(sqrt(x, v)\big)^2 \leq x \wedge$ $x < \big(sqrt(x, v) + v\big)^2$		

This now declares that the square-root program behaves properly for any pairs of inputs less than the original inputs under $\prec_w$. The well-founded relation applies to two pairs, that is, 2-tuples, rather than to two individual nonnegative rationals. ▮

Recursion Formation

The induction hypothesis introduced by application of the induction rule contains occurrences of the function symbol f, which denotes the function we are trying to compute. If the induction hypothesis is used in the proof, it can happen that terms of form $f(t)$ will be introduced into the output column and hence into the derived program. This is the mechanism by which recursive calls are introduced into the program.

Example (recursion formation)

We have applied the induction rule to the initial goal of the *front-last* derivation, introducing the induction hypothesis

$\mathit{if}\ x \prec_w s$ $\mathit{then}\ \mathit{if}\ \neg(x = \Lambda)$ $\mathit{then}\ char\big(last(x)\big) \wedge$ $\Big(x = \boxed{front(x) * last(x)}\Big)^-$			

This induction hypothesis contains occurrences of the function symbols *front* and *last*, which we are trying to compute. Suppose we have also derived the following goal:

	$char(u) \wedge$ $char(z_2) \wedge$ $s = u \cdot \boxed{\hat{z}_1 * z_2}$	$u \cdot \hat{z}_1$	z_2

Note that the boxed subterms of the two rows are unifiable, with most-general unifier $\{\hat{z}_1 \leftarrow front(x),\ z_2 \leftarrow last(x)\}$. By application of the right-to-left version of the equality rule, we obtain, after simplfication, the goal

	$x \prec_w s \wedge$ $\neg(x = \Lambda) \wedge$ $char(u) \wedge$ $char\big(last(x)\big) \wedge$ $s = u \cdot x$	$u \cdot front(x)$	$last(x)$

By using the induction hypothesis, we have introduced the terms $front(x)$ and $last(x)$ into the output column. This will result in the formation of recursive calls in the final program.

The condition $x \prec_w s$ in the goal has the effect of ensuring that these recursive calls will not cause a nonterminating computation of the final program. If there were an infinite sequence of calls to either program *front* or *last*, the corresponding arguments would constitute an infinite sequence of strings decreasing with respect to $\prec_w$, which contradicts the well-foundedness of $\prec_w$.

The condition $\neg(x = \Lambda)$ in the goal guarantees that the argument to the recursive calls is a legal input, that is, that it is nonempty.

The relation $\prec_w$ to be used in the proof has not been determined; it may be any well-founded relation. ▮

We illustrate recursion formation with another example.

Example (recursion formation)

In the derivation of the rational square-root program, suppose we have derived the assertion

if $\langle x, v\rangle \prec_w \langle r, \epsilon\rangle$ *then if* $v > 0$ *then* $\boxed{\begin{array}{l}\big(sqrt(x,v)\big)^2 \le x \wedge \\ \neg\big[\big(sqrt(x,v)+v\big)^2 \le x\big]\end{array}}^{-}$		

This is an immediate consequence of our induction hypothesis.

We have earlier obtained the goal

	$\boxed{z^2 \le r \wedge \neg\left[(z+2\epsilon)^2 \le r\right]}^{+}$	*if* $(z+\epsilon)^2 \le r$ *then* $z+\epsilon$ *else* z

This was obtained by an application of the resolution rule in a previous example. The boxed subsentences of the two rows are unifiable, with most-general unifier $\{x \leftarrow r,\ v \leftarrow 2\epsilon,\ z \leftarrow sqrt(r, 2\epsilon)\}$. By application of the resolution rule, we obtain, after simplification, the goal

	$\langle r, 2\epsilon\rangle \prec_w \langle r, \epsilon\rangle \wedge$ $2\epsilon > 0$	*if* $(sqrt(r,2\epsilon)+\epsilon)^2 \le r$ *then* $sqrt(r,2\epsilon)+\epsilon$ *else* $sqrt(r,2\epsilon)$

By using the induction hypothesis in the proof, we have introduced three occurrences of the recursive call $sqrt(r, 2\epsilon)$ into the output column. The condition $\langle r, 2\epsilon\rangle \prec_w \langle r, \epsilon\rangle$ in the goal guarantees that these recursive calls do not lead to a nonterminating computation. The condition $2\epsilon > 0$ guarantees that the arguments r and 2ϵ of the recursive calls are legal inputs, that is, that 2ϵ is positive. The well-founded relation $\prec_w$ is yet to be determined. ∎

Choice of a Well-Founded Relation

There are many well-founded relations that can serve as the basis for an induction proof. Until the proof is well underway, it may be difficult to determine which relation will be most convenient to use. Rather than attempting to choose a relation at the beginning of the proof, we prefer to start the proof with an unspecified relation $\prec_w$, so that we can discover what properties the relation is required to satisfy.

We assume that a number of relations are given in advance to be well-founded, with certain known properties. In addition, there are mechanisms for constructing new well-founded relations from old ones, to satisfy certain properties. When the required properties of the unspecified relation $\prec_w$ match the properties of a known or constructed relation $\prec_r$, we can choose $\prec_w$ to be that relation $\prec_r$.

Example (choice of a well-founded relation)

In the theory of strings, the proper substring relation $\prec_{string}$ is given to be well-founded and known to have the property that the tail of a nonempty string is its proper substring, that is,

if $\neg(y = \Lambda)$ *then* $\boxed{tail(y) \prec_{string} y}^{-}$		

In a derivation of the *front-last* program, we obtain the goal

	$\boxed{tail(s) \prec_w s}^+ \wedge$ $\neg\,(tail(s) = \Lambda)$	$head(s) \cdot$ $front(tail(s))$	$last(tail(s))$

This suggests that we take the relation $\prec_w$ to be the proper substring relation $\prec_{string}$. We can then apply the resolution rule to these two rows, with most-general unifier $\{y \leftarrow s\}$, to obtain, after simplification, the goal:

	$\neg\,(s = \Lambda) \wedge$ $\neg\,(tail(s) = \Lambda)$	$head(s) \cdot$ $front(tail(s))$	$last(tail(s))$

∎

The selection of the well-founded relation may be regarded as an extra-logical step, to be performed by an external mechanism. Alternatively, we can extend our theories to include well-founded relations as objects. We may then regard $x \prec_w y$ as an abbreviation for $\prec(w, x, y)$, where w is a variable that ranges over well-founded relations. In the above resolution step, when we unified $tail(y) \prec_{string} y$ with $tail(s) \prec_w s$, the unification algorithm would then include $w \leftarrow string$ as a replacement in the most-general unifier. In other words, the choice of well-founded relation would be a by-product of the proof process. ∎

EXAMPLES

In this section we give some examples of the derivation of specific programs.

The Front-Last Derivation

We have not given all the rules in the system, but we have shown enough to illustrate a full derivation of the front-last program. This program, the reader will recall, is to find, for a given nonempty string s, two outputs: the last character $last(s)$ of s and the string $front(s)$ of all but the last character of s. The specification is

$$\langle front(s), last(s)\rangle \Leftarrow \begin{cases} \textit{find}\ \langle z_1, z_2\rangle \text{ such that} \\ \quad \textit{if}\ \neg(s = \Lambda) \\ \quad \textit{then}\ char(z_2)\ \wedge\ s = z_1 * z_2 \end{cases}$$

in the theory of strings. Our initial goal is thus

assertions	goals	$front(s)$	$last(s)$
	1. *if* $\neg(s = \Lambda)$ *then* $char(z_2)\ \wedge\ s = z_1 * z_2$	z_1	z_2

Properties of the theory of strings, expressed as assertions, are present in the initial tableau, and will be mentioned as we use them.

By the *if-split* rule, we may decompose our goal into its antecedent and consequent

2. $\neg(s = \Lambda)$			
	3. $char(z_2) \land s = \boxed{z_1 * z_2}$	z_1	z_2

The output entries z_1 and z_2 have been dropped from the row 2, because these variables do not occur free in the assertion. We have annotated goal 3 in anticipation of a future step.

The Base Case

By the equality rule, applied to an axiom for concatenation

$\left(\boxed{\Lambda * y} = y\right)^-$			

and the most recent goal, with most-general unifier $\{z_1 \leftarrow \Lambda,\ y \leftarrow z_2\}$, we obtain the goal

	4. $char(z_2) \land \boxed{s = z_2}^+$	Λ	z_2

Note that the first output entry has been instantiated.

By the resolution rule, applied to the *reflexivity* axiom

$\boxed{x = x}^-$			

and the goal, with most-general unifier $\{x \leftarrow s,\ z_2 \leftarrow s\}$, we obtain

	5. $char(s)$	Λ	s

Now both output entries have been instantiated. The intuitive content of this row is that in the case in which the input string s consists of a single character, *front*(s) may be taken to be Λ and *last*(s) to be s itself. This will lead to the base case for the program we are constructing. Let us set it aside for awhile and turn our attention to the recursive case.

The Recursive Case

We have earlier developed the goal

	3. $char(z_2) \wedge$ $s = \boxed{z_1 * z_2}$	z_1	z_2

By the equality rule, applied to an axiom for concatenation,

if $char(u)$ *then* $\left(\boxed{(u \cdot y_1) * y_2} = u \cdot (y_1 * y_2)\right)^-$			

and the goal, with most-general unifier $\{z_1 \leftarrow u \cdot y_1, z_2 \leftarrow y_2\}$, we obtain

	6. $char(u) \wedge$ $char(y_2) \wedge$ $s = u \cdot \boxed{y_1 * y_2}$	$u \cdot y_1$	y_2

By the induction rule, applied as always to the initial goal, we may assume the induction hypothesis

7. *if* $x \prec_w s$ *then if* $\neg(x = \Lambda)$ *then* $char(last(x)) \wedge$ $\left(x = \boxed{front(x) * last(x)}\right)^-$			

By the equality rule, applied right-to-left to assertion 7 and goal 6, with most-general unifier $\{y_1 \leftarrow front(x), y_2 \leftarrow last(x)\}$, we obtain the goal

	8. $x \prec_w s \wedge$ $\neg(x = \Lambda) \wedge$ $char(u) \wedge$ $\boxed{char(last(x))}^+ \wedge$ $s = u \cdot x$	$u \cdot front(x)$	$last(x)$

Note that, by use of the induction hypothesis, recursive calls $front(x)$ and $last(x)$ have been introduced into the output columns.

We shall next apply the resolution rule, again to the induction hypothesis and the goal. Because these rows have the variable x in common, we rename the variable in the induction hypothesis:

7. $if\ x' \prec_w s$ $then\ if\ \neg(x' = \Lambda)$ $then\ \boxed{char(last(x'))}^{-}\ \wedge$ $x' = front(x') * last(x')$			

Applying the rule, with most-general unifier $\{x' \leftarrow x\}$, we obtain

	9. $x \prec_w s\ \wedge$ $\neg(x = \Lambda)\ \wedge$ $char(u)\ \wedge$ $\boxed{s = u \cdot x}^{+}$	$u \cdot front(x)$	$last(x)$

By the resolution rule, applied to the *decomposition* property for strings,

$if\ \neg(y = \Lambda)$ $then\ \boxed{y = head(y) \cdot tail(y)}^{-}$			

and the goal, taking most-general unifier $\{y \leftarrow s,\ u \leftarrow head(s),\ x \leftarrow tail(s)\}$, we obtain

	10. $\neg(s = \Lambda)\ \wedge$ $\boxed{tail(s) \prec_w s}^{+}\ \wedge$ $\neg\big(tail(s) = \Lambda\big)\ \wedge$ $char\big(head(s)\big)$	$head(s) \cdot front\big(tail(s)\big)$	$last\big(tail(s)\big)$

Note that at this stage the output entries are fully instantiated. It remains to select the well-founded relation and to combine the base case and the recursive calls.

Choice of a Well-Founded Relation

Let us assume we know the following property of the proper substring relation $\prec_{string}$:

$if\ \neg(y = \Lambda)$ $then\ \boxed{tail(y) \prec_{string} y}^{-}$			

By application of the resolution rule, taking $\prec_w$ to be the $\prec_{string}$ and taking the most-general unifier to be $\{y \leftarrow s\}$, we obtain

	11. $\neg(s = \Lambda)\ \wedge$ $\neg\big(tail(s) = \Lambda\big)\ \wedge$ $\boxed{char\big(head(s)\big)}^{+}$	$head(s) \cdot$ $front\big(tail(s)\big)$	$last\big(tail(s)\big)$

This step has suggested that the well-founded relation be taken to be the proper substring relation.

The Final Steps

By the resolution rule, applied to the axiom

if $\neg(y = \Lambda)$ *then* $\boxed{char(head(y))}^{-}$			

and the goal, with most-general unifier $\{y \leftarrow s\}$, we obtain

	12. $\neg(s = \Lambda) \wedge \boxed{\neg(tail(s) = \Lambda)}^{+}$	$head(s) \cdot front(tail(s))$	$last(tail(s))$

By the resolution rule, applied to the *trichotomy* property of strings

$y = \Lambda \vee char(y) \vee \boxed{\neg(tail(y) = \Lambda)}^{-}$			

with most-general unifier $\{y \leftarrow s\}$, we obtain

	13. $\boxed{\neg(s = \Lambda)}^{+} \wedge \neg\, char(s)$	$head(s) \cdot front(tail(s))$	$last(tail(s))$

By the resolution rule, applied to our earlier assertion

2. $\boxed{\neg(s = \Lambda)}^{-}$			

and the goal, we obtain

	14. $\neg \boxed{char(s)}^{-}$	$head(s) \cdot front(tail(s))$	$last(tail(s))$

Finally, by the resolution rule applied to this goal and the one we set aside for the base case,

	5. $\boxed{char(s)}^{+}$	Λ	s

we obtain the goal

	15. *true*	*if* $char(s)$ *then* Λ *else* $head(s) \cdot front(tail(s))$	*if* $char(s)$ *then* s *else* $last(tail(s))$

By this step, conditional terms have been introduced into our program.

Because we have obtained the goal *true* with primitive output entries, we can take this to be the final goal of our tableau. The program we extract from the proof is

$$front(s) \Leftarrow \begin{cases} \textit{if } char(s) \\ \textit{then } \Lambda \\ \textit{else } head(s) \cdot front\big(tail(s)\big) \end{cases}$$

$$last(s) \Leftarrow \begin{cases} \textit{if } char(s) \\ \textit{then } s \\ \textit{else } last\big(tail(s)\big) . \end{cases}$$

The Final Square-Root Program

We do not give the full derivation for the square-root program we have been using as an example; it is described in Manna and Waldinger [MW87a]. The final program we obtain is

$$sqrt(r, \epsilon) \Leftarrow \begin{cases} \textit{if } max(r, 1) < \epsilon \\ \textit{then } 0 \\ \textit{else if } \big(sqrt(r, 2\epsilon) + \epsilon\big)^2 \leq r \\ \qquad \textit{then } sqrt(r, 2\epsilon) + \epsilon \\ \qquad \textit{else } sqrt(r, 2\epsilon) \end{cases}$$

Let us explain this somewhat odd program, because it illustrates a more general phenomenon.

Recall that the program is intended to find a rational approximation $sqrt(r, \epsilon)$ that is within ϵ less than the exact square root of r; that is, $\sqrt{r}$ is to belong to the half-open interval $\big[sqrt(r, \epsilon),\ sqrt(r, \epsilon) + \epsilon\big)$.

In the case in which the error tolerance is quite large, that is, $max(r, 1) < \epsilon$, it turns out that $r < \epsilon^2$, that $\sqrt{r}$ belongs to the interval $[0, \epsilon)$, and hence that 0 is a good enough approximation to the square root of r.

Otherwise, we double our error tolerance and recursively find an approximation $sqrt(r, 2\epsilon)$ that is within 2ϵ less than the square root of r; that is, $\sqrt{r}$ belongs to the interval $\big[sqrt(r, 2\epsilon),\ sqrt(r, 2\epsilon) + 2\epsilon\big)$. The program then asks whether or not $\big(sqrt(r, 2\epsilon) + \epsilon\big)^2 \leq r$, that is, whether $\sqrt{r}$ is in the right or the left half of our interval.

In the case in which $\sqrt{r}$ is in the right half $\big[sqrt(r, 2\epsilon) + \epsilon,\ sqrt(r, 2\epsilon) + 2\epsilon\big)$, we can take $sqrt(r, 2\epsilon) + \epsilon$ to be our approximation to the square root; it is certain to be within ϵ less than $\sqrt{r}$.

In the alternative case, in which $\sqrt{r}$ is in the left half $\big[sqrt(r, 2\epsilon),\ sqrt(r, 2\epsilon) + \epsilon\big)$, we can take $sqrt(r, 2\epsilon)$ itself to be our approximation. In either case, the conditional expression will yield an approximation within ϵ less than $\sqrt{r}$.

This recursive program uses a binary-search technique but it does not resemble conventional iterative binary-search algorithms. Usually a binary-search algorithm will begin with a very large interval containing the desired output. It will divide the interval in half at each iteration and retain the half that contains the output. The process continues until the interval is sufficiently small, that is, shorter than a given error tolerance.

Rather than dividing an interval in half at each iteration, our derived program doubles its error tolerance at each recursion, until the tolerance is quite large. At this point, it can form a large interval which contains the desired output. As it unwinds from the recursion, it implicitly divides this interval in half, just like a conventional algorithm. Similar recursive binary-search programs may be obtained for division and other numerical problems.

This program was first derived by purely formal manipulation of the rules of the system, to explore the search space, without any expectation of finding a program of this form. When the program was obtained, we did not understand it and thought we had made an error in the derivation.

The program as derived is quite inefficient, as it contains several occurrences of the same recursive call $sqrt(r, 2\epsilon)$. These can be replaced by a single recursive call by ordinary common-subexpression elimination. More sophisticated program-transformation techniques (Harrison and Khoshnevisan [HK86]) have been applied to transform the program into a linear iterative form.

The Slowsort Program

Another example of a program obtained by formal manipulation is this sorting program obtained by Traugott [Tr 89].

$$sort(l) \Leftarrow \begin{cases} \textit{if } l = \langle\,\rangle \\ \textit{then } \langle\,\rangle \\ \textit{else if } tail(l) = \langle\,\rangle \\ \qquad \textit{then } l \\ \qquad \textit{else if } head(l) \leq head\big(sort\big(tail(l)\big)\big) \\ \qquad\qquad \textit{then } head(l) \cdot sort\big(tail(l)\big) \\ \qquad\qquad \textit{else } head\big(sort\big(tail(l)\big)\big) \cdot \\ \qquad\qquad\qquad sort\big(head(l) \cdot tail\big(sort\big(tail(l)\big)\big)\big) \end{cases}$$

Here l is a list of numbers and $\langle\,\rangle$ is the empty list. No particular claims are made for the efficiency of this program. For example, to find the minimum element of $tail(l)$, it sorts it and throws away all but the first element. The program is unusual in that it sorts the list without invoking any auxiliary programs, only basic list-processing primitives. Traugott derived other sorting programs with this property too.

SUBPROGRAMS

Once we have derived a program f, we can use it as a subprogram in future derivations. We do this by including in the tableau for these derivations an assertion stating that the derived program f does indeed meet its specification.

More precisely, suppose we have derived a program $f(a) \Leftarrow t$ to meet a specification

$$f(a) \;\Leftarrow\; \textit{find } z \textit{ such that } \mathcal{Q}[a, z].$$

Then in the initial tableau for the derivation of a new program g, we may include the assertion

$(\forall a)\mathcal{Q}[a, f(a)]$		

which states that f does satisfy its specification. If this assertion is used in the proof, the new program g may invoke the earlier program f. (The function symbol f is included in the primitive list for the derivation of g.)

If we choose, we may include the program f itself as an assertion in the derivation for g. That is, we may include the assertion

$(\forall a)[f(a) = t]$		

in the initial tableau for g. If we do this we have lost a certain degree of modularity, because the program for g may depend on the particular implementation for f. We are no longer free to replace the program for f with a different program meeting the same specification.

Program Transformation

At this point we give an example that illustrates both the formation of subprograms and the application of deductive methods to program transformation problems.

We suppose we are given a program

$$\textit{reverse}(s) \;\Leftarrow\; \begin{cases} \textit{if } s = \langle\,\rangle \\ \textit{then } s \\ \textit{else } \textit{reverse}\big(\textit{tail}(s)\big) * \textit{head}(s) \end{cases}$$

for reversing the characters of a given string s. The program is inefficient because, in executing successive recursive calls to *reverse*, it will be computing the concatenation function many times.

To transform a given program, we may regard that program as the specification for a new program. For this example, the new specification is

$$\textit{reverse1}(s) \;\Leftarrow\; \textit{find } z \textit{ such that } z = \textit{reverse}(s).$$

Of course the *reverse* program itself will satisfy this specification, but different derivations will yield different programs, some of them more efficient than others.

Looking ahead, the particular program we shall derive is

$$reverse1(s) \Leftarrow reverse2(s, \Lambda)$$

where

$$reverse2(s,t) \Leftarrow \begin{cases} \textit{if } s = \Lambda \\ \textit{then } t \\ \textit{else } reverse2\big(tail(s),\ head(s) \cdot t\big) \end{cases}$$

The auxiliary subprogram $reverse2(s,t)$ may be regarded as a generalization of *reverse*. It meets the specification

$$reverse2(s,t) \Leftarrow \textit{find } z \textit{ such that } z = reverse(s) * t.$$

In other words, it reverses s and concatenates the result with t. To complete the derivation of *reverse1*, it is necessary to derive *reverse2*.

Derivation of Reverse2

We shall not give the full derivation of *reverse2*, but let us present those steps that will be relevant to our present discussion.

We begin with the initial tableau

assertions	goals	$reverse2(s,t)$
	$z = reverse(s) * t$	z

The function symbol *reverse* is excluded from the primitive list, so that it may not occur in the derived program. By the well-founded induction rule, we may assume the induction hypothesis

if $\langle x,\ y\rangle \prec_w \langle s,\ t\rangle$ *then* $\boxed{reverse2(x,y) = reverse(x) * y}^{-}$		

From our initial goal, the definition of *reverse*, and some properties of strings, we eventually obtain the goal

	$\neg(s = \Lambda)\ \wedge$ $\boxed{z = reverse\big(tail(s)\big) * \big(head(s) \cdot t\big)}^{+}$	z

The boxed subsentences of the assertion and the goal unify, with most-general unifier $\{x \leftarrow tail(s),\ y \leftarrow head(s) \cdot t,\ z \leftarrow reverse2\big(tail(s),\ head(s) \cdot t\big)\}$. By the resolution rule, we obtain

	$\langle tail(s),\ head(s) \cdot t \rangle \prec_w \langle s,\ t \rangle \wedge$ $\neg (s = \Lambda)$	$reverse2\big(tail(s),\ head(s) \cdot t\big)$

Use of the induction hypothesis has accounted for the introduction of the recursive call $reverse2\big(tail(s), head(s) \cdot t\big)$ into the output entry and, ultimately, into the *reverse2* program.

Derivation of Reverse1

Once we have derived *reverse2*, the derivation of the program *reverse1* is simple. We begin with the initial tableau.

assertions	goals	$reverse1(s)$
	$\boxed{z = reverse(s)}^{+}$	z

We may include in our tableau the assertion that, for all x and y, the program $reverse2(x, y)$ does meet the specification from which it was derived:

$reverse2(x, y) = \boxed{reverse(x) * y}$		

We assume that we have the following property of concatenation:

$\left(\boxed{\widehat{x} * \Lambda} = \widehat{x}\right)^{-}$		

By the equality rule applied to this and the previous assertion, with most-general unifier $\{\widehat{x} \leftarrow reverse(x),\ y \leftarrow \Lambda\}$, we obtain

$\boxed{reverse2(x, \Lambda) = reverse(x)}^{-}$		

By the resolution rule, applied to this assertion and the initial goal, with most-general unifier $\{x \leftarrow s,\ z \leftarrow reverse2(s, \Lambda)\}$, we obtain the final goal

	true	$reverse2(s, \Lambda)$

From this proof, we extract the program

$$reverse1(s) \Leftarrow reverse2(s, \Lambda).$$

The Need for Generalization

This derivation illustrates a phenomenon in program synthesis that reflects a corresponding observation in theorem proving. It has been remarked that in proving a theorem by induction, it is often necessary to prove a more general theorem, so as to have the benefit of a more general induction hypothesis. (This fact has been exploited by the Boyer-Moore theorem prover [BM79].) Similarly, deriving a program, it is sometimes necessary to derive a more general program, so as to have the benefit of a more general recursive call.

To illustrate this phenomenon, let us see what would have happened had we begun the derivation of the *reverse1* program without first deriving *reverse2*. We begin with the goal

assertions	goals	*reverse1*(s)
	$z = reverse(s)$	z

By the well-founded induction rule, we may assume the induction hypothesis

if $x \prec_w s$ *then* $\boxed{reverse1(x) = reverse(x)}^{-}$		

As in the derivation of the *reverse2* program, we may obtain from the initial goal, the definition of *reverse*, and properties of strings the new row

	$\neg(s = \Lambda) \ \wedge$ $\boxed{z = reverse\big(tail(s)\big) * head(s)}^{+}$	z

This time, however, the boxed subsentence of the goal fails to unify with the boxed subsentence of the induction hypothesis. Because the specification for *reverse1* is less general than the specification for *reverse2*, its induction hypothesis is also less general; in fact, the induction hypothesis is not general enough to unify with the desired goal.

Motivation for Generalization

In our successful derivation for *reverse1*, we have assumed that we were clever enough to first derive *reverse2*. If we were not given the specification for *reverse2*, could we, or perhaps a system, be led to discover it? In general, automatic generalization of this sort is a difficult problem. We speculate that appropriate generalizations may be discovered by observing regularities in the structure of a derivation attempt.

For example, in the attempted derivation of *reverse1* (assuming that *reverse2* has not yet been developed), we begin with the goal

$$z = reverse(s)$$

and obtain the subsentence

$$z = reverse(tail(s)) * head(s).$$

If we apply the same steps to this subsentence as we did to the original goal, we obtain the subsentence

$$z = reverse(tail(tail(s))) * head(tail(s)) * head(s).$$

If we can observe the regularity in these goals, we may be inspired to construct a subprogram to satisfy instead the input-output condition

$$z = reverse(\widehat{s}) * \widehat{t}.$$

This is the specification for the auxiliary subprogram *reverse2*. Each of the above three subsentences is equivalent to an instance of this condition, taking $\widehat{s}$ to be s, $tail(s)$, and $tail(tail(s))$, respectively, and $\widehat{t}$ to be Λ, $head(s)$, and $head(tail(s)) * head(s)$, respectively.

Some generalizations, however, are more difficult to motivate. For example, in the derivation of a unification algorithm (Manna and Waldinger [MW81], Nardi [Na89]), we begin with a specification

$$unify(e_1, e_2) \Leftarrow \left\{ \begin{array}{l} \textit{find } \theta \textit{ such that} \\ \quad \left[\begin{array}{l} e_1\theta = e_2\theta \ \wedge \\ (\forall \phi) \left[\begin{array}{l} \textit{if } e_1\phi = e_2\phi \\ \textit{then } (\exists \lambda)[\phi = \theta\lambda] \end{array} \right] \end{array} \right] \\ \qquad\qquad \vee \\ \quad \left[\begin{array}{l} (\forall \phi) \neg (e_1\phi = e_2\phi) \ \wedge \\ \theta = nil \end{array} \right] \end{array} \right.$$

In other words, we wish the program to return a substitution θ that is a unifier of e_1 and e_2 and that is more general than any other unifier ϕ. In the case in which e_1 and e_2 are not unifiable, the program is to return the special object *nil*, which is not a substitution.

The details of this derivation are outside the scope of this discussion. For our derivation proof to go through, we found it necessary to add to the first disjunct of the specification the new condition $\theta\theta = \theta$, that is, θ is idempotent under composition. Nonidempotent most-general unifiers are unintuitive; for example, $\{x \leftarrow y\}$ and $\{y \leftarrow x\}$ are idempotent most-general unifiers of x and y but $\{x \leftarrow z,\ y \leftarrow z,\ z \leftarrow x\}$ is a nonidempotent most-general unifier. On the other hand, idempotence had not been studied in connection with unification, so we were surprised to have to introduce it into the specification. (Idempotence has been studied independently in the work of Eder [Ed85].)

SPECIALIZED INFERENCE RULES

Progress in program synthesis depends on the development of techniques for automated deduction, both interactive and automatic. The inference rules we have introduced so far are very general—they apply to proving theorems in any theory. If we are satisfied with a more specialized system, one which is competent in a particular theory, such as the strings, we may be able to devise more powerful inference rules whose applicability is limited to that theory. Such rules may be able to achieve in a single step inferences that would otherwise require several steps.

The first benefit of this is to shorten proofs. This is a clear advantage in an interactive system, in which each step of the proof requires some effort on the part of the user. For an automatic system, a shorter proof may be an advantage if it can be found more easily. Because introducing new inference rules gives us more choices at each stage, it can actually increase the search space. Although the proof is shorter, it may be more difficult to discover.

A new inference rule may pay for itself, however, if, in addition to shortening the proof, it allows us to discard from the initial tableau some of the assertions that represent valid properties of the theory. We can do this only if the rule has certain completeness properties, which guarantee that in discarding the assertions we are not losing any opportunity to complete a proof. If so, the rule may reduce the number of choices at each stage and hence contract the search space.

Associative-Commutative Unification and E-Unification

One way to increase the power of an inference rule is to extend the unification algorithm to take properties of the theory into account. For example, the associative and commutative properties of operators, such as the addition and multiplication functions in the theories of numbers or the conjunction and disjunction connectives in any logicial theory, may be incorporated into an ***associative-commutative*** (AC) ***unification*** algorithm (Stickel [St81]). While the ordinary unification algorithm would not be able to unify the two terms $a+(x+b)$ and $(c+a)+b$, the AC algorithm would, returning the unifier $\{x \leftarrow c\}$.

Completeness results for the algorithm have been established; that is, if the algorithm is adopted, we may discard the ***associativity*** property

$$(u+v)+w = u+(v+w)$$

and the ***commutativity*** property

$$u+v = v+u$$

from the initial tableau.

Unlike ordinary unification, which always returns a single most-general unifier, the AC unification algorithm may return a finite number of distinct unifiers. For example, for an associative-commutative function f, the result of unifying the two terms $f(a,x)$ and $f(b,y)$ can be either $\{x \leftarrow b,\ y \leftarrow a\}$ or $\{x \leftarrow f(b,u),\ y \leftarrow f(a,u)\}$, where u is a new variable.

Special unification algorithms have been devised (Siekmann [Si89]) for treating operators with various combinations of properties, including associativity, commutativity, identity, and idempotence. More general ***E-unification*** algorithms (e.g., Fay [Fa79]) treat operators with properties defined by a set of equations supplied by the user. Some of these algorithms produce multiple most-general unifiers or even an infinite stream of unifiers; some are not guaranteed to terminate, whether they produce an infinite stream or not.

Sorted Unification

Some unification algorithms have been devised for dealing with sort relations; these are the unary relations, such as $integer(x)$, $string(x)$, or $char(x)$, that serve to categorize our set of objects. *Sorted unification* algorithms (e.g., Schmidt-Schauss [Sc88]) allow us to provide a declaration that associates with each variable and term a particular sort relation. Thus we might declare that x is of sort *integer* and s is of sort *string*. The sorted unificaiton algorithm will produce only substitutions of replacement pairs $x \leftarrow t$ such that x and t are of the same sort.

An advantage of using sorted unification is that we can then drop from our assertions and goals all subsentences $p(t)$, where p is a sort relation. For instance, if we have declared x to be of sort *string* and y to be of sort *integer*, the sentence

$$(\forall x)(\exists y)\, q(x, y)$$

will be understood to mean

$$(\forall x) \begin{bmatrix} \textit{if } string(x) \\ \textit{then } (\exists y)[integer(y) \ \wedge \ q(x, y)] \end{bmatrix} .$$

Some assertions may disappear completely. Use of sorted unification has achieved dramatic reduction of the search space for some problems.

Extended unification algorithms may replace ordinary unification in the resolution and other inference rules of a deductive system. Where the algorithm returns multiple unifiers or may fail to terminate, the control for the rule must be adapted accordingly.

Special Inference Rules

Another way to specialize a deduction system to a particular theory is to introduce entirely new inference rules. We have already seen how paramodulation (our equality rule) allows us to give special treatment to the equality relation, and thereby eliminate such axioms as *transitivity* and the *functional-substitutivity* of equality from our initial tableau. Manna and Waldinger [MW86] introduce an analogous rule for dealing with ordering relations; adopting this rule allows us to give special treatment to the ordering relation, and eliminate the *transitivity* and other so-called *monotonicity* properties of these relations. Bledsoe and Hines give special inference rules for the real numbers [BH80] and set theory [Hi90].

We have seen that we can specialize a rule to a particular theory or subtheory if we have a special unification algorithm for that theory. Stickel [St85] has shown that we can also specialize a rule if we are given a procedure for determining the validity of sentences in a subtheory. The specialized rule can then be used to perform derivations in a combination of the subtheory with other theories.

For example, suppose we have two goals:

	$\boxed{z_1 \succeq b}$	z_1
	$\boxed{b \succeq a}$	b

where $\succeq$ is a total reflexive relation. The ordinary resolution rule cannot be applied because the boxed subsentences are not unifiable. If, however, we have a procedure capable of determining that, if z_1 is taken to be a, the disjunction of the instances

$$a \succeq b \vee b \succeq a$$

is valid in the total reflexive theory, the *theory resolution* rule is able to deduce the final row

	true	*if* $a \succeq b$ *then* a *else* b

Stickel formulates completeness results that allow us to remove axioms from the initial tableau, such as the *totality* axiom

$$u \succeq v \vee v \succeq u .$$

Analogous *theory* extensions may be formulated for the equality rule and other inference rules. Such rules have been found to achieve sizable reductions in the search space.

DISCUSSION

There are of course many aspects of program synthesis that have not been discussed in this paper, both because of space restrictions and because many of these topics are still being developed.

We have limited ourselves to discussing the synthesis of applicative programs, which return an output but produce no side effects. Some work on the deductive synthesis of imperative programs, which may alter data structures and produce other side effects, is discussed in [MW87b]. We have also disregarded the synthesis of concurrent, real-time, and reactive programs, which may interact with their environments [e.g., PR88].

We have considered specifications only in the form of first-order input-output relations. In general, it is necessary to deal with higher-order specifications that describe properties other than input-output relations. For example, if we are constructing a pair of programs, we should be able to say that one is the inverse of the other.

We have for the most part ignored the efficiency of the programs we construct; in fact, automatically synthesized programs are often wantonly wasteful of time and space. One way of treating this is to include performance criteria as part of the program's specification; the synthesized program would then be forced to meet these criteria. Another approach is to maintain a crude performance estimate for each output entry, in a separate column. Performance estimates could be taken into account in directing the search for a program. Furthermore, once a program was constructed, the search could continue for programs with better performance estimates, because of a better algorithm or data structure, for instance.

Finally, we have concentrated on program synthesis to the exclusion of the use of deductive techniques in collaboration with other software production methods, e.g., deductive testing, debugging, verification, modification, and maintenance.

At present, progress in program synthesis is limited by the power of automated proof systems. Derivation proofs are an appealing and challenging area of application for both automatic and interactive theorem proving.

For automatic systems, program synthesis has an advantage over mathematics as an application area. To make a contribution to mathematics, a system must be able to prove theorems that a human mathematician cannot. For this reason, theorem-proving systems such as Argonne's have had their greatest successes in areas in which human intuition is weak, such as combinatory logic and ternary boolean algebras, so that the machine can compete on a more equal footing. For program synthesis, there is great utility in a system that can reliably be expected to prove routine and mathematically naive results, because from these results we can extract correct programs. The challenge is that many such proofs are still outside the reach of current automatic deductive technology.

To construct an interactive, rather than an automatic, program synthesis system is closer to an engineering feat today (e.g., [BMW90]). Such a system would rely on human intuition to guide the upper levels of the proof search, but would complete the automatable details itself. Errors in human guidance could delay the discovery of a program but could never cause the system to construct an incorrect program. The challenge in designing an interactive system is to phrase the interaction in terms that the human guide could understand.

REFERENCES

[An81] P. B. Andrews, Theorem Proving via General Matings, *Journal of the ACM* 28: 2 (1981), 193–214.

[Bi83] W. Bibel, Matings in Matrices, *Communications of the ACM* 26 (1983), 844–852.

[BD77] R. M. Burstall and J. Darlington, A Transformation System for Developing Recursive Programs, *Journal of the ACM* 24: 1 (1977), 44–67.

[BH80] W. W. Bledsoe and L. Hines, Variable Elimination and Chaining in a Resolution-Based Prover for Inequalities, *Proceedings of the Fifth Conference on Automated Deduction* (1980), 281–292.

[BM79] R. S. Boyer and J S. Moore, *A Computational Logic*, Academic Press, 1979.

[BMW90] R. Burback, Z. Manna, R. Waldinger, et al, Using the Deductive Tableau System, in *MacIntosh Educational Software Collection*, Chariot Software Group, San Diego, 1990.

[Con86] R. Constable et al., *Implementing Mathematics with the NuPrl Proof Development System*, Prentice-Hall, 1986.

[CH88] T. Coquand and G. Huet, The Calculus of Constructions, *Information and Control* 76: 2/3 (1988), 95–120.

[De83] N. Dershowitz, *The Evolution of Programs*, Birkhäuser, 1983.

[Dij76] E. Dijkstra, *A Discipline of Programming*, Prentice-Hall, 1976.

[Ed85] E. Eder, Properties of Substitutions and Unifications, *Journal of Symbolic Computation* 1 (1985), 31–46.

[Fa79] M. Fay, First-Order Unification in an Equational Theory, *Proceedings of the Fourth Conference on Automated Deduction* (1979), 161–167.

[FM88] A. Felty and D. Miller, Specifying Theorem Provers in a Higher-Order Logic Programming Language, *Ninth International Conference on Automated Deduction* (1988), 61–80.

[GMW79] M. J. Gordon, A. J. Milner, C. P. Wadsworth, *Edinburgh LCF: A Mechanised Logic of Computation*, Springer-Verlag, Berlin, 1979.

[GH80] J. Guttag and J. Horning, Formal Specification as a Design Tool, *Seventh ACM Symposium on Principles of Programming Languages* (1980), 251–261.

[HK86] P. Harrison and H. Khoshnevisan, Efficient Compilation of Linear Recursive Functions into Object-Level Loops, *Proceedings of the SIGPLAN '86 Symposium on Compiler Constructions* (1986), 207–218.

[HN88] S. Hayashi and H. Nakano, *PX: A Computational Logic*, MIT Press, Cambridge, MA, 1988.

[Hi90] L. Hines, Str$\dot{+}$ve $\subseteq$: The Str$\dot{+}$ve-based Subset Prover, *Tenth International Conference on Automated Deduction* (1990), 193–206.

[KN84] D. Kapur and P. Narendran, An Equational Approach to Theorem Proving in the First-Order Predicate Calculus, Report, General Electric Research and Development Center, Schenectady, New York, April 1984.

[Kow74] R. Kowalski, Predicate Logic as a Programming Language, *IFIP Congress 74* (1974), 569–544.

[Mc89] W. W. McCune, OTTER 1.0 Users' Guide, Mathematics and Computer Science Division, Argonne National Laboratory, 1989.

[MW80] Z. Manna and R. Waldinger, A Deductive Approach to Program Synthesis, *ACM Transactions on Programming Languages and Systems* 2: 1 (1980), 90–121.

[MW81] Z. Manna and R. Waldinger, Deductive Synthesis of the Unification Algorithm, *Science of Computer Programming* 1 (1981), 5–48.

[MW85] Z. Manna and R. Waldinger, *The Logical Basis for Computer Programming*, Volume 1: *Deductive Reasoning*, Addison-Wesley, 1985.

[MW86] Z. Manna and R. Waldinger, Special Relations in Automated Deduction, *Journal of the ACM* 33: 1 (1986), 1–59.

[MW87a] Z. Manna and R. Waldinger, The Origin of a Binary-Search Paradigm, *Science of Computer Programming* 9 (1987), 37–83.

[MW87b] Z. Manna and R. Waldinger, The Deductive Synthesis of Imperative LISP Programs, *Sixth AAAI National Conference on Artificial Intelligence* (1987), 155–160.

[MW90] Z. Manna and R. Waldinger, *The Logical Basis for Computer Programming*, Volume 2: *Deductive Systems*, Addison-Wesley, 1990.

[M-L82] P. Martin-Löf, Constructive Mathematics and Computer Programming, *Sixth International Congress for Logic, Methodology, and Philosophy of Science* (1982), 153–175.

[Mu82] N. Murray, Completely Nonclausal Theorem Proving, *Artificial Intelligence* 18: 1 (1982), 67–85.

[Na89] D. Nardi, Formal Synthesis of a Unification Algorithm by the Deductive-Tableau Method, *Journal of Logic Programming* 7 (1989), 1–43.

[NPS90] B. Nordström, K. Petersson, and J. M. Smith, *Programming in Martin-Löf's Type Theory: An Introduction*, Oxford University Press, Oxford, England, 1990.

[Pa89] L. C. Paulson, The Foundation of a Generic Theorem Prover, *Journal of Automated Reasoning* 5: 3 (1989), 363–398.

[PR88] A. Pnueli and R. Rosner, A Framework for the Synthesis of Reactive Modules, *Proceedings of Concurrency 88* (1988), 4–17.

[Ro65] J. A. Robinson, A Machine-Oriented Logic Based on the Resolution Principle, *Journal of the ACM* 12: 1 (1965), 23–41.

[Sa79] M. Sato, Towards a Mathematical Theory of Program Synthesis, *Proceedings of the Sixth International Joint Conference on Artificial Intelligence* (1979), 757–762.

[Sc88] M. Schmidt-Schauss, Computational Aspects of an Order Sorted Logic with Term Declarations, SEKI Report SR–88–10, Universität Kaiserslautern, 1988.

[Si89] J. Siekmann, Unification Theory, *Journal of Symbolic Computation* 7: 3/4 (1989), 207–274.

[Sm85] D. R. Smith, Top-Down Synthesis of Divide-and-Conquer Algorithms, *Artificial Intelligence* 27: 1 (1985), 43–96.

[Sn85] N. Shankar, Checking the Proof of Gödel's Incompleteness Theorem, Institute for Computing Science, University of Texas at Austin, 1985.

[Sp83] E. Shapiro, *Algorithmic Program Debugging*, MIT Press, 1983.

[St81] M. E. Stickel, A Unification Algorithm for Associative-Commutative Functions, *Journal of the ACM* 28: 3 (1981), 423–434.

[St85] M. E. Stickel, Automated Deduction by Theory Resolution, *Journal of Automated Reasoning* 1: 4 (1985), 333–355.

[Tr89] J. Traugott, Deductive Synthesis of Sorting Programs, *Journal of Symbolic Computation* 7 (1989), 533–572.

[WR69] L. Wos and G. Robinson, Paramodulation and Theorem Proving in First-Order Theories with Equality. In B. Meltzer and D. Michie (editors), *Machine Intelligence 4*, American Elsevier (1969), 135–150.

[WW83] L. Wos and S. Winker, Open Questions Solved with the Assistance of AURA, in W. W. Bledsoe and D. W. Loveland (editors), *Automated Theorem Proving: After 25 Years*, American Mathematical Society (1983), 73–88.

Notes on resolution

J. A. Robinson

Syracuse University

Introduction. These notes are a revised and extended version of [18]. They offer a brief but reasonably complete account of the main ideas underlying the *resolution* formulation of first order predicate logic, with computational issues receiving the lion's share of attention. The *unification* algorithm is discussed in considerable detail, with logical formulas regarded simply as certain data structures. The aim is to explain resolution in general in such a way that the important special case of *Horn clause resolution* can be properly understood within the broader setting. Horn clause resolution is the theoretical framework for the kind of logic programming which is done by users of PROLOG. However, although many of the ideas we shall discuss are concretely realized in various versions of that programming language, we shall not be explicitly concerned with it. Surface details differ markedly from version to version and often obscure the relatively simple underlying conceptual system. We therefore try to present that system directly.

Expressions as data structures. Computational logic is concerned with specifying logical processes as formal symbolic computations, that is, as computations which operate on formal symbolic expressions as data objects. Logicians have long dealt with expressions in this formal way: in formal logic one must be able to treat logical expressions in abstraction from their meaning, as mere objects to be disassembled and constructed according to combinatorially precise procedures. In computer science, also, it is necessary to adopt a similar approach in treating programming languages rigorously.

Thus in particular in the *automatic theorem proving problem for the predicate calculus* one seeks an algorithm (a *proof procedure*) which, when given as input a sentence S will construct as output a *proof* P of S (provided that S is logically true).

NATO ASI Series, Vol. F 79
Logic, Algebra, and Computation
Edited by F. L. Bauer

Such an algorithm may sometimes (but not always, because of the demonstrable impossibility of a *decision* procedure for logical truth in the predicate calculus) also be able to indicate correctly that S is *not* logically true if that is indeed the case. In this problem both S and P have to be treated merely as data structures. Many different such proof procedures are now known, but in these notes we shall be concerned only with proof procedures based on the resolution principle.

In most theorem proving or logic programming applications the formal expressions will also of course have an informal (perhaps indeed even a formal) *semantics* - a logic programmer seeking to generate a set of "answer expressions" or a mathematician seeking to demonstrate a theorem will usually mean something by them. In *formal* theorem proving no "official" notice is taken at all of the semantics of the expressions. Only their (abstract) *syntactical form* is used as a basis for both their analysis and synthesis, and their semantics is ignored.

There are, however, no theoretical reasons why semantic proof procedures should not be pursued, only practical ones. Presumably, humans take into account the meanings of formulas when trying to prove a theorem. Indeed the resolution system, (and all other proof procedures) must be in very close harmony with the semantics of first order logic; the whole point of a proof procedure is that a semantic property of formulas (that of being logically true) can be exactly mirrored by a semicomputable syntactical property (that of being provable). The motivation for much of the detailed internal design of the resolution principle is semantic: resolution can be intuitively viewed as the organizing principle of a systematic search for a counterexample (an interpretation which makes a formula come out false) for a formula; a resolution proof of the formula is then just a record of the failure of the search to find any counterexample for it. This point of view is worked out at length in [19].

Proper names, variables and conses. LISP's simple and convenient symbolic universe of discourse is our model in setting up the following syntactic framework. We take expressions to be the objects generated from two countably infinite, disjoint sets, the set of *proper names* and the set of *variables*, by the single binary operation, *dot.* The proper names and the variables are also known as the *atoms.* They are concretely represented by finite strings over some suitable alphabet of characters. It does not much matter what these characters are, but in order to avoid confusion they include neither the various *bracketing characters* < > { } ( ) [ ] nor the *space* nor the the *comma* nor the character reserved for the dot operation: • . In these notes we write variables as nonempty strings of *lower case* letters, possibly subscripted, as for example: x, reverse, y_2. All other strings are proper names (this includes numerals, arithmetic operators, strings containing upper case letters, and so on).

Proper names and variables are the all those expressions which are not conses, in the sense of the following definition. If A and B are any expressions, then there is a unique expression P, called the *cons* of A and B, and whose *car* is A and whose *cdr* is B. We introduce the basic notations:

P = [• A B]	("P is the cons of A and B")
aP = A	("the car of P is A")
dP = B	("the cdr of P is B").

Remark. In the speech of the LISP community "cons" rhymes with "once" and is singular, while "cdr" is pronounced "kidder" by some and "kudder" by others. The point of our using in these notes such odd-sounding coinages (which are vestiges of the first LISP implementation at MIT on the IBM 704) is that they have no other meanings and hence allow us to preserve such words as "pair", "head", tail" for other purposes, and to mark and enforce the distinction between the *general* syntax of expressions-as-such and the *particular* syntax (to which we shall soon come) of the predicate calculus.

The original LISP convention was to write the dot *infixed,* [A • B] rather than [• A B]. In general [• A B] is the same expression as [• B A] only when A and B are the same expression. The expressions [•A[• B C]] and [•[• A B]C] are never the same.

Lists. This general syntax also provides the notion of a *list*. The proper name **nil** is given a special role as the *empty list* in the following definition, in which several notions relating to lists are defined simultaneously:

- **nil** is a *list*; moreover it is the (only) *empty list*, and it has *length* 0;

- for any list L of length $n \geq 0$ and any expression A, the cons [• A L] is a list of length n+1; the 1st *element* of [• A L] is A and the (j+1)st element of [• A L] is the jth element of L, for j=1, . . ., n.

According to this definition **nil** is the only list which is an atom, and all other lists, that is, all lists of nonzero length, are conses. However, not all conses are lists: Only those conses whose cdrs are lists are themselves lists.

The dot notation for conses in general is in principle completely adequate for all purposes, but since we shall mostly be dealing with lists we shall make use of LISP's more intuitive and compact *list notation*. This is based on the additional convention that allows a nested cons of the form

$$[\bullet\ A_1[\bullet\ A_2\ \ldots [\bullet\ A_n\ \textbf{nil}] \ldots]],$$

namely a *nonempty list*, to be written without its dots and interior brackets, thus:

$$[A_1\ A_2 \ldots A_n].$$

The list notation also allows **nil** to be written alternatively and intuitively as: [ ].

To sum up the general syntax of expressions we have:

expressions = proper names + variables + conses.

Substitutions. We shall have much to do with certain mappings of expressions to expressions, called *substitutions.* Abstractly, a substitution is any mapping θ of the expressions which sends each proper name onto itself and each cons [• A B] onto the cons [• Aθ Bθ]. Intuitively, the image Eθ of an expression E under the substitution θ is the expression which we get by replacing every occurrence of a variable within E by another given expression (possibly by another variable, possibly by the same variable). Each occurrence of the same variable is replaced by an occurrence of the same expression (called the *value* of that variable *under* θ). Thus the effect of a substitution θ on every expression is completely determined by its effect on the variables.

Instantiation. For each expression E and each substitution θ, the expression Eθ onto which θ maps E is called *the instance of* E *by* θ. The following *instantiation rule* then says, intuitively, that the expression Eθ is the result of simultaneously replacing each variable in E by the expression which is the instance of that variable by θ. The proper names in E are left alone. The rule is simple and has three cases:

(1)	Eθ	=	E	when E is a proper name
(2)		=	[• Aθ Bθ]	when E is the cons [• A B]
(3)		=	the instance of E under θ	when E is a variable.

Descriptions of substitutions. If a variable is sent onto itself by a substitution, it will be treated by part (3) of the rule exactly as part (1) treats a proper name. What this means in practice is that the instantiation rule allows us to construct Eθ, for any expression E, if we know only what the instance Xθ of each variable X is whose instance by θ is *different* from the variable itself. It is these variables (and only these) which are said to be *bound by* θ. For each variable X which is bound by θ the couple < X Xθ > is a *binding* of θ. In order to have a smooth, flexible notation for describing substitutions, we say that *any* set of couples of the form $< X\ X\theta >$ (where X is a variable) *which contains all the bindings of* θ, is a *description* of θ. A couple in a description of θ which is not a binding of θ must therefore have the trivial form $< X\ X >$. If we discard such trivial couples from any description of θ we get the (unique) *minimal description* of θ.

It is a standard convention, which we have been following above, to use lower case Greek letters to denote substitutions, and to indicate their application to expressions by juxtaposition on the right.

We shall frequently indulge in a harmless abuse of the description notation, in which a substitution is identified with a description of it. Thus we shall write, for example,

$$\theta = \{<y\ [A\ x]>,\ \ <x\ 5>,\ \ <z\ [B\ y]>\}$$

to mean that θ is the substitution *described by* the right hand side. If we are given a substitution in this way we can then for example easily verify that

$$[A\ [B\ x\ y]\ [C\ x\ y\ z\ w]]\theta = [A\ [B\ 5\ [A\ x]]\ [C\ 5\ [A\ x]\ [B\ y]\ w]].$$

We shall also find it convenient to extend the instantiation notation to sets and write:

$$S\theta \quad = \quad \{X\theta \mid X \text{ in } S\}.$$

It is then important to realise that *the set* Sθ *may have fewer members than the set* S, since it can happen that Xθ and Yθ are the same expression even though X and Y are different expression. Indeed, as we shall see shortly, this is exactly what is meant by saying that the expressions X and Y are *unified* by θ.

Products of substitutions. The *product* θλ of two substitutions θ and λ is simply their composition as mappings. That is, θλ is the substitution which maps each expression E onto the expression (Eθ)λ. θ maps E onto Eθ and then λ maps

$E\theta$ onto $(E\theta)\lambda$. Thus the product operation is associative. It also has an identity element, namely the substitution which maps every expression onto itself. The standard convention, which we shall follow, is to denote the identity substitution by the lower case Greek letter ε. Note that the empty set { } is the minimal description of ε.

Not all substitutions are bijections, of course, but when θ *is* a bijection we shall write θ^{-1} for its inverse. We then have $\theta\theta^{-1} = \theta^{-1}\theta = \varepsilon$. A bijective substitution is called a *change of variables*.

Descriptions of products. Since in computations substitutions are represented by their descriptions, we shall often need to construct a description of a product substitution $\theta\lambda$, given only descriptions of θ and λ. We can readily do this by using the following easily verified fact:

- if V and W are the sets of variables bound by θ and λ respectively, then the set of all couples

  $$\{ < X\ (X\theta)\lambda > \mid X \text{ in } (V \cup W) \}$$

  is a description of $\theta\lambda$.

This set is not necessarily the minimal description of $\theta\lambda$. In order to write the minimal description of $\theta\lambda$ we would first construct this set and then drop from it trivial couples (if any) of the form: <V V>. So, for example, the product of the substitution

$$\theta = \{ < x\ [A\ x\ y] >, < v\ w > \}$$

with the substitution

$$\lambda = \{ < x\ [C\ y] >, < y\ [B\ z\ x] >, < w\ v > \}$$

is (dropping the trivial couple < v v >) the substitution

$$\theta\lambda = \{ < x\ [A\ [C\ y]\ [B\ z\ x]] >, < y\ [B\ z\ x] >, < w\ v > \}$$

and their product in the reverse order is (dropping the trivial couple < w w >) the substitution

$\lambda\theta$ = {<x [C y]>, <y [B z [A x y]]><v w>}.

As this example illustrates, substitutions do not in general commute.

Unification. A central concept of the theory of resolution is that of *unification.* Consider the following *unification problem*:

- Given any finite set S of expressions, find (if one exists) a substitution θ which maps every expression in S onto the same expression, or else show (if no such substitution exists) that no such substitution exists.

A positive solution θ of this problem is said to *unify*, or to be a *unifier* of, the set S, and S is said to be *unifiable.* The expression in the singleton Sθ is thus *a common instance of all the expressions in* S.

A unifiable set of expressions may have many, possibly infinitely many, different unifiers. However, if in the above unification problem we ask that the substitution θ be not just *a* unifier of S, but a *most general* unifier of S, then the solution is essentially unique if there is one. A substitution σ is a *most general* unifier of a set S of expressions if it is a unifier of S with the extra property that *every* unifier θ of S satisfies an equation $\theta = \sigma\lambda$ for some substitution λ. Intuitively: any unifier of S is simply an instance (or special case) of a most general unifier of S. It turns out that every unifiable set has a most general unifier.

Of all the ideas needed for an understanding of resolution and logic programming the idea of *most general unification* is probably the most important.

Examples. The set { A, B } whose two members are the expressions

A = [[P x y u] [P y z v] [P x v w] [P [K t] t [K t]]]

and

B = [[P [G r s] r s] [P a [H a b] b] [P x y u] [P u z w]]

is unified by the substitution

σ = { <x [G r [K [H r r]]]> <y r> <z [H r r]>
<u [K [H r r]]> <v r> <w [K [H r r]]>
<a r> <b r> <s [K [H r r]]> <t [H r r]>},

as may be readily verified by applying σ to A and B and comparing the results. Indeed σ maps both A and B onto the expression:

[[P [G r [K [H r r]]] r [K [H r r]]]
[P r [H r r] r]
[P [G r [K [H r r]]] r [K [H r r]]]
[P [K [H r r]] [H r r] [K [H r r]]]].

Now, as it happens, σ is also a most general unifier of {A B}: any unifier θ of {A B} is a product σ{< r E >} of σ with a substitution which maps the variable r onto any expression E whatsoever. In particular, if E is a variable, then θ will also be a most general unifier of A and B.

On the other hand, for example, the set

{ [P x y u], [Q a b c] }

is not unifiable. It is not difficult to see this: any unifier of this set would have to unify the two proper names P and Q, which is impossible.

Again, the set

{ [F x], x }

is not unifiable: any unifier of this set would have to map the variable x onto an expression which contained itself as a proper subexpression, which is impossible.

A unification algorithm. We next give a simple *binary unification algorithm* which finds a most general unifier for any unifiable set of *two* expressions, and detects the nonunifiability of any set which is not unifiable. Finite sets with more than two members are handled by iterating this algorithm, using the (easily verified) fact that

- $\theta\lambda$ is a most general unifier of a set S of $n \geq 2$ expressions among which are A and B, if θ is a most general unifier of the set {A, B}, and λ is a most general unifier of the set $S\theta$.

Note that since the subset $\{A, B\}\theta$ of $S\theta$ is a singleton, $S\theta$ has at most $n-1$ elements, and thus the iteration halts after no more than $n-1$ steps.

Differences between expressions. In formulating the simple and intuitive (but inefficient) binary unification algorithm we shall use the notion of the *difference* $\Delta(X,Y)$ between two expressions X and Y.

Intuitively, we describe the difference between X and Y by giving all those (unordered) pairs of expressions which occur opposite each other at corresponding positions in X and Y where X and Y are not the same. This set of pairs can be computed recursively by means of the following equations:

$\Delta(X,Y) = \{\}$	if X and Y are the same expression,
$= \Delta(aX, aY) \cup \Delta(dX,dY)$	if A and B are both conses,
$= \{\{X,Y\}\}$	otherwise.

Negotiability. Reductions. A difference is said to be a *negotiable* one if

(1) it is nonempty;

(2) each unordered pair in it has the following two properties:

- *at least one of its members is a variable*
- *neither member occurs in the other.*

Notice that, because of the first property under (2), for *every* couple {U, V} in a negotiable difference, at least one (and possibly both) of {<U V>} or {<V U>} must describe a substitution. These substitutions are called *reductions* of the difference. Thus *every negotiable difference has at least one reduction.*

Examples of differences and their reductions. The difference

$$\Delta(\,[P\ x\ y\ z],\ [Q\ a\ b\ c]\,)$$

is the set

$$\{\, \{P, Q\},\ \{x, a\},\ \{y, b\},\ \{z, c\} \,\}.$$

This is *not* negotiable, because it contains the pair $\{P,Q\}$ and neither of P, Q is a variable. Nor is the difference

$$\Delta([F\ x], x) \quad = \quad \{\, \{[F\ x], x\} \,\}$$

negotiable, because although at least one of the two expressions x, [F x], *is* a variable (namely x), it occurs in the other (namely in [F x]). On the other hand, the difference

$$\Delta([P\ x], [y\ Q]) \;=\; \{\, \{P, y\}\ \{x, Q\}\}$$

is negotiable, and has two reductions, namely:

$$\mu = \{ < y \quad P > \} \text{ and } \nu = \{ < x \quad Q > \}.$$

Unifiers remove differences. The intuitive content of the following proposition is then that *the difference between distinct but unifiable expressions is always eliminable*:

- **Negotiability Lemma.** If A and B are distinct expressions, and θ unifies $\{A, B\}$, then $\Delta(A, B)$ is negotiable, and θ unifies each pair in $\Delta(A, B)$.

In the following algorithm we repeatedly reduce the difference between two given expressions by applying to them an arbitrary reduction of the difference, and accumulate the product of these reductions. This process eventually halts when the difference is no longer negotiable, at which point the outcome depends on whether the difference is empty or nonempty.

Here then is the

- **Binary Unification Algorithm,** for two expressions A, B as input:

  1 $\sigma \leftarrow \varepsilon$;

  2 **while** $\Delta(A\sigma,B\sigma)$ is negotiable **do** $\sigma \leftarrow \sigma\mu$

  where μ is any reduction of $\Delta(A\sigma,B\sigma)$;

  3 **return** (**if** $\Delta(A\sigma,B\sigma)$ is empty **then** σ **else** "FAIL").

We are assured that this algorithm always terminates and yields the correct result, by the

> **Binary Unification Theorem.** Let A and B be any two expressions. Then {A,B} is unifiable *if and only if* the Binary Unification Algorithm terminates, when applied to A and B as input, without returning "FAIL". The substitution σ it returns is then a most general unifier of {A, B}.

The proof of this theorem rests on two ideas, one dealing with the *termination* of the algorithm and the other dealing with its *correctness.*

Why the algorithm terminates. Obviously, the algorithm terminates if and only if its looping step 2 terminates. To see that step 2 always terminates we need only note that after each repetition of the loop the number of distinct variables in the two expressions $A\sigma$, $B\sigma$ will have decreased by 1 (since a reduction {<U V>} *eliminates the variable* U because of the fact that U does not occur in V). Hence step 2 must terminate after no more repetitions than there are distinct variables in the *input* expressions A and B.

Why the algorithm is correct. If, at step 3 of the algorithm, $\Delta(A\sigma, B\sigma)$ is empty, then (obviously) the σ then returned as output is *a* unifier of A and B, but we have to show that it is a *most general* unifier of A and B. That is, we have to show that for *every* unifier θ of A and B we will have that $\theta = \sigma\lambda$ for some λ. Well: we shall show that if θ unifies {A,B} then *the equation* $\theta = \sigma\theta$ *is an invariant of the computation.* This equation is (trivially) true after step 1. That its truth is preserved throughout step 2 follows from the fact that *any* reduction μ of $\Delta(A\sigma,B\sigma)$ satisfies the equation $\theta = \mu\theta$. This allows us to calculate (writing σ_i for the substitution σ after the ith repetition and μ_i for the reduction applied at the ith repetition):

$$\theta = \sigma_i\,\theta = \sigma_i\,(\mu_i\theta) = (\sigma_i\mu_i)\theta = \sigma_{i+1}\theta.$$

Since the final output σ is σ_i for some i, the equation $\theta = \sigma\theta$ will still be true after the termination of step 2. Thus for any unifier θ of {A,B} we have $\theta = \sigma\lambda$, with $\lambda = \theta$.

Faster binary unification algorithms. This binary unification algorithm has the pedagogical and theoretical advantage of being both concise and intuitive, and of having a termination and correctness proof which is short and easy to understand. However, it is not a practical algorithm, being *very* inefficient in terms of both time and space. For example, to unify with this algorithm the two expressions

$$[\ [F\ x_0\ x_0]\ \ [F\ x_1\ x_1]\ \ [F\ x_2\ x_2]\ \dots\ [F\ x_{n-1}\ x_{n-1}]\ ]$$
$$[\ x_1\ \ x_2\ \ x_3\ \dots\ x_n\ ]$$

requires time and space proportional to 2^n. It is easy to see that the most general unifier is

{<x_1 [F x_0 x_0]>

<x_2 [F [F x_0 x_0] [F x_0 x_0]]>

<x_3 [F [F [F x_0 x_0] [F x_0 x_0]] [F [F x_0 x_0] [F x_0 x_0]]]>

:

<x_n (an expression containing "F" 2^n -1 times and "x_0" 2^n times) > } .

Fortunately it turns out to be possible to exploit other aspects of unification in such a way that much faster unification algorithms can be devised. The idea behind them is to focus attention on *the partition which is induced in a set of expressions by a substitution.*

Substitution partitions. For any set S of expressions let us define S^* to be the smallest set of expressions which includes S and is closed under the taking of cars and cdrs. For example, if S is

{ [F [G x]], [F y] }

then S^* is:

{ [F [G x]], [F y], F, [[G x]], [G x], G, [x], x, [y], y, [ ] }. (1)

Then, for any set S of expressions and any substitution θ we can consider the partition induced in S^* by θ. By this we mean that partition of S^* in which two expressions lie in the same block if and only if they are unified by θ. For example, the partition induced in (1) by the substitution

{<y [G x]>}

is:

{ { [F [G x]], [F y] },
{ [[G x]], [y] },
{ [G x], y },
{ F },
{ G },
{ x },
{ [ ] } }.

Such partitions are called *substitution partitions*. For any substitution partition P we define the following two relations:

$X \equiv_P Y$ iff X and Y lie in the same block of P,

$X <_P Y$ iff for some A, B: $X \equiv_P A$ and $Y \equiv_P B$ and A occurs in B.

We note that for any substitution partition P of S^* we will have the following two properties:

no cycles for no $X_1, \ldots, X_n$ do we have $X_1 <_P \ldots <_P X_n <_P X_1$

structure for all distinct X, Y in S^*:

$X \equiv_P Y$ if and only if

either at least one of X, Y is a variable

or X, Y are both lists and **a**$X \equiv_P$ **a**Y and **d**$X \equiv_P$ **d**Y.

The converse also holds: any partition P of S^* which satisfies ***no cycles*** and ***structure*** is a substitution partition. A proof of this proposition and indeed a full discussion of these ideas can be found in Gerard Huet's thesis [7].

Set union algorithm. The idea, then, of the various known fast binary unification algorithms is to seek to construct, given distinct expressions A and B, a substitution partition on $\{A, B\}^*$, that is, one satisfying the properties ***no cycles*** and ***structure***. First, one constructs, if possible, a partition P satisfying ***structure*** for which $A \equiv_P B$. This is done by imitating the method of the very fast (almost linear in time, and linear in space) *set union algorithm* analyzed by R. E. Tarjan [14]. This partition P is then checked for the ***no cycles*** property by the (linear in time and space) *topological sort* method described by Knuth [9].

Gerard Huet was the first to describe this idea in [7], and there are also papers by Paterson and Wegman [11], and by Martelli and Montanari [10], which give strictly linear algorithms. A recent paper by J. S. Vitter and R. A. Simmons [16] contains a detailed discussion of the 'almost linear' unification process based on the very fast set union algorithm, and also airs the possibility of speeding it up by taking advantage of the opportunities it offers for parallel computation. The following brief sketch will give a feel for the general set union approach to unification. This elegant version of the algorithm is due to Lockwood Morris [20].

Partitions represented by forests. In the set union algorithm, the blocks of a substitution partition are represented as *trees* in a *forest*. The substitution is determined as explained below by means of a set E of ordered couples, no two ordered couples in E having the same left hand side. If E contains the ordered couple <U V> we say that E *equates* U *to* V, and we think of U and V as being nodes in the same tree, with V the *parent* of U. The following function ROOT is defined for any expression A and any such collection E of ordered couples as arguments, and finds the root of the tree in which A lies:

ROOT A E = **if** E equates A to B **then** ROOT B E **else** A.

The partition P represented by E is then that for which

$$X \equiv_P Y \quad \textbf{iff} \quad \text{ROOT X E} = \text{ROOT Y E}.$$

The ***structure*** part of the algorithm is embodied in two cooperating functions EQUIV and EQUATE.

The function EQUIV, when given expressions A, B, as first and second arguments and a collection E of equations as third argument, returns E itself, if ROOT A E is the same as ROOT B E. If ROOT A E is not the same as ROOT B E then EQUIV returns a set E^+ of ordered couples obtained by adding to E such further ordered couples as are required both for E^+ to satisfy ***structure*** and for ROOT A E^+ and ROOT B E^+ to be the same. If no such E^+ exists, then EQUIV returns the message "FAIL".

The function EQUATE behaves in just the same way as EQUIV, but it is only ever called when its third argument E is *not* the message "FAIL" and when its first and second arguments A and B *are* roots of E.

The function UNIFY, when given expressions A and B as arguments, returns a set of ordered couples (if A and B are unifiable), or returns "FAIL" if A and B are not unifiable. Let τ be the set of ordered couples in (UNIFY A B) whose first component is a variable:

$$\tau = \{ \langle V, W\rangle \mid \langle V, W\rangle \text{ is in (UNIFY A B) and V is a variable } \}.$$

The set τ describes a substitution from which we can construct a most general unifier of {A, B}.

The substitution σ constructed from τ by the following algorithm is a most general unifier of {A, B}:

$\sigma \leftarrow \tau$;

while $\sigma\tau \neq \sigma$ **do** $\sigma \leftarrow \sigma\tau$;

return σ .

This is indeed an algorithm - the **while** loop always terminates (see [20], p. 14, for a proof).

The function UNIFY is given by the equation:

UNIFY A B = EQUIV A B { }

where

EQUIV A B E =	**if**		E is "FAIL" **then** "FAIL"	
	else		EQUATE (ROOT A E) (ROOT B E) E	

and

EQUATE A B E =		**if**	A = B **then** E	1
	else	**if**	{A,B} contains a variable **then** MERGE A B E	2
	else	**if**	{A,B} contains a proper name **then** "FAIL"	3
	else		EQUIV **a**A **a**B (EQUIV **d**A **d**B (MERGE A B E))	4.

Since EQUATE is called only by EQUIV, in the context shown, the input assumptions for EQUATE are always true, namely that E is not the message "FAIL" and that A and B are roots of E. The action in line 3 of EQUATE is appropriate to having discovered that ***structure*** cannot be satisfied, since in view of the tests in lines 1, 2 and 3, one of A, B must be a proper name while the other must be either a (distinct) proper name or else a cons. The action in lines 2 and 4 is to merge the two trees (i.e. blocks of the partition) into a single tree (block) by equating the root of one tree to the root of the other. The action in line 4 is appropriate to having found that both A and B are conses, so that not only are the two trees (blocks) merged, but the resulting forest (partition) is further modified by whatever mergings of blocks as may be necessary to satisfy ***structure***.

If we use the following simple definition for MERGE

MERGE A B E ≡ **if** A is a variable **then** {<A B>} ∪ E **else** {<B A>} ∪ E

then the calls of MERGE from EQUATE preserve the property that

- if a tree representing a block has a variable as its root, then *all* expressions in that block are variables.

Balancing and path compression. However, this simple method of merging foregoes the opportunity to imitate fully the set union algorithm, which, when merging two trees, *balances* the resulting tree by making the root of the *taller* tree the parent of the root of the *shorter* tree. The set union algorithm also *compresses* the trees, as follows. Each call (ROOT A_0 E) returns a root A_n by accessing a sequence of ordered couples

$$<A_i \; A_{i+1}>, \quad (i = 0, \ldots, n-1)$$

in E (a "path"). To compress the tree along this path, each of these couples is *replaced* by the couple $<A_i = A_n>$, thus making the root A_n of the tree the immediate parent of each node A_i, $(i = 0, \ldots, n-1)$ instead of merely an ancestor. By complicating the fast unification algorithm slightly to reflect these two refinements, an essentially linear cost performance can be achieved.

Logic. After these preliminaries we can now get to our main topic and describe the *clausal predicate calculus.* This is a special machine-oriented version of the predicate calculus, in which one deals principally with only one form of proposition (the clausal sequent) and uses only one inference principle by which to prove the truths among such propositions (the resolution principle).

Clausal sequents. A clausal sequent is an intuitively meaningful proposition which asserts that a given set P of *universal disjunctive clauses* logically entails a given set Q of *existential conjunctive clauses.*

Sequent notation. We write the sequent by joining the two sets of clauses with a *sequent arrow*, thus:

$$P \Rightarrow Q$$

and we say that P is the *antecedent*, and Q the *succedent*, of the sequent. We read the sequent as "P logically entails Q" or as "Q logically follows from P". When Q is the empty set we can write the sequent simply as

$$P \Rightarrow$$

and read it as "P is logically unsatisfiable". We often write the antecedent and succedent sets by simply listing their clauses in some order (the order being irrelevant) and omitting the external set brackets. Thus, we can write

$$A, B, C \Rightarrow M, N$$

rather than

$$\{A, B, C\} \Rightarrow \{M, N\}.$$

Meaning of a clausal sequent. Pending the more exact definitions to be given below, we can say that the sequent $P \Rightarrow Q$ expresses the claim that there is no *interpretation* of the predicate symbols and function symbols of the clauses in P and Q under which all the clauses in P are true and all those in Q are false.

When Q is a singleton {C} this accords well with our everyday intuitive understanding of what it means to say that C logically follows from (the sentences in) P: that C must be true whenever all the sentences in P are true. When Q is not a singleton, however, the sequent $P \Rightarrow Q$ expresses a proposition which has no natural intuitive counterpart in everyday language. Its meaning is then something like: if all the sentences in P are true then not all of the sentences in Q are false.

Universal disjunctive and existential conjunctive clauses. A universal disjunctive (u.d.) clause is a sentence of the form:

$$\forall x_1 \ldots x_k\ (\sim A_1 \vee \ldots \vee \sim A_p \vee B_1 \vee \ldots \vee B_q).$$

We may note that this sentence is is equivalent to the sentence

$$\forall x_1 \ldots x_k \; ((A_1 \wedge \ldots \wedge A_p) \rightarrow (B_1 \vee \ldots \vee B_q)).$$

An existential conjunctive (e.c.) clause is one of the form:

$$\exists x_1 \ldots x_k \; (A_1 \wedge \ldots \wedge A_p \wedge \sim B_1 \wedge \ldots \wedge \sim B_q).$$

This sentence is equivalent to the sentence

$$\exists x_1 \ldots x_k \; ((A_1 \wedge \ldots \wedge A_p) \wedge \sim (B_1 \vee \ldots \vee B_q)).$$

In each clause, the expressions $A_1, \ldots, A_p$ and $B_1, \ldots, B_q$ are predications (defined below), and the $x_1, \ldots, x_k$ are all the distinct variables which occur in them.

For both kinds of clause, we say that

- the set $\{x_1 \ldots x_k\}$ of variables is the *prefix* of the clause,
- the set $\{A_1 \ldots A_p\}$ of predications is the *body* of the clause, and
- the set $\{B_1 \ldots B_q\}$ of predications is the *head* of the clause.

A clause with an empty prefix (i.e. a clause which contains no variables) is called a *ground clause*. Note that a clause contains no *free* variables. Previous treatments of resolution have followed the convention that all variables in u.d. clauses are free, but are to be given the universal interpretation. Thus the present formulation is not essentially different in this respect.

Empty clauses. When both the head and the body of a clause are empty then (necessarily) also the prefix is empty. The clause itself is then said to be *the empty clause* of the one kind or the other. The empty u.d. clause is logically equivalent to the sentence **false,** and the empty e.c. clause is logically equivalent to the sentence **true**. The sentence **false** is false in every interpretation (see below), and the sentence **true** is true in every interpretation.

Duality; kernels of clauses. A u.d. clause with a given prefix, body and head is changed into its *dual*, that is, into the e.c. clause which has the same prefix, body and head, by interchanging $\forall$ with $\exists$, $\wedge$ with $\vee$, and (for each predication P) $\sim$P with P. We find it useful to define the basic operations and notions of resolution in terms of these invariants of the duality transformation. We call the triple

$< \{x_1 \ldots x_k\} \; \{A_1 \ldots A_p\} \; \{B_1 \ldots B_q\} >$

consisting of the prefix, body and head of a clause C, the *kernel* of C, regardless of whether C is a u.d. clause or an e.c. clause. Notice that each of two clauses which are duals of each other, (that is, a u.d. clause and an e.c. clause having the same kernel) is logically equivalent to the *negation* of the other.

It has hitherto been the usual practice in clausal predicate calculus to deal only with u.d. clauses, and to call these simply *clauses* without qualification. However, this practice leads to unnecessary awkwardness in the later development and applications of the resolution principle. By working with kernels wherever possible we are able to enjoy the conceptual economy of the usual treatment without giving up the richer and more natural means of expression which the availability of both kinds of clause provides.

Predications and terms. Herbrand Universes and Bases. In any particular application, we shall regard clauses and clausal sequents as data objects all built ultimately out of the variables and certain proper names, each of which is classified as a *predicate symbol* or a *function symbol* and assigned an *arity*. (The arity of a proper name may be any natural number. When the arity of a function symbol is 0 the symbol is also called an *individual symbol*). For any fixed set L of predicate and function symbols with assigned arities (such a set is called a *lexicon*) we define certain expressions to be the *terms over* L and the *predications over* L, and when these expressions contain no variables we say they are the *ground* terms over L and the *ground* predications over L. The definitions are:

- The terms over L are the variables, and those lists $[F \; t_1 \ldots t_n]$ in which F is a function symbol of arity n in L, and the t_i are terms over L. The set of all ground terms over L is called the *Herbrand Universe over* L.

- The predications over L are those lists $[P \; t_1 \ldots t_n]$ in which P is a predicate symbol of arity n in L, and the

t_i are terms over L. The set of all ground predications over L is called the *Herbrand Base over* L.

Substitutions extended to clauses. It is useful to extend the substitution notation not only (as was done earlier) to sets of expressions, but also to *kernels*, as follows: the instance of the kernel $< X\ A\ B >$ by the substitution θ is the kernel $< Y\ A\theta\ B\theta >$, where Y is the set of all variables occurring in expressions in $X\theta$; and we write it as:

$$< X\ A\ B >\theta.$$

It is then easy to apply substitutions to clauses. The instance by θ of the u.d. (resp. e.c.) clause C whose kernel is K is the u.d. (resp. e.c.) clause whose kernel is $K\theta$. We write it as $C\theta$. For example, the instance by the substitution

$$\theta = \{ < x\ [K\ u] >, < y\ [M\ v] > \}$$

of the of the u.d. clause

$$\forall x\ y\ z\ (\sim[P\ x\ y] \vee \sim[Q\ y\ z] \vee [R\ x\ y\ z])$$

whose kernel is

$$< \{x, y, z\}\ \{[P\ x\ y], [Q\ y\ z]\}\ \{[R\ x\ y\ z]\} >$$

is the u.d. clause

$$\forall u\ v\ z\ (\sim[P\ [K\ u]\ [M\ v]] \vee \sim[Q\ [M\ v]\ z] \vee [R\ [K\ u]\ [M\ v]\ z])$$

whose kernel is

$$< \{u, v, z\}\ \{[P\ [K\ u]\ [M\ v]], [Q\ [M\ v]\ z]\}\ \{[R\ [K\ u]\ [M\ v]\ z]\} >$$

since $\{u, v, z\}$ is the set of all variables occurring in expressions in

$$\{x, y, z\}\theta = \{[K\ u], [M\ v]\ z\}.$$

Note that in forming an instance of a clause, the substitution operates on the variables of the prefix even though they are from the logical point of view *bound* variables of the clause. Our substitution operations know nothing about the meanings, if any, that are associated with expressions.

In particular a *variant* of a clause C is an instance of C by a change of variables. Every clause is obviously a variant of itself (with the change of variables ε). Note also that if $C\theta$ is a variant of C then C is a variant of $C\theta$ since θ^{-1} is a change of variables and

$$(C\theta)\theta^{-1} = C(\theta\theta^{-1}) = C\varepsilon = C.$$

Finally, if A is a variant of B and B is a variant of C then A is a variant of C. Thus 'being a variant of' is an equivalence relation on clauses.

If the prefix of an instance D of a clause C is empty, then D is a *ground instance* of C.

Separated clauses. Two clauses are said to be *separated* if their prefixes are disjoint. This notion comes up later, in the definition of resolvents.

Semantics: Herbrand interpretations of lexicons. Herbrand was able in effect to reduce the semantics of a lexicon L to its syntax by the ingenious idea of choosing as the domain D of individuals the set of all ground terms over L. An Herbrand interpretation J of a lexicon L is then given by specifying, for each predication in the Herbrand Base of L, whether it is *true in* J or *false in* J. This in effect associates an n-ary relation P* with every n-ary predicate symbol P in L, namely, the set of all n-tuples:

$$P^* = \{ < t_1 \dots t_n > | \text{ the predication } [\, P\, t_1 \dots t_n \,] \text{ is true in J } \}.$$

Such a specification of truth or falsehood for each ground predication is then automatically extended to clauses, since in every Herbrand interpretation the variables in clauses over L range over the Herbrand Universe of L as the domain of individuals of the interpretation. Thus

- a u.d. (respectively, e.c.) clause is true (respectively, false) in J if and only if all its ground instances are true (respectively, false) in J; and
- a u.d. (respectively, e.c.) ground clause with body A and head B is false (respectively, true) in J if and only if each member of A is true in J and each member of B is false in J.

Counterexamples of clausal sequents. Let S be a clausal sequent over the lexicon L. Then an interpretation J of L is a *counterexample* to S if and only if every clause in the antecedent of S is true in J, and every clause in the succedent of S is false in J. A counterexample to S is a witness to the fact that S is not true.

Truth of clausal sequents. A clausal sequent is *true* if and only if there are no counterexamples to it.

Equivalence of clausal sequents. Two sequents over L are *equivalent* if and only if every counterexample to one of them is also a counterexample to the other.

Being true is a semidecidable (but only a semidecidable) property of clausal sequents. There is no general algorithm for detecting the falsehood of every false sequent, but there are, as we shall see, sound and complete proof procedures for clausal sequents. Such proof procedures for clausal sequents are algorithmic methods of *recognizing the truth of any clausal sequent which is in fact true.* The resolution principle is intended as the basis for designing such proof procedures in a reasonably efficient form.

Obvious sequents. Some true clausal sequents can be immediately recognized as being true: for example, those which contain **false** in the antecedent and those which contain **true** in the succedent.

Why are these sequents obvious? Clearly, if a set of sentences contains the sentence **false** then there can be no interpretation under which every sentence in the set is true (since **false** is false in every interpretation), hence any sequent whose antecedent contains **false** can have no counterexample. Similarly, if the succedent contains the sentence **true**.

So these sequents indeed are obvious. In general, however, clausal sequent can be true without being obvious, and so we need some other means of establishing its truth. This is where the resolution principle comes in.

Proving true clausal sequents by means of resolution. The resolution principle is based upon the idea of a *resolvent* of of two separated clauses.

Resolvents. Let E, F be two separated clauses with kernels < X A B >, < Y C D > respectively. Then a clause R with kernel < Z M N > is a *resolvent of* E *with* F *on* K if and only if

- K is a unifiable subset of $B \cup C$
- $K \cap B$ and $K \cap C$ are nonempty
- σ is the most general unifier of $B \cup C$
- Z = the set of variables occurring in expressions in $(X\sigma \cup Y\sigma)$
- $M = A\sigma \cup (C\sigma — K\sigma)$
- $N = (B\sigma — K\sigma) \cup D\sigma$.

Note that according to this definition both the u.d. clause R with kernel < Z M N > and the e.c. clause R with kernel < Z M N > are resolvents of E with F on K. Note also that we do not specify whether the clauses E and F are both u.d. clauses or both e.c. clauses, or one of each kind. The definition applies to all eight combinations:

E	F	R
u.d.	u.d.	u.d.
u.d.	u.d.	e.c.
u.d.	e.c.	u.d.
u.d.	e.c.	e.c.
e.c..	u.d.	u.d.
e.c.	u.d.	e.c.
e.c.	e.c.	u.d.
e.c	e.c.	e.c.

instead of only to the first one.

In general we say that R is a resolvent of E and F if R is a resolvent of E and F on K, for some K.

The resolution principle. The resolution principle is an inference principle for clausal sequents. This is a somewhat different and more general point of view from the original one [12]. In the original treatment (which is the one generally described in the literature) resolution is formulated as an inference principle only for u.d. clauses, and it allows us to infer only a u.d. clause as conclusion from two u.d. clauses as premises. In other words, in the original treatment we are restricted to the first of the eight possible combinations listed above. By restating the principle in the present form, however, we uncover more of the power and flexibility of the underlying idea, and allow e.c. clauses to enter into resolution reasoning in a natural way. In particular we get rid of the idea that resolution is somehow essentially tied up with the notions of refutation and unsatisfiability (which was the narrower view taken in [12] and which has been the prevailing view ever since).

We state the *resolution principle* as follows:

- from a clausal sequent S one may infer the clausal sequent S + R if
  (1) S is not obvious,
  (2) R is a resolvent of two variants of clauses in S,
  (3) R is not a variant of any clause in S, nor of the dual of any clause in S.

By S + R we mean the clausal sequent obtained by adding the clause R to the antecedent of S, if it is a u.d.clause, or to the succedent of S, if it is an e.c. clause. Any such sequent S + R is then said to be a *resolution* of the sequent S.

Comment on the definition. Provision (1) removes the possibility of continuing to look for a proof when the proof is in fact complete. This point will become clearer in the next paragraph. Provision (2) allows the resolved clauses to be unseparated.

The process of forming their resolvent includes separating their variables if need be. Indeed, it allows a clause to be resolved with itself. Provision (3) eliminates an undesirable source of redundancy (namely, adding a resolvent to a sequent which already contains it) and ensures that a sequent has only finitely many essentially different resolutions.

Resolution series; resolution proofs. A series $S_0, \ldots, S_r$ of clausal sequents in which S_{i+1} is a resolution of S_i, $0 \leq i < r$, is called a *resolution series.* A resolution series is a *resolution proof* (of its initial sequent) if and only if its final sequent is obvious.

Soundness of the resolution principle. The logical justification of the resolution principle rests on the fact (not difficult to prove) that any resolution of a sequent is equivalent to it. Thus the resolution principle is *sound.* We cannot infer a false sequent from a true sequent. Indeed, since *all sequents in a resolution series are equivalent,* all sequents in a resolution proof are true: they are equivalent to its final sequent which, being by definition an obvious sequent , is (obviously) true. In particular, then, the initial sequent of a resolution proof is true, and so the proof really is a witness to the truth of its initial sequent.

We can thus simply read the resolution proof *backwards,* starting with its terminal (obviously true) sequent, and proceed in truth-preserving steps (passing from a sequent we know to be true to one which must also be true) until we arrive at the initial sequent, with the knowledge that it too must be true.

The significance of the resolution principle for computational logic then rests on two further properties of resolution, *local finiteness* and *completeness,* one of which is easy to see, the other of which is not.

Local finiteness. A clausal sequent has only *finitely* many resolutions, all of which can be effectively constructed. There are only finitely many clauses in the sequent: hence only finitely ways of constructing a resolvent from two variants of clauses in it. Hence we can effectively find all finite resolution series starting with a given sequent, and therefore all resolution proofs of that sequent, if it has any.

Completeness of the resolution principle. Every true sequent has at least one resolution proof. This fact is nontrivial, but the proof is omitted here. It can be found in [19].

Horn sequents. The general resolution-proof-finding procedure as sketched above is rather attractive compared with older proof-finding systems for the predicate calculus. However, it is not yet in an efficient enough form for what we now call logic programming. There are just too many resolutions of a given sequent for it to be feasible to search out exhaustively all resolution proofs of a given true clausal sequent. Nevertheless Green [5] was able to use the general procedure to lay out and motivate the main ideas of logic programming as we know them today. Soon thereafter Colmerauer [4] with his PROLOG system, and Kowalski [8] in a more abstract form, showed that by sacrificing some generality one can achieve a remarkably useful system of logical computation. Their key idea was to work with a restricted system of clausal predicate calculus rather than the full system. In the restricted system one considers only Horn sequents, rather than clausal sequents in general.

Horn sequents; procedure clauses; goal clauses. Kowalski calls a clause is a *procedure* clause if its head is a singleton, and a *goal* clause if its head is empty. Both procedure clauses and goal clauses are called *Horn* clauses. In both kinds of clauses the predications in the body of the clause are called *goals.* A Horn sequent is a then a clausal sequent whose antecedent contains only procedure clauses and whose succedent contains only goal clauses. A *minimal* Horn sequent is one which contains only *one* goal clause.

Logic programs = minimal Horn sequents. A minimal Horn sequent S thus has the form

$$P \Rightarrow (\exists Y)(G_1 \wedge \ldots \wedge G_n)$$

where the antecedent P is a set of procedure clauses, and $\{G_1, \ldots, G_n\}$ is a nonempty set of goals. The various resolution proofs (if any) of S will include the very special LUSH resolution proofs which we define below, but also a great many more. It turns out that if S is true then not only does it have a resolution proof (which we already know) but bacuase of its special form it even has a LUSH resolution proof. This fact is crucial for logic programming purposes. It means that a logic programming engine need only search through the very much sparser space of LUSH resolution series starting with a true minimal Horn sequent S, in order to be sure of finding a proof of S.

Selection and removal functions. To help in describing LUSH resolution we need the idea of a *selection* function, namely, a function $\uparrow$ defined on every nonempty set M, whose value $\uparrow M$ at M is some member of M. For each such selection function there is a corresponding *removal* function $\downarrow$, which when applied to a nonempty set M returns the the result of removing the element $\uparrow M$ from M.

LUSH resolution series. Let $\uparrow$ be any selection function. Given the resolution series

$$S_0, S_1, \ldots, S_i, \ldots,$$

let us write

$$S_{i+1} = S_i + R_{i+1} \qquad (i \geq 0)$$

to show that at each step the sequent S_{i+1} is obtained by adding the resolvent R_{i+1} to the sequent S_i. Then the series is a *LUSH resolution series controlled by* $\uparrow$ provided that

- each S_i $(i \geq 0)$ is a Horn sequent,
- S_0 is a *minimal* Horn sequent with goal clause R_0,
- R_{i+1} $(i \geq 0)$ is a resolvent with R_i, on the set $\{H_i, \uparrow C_i\}$, of (a variant) $\forall X_i\,(A_i \rightarrow H_i)$ of some procedure clause in S_i, where C_i is the body of R_i.

The clause R_i is called the *active clause* of the sequent S_i. Thus the active clause of each sequent in the series after the first is the resolvent of some procedure clause in the preceding sequent with the active clause of the preceding sequent. The active clause in the initial sequent is the (only) goal clause in that sequent. Note that in a LUSH resolution series the resolvents are always goal clauses: *no procedure clauses are ever generated as resolvents.* Note also that a procedure clause can produce *at most one* resolvent with the active clause, and will do so if, but *only* if, its head can be unified with the goal selected by $\uparrow$ from the body of the active clause. The rapidity with which this can be decided will depend on the complexity of the particular function $\uparrow$ being used, on the method of representation of the sets of goals to which it is applied, and on the method of representing the procedure clauses in the initial sequent. PROLOG uses particularly efficient methods, in which an order is imposed on the goals. PROLOG's $\uparrow$ respects this ordering in the sense that $\uparrow M$ is always the first (in that ordering) of the most recently added elements of M. The details are too many for further discussion here.

The idea of LUSH resolution is originally due to Kowalski [8] but its acronymic label is due to Hill [6]. It is intended to suggest: **L**inear resolution with **U**nrestricted **S**election function for **H**orn clauses. The LUSH resolution proofs of a true minimal Horn sequent are far fewer in number than the ordinary resolution proofs of it. Many people, following the example of van Emden and Kowalski [15] and Lloyd [17], prefer to use the acronym SLD (**S**elected **L**inear resolution for **D**efinite clauses) instead of LUSH. However, we feel that acronyms should not be multiplied, or even added, beyond necessity.

LUSH resolution is complete in the following very strong sense: if S is a true minimal Horn sequent, then for <u>all</u> selection functions $\uparrow$ there is at least one LUSH resolution proof of S which is controlled by $\uparrow$.

Lush resolution series as computations. We can associate with the LUSH resolution series $S_0, S_1, \ldots, S_i, \ldots$, controlled by $\uparrow$, the *computation* which is the series $C_0, C_1, \ldots, C_i, \ldots$, of *states* each of which is the body of the active clause R_i of the corresponding sequent S_i. Consider, then, a LUSH resolution series whose initial Horn sequent $P \Rightarrow (\exists Y)C$ provides the initial state C. The relationship between successive states corresponding to the successive Horn sequents in the series exhibits a repetitive pattern: each successive state C_{i+1} is obtained from its predecessor C_i by the same "computation cycle".

The computation cycle. To obtain a successor of the nonempty state C_i we take some procedure clause in P such that, for some suitable variant

$$<X_i\ A_i\ \{H_i\}>\theta_i\,,$$

of its kernel $<X_i\ A_i\ \{H_i\}>$, the set $\{\uparrow C_i, H_i\theta_i\}$ is unifiable with most general unifier σ_i. If no such procedure clause exists, then the state C_i is a "failure" and has no successors. Otherwise, a new state C_{i+1} is formed by the construction

$$C_{i+1} = A_i\theta_i\sigma_i \cup (\downarrow C_i)\,\sigma_i\,.$$

In general there may be more than one procedure clause for which this construction can be made. *In other words, a state may have more than one successor.*

Stacklike behavior of successive states. This computation cycle does indeed abstractly resemble the basic cycle of a simple stack-oriented computer, if we think of $\uparrow$ as returning the top element of the stack, and of $\downarrow$ as removing the top element of the stack and returning the rest. The states C_i are then the successive contents of the stack. The cycle thus consists, partly, of popping the top goal from the stack and pushing the goals of the procedure body onto the stack. This is what PROLOG actually does. Unfortunately this simple analogy does not account for the application, to each goal in the stack, of the most general unifier σ_i. However, by means of an idea due originally to Boyer and Moore in their Edinburgh resolution theorem prover [1], we can find a natural computational role in this analogy for the substitutions θ_i and σ_i of this basic cycle: that of the *environment of bindings* for the variables.

Implicit representation of expressions. The idea of Boyer and Moore is to represent an expression $E\theta$ *implicitly* as the ordered pair $<E\ \theta>$ instead of actually carrying out the work of applying θ to E. This ordered pair can be thought of as a "closure" or a "delayed evaluation". The ordered pair $<E\ \theta>$ can be treated in all respects as though it were the explicit expression $E\theta$ that it implicitly represents: for example the result $E\theta\lambda$ of applying λ to the expression represented by the ordered pair $<E\ \theta>$ is itself represented by the ordered pair

$$<<E\ \theta>\ \lambda>$$

and so on. When pairs are nested to the left like this we follow an "association to the left" convention and drop the inner brackets. In general, the expression implicitly represented by $<E\ \theta_1\ \ldots\ \theta_n>$ can be found simply by carrying out the "delayed" work of applying the successive substitutions, to yield the explicit expression: $((E\theta_1)\ \ldots\ \theta_n)$. It is straightforward to adapt the unification procedures UNIFY, EQUIV, EQUATE, ROOT, MERGE, etc., to this method of implicit representation of expressions. We can then use the Boyer-Moore idea to represent the successive states of a LUSH computation. Instead of actually explicitly constructing the state C_{i+1} by applying the most general unifier σ_i to the set $(A_i\theta_i \cup \uparrow C_i)$ we can simply represent C_{i+1} by pairing this set with σ_i:

$$C_{i+1} = <(A_i\theta_i \cup \uparrow C_i)\ \ \sigma_i>.$$

The set $A_i\theta_i$ can also be represented in the same way, so that the transition equation can be written

$$C_{i+1} = <(<A_i\ \ \theta_i> \cup \uparrow C_i)\ \ \sigma_i>.$$

Computations on the machine will of course use this implicit representation wherever possible. In particular the successive unification substitutions will be separate components of each state. The successive states of the computation corresponding to a LUSH resolution proof of the minimal Horn sequent

$P \Rightarrow (\exists Y)C$

are then the following (in Boyer-Moore form):

$$
\begin{array}{lll}
C_0 & = & < C \;\; \varepsilon > \\
C_1 & = & < (\{<A_1 \;\; \theta_1>\} \;\; \cup \uparrow C_0) \;\; \sigma_1 > \\
C_2 & = & < (\{<A_2 \;\; \theta_2>\} \;\; \cup \uparrow C_1) \;\; \sigma_1 \, \sigma_2 > \\
 & \vdots & \\
C_{i+1} & = & < (\{<A_{i+1} \, \theta_{i+1}>\} \;\; \cup \uparrow C_i) \;\; \sigma_1 \, \sigma_2 \, \ldots \, \sigma_{i+1}> \\
 & \vdots & \\
C_t & = & < \{\} \qquad\qquad\qquad \sigma_1 \, \sigma_2 \, \ldots \;\; \ldots \sigma_t >
\end{array}
$$

with the successive kernels $<X_1 \, A_1 \, \{H_1\}>, \ldots, <X_t \, A_t \, \{H_t\}>$ of (not necessarily different) procedure clauses in P supplying the sets A_i of new goals at each step, and with each substitution σ_j satisfying the equation:

$$\sigma_i = (\text{UNIFY } H_i\theta_i \; (\uparrow C_i) < \sigma_1 \ldots \sigma_{i-1} >), \qquad (i \geq 1).$$

Throughout this LUSH computation no expression need actually be constructed explicitly. At termination, the substitution $(\sigma_1 \, \sigma_2 \ldots \sigma_t)$ is available to construct the output of the computation.

The computation tree. The procedure clause chosen at each step of the computation is one of the only finitely many comprising the antecedent P of the sequent. This gives rise to a *computation space* which is a finitary (but not necessarily finite) tree. The various branches of the tree correspond to the various LUSH resolution series starting with the given initial sequent. The root of the tree is the body of the goal clause (= the active clause) of the initial sequent, and in general each node of the tree is *either*

- empty, and a leaf of the tree, (a "success") *or*
- nonempty but with no successors, and a leaf of the tree, (a "failure") *or*
- nonempty and with one or more successors.

The "success" branches (if any) of the computation tree are the completed computations corresponding to the various LUSH resolution proofs of the initial minimal sequent $P \Rightarrow (\exists Y)C$, the body C of whose active clause $(\exists Y)C$ is the root of the tree.

Each completed computation yields an output. Recall that the original program was

$$P \Rightarrow (\exists Y)(G_1 \wedge \ldots \wedge G_n).$$

It is then natural to view the equation

$$Y = Y\sigma_1\sigma_2 \ldots \sigma_t$$

as the *output* of the completed computation, where the state $< \{\} \quad \sigma_1\ \sigma_2\ \ldots\ \sigma_t >$ is its terminal node. The term(s) $Y\sigma_1\sigma_2 \ldots \sigma_t$ are expressions constructed stepwise by the successive unifiers $\sigma_1, \sigma_2, \ldots, \sigma_t$.

Nondeterminacy. For every *true* initial sequent S there is at least one such computation since the sequent has at least one LUSH resolution proof. S may, however, have many, even infinitely many, such proofs; and for the computation corresponding to each proof there will be a possibly different output. It is the responsibility of a logic programming engine, such as a PROLOG machine, to obtain all such outputs when given a true minimal Horn sequent. This it does by making a complete exploration of the tree which is the search space. If the search space is infinite then it may contain infinitely many success branches (and it may also contain infinitely many failure branches). A properly designed logic programming engine should presumably generate the search tree *fairly*, i.e., in such a way (and there are many options) as to reach any given node in the tree after only finitely much time. Unfortunately, most PROLOG engines are designed unfairly (for the sake of speed) as "depth-first backtracking" devices.

Example. The minimal Horn sequent sequent

- $\forall x($ [NUMBER x] $\rightarrow$ [NUMBER [1+ x]])
- [NUMBER 0]

$\Rightarrow$ $\exists y$: [NUMBER y]

is true, and has infinitely many LUSH resolution proofs.

The outputs of the corresponding computations are the equations:

y = 0
y = [1+ 0]
y = [1+ [1+ 0]]

and so on.

At the jth step (j > 0) the active clause is (understanding y_0 as just y itself):

$\exists y_j$: [NUMBER [1+ . . . [1+ y_j] . . .]] (with j 1+'s)

and the next resolution can be obtained by choosing either the second procedure clause

(*) [NUMBER 0]

or the [j+1]st variant

(**) $\forall y_{j+1}($ [NUMBER y_{j+1}] $\rightarrow$ [NUMBER [1+ y_{j+1}]])

of the first procedure clause. The choice of (*) will yield an obvious sequent, and the output will be

$$y = [1+ \ldots [1+0] \ldots] \qquad \text{(with } j \text{ 1+'s)}$$

The choice of (**) will produce the new active clause

$$\exists y_{j+1}: [\text{NUMBER } [1+ \ldots [1+ y_{j+1}] \ldots]] \qquad \text{(with } j+1 \text{ 1+'s).}$$

A depth-first backtracking device which chose (**) before (*) at each level would therefore delay permanently the choice of (*) at the first level, and never reach even the first success node in the search tree! Merely reversing the order of the choice would produce a complete (although of course nonterminating) traversal of the search tree.

Example. The two procedure clauses

P_1: $\forall x$: [CAT nil x x]

P_2: $\forall x,a,b,c$: ([CAT a b c] $\rightarrow$ [CAT [• x a] b [• x c]])

intuitively define [CAT x y z] to mean that the list z is the result of conCATenating the lists x and y in that order. The goal clause

Q: $\exists p,q$: [CAT p q [• 1 [• 2 [• 3 nil]]]]

intuitively says that the list [1 2 3] is p ++ q, i.e., the result of concatenating two lists, p and q, in that order. The minimal Horn sequent $P_1, P_2 \Rightarrow Q$ is true, and has *four* different LUSH resolution proofs, the simplest of which is obtained by adding just one resolution invoking the procedure clause P_1. The corresponding computation has the output

$[p\ q] = [\ \mathbf{nil}\ \ [1\ 2\ 3]\ ]$

describing the construction $[1\ 2\ 3] = \mathbf{nil} +\!+ [1\ 2\ 3]$. The other three proofs consist respectively of one, two and three successive invocations of the procedure clause P_2 followed by an invocation of the procedure clause P_1. The outputs of the corresponding computations describe respectively the constructions

$$[1\ 2\ 3] = [1]+\!+[2\ 3],\quad [1\ 2\ 3] = [1\ 2]+\!+[3],\quad [1\ 2\ 3] = [1\ 2\ 3]+\!+\ \mathbf{nil}.$$

Another view of LUSH resolution proofs. Clark [3] has pointed out an interesting alternative way to view a computation with output $Y = Y\theta$ corresponding to a LUSH resolution proof of a true minimal Horn sequent

$$P \Rightarrow (\exists Y)(G_1 \wedge \ldots \wedge G_n).$$

Namely, it can be interpreted as the construction of *n separate hyper-resolution proofs* by which n different unconditional procedure clauses

$$(\forall X_1)H_1,\ \ldots,\ (\forall X_n)H_n$$

are simultaneously deduced from the procedure clauses in P, and which are such that the expressions

$$[\,H_1\ \ldots\ H_n\,] \text{ and } [\,G_1\ \ldots\ G_n\,]$$

are unifiable with most general unifier θ.

Hyper-resolution. Hyper-resolution is an inference pattern for u.d. clauses which requires a *conditional* procedure clause

(1) $(\forall X)(A_1 \wedge \ldots \wedge A_k \rightarrow B)$

with $k \geq 1$ goals $A_1, \ldots, A_k$, as *major premise*, and k *unconditional* procedure clauses

(2) $(\forall X_1)B_1, \ldots, (\forall X_k)B_k$,

separated from it and from each other, as *minor premises*, such that the expressions

$[A_1 \ldots A_k]$ and $[B_1 \ldots B_k]$

are unifiable with most general unifier σ. The *conclusion* of the hyper-resolution inference is then the unconditional procedure clause

(3) $(\forall Z)C$

where $C = B\sigma$, and $Z =$ the set of variables occurring in expressions in the set $(X \cup X_1 \cup \ldots \cup X_k)\sigma$. It is straightforward to verify that the clause (3) is indeed a logical consequence of the clauses (2) together with the clause (1).

One may use this inference pattern to obtain, from a set P of procedure clauses, a series P_0, P_1, ..., of sets of procedure clauses all of which are logical consequences of P, as follows. The set P_0 is the set of unconditional procedure clauses in P (sometimes called "facts", although they need not be ground clauses). The set P_{j+1} is the result of adding to the set P_j all of the unconditional procedure clauses which can be inferred by a single application of hyper-resolution with major premise in the set P and minor premises in the set P_j. Clearly the set P_k contains all the unconditional procedure clauses deducible from P by no more than k steps of hyper-resolution. It is natural to organize a deduction from P of such an unconditional procedure clause as a tree with that clause as its root and with members of P_0 as leaves. Each nonleaf node in the tree is the immediate consequence, by hyper-resolution, of its immediate descendents in the tree (as minor premises) and some conditional clause in P as major premise.

LUSH resolution and hyper-resolution. Now consider again a LUSH resolution proof whose initial sequent is $P \Rightarrow (\exists Y)(G_1 \wedge \ldots \wedge G_n)$ and whose output is $Y = Y\theta$. Let

$C_0, \ldots, C_t$

be the successive states of the corresponding computation and let

$D_0, \ldots, D_t$

be the successive procedure clauses D_i such that D_i is used with C_i to get C_{i+1}. Note that

$C_0 = \{G_1, ..., G_n\}$.

Clark observes that the ith step of the computation can be interpreted as attaching the goals in the body of D_i as immediate successors to the goal $\uparrow C_i$. These goals then become, together with those in the set $\downarrow C_i$, the goals in the set C_{i+1}. In this manner n trees are grown, each nonleaf node in which follows, by a single application of hyper-resolution, from its immediate successors (as minor premises) and some procedure clause in P (as major premise). Clark's interpretation is most illuminating, for it shows where the extraordinary freedom of the LUSH resolution scheme (to use any selection function whatsoever) comes from. It comes from the fact that it does not matter in what order the nodes of the n hyper-resolution trees are treated as these trees are grown.

Negation as failure. The preceding discussion has dealt only with the pure Horn clause case of logic programming. In practice, one can work (as first explained in Clark [2]) with *pseudo-Horn clauses* by generalising the definition of Horn clauses in the following way. Instead of restricting the bodies to be sets of predications (= unnegated atomic sentences) we can allow them to be *literals*, that is, either predications or negated predications. However, both the definition of procedure clauses as having a head containing exactly one predication, and the definition of goal clauses as having an empty head, are retained. The basic computation cycle is then extended to deal with the case that the goal $\uparrow C_i$ can now be a negated predication, say, ~G (hence not unifiable with the head of any procedure clause). If G is a ground expression an attempt is made to prove the sequent $P \Rightarrow G$. Such an attempt can have only three outcomes:

1 it can terminate with a proof, in which case ~G is treated as "disproved" and C_i has no successors

2 it can terminate but without finding a proof ("finite failure"), in which case ("negation by failure") ~G is treated as "proved" and C_i has the successor $\downarrow C_i$

3 it can fail to terminate, in which case the attempt must be eventually aborted without any decision as to the provability of G from P.

Evidently, **1** assumes the simple consistency of P. All that has been shown is that G is a logical consequence of P, and in general it might well also be that –G is also a logical consequence of P. However, if P is simply consistent it cannot have both G and –G as logical consequences.

On the other hand, **2** appears to assume something like the adequacy of the axiom system P as a characterization of all the ground literals which are intended to be taken as true in the intended interpretation of the lexicon ("if G were true in our intended interpretation then it would be provable from P").

We cannot discuss further here the important and interesting way of handling negation in logic programming systems. There is an abundant recent research literature on the subject. We refer the reader to Clark [2] and Lloyd [17] for further reading.

References

[1] Boyer, R.S. and Moore, J S.
The sharing of structure in theorem proving programs.
Machine Intelligence 7, Edinburgh University Press, 1972, 101 - 116.

[2] Clark, K.L.
Negation as failure.
In Logic and Databases, edited by Gallaire and Minker, Plenum Press, 1978, 293 - 322.

[3] Clark, K.L.
Predicate logic as a computational formalism.
Ph.D. Thesis, Imperial College, London, 1979.

[4] Colmerauer, A., et al.
Un systeme de communication homme-machine en francais. Groupe d'Intelligence Artificielle, U.E.R. de Luminy, Universite d'Aix-Marseille, Luminy, 1972.

[5] Green, C.C.
Application of theorem proving to problem solving.
Proceedings of First InternationalJointConference on Artificial Intelligence, Washington D.C., 1969, 219 - 239.

[6] Hill, R.
LUSH resolution and its completeness.
DCL Memo 78, University of Edinburgh Department of Artificial Intelligence, August 1974.

[7] Huet, G.
Resolution d'equations dans les langages d'ordre 1, 2, . . . , ω.
These d'Etat, Universite Paris VII (1976).

[8] Kowalski, R.A.
Predicate logic as programming language.
Proceedings of IFIP Congress 1974, North Holland, 1974, 569 - 574.

[9] Knuth, D. E.
The Art of Computer Programming, Volume 1, Addison-Wesley 1969, 258 - 268.

[10] Martelli, A. and Montanari, U.
Unification in linear time and space.
Technical Report B76-16, University of Pisa, Italy, 1976.

[11] Paterson, M. S., and Wegman, M. N.
Linear unification. Journal of Computer and Systems Sciences 16, 1978, 158 - 167.

[12] Robinson, J. A.
A machine-oriented logic based on the resolution principle.
Journal of the Association for Comp[uting Machinery 12, 1965, 23 - 41.

[13] Robinson, J. A.
Automatic deduction with hyper-resolution.
International Journal of Computer Mathematics 1, 1965, 227 - 234.

[14] Tarjàn, R. E.
Efficiency of a good but not linear set union algorithm.
Journal of the Association for Computing Machinery 22, 1975, 215 - 225.

[15] van Emden, M.H., and Kowalski, R.A.
The semantics of predicate logic as a programming language.
J.ournal of the Association for Computing Machinery 23, 1976, 733 - 742.

[16] Vitter, J. S., and Simmons, R. A.

New classes for parallel complexity: a study of unification and other complete problems for ***P***.

I.E.E.E.Transactions on Computers, Vol C-35, May 1986, 403 - 417.

[17] Lloyd, J. W.

Foundations of logic programming.

Springer-Verlag, 1984. Second, extended edition, 1987.

[18] Robinson, J. A.

Notes on Logic Programming.

Logic of Programming and Calculi of Discrete Design, Springer-Verlag, 1987, 109 - 144.

[19] Robinson, J. A.

Logic: form and function.

Edinburgh University Press, 1979.

[20] Morris, F.L.

On list structures and their use in the programming of unification.

School of Computer and Information Science Report 4-78, Syracuse University, 1978.

Introduction au λ-calcul pur

(Introduction to Pure λ-Calculus)

Gérard Huet
INRIA
Rocquencourt, France

Abstract

These course notes introduce pure λ-calculus, the basic formal system underlying functional programming languages. Section 1 discusses notations for functions in mathematics and introduces Church's λ-notation. Section 2 introduces de Bruijn's abstract syntax for λ-terms, and gives the basic definitions constructively as executable CAML programs. Section 3 introduces the notions of a redex and of β-reduction, and gives a table of some commonly used notations for λ-terms. Section 4 gives a proof using Tait's technique of parallel reduction that β-reduction satisfies the diamond property; the Church–Rosser theorem follows as a corollary. A systematic proof method based on annotated terms is introduced here. Section 5 proves the theorem of finite developments, and Section 6 proves its corollary the standardisation theorem, with a method due to Klop. Section 7 introduces the notion of a Böhm tree. Section 8 presents some variations of the theory of β-reduction of pure λ-K-calculus. Section 9 proves Böhm's theorem and discusses its semantic consequences. Section 10 contains a short guide to the literature.

Table des matières

NATO ASI Series, Vol. F 79
Logic, Algebra, and Computation
Edited by F. L. Bauer

Introduction

Ces notes de cours introduisent la notion de fonction calculable en étudiant le formalisme du λ-calcul pur, plus précisément le λ-K-calcul pur avec la β-conversion.

Ce formalisme est traité de façon homogène, avec l'usage systématique de la structure abstraite de de Bruijn. Les définitions sont exprimées de manière complètement constructive et même exécutable, dans le langage CAML développé à l'INRIA. Une méthode uniforme de marquage permet de dégager simplement l'idée directrice des preuves.

Les principaux théorèmes présentés sont le théorème du losange et sa conséquence la confluence des calculs, le théorème des développements finis et sa conséquence le théorème de standardisation, enfin le théorème de Böhm et ses conséquences sémantiques.

Un deuxième chapitre de ces notes de cours, en préparation, traitera de différents calculs typés, modélisant des langages fonctionnels divers, ou des systèmes d'inférence en déduction naturelle, suivant la correspondance de Curry-Howard.

1. La lambda notation

Il n'y a pas de notation généralement admise en mathématiques pour décrire une expression fonctionnelle, c'est à dire dénotant une fonction. A vrai dire, la notion même de fonction est relativement récente en mathématiques. Jusqu'au XIXème siècle, les fonctions n'étaient pas des objets mathématiques à part entière, mais juste une manière de parler pour désigner des procédés permettant de calculer des valeurs mathématiques à partir d'autres. Par exemple, à partir des fonctions sin et cos, on peut fabriquer la fonction qui aux réels x et y, fait correspondre le réel $\sin(x)+\cos(y)$. On désigne généralement cette fonction par la notation :

$$x,y \mapsto \sin(x)+\cos(y) \qquad (1)$$

qui ne peut pas être employée comme expression, à l'intérieur d'une autre expression par exemple.

Les fonctions sont devenues des objets mathématiques à part entière depuis l'introduction par Cantor de la théorie des ensembles. Les fonctions de la théorie des ensembles sont identifiées à leur graphe, c'est à dire à l'ensemble des couples argument-résultat. Il faut bien remarquer que cette nouvelle définition de fonction, extensionnelle, est fondamentalement différente de la notion intentionnelle d'algorithme, ou règle de calcul d'un résultat à partir d'une valeur de l'argument.

Nous allons dans ces notes nous attacher à la notion d'algorithme. Une fonction calculable sera une fonction qui peut être définie par un algorithme. La notion d'algorithme va s'appuyer sur un langage permettant d'écrire des expressions dénotant des valeurs ou des fonctions, et sur des règles de calcul spécifiant comment une expression peut expliciter une valeur, par évaluation progressive. Plusieurs difficultés surgissent.

a) Faut il que la notation distingue une valeur concrète, par exemple un entier, d'une valeur fonctionnelle? On distingue ici les langages typés (par exemple, CAML), et les langages non typés (par exemple, le λ-calcul que nous allons étudier). Certains langages sont intermédiaires, comme LISP, qui distingue les (C-VAL) des fonctions (F-VAL).

b) Le langage et ses règles de calcul garantissent-ils que l'évaluation d'une expression termine toujours? Nous verrons qu'une telle contrainte est nécessairement restrictive, et que les algorithmes ne définissent donc en général que des fonctions partielles, non définies sur certaines valeurs.

c) Comment peut-on désigner de manière non-ambigue l'argument d'un algorithme? Dans la notation ci-dessus, les meta-variables x et y sont utilisées pour cet usage. Cette utilisation traditionnelle de noms pour les variables liées, ou muettes, qui dénotent l'argument d'une fonction, est pratique pour le mathématicien, qui joue sur l'ambiguité entre l'algorithme et la valeur résultat sin(x)+cos(y) pour des valeurs données des arguments x et y. Mais les difficultés surgissent dès qu'on essaye d'étendre cette confusion à des fonctionnelles, c'est à dire des expressions qui prennent en argument ou retournent en résultat des fonctions. Les notations usuelles $\int f(x)dx$, $\partial f/\partial x$, $\Sigma_{i=1,10}E(i)$, $\forall x \exists y\ P(x,y)$ montrent qu'une notation fonctionnelle uniforme est désirable.

Nous donnons ci-dessous quelques notations concrètes pour l'expression fonctionnelle (1), dans différents langages.

λx. λy. ((+ (sin x)) (cos y))	λ-calcul de Church
function x → function y → sin(x)+cos(y)	langage CAML
[x][y] <<y>cos><<x>sin>+	AUTOMATH
[x,y] (+ (sin x) (cos y))	

Cette dernière notation sera utilisée dans ces notes de cours.

Toutes les notations concrètes ci-dessus souffrent du défaut d'avoir à utiliser un nom, c'est à dire une chaîne de caractères spécifique, pour noter les variables muettes. Elles ne rendent donc pas compte du concept abstrait d'algorithme, pour lequel le nom de ces variables n'est pas pertinent. Par exemple, [x,y] (+ (sin x) (cos y)) et [u,v] (+ (sin u) (cos v)) dénotent le même algorithme. Pourtant, il n'est pas aisé de spécifier rigoureusement, mais sans lourdeur, la congruence de renommage. Par exemple, il convient de ne pas confondre ci-dessus les noms x et y. La congruence de renommage est appelée traditionnellement α-conversion dans la notation du λ-calcul de Church. Nous éviterons ici ces difficultés en considérant un langage plus abstrait qui contourne cette difficulté.

2. Notation abstraite, substitution

2.1. Quelques propositions de notation abstraite

Un certain nombre de propositions de représentations des λ-termes modulo renommage des variables liées ont été faites. Nous en discutons certaines sur l'exemple:

[x](x [y](x y)).

Remarquons que cette syntaxe utilise la notation (f x) pour indiquer l'application de la fonction f à l'argument x.

Une première proposition de syntaxe abstraite, n'utilisant pas de noms de variables, a été évoquée (mais non vraiment utilisée!) par Bourbaki :

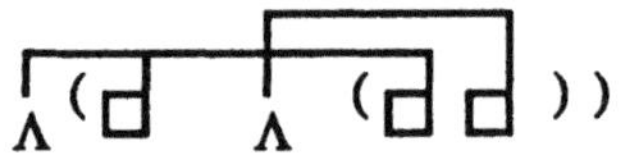

Une autre proposition, liée à une implémentation sur ordinateur du λ-calcul, a été faite par C. Wadsworth. Elle utilise un formalisme de graphes :

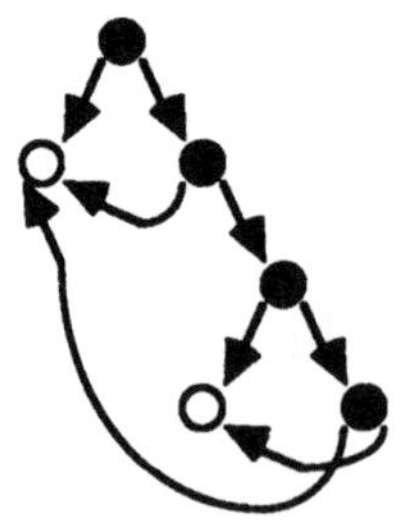

Les deux propositions ci-dessus sont en fait très proches. Celle de Wadsworth présente l'avantage d'autoriser d'exprimer certains partages de sous-expressions. Elle ont toutes deux un défaut essentiel, celui d'être basées sur un langage bi-dimensionnel. De plus, elles ne permettent pas de parler facilement de sous-expressions, dans lesquelles certaines occurrences de variables peuvent être libres. Enfin, la représentation de Wadsworth n'exprime pas la bonne notion de partage.

La bonne notation abstraite pour les λ-termes est due à N. de Bruijn. Elle utilise l'idée d'indice de liaison: chaque occurrence de variable liée est représentée par un entier, qui est la profondeur relative du lieur de la variable. Sur notre exemple, on obtient :

[] (1 [] (2 1))

Ici les crochets de liaison [] doivent être lus comme un symbole unaire, l'abstraction. Les expressions parenthésées représentent l'application (ici opérateur binaire). Les entiers sont les indices représentant les occurrences de variables.

Remarquons tout de suite que cette représentation abstraite n'est pas une notation lisible, au contraire de la représentation concrète : les deux occurrences de la variable x sont représentées par deux indices différents, alors qu'inversement les deux indices "1" représentent des occurrences de variables distinctes x et y.

2.2. Définitions précises des structures

Nous avons donc besoin des deux langages, correspondant aux deux structures définies ci-dessous en CAML :

```
type lambda =                   (* syntaxe abstraite *)
      Ref of num                      (* variable *)
    | Abs of lambda                   (* abstraction *)
    | App of lambda * lambda;;        (* application *)

type concrete =                 (* syntaxe concrète *)
      Var of string                   (* x *)
    | Lambda of string * concrete     (* [x] E *)
    | Apply of concrete * concrete;;  (* (E1 E2) *)

let closed = valid 0
    where rec valid n = function
      Ref(m)      → n ≥ m
    | Abs(e)      → valid (n+1) e
    | App(e1,e2) → valid n e1 & valid n e2;;
```

On écrit Λ_n pour l'ensemble des λ-termes M valides dans un contexte de n variables libres, c'est à dire tels que valid n M = true. Les termes fermés (dans Λ_0) n'ont pas de variables libres.

A tout terme M de Λ_n correspond sa fermeture $\underline{M}$ = closure M n, avec :

```
let rec closure e = function
      0  → e
    | n  → closure (Abs e) (n-1);;
```

Voici maintenant l'algorithme calculant l'indice correspondant à un nom de variable dans un environnement donné, représenté par une liste de chaînes de caractères :

```
let index name = search 1
     where rec search n = function
        []          → Erreur "variable non liée"
      | first::rest → if first=name then n else search (n+1) rest;;
```

On donne maintenant l'algorithme traduisant un λ-terme concret fermé en la structure abstraite correspondante :

```
(* parser : concrete →  lambda *)
let parser = parse_env []
      where rec parse_env env = abstract
        where rec abstract = function
        Var(name)          → Ref(index name env)
      | Lambda(name,c)     → Abs(parse_env (name::env) c)
      | Apply(c1,c2)       → App(abstract c1, abstract c2);;
```

On peut maintenant définir complètement l'interface entre syntaxe abstraite et syntaxe concrète par une grammaire et un formateur :

```
grammar lambda =
rule entry concrete = parse
          lambda e                      → parser(e)
and lambda = parse
          "("; lambda g; lambda d; ")"  → Apply(g,d)
      |   IDENT x                       → Var(x)
      |   " ["; binder b; "]"; lambda e → list_it (curry Lambda) b e
      |    "("; lambda e; ")"           → e
and binder = parse
          IDENT x                       → x
      |   IDENT x; binder b             → x::b;;
```

Nous laissons l'écriture du formateur en exercice.

2.3 Principe de récurrence contextuelle

Les définitions récursives sur les λ-termes suivent toutes le même schéma : on appelle une fonction d'un entier n avec la valeur 0 sur une expression fermée; la fonction s'appelle récursivement sur les sous-expressions, l'entier n étant incrémenté à chaque abstraction.

A ce style de définition correspond un style de preuve par récurrence, qui est l'analogue de la récurrence structurelle traditionnelle des structures libres, mais tenant compte du fait que l'opérateur d'abstraction lie une nouvelle variable.

Principe de récurrence contextuelle.

Soit P_n une propriété des lambda-termes, indicée par un entier naturel n, et vérifiant les conditions de fermeture suivantes.

(a) $P_n(M) \;\&\; P_n(N) \Rightarrow P_n(App(M,N))$

(b) $P_{n+1}(M) \Rightarrow P_n(Abs(M))$

(c) $0<m\leq n \Rightarrow P_n(Ref(m))$

Alors $P_n(M)$ est vrai pour tout M dans Λ_n, avec n quelconque.

Au principe de récurrence contextuelle correspond le procédé de définition par <u>récursion contextuelle</u>. Par exemple, donnons les algorithmes de substitution.

2.4. La substitution

Voici l'algorithme qui recalcule les indices des variables libres d'un terme à travers k niveaux d'abstraction :

```
let lift k = lift_rec 1
where rec lift_rec n = function
          Ref(m)    →  if m<n then Ref(m)          (* variable liée  *)
                       else Ref(m+k)               (* variable libre *)
        | Abs(e)    →  Abs(lift_rec (n+1) e)
        | App(g,d)  →  App(lift_rec n g, lift_rec n d);;
```

La fonction (lift k) injecte Λ_n dans Λ_{n+k}, avec n quelconque. Si M=lift k N, on écrit $N=lift^{-1}$ k M. On a ($lift^{-1}$ k M) = lift (-k) M, sauf qu'il faut vérifier maintenant m+k>0 dans le cas variable libre, et échouer dans le cas contraire.

Voici maintenant l'algorithme de substitution. Si M est un lambda construit dans un environnement $x::\Gamma$ et N est un lambda construit dans l'environnement Γ, alors $M\{x \leftarrow N\}$ (notation concrète) ou $M\{N\}$ (notation abstraite) est défini comme le terme subst N M, avec subst défini comme suit. On vérifie que ce terme est bien construit dans l'environnement Γ.

```
let subst lam = subst_rec 1
where rec subst_rec n = function
        Ref(m)      →  if m=n then (lift (n-1) lam)
                       if m<n then Ref(m)
                       else Ref(m-1)
      | Abs(e)      →  Abs(subst_rec (n+1) e)
      | App(g,d)    →  App(subst_rec n g, subst_rec n d);;
```

Exemple.
[y](y x){x ← [z]z} = subst (Abs(Ref(1))) (Abs(App(Ref(1),Ref(2)))
= Abs(App(Ref(1),Abs(Ref(1)))) = [y](y [z]z).

3. Réduction, conversion

On définit maintenant la relation de réduction ⇨ entre λ-termes, comme suit.

$$([x]M\ \ N) \ ⇨\ M\{x \leftarrow N\} \qquad (\beta)$$

On dit que le terme ([x]M N) est un radical, qui se réduit en $M\{x \leftarrow N\}$. On étend la relation ⇨ par congruence par rapport à la structure de λ-terme :

$$M ⇨ M' \Rightarrow []M ⇨ []M' \qquad (\xi)$$
$$M ⇨ M' \Rightarrow (M\ \ N) ⇨ (M'\ \ N)$$
$$M ⇨ M' \Rightarrow (N\ \ M) ⇨ (N\ \ M')$$

Historiquement, la fermeture réflexive-transitive ⇨* de la relation ⇨ s'appelle la β-réduction.

La β-réduction est la règle de calcul du langage algorithmique lambda. Elle est non-déterministe, dans la mesure où un terme possédant plusieurs occurrences de radicaux peut se calculer de manières différentes. Pourtant, le langage est intrinsèquement déterministe, dans la mesure où le résultat final d'un calcul est unique, comme le montrera le théorème du losange ci-dessous. Donnons tout d'abord un exemple de calcul.

Exemple.
On calcule, en soulignant les radicaux à chaque étape :

M = ((([x] [y] (x (y x)) [u] (u u)) [v] [w] v).
⇨ (([y] ([u] (u u) (y [u] (u u))) [v] [w] v)
⇨ ([u] (u u) ([v] [w] v [u] (u u)))
⇨ ([u] (u u) [w] [u] (u u))
⇨ ([w] [u] (u u) [w] [u] (u u))
⇨ [u] (u u) = Δ.

Δ est une forme irréductible, car ce terme ne possède pas de radical. On dit qu'un tel terme est en forme normale. On dit aussi que Δ est la forme normale de M. Cette terminologie est justifiée par la prochaine section. Mais remarquons tout d'abord qu'un terme peut ne pas posséder de forme normale, car ses calculs peuvent ne pas terminer; ainsi :

Ω = (Δ Δ) ⇨ (Δ Δ) ⇨ (Δ Δ) ⇨ .... boucle!

Remarque. Dans la terminologie anglo-saxonne, on utilise redex, abréviation de "reducible expression", pour radical.

Nous finissons cette section par une table donnant quelques lambdas usuels :

I	=	[x]x	
K	=	[x, y]x	
S	=	[x, y, z](x z (y z))	
A	=	[x, y](x y)	
B	=	[x, y, z](x (y z))	
C	=	[x, y, z](x z y)	
P	=	[x, y, z](z x y)	
T	=	[x, y]x	(= **K**)
F	=	[x, y]y	
0	=	[f, x]x	(= **F**)
1	=	[f, x](f x)	(= **A**)
n	=	[f, x](f (f ... (f x)))	(n fois f)
Y	=	[f]([x](f x x) [x](f x x))	
Θ	=	([x, y](y (x x y)) [x, y](y (x x y)))	
Δ	=	[x](x x)	
Ω	=	([x](x x) [x](x x))	(= (Δ Δ))

4. Le théorème du losange

4.1 Enoncé du théorème et corollaires

Théorème du losange.
Pour tous termes M, M_1 et M_2 tels que $M \Rightarrow^* M_1$ et $M \Rightarrow^* M_2$, il existe un terme N tel que $M_1 \Rightarrow^* N$ et $M_2 \Rightarrow^* N$.

Autrement dit, la relation $\Rightarrow$ est <u>confluente</u>, ou de manière équivalente la β-réduction $\Rightarrow^*$ est <u>fortement confluente</u>, avec les définitions de ces notions exprimées par les diagrammes ci-dessous :

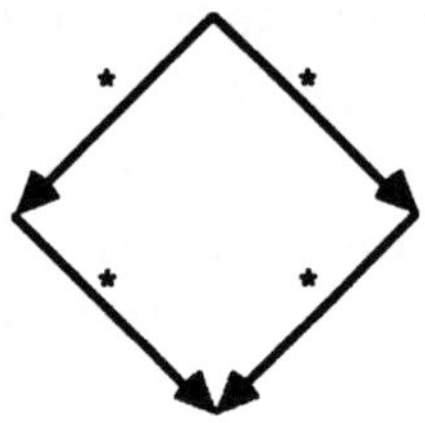

confluence

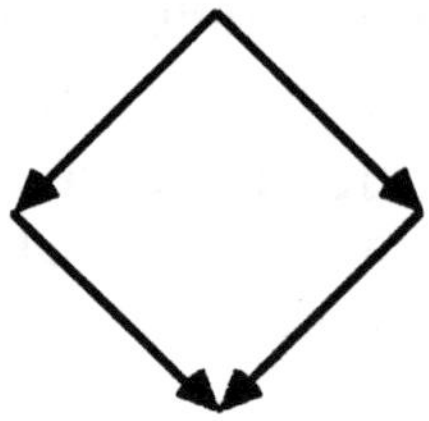

confluence forte

Dans de tels diagrammes, la partie supérieure, en traits pleins, est universellement quantifiée, alors que la partie inférieure, en pointillés, est existentiellement quantifiée.

Définition. On appelle <u>β-conversion</u> l'équivalence $\cong$ engendrée par $\Rightarrow$.

Enonçons maintenant quelques conséquences importantes du théorème.

Corollaire 1 : <u>Propriété de Church et Rosser</u>. Pour tous termes M et M' convertibles : $M \cong M'$, il existe un terme commun N résultat de calculs effectués à partir de M et M' respectivement : $M \Rightarrow^* N$ et $M' \Rightarrow^* N$.

Corollaire 2 : <u>Unicité des formes normales</u>. La forme normale de tout terme, si elle existe, est unique.

On appelle normalisable un terme possédant une forme normale, et fortement normalisable un terme tel que tout calcul issu de ce terme se termine. Par exemple, le terme ([w][x]x Ω) est normalisable, de forme normale **I** = [x]x, mais non fortement normalisable, à cause de son sous-terme non normalisable Ω.

4.2 Occurrences et sous-termes

Avant de faire la preuve du théorème du losange, il convient de développer les notions fondamentales permettant d'exprimer les transformations progressives d'une expression par calcul.

La première notion est celle d'occurrence. Jusqu'ici nous avons parlé de manière informelle d'occurrences de variables. De manière plus formalisée, les occurrences vont désigner l'emplacement des sous-expressions d'une expression. Une étape de calcul sera alors déterminée par l'occurrence à laquelle se trouve le radical en question.

Définition. Soit M une λ-expression. On définit l'ensemble de ses occurrences dom(M) par récurrence comme suit.

```
type sibling = A | B | C
and occurrence == sibling list
and domaine == occurrence list;;

let top = []                  (* l'occurrence principale de tout terme *)
and sons s = map (cons s);;        (* préfixage par le sibling s *)

let rec dom = function        (* dom : lambda → domaine *)
      Ref(_)      → [top]
    | App(g,d)    → top :: ((sons A (dom g)) @ (sons B (dom d)))
    | Abs(e)      → top :: (sons C (dom e));;
```

Exemple.
dom([x](x [y](x y))) = [[]; [C]; [C;A]; [C;B]; [C;B;C]; [C;B;C;A]; [C;B;C;B]].

Deux occurrences u et v sont dites cohérentes si elles peuvent appartenir à un même terme. Dans le cas contraire on écrit u#v.

Les occurrences sont ordonnées naturellement par l'ordre préfixe. On dit que u est au dessus de v, et on écrit u<v si u est un préfixe (sous-liste initiale) de v. On dit que u et v sont disjointes, et on écrit u|v, si u et v sont des occurrences cohérentes non comparables. L'algorithme suivant donne tous les cas de comparaison :

```
let rec compare = function
        ([],[])        →  "="
      | ([],_)         →  "<"
      | (_,[])         →  ">"
      | (A::u,A::v)    →  compare(u,v)
      | (B::u,B::v)    →  compare(u,v)
      | (C::u,C::v)    →  compare(u,v)
      | (C::_,_)       →  "#"
      | (_,C::_)       →  "#"
      | _              →  "|";;
```

L'algorithme factor calcule le préfixe commun maximal de deux occurrences, avec les suffixes correspondants :

```
let factor = fact_prefix []
where rec fact_prefix w = function
        (A::u,A::v)    →  fact_prefix (A::w) (u,v)
      | (B::u,B::v)    →  fact_prefix (B::w) (u,v)
      | (C::u,C::v)    →  fact_prefix (C::w) (u,v)
      | suffixes       →  (rev w, suffixes);;
```

Remarque. Un ensemble E fini non vide d'occurrences deux à deux cohérentes peut être ordonné en un domaine d'occurrences de terme si et seulement s'il vérifie les deux conditions suivantes :

a) $u @ [s] \in E \Rightarrow u \in E$ (fermé vers le haut)

b) $u @ [B] \in E \Leftrightarrow u @ [A] \in E.$ (complet pour App)

On définit le transfert d'une occurrence comme le nombre d'abstractions qu'elle traverse :

```
let transfert u = trans (0,u) where rec trans = function
        (n, C::u)    →  trans(n+1,u)
      | (n, _::u )   →  trans(n,u)
      | (n, [])      →  n;;
```

On définit maintenant le sous-terme de M à l'occurrence u∈dom(M), noté M/u, et calculé comme sub(M,u), avec l'algorithme sub ci-dessous :

```
let rec sub = function
    (e , [])              → e
  | (App(g,_) , A::u)     → sub(g,u)
  | (App(_,d) , B::u)     → sub(d,u)
  | (Abs(e) , C::u)       → sub(e,u)
  | _                     → Erreur "Occurrence hors du domaine";;
```

Remarque. Une occurrence est une représentation d'un terme "filiforme" dénotant un chemin d'accès à un sous-terme.

On définit l'égalité des deux sous-termes u et v du terme M comme suit:

let eq_sub M (u,v) =
let (_,u',v') = factor(u,v) in
lift^{-1} (length u') (sub(M,u)) = lift^{-1} (length v') (sub(M,v));;

On appelle radical du terme M toute occurrence d'un sous terme de M de la forme App(Abs(_),_). C'est un endroit immédiatement réductible par substitution. Formellement, on définit :

```
let radical u M = match sub(e,M) with
     App(Abs(_),_) → true
   | _             → false;;
```

Autrement dit : radical u M $\Leftrightarrow$ u @ [A;C] $\in$ dom(M).

Notation. On note R(M) l'ensemble des occurrences de radicaux dans le terme M : R(M) = {u$\in$ dom(M) | radical u M}.

```
let rec R = function
        Ref(_)              → []
      | Abs(e)              → sons C (R e)
      | App(Abs(e),d)       → top :: (sons A (sons C (R e))) @ (sons B (R d))
      | App(g,d)            → (sons A (R g)) @ (sons B (R d));;
```

Le remplacement du sous-terme de M à l'occurrence u par le terme N, que nous noterons M[u←N], se calcule par remplace N (M,u), avec remplace défini ci dessous :

```
let remplace N = remp 0
where rec remp n = function
      (_ , [])             →  lift n N
    | (App(g,d) , A::u)    →  App(remp n (g,u), d)
    | (App(g,d) , B::u)    →  App(g, remp n (d,u))
    | (Abs(e) , C::u)      →  Abs(remp (n+1) (e,u))
    | _                    →  Erreur "Occurrence hors du domaine";;
```

On peut maintenant calculer formellement le terme N obtenu par une étape de réduction $M \Rightarrow_u N$ réduisant le radical u de M, comme suit :

```
let réduit M u = match sub(M,u) with
      App(Abs(body),arg)  → remplace (subst arg body) (M,u)
    | _                   → Erreur "Réduction d'un non-radical";;
```

On écrit simplement $M \Rightarrow N$ pour $\exists\, u \in R(M)\;\; M \Rightarrow_u N$.

4.3 Résidus

On définit maintenant une notion fondamentale : celle de <u>résidu</u>. Intuitivement, les sous-termes d'un terme progressivement calculé proviennent de morceaux recombinés du terme d'origine. On trace cet héritage à l'aide de la notion de résidu d'occurrence.

Tout d'abord, considérons un radical R = ([x]M N). L'étape de calcul réduisant R calcule : $R \Rightarrow M'$, avec $M' = M\{x \leftarrow N\}$. Les occurrences de M' proviennent de deux sources : les occurrences de N qui remplacent d'éventuelles occurrences de x dans M, et les autres occurrences de M. Calculons tout d'abord l'ensemble X des occurrences de x dans M à l'aide de X = locaux M, avec locaux défini récursivement par :

```
let locaux = loc 1
where rec loc n = function
      Ref(m)    → if m=n then [top] else []
    | App(g,d)  → (sons A (loc n g)) @ (sons B (loc n d))
    | Abs(e)    → sons C (loc (n+1) e);;
```

Maintenant soit T un terme possédant à l'occurrence u le radical T/u = R. On a T $\Rightarrow_u$ T', avec T' = T[u←M']. A chaque occurrence v de T on fait correspondre un ensemble d'occurrences V de T', dites <u>résidus de v selon u</u>, comme suit. Tout d'abord, si v<u ou v|u, V={v}. Ensuite, les occurrences du radical proprement dit, u et u@[A], n'ont pas de résidu. Une occurrence u@(A::C::w) dans M a pour résidu l'occurrence u@w, sauf pour les occurrences de la variable substituée x, qui n'ont pas de résidu. Enfin, une occurrence u@(B::w) dans N a pour résidus tous les u@x@w, pour x∈X quelconque.

Formellement, les résidus de v selon u dans le terme M se calculent par résidus M u v, avec l'algorithme résidus ci dessous :

```
let résidus M u =
   let X = locaux(sub(M,u@[A;C])) in function v → res(u,v)
(* u∈R(M)   v∈dom(M)  *)
   where rec res = function
      ([],[])         → []
    | ([],[A])        → []
    | ([],A::C::w)    → if mem w X then [] else [u@w]
    | ([],B::w)       → let wrap x = u@x@w in map wrap X
    | ([],_)          → Erreur "Occurrence hors du domaine"
    | (_,[])          → [v]
    | (A::w,B::w')    → [v]
    | (B::w,A::w')    → [v]
    | (A::w,A::w')    → res(w,w')
    | (B::w,B::w')    → res(w,w')
```

```
| (C::w,C::w') → res(w,w')
| _            → Erreur "Occurrence hors du domaine";;
```

Proposition. Soit M $\Rightarrow_u$ N.

1. $\forall v \in dom(N)\ \exists! w \in dom(M)$: $v \in$ résidus M u w.
   On écrit : w = résidus^{-1} M u v.
2. $\forall w \in dom(M)$, $\forall v_1, v_2 \in$ résidus M u w : eq_sub N (v_1,v_2).
3. $\forall w \in R(M)$: (résidus M u w) $\subseteq$ R(N).

Nous laissons au lecteur la preuve de cette proposition. Toute occurrence dans le terme réduit est donc résidu d'une occurrence unique. Les sous-termes résidus d'une même occurrence sont égaux. Les résidus d'une occurrence de radical sont des occurrences de radicaux, qui intuitivement partagent le même "grain de calcul". On appelle radical résidu dans N une occurrence de R(N) résidu d'une occurrence de R(M). Par contre, certains radicaux de N proviennent d'un nouvel assemblage App(Abs(g),d). On dit que ces radicaux sont créés par le calcul. Formellement, on définit RR et RC comme suit :

RR(M,u) = $\{v \in R(N) \mid$ résidus^{-1} M u $v \in R(M)\}$, RC(M,u) = R(N) - RR(M,u), avec N=réduit M u.

Remarque. Le radical R = ([x]M N) peut par sa réduction créer un radical de 3 manières différentes :

- vers le bas : à toutes les occurrences dans M de la forme (x P), si N est une abstraction.
- vers le haut : si R apparaît en partie gauche d'une application dans le terme à réduire : (R P), avec :
  - soit M une abstraction
  - soit M=x, et N une abstraction.

Remarquez que les deux premiers cas peuvent se cumuler.

Les résidus d'un ensemble d'occurrences V sont l'union des résidus de chacune des occurrences de V. On les calcule par (Résidus M u V), avec :

```
let Résidus M u = set_extension (résidus M u);;
(* where set_extension f = it_list (fun set x → union set (f x))  *)
```

Finalement, les résidus s'itèrent le long des séquences de réduction. Ainsi, si on définit une dérivation $M = M_1 \Rightarrow M_2 \Rightarrow \ldots M_n \Rightarrow N$ de longueur n par une suite d'occurrences de radicaux : $D = [u_1 ; \ldots ; u_n]$, alors on calcule les résidus de l'ensemble d'occurrences V de M comme l'ensemble d'occurrences de N :

$$W = \text{Résidus } M_n\ u_n\ (\text{Résidus } M_{n-1}\ u_{n-1} \ldots\ (\text{Résidus } M_1\ u_1\ V) \ldots).$$

Voici l'algorithme trace qui, étant donné M, V et D, calcule N et W :

```
let rec trace (M,V) = function
  []    →  (M,V)
 | u::D  →  let N = réduit M u and W = Résidus M u V in trace (N,W) D;;
```

Exemple. Voici un exemple de dérivation D = [[] ; []] à partir d'un terme L:

```
L =     ([u](u u)   [v]([x]v  y))
                       ‾‾‾‾‾‾‾‾‾‾ 0

⇨     ([v]([x]v  y)   [v]([x]v  y))  = R
          ‾‾‾‾‾‾‾‾ 1      ‾‾‾‾‾‾‾‾ 2

⇨        ([x][v]([x]v  y) 4 y)
                ‾‾‾‾‾‾‾‾
         ‾‾‾‾‾‾‾‾‾‾‾‾‾‾‾‾‾‾‾‾‾‾ 3
```

Figure 1

Après la première étape le radical marqué 0 a pour résidus les deux radicaux marqués 1 et 2. Après la 2ème étape le radical 1 (resp. 2) a pour unique résidu le radical 3 (resp. 4). Les résidus de 0 par D sont donc 3 et 4. Formellement, on calcule : trace (L,[[B; C]]) D = (L',[[] ; [A; C; C]]), avec L'= ([x][v]([x]v y) y). On remarque que le radical R ci-dessus n'est pas résidu d'un radical de L, mais est créé par la dérivation D.

Remarque. Les résidus donnent une idée de "partage" : les sous-termes résidus d'un sous-terme donné peuvent être "partagés" avec ce sous-terme, plutôt que copiés. Toutefois, ce partage ne rend pas compte d'un partage plus complexe de "contextes", ou portions de termes non nécessairement sous-termes. De plus, la représentation de de Bruijn ne rend pas compte de ce partage, dans la mesure où les variables globales des résidus doivent être éventuellement recalées (par l'opérateur lift).

4.4 Réductions parallèles.

On cherche à prouver le lemme du losange, c'est à dire à montrer que la relation ⇨ est confluente. Une idée naïve consiste à vouloir prouver que ⇨ est fortement confluente. Ceci échoue, à cause du problème de duplication :

Avec M = (Δ (**I** **I**)) , on a M ⇨ N1 = (Δ **I**) et M ⇨ N2 = ((**I** **I**) (**I** **I**)), mais la seule manière de fermer le diagramme est N1 ⇨ (**I** **I**), alors que N2 ne conduit à (**I** **I**) qu'après 2 étapes de réduction.

Par contre, il est vrai que la relation ⇨ est localement confluente. Cette propriété est définie graphiquement par :

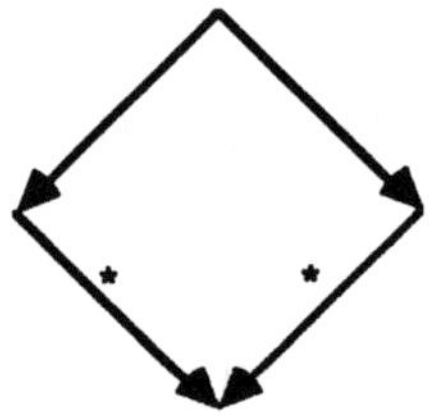

confluence locale

Toutefois, ceci ne suffit pas pour prouver directement la confluence, car en général les calculs peuvent ne pas terminer, et on ne peut donc pas employer de récurrence permettant d'utiliser l'idée d'un progrès effectué par la fermeture du diagramme de confluence locale. Remarquez par exemple que le graphe ci-dessous (paradoxe d'Escher-Newman) détermine une relation acyclique localement confluente, mais non confluente :

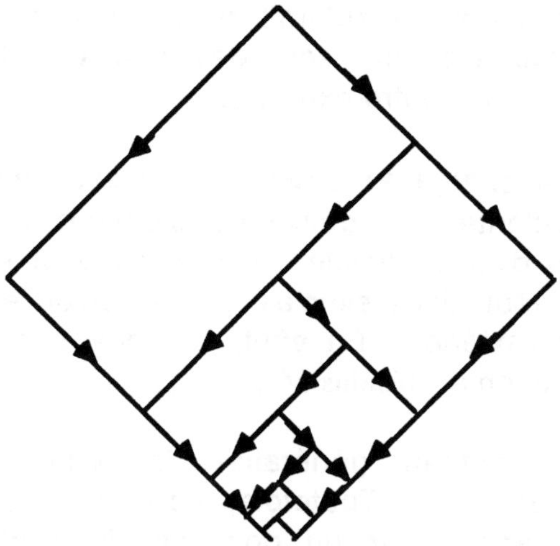

Figure 2

Il y a en gros deux manières pour sortir de cette difficulté. La première consiste à persévérer dans l'idée ci-dessus, en remarquant que la partie inférieure du diagramme de confluence locale n'est pas arbitraire, mais consiste à réduire des résidus des radicaux réduits dans sa partie supérieure. Le progrès peut alors s'exprimer sous la forme du théorème de finitude des développements que nous verrons plus loin. La deuxième manière, due à Tait, consiste à prouver la confluence forte pour une relation $\twoheadrightarrow$ intermédiaire entre $\Rightarrow$ et $\Rightarrow^*$. Nous allons maintenant poursuivre cette idée.

Définition.

La <u>réduction parallèle</u> $\twoheadrightarrow$ est définie comme la plus petite relation entre λ-termes vérifiant :

$$M \twoheadrightarrow M' \,\&\, N \twoheadrightarrow N' \;\Rightarrow\; ([x]M\ N) \twoheadrightarrow M'\{x \leftarrow N'\}$$
$$M \twoheadrightarrow M' \;\Rightarrow\; []M \twoheadrightarrow []M'$$
$$M \twoheadrightarrow M' \,\&\, N \twoheadrightarrow N' \;\Rightarrow\; (M\ N) \twoheadrightarrow (M'\ N')$$
$$x \twoheadrightarrow x$$

Remarque. La relation $\twoheadrightarrow$ correspond à une idée de réduction en parallèle de radicaux dont les occurrences peuvent ne pas être disjointes. En effet, la notion de réduction en parallèle d'occurrences deux à deux disjointes ne suffirait pas, car il n'est pas vrai que les résidus de radicaux disjoints sont des radicaux disjoints. La figure 1 ci-dessus en donne un contre-exemple.

Il est clair que $\Rightarrow \subseteq \twoheadrightarrow$ et $\twoheadrightarrow \subseteq \Rightarrow^*$, et donc que $\twoheadrightarrow^* = \Rightarrow^*$. La première assertion est évidente. La deuxième s'obtient en constatant que la réduction parallèle d'un ensemble de radicaux peut être séquentialisée en une séquence de réductions simples, pourvu qu'on effectue les réductions de l'intérieur vers l'extérieur. En effet, si u et v sont des occurrences de M, avec v<u ou v|u, on a : résidus M u v = [v].

Nous allons maintenant généraliser la notion d'occurrence à celle d'ensemble d'occurrences. Toutefois, nous ne représenterons pas un ensemble d'occurrences par un domaine (liste d'occurrences) comme précédemment défini, car cette représentation n'a pas la bonne structure pour partager les préfixes communs. On choisit plutôt de généraliser la notion d'occurrence vue comme "terme filiforme" en un terme dont certains des symboles sont marqués. Comme nous nous intéresserons à ces ensembles d'occurrences essentiellement pour marquer certains radicaux, on ne marque que les opérateurs d'application.

4.5 Termes marqués

Voici tout d'abord la structure de λ-terme avec applications marquées.

```
type marqué =
     MRef of num                          (* variable *)
   | MAbs of marqué                       (* abstraction *)
   | MApp of bool * marqué * marqué;;     (* application marquée *)
```

Voici comment traduire un lambda en un terme marqué, étant donné un domaine U :

```
let mark U = markrec []
where rec markrec u = function
  Ref(m)      → MRef(m)
```

```
| Abs(e)    → MAbs(markrec (C::u) e)
| App(g,d)  → MApp(mem u U, markrec (A::u) g, markrec (B::u) d);;
```

Inversement, unmark enlève les marques :

```
let rec unmark = function
    MRef(m)      → Ref(m)
  | MAbs(e)      → Abs(unmark e)
  | MApp(_,g,d)  → App(unmark g, unmark d);;
```

On suppose partout dans la suite que les marques ne valent true que sur les occurrences de radicaux : M = mark U L ⇒ U ⊆ R(L). On dit qu'un terme marqué M est un <u>marquage</u> du lambda L lorsque L = unmark M. On écrit aussi M ↑ L. On dit que les termes marqués M et M' sont <u>compatibles</u> s'ils sont les marquages d'un même lambda, et on écrit alors M ↑ M'. On écrit de même U ↑ L lorsque U est un ensemble d'occurrences de radicaux du terme L.

Les termes marqués compatibles ont naturellement une structure d'algèbre Booléenne héritée de la structure de l'ensemble des marques. On écrit par exemple M ⊆ M', M ∪ M' etc... avec la signification évidente. Par exemple, M ⊆ M' ssi M = mark U L, M' = mark U' L, et U ⊆ U'. Cette notation s'étend aussi aux termes, en identifiant L à mark [] L.

Donnons maintenant l'algorithme dérivé qui réduit un terme marqué M suivant toutes les marques d'un terme marqué compatible N. On suppose ci-dessous que l'algorithme subst a été étendu aux termes marqués en l'algorithme msubst de la manière naturelle, c'est à dire préservant la valeur des marques.

Définition : Dérivation.
On suppose que M et N sont des termes marqués compatibles. Le terme marqué P = dérivé(M,N) représente l'effet de réduire M en tous les radicaux marqués par N, les réductions ayant lieu de l'intérieur vers l'extérieur.

```
let rec dérivé = function
    (MRef(k),_)          → MRef(k)
```

```
| (MAbs(e),MAbs(m))→ MAbs(dérivé(e,m))
| (MApp(_,MAbs(e),e'),MApp(true,MAbs(m),m'))
                    → let body=dérivé(e,m) and arg=dérivé(e',m')
                      in msubst arg body
| (MApp(b,e,e'),MApp(false,m,m'))
                    → MApp(b,dérivé(e,m),dérivé(e',m'))
| _                 → Erreur "Termes non cohérents";;
```

Par la suite, on notera M\N pour dérivé(M,N) :

```
#infix "\\";;
let x\y = dérivé(x,y);;
```

Proposition. L ⋙ L' si et seulement si il existe un marquage N de L tel que L' = unmark(M\N), avec M = mark [] L.

Nous laissons au lecteur la preuve de la proposition. En fait l'algorithme dérivé nous donne beaucoup plus d'information que la relation ⋙, car les marquages permettent de tracer les résidus de radicaux. Remarquons par exemple que si U ↑ L, alors plus généralement on a L ⋙ L'=unmark(P), pour P= M\N, avec M = mark U L. Mais de plus P est la marque des résidus de U par le calcul L ⋙ L'. Autrement dit, l'algorithme dérivé joue simultanément les rôles des algorithmes réduit et résidu ci-dessus. Remarquez par exemple que si V ↑ L, et si D_V est une dérivation représentant V de l'intérieur vers l'extérieur, alors pour tout U ↑ L, avec M = mark U L et N = mark V L, on a M\N = mark W L', avec (L',W) = trace (L,U) D_V.

On étend la notation aux λ-termes L et aux ensembles de radicaux U ⊆ R(L) en écrivant L\U pour (mark [] L)\(mark U L). De même on écrit M\U pour M\(mark U (unmark M)). De plus, lorsque le terme L est sous-entendu par le contexte, on écrit V\U pour (mark V L)\(mark U L).

Remarque. A tout terme marqué M correspond de manière unique U et L tels que M = mark U L. Inversement, à L correspond l'ensemble (fini) de tous ses marquages. Mais à U correspond un ensemble infini de termes marqués M compatibles.

4.6. Démonstration du théorème du losange

On montre d'abord un résultat technique, expliquant que la substitution et la dérivation distribuent :

Lemme de substitution.
Pour tous termes marqués M,M',N,N', tels que M ↑ M' et N ↑ N', on a :
msubst (N\N') (M\M') = (msubst N M) \ (msubst N' M').

Démonstration. Nous laissons au lecteur la preuve de cette propriété, par récurrence sur M.

Le résultat fondamental s'exprime maintenant comme suit.

Lemme de monotonie des résidus. Soit L un λ-terme, U ↑ L, W ↑ L avec U ⊆ W. Pour tout M ↑ L, on a (M\U)\(W\U) = M\W.

Démonstration.
Récurrence sur M.
Le seul cas intéressant est lorsque M = MApp(b,MAbs(M1), M2). On pose M' = M\U et W' = W\U. Avec U1 = {u | A::C::u∈U}, U2 = {u | B::u∈U}, et notations similaires pour W, on a M1\U1 = M'1, M2\U2= M'2, W1\U1 = W'1 et W2\U2 = W'2 tels que, par récurrence : M'1\W'1 = M1\W1 et M'2\W'2 = M2\W2. Maintenant il y a 2 cas:

a) top ∈ U. Alors top ∈ W aussi, et donc M\W = msubst (M2\W2) (M1\W1), et M' = msubst M'2 M'1. De la même manière W' est un marquage de L\U de la forme msubst W'2 W'1 (avec un abus de notation évident) et donc par le lemme de substitution M'\W' = msubst (M'2\W'2) (M'1\W'1) = M\W.

b) top ∉ U. Alors M' = MApp(b, MAbs(M'1), M'2). Soit W'0 = {A::C::w | w∈W'1} ∪ {B::w | w∈W'2}. Il y a de nouveau 2 cas :

b1) top ∈ W. Alors W' = W'0 ∪ {top}, et de nouveau :
M\W = msubst (M2\W2) (M1\W1) = msubst (M'2\W'2) (M'1\W'1) = M'\W'.

b2) top ∉ W. Alors W' = W'0, et de même :
M\W = MApp(b, MAbs(M1\W1), M2\W2) = M'\W'.

Corollaire : Lemme du cube.
Soit L un λ-terme, U et V des sous-ensembles de R(L). Pour tout M ↑

L, on a $(M\backslash U)\backslash(V\backslash U) = (M\backslash V)\backslash(U\backslash V)$.

Démonstration. On applique le lemme ci-dessus, en prenant $W = U \cup V = V \cup U$.

Corollaire : Théorème du losange.

Démonstration. Le lemme du cube donne la confluence forte de ⤀, et donc la confluence de ⇨.

En fait on obtient beaucoup plus. En particulier, on aurait pu faire converger le diagramme de forte confluence plus simplement sur $L\backslash R(L)$. Mais cette preuve n'aurait pas exprimé que ce diagramme peut être fermé d'une manière minimale. Nous allons voir maintenant comment le lemme du cube peut s'exprimer comme propriété catégorique.

4.7. Structure des dérivations

Soit M et N des termes marqués. On définit récursivement la notion de dérivation parallèle D de M vers N, notée D : M ⤀* N, comme une paire (M,S), où S est une suite d'ensembles d'occurrences, vérifiant :

a) (M,[]) : M ⤀* M, pour tout M.

b) si (N,S) : N ⤀* N' et U ↑ M, avec $M\backslash U = N$, alors (M, U::S) : M ⤀* N'.

Si (M,S) est une dérivation parallèle de M vers N, alors N est unique. On écrit $N = M\backslash S$. Si M=mark U L et N=mark V L, on écrit de même $V = U\backslash S$.

Remarque. On commet ici un abus de notation : la relation ⤀ était définie plus haut pour les λ-termes, et ici la relation ⤀ est définie entre termes marqués. Cet abus est justifié en remarquant que M ⤀ N ssi M = mark U L, L ⤀ L' par une dérivation D comme définie précédemment, et N = mark V L' avec (L',V) = trace (L,U) D.

Réciproquement, on définit une dérivation parallèle de L vers L' comme une paire (L,S), avec (M,S) : M ⤀* M' par une dérivation parallèle comme défini ci-dessus, avec M = mark [] L et L' = unmark M'. On dit que deux séquences de calcul S et S' sont équivalentes en L ssi $M\backslash S = M\backslash S'$ pour tout marquage M de L. On écrit alors $(L,S) \equiv (L,S')$.

Remarque. L'équivalence entre dérivations $\equiv$, appelée équivalence de permutations, distingue des dérivations aboutissant au même terme. Par exemple, les deux dérivations qui réduisent (**I** (**I** M)) en (**I** M) ne sont pas équivalentes (pourquoi?). Par contre, toutes les dérivations issues d'un terme normalisable et aboutissant à sa forme normale sont équivalentes.

On dit que l'étape de dérivation est vide lorsque l'ensemble de radicaux réduits à cette étape est vide.

On appelle développement de M une dérivation parallèle issue de M telle qu'à chaque étape b) ci-dessus on ait $U \subseteq M$. Autrement dit, un développement est une dérivation parallèle qui ne réduit pas de radicaux créés, les seuls radicaux contractés étant des résidus des radicaux marqués initialement dans M. Le développement D est dit complet s'il aboutit à un terme n'ayant plus de radicaux marqués.

Nous allons maintenant nous intéresser à la structure des dérivations. La première opération est la concaténation.

Soit $D : M \twoheadrightarrow^* N$ par la suite S et $D' : N \twoheadrightarrow^* N'$ par la suite S', on définit la concaténation $D @ D' : M \twoheadrightarrow^* N'$ comme donnée par la suite $S @ S'$.

On étend maintenant la notion de résidu à une opération entre dérivations. Tout d'abord, soit S une suite de calcul et U un ensemble d'occurrences. On définit $S\backslash U$ par récurrence sur la longueur de S comme suit.

a) $[] \backslash U = []$
b) $(V::S) \backslash U = (V\backslash U) :: (S\backslash(U\backslash V))$.

Cette dernière clause suppose bien sûr $U \uparrow V$. On laisse au lecteur le soin de vérifier que si $D = (M,S) : M \twoheadrightarrow^* N$ avec $M \uparrow U$, alors $D\backslash U = (M\backslash U, S\backslash U)$ définit une dérivation parallèle $M\backslash U \twoheadrightarrow^* N\backslash U$.

Soient maintenant $D = (M, S)$ et $D' = (M', S')$ deux dérivations, avec $M \uparrow M'$. On définit $D'\backslash D$ comme la dérivation $(M\backslash S,\ S'\backslash S)$, avec l'opération $\backslash$ définie par récurrence sur la longueur de S' par :

a) $[] \backslash S = []$
b) $(U::S') \backslash S = (U\backslash S) :: (S'\backslash(S\backslash U))$.

Nous laissons au lecteur le soin de vérifier que cette définition est bien fondée.

Soit D : M $\twoheadrightarrow^*$ N et D' : M $\twoheadrightarrow^*$ N' deux dérivations issues du même terme marqué M. On dit que D calcule moins que D', et on écrit D ≤ D', si et seulement si on peut prolonger D en D', c'est à dire s'il existe une dérivation N $\twoheadrightarrow^*$ N'.

L'ensemble des dérivations issues d'un même terme marqué M forme un sup demi-treillis, avec l'ordre préfixe : (M, S) ≤ (M, S') ssi S ≤ S'. On laisse au lecteur le soin de vérifier que l'opération ∪ est la borne supérieure, avec ∪ définie par D ∪ D' = D @ (D'\D).

Il est clair que la notion de dérivation issue d'un terme marqué n'est qu'un instrument technique pour étudier les dérivations à partir de λ-termes. Pour celà, on remarque que toutes les notions définies ci-dessus s'étendent aux dérivations sur les λ-termes (non marqués), et que les deux opérations @ et \ préservent l'équivalence ≡ ci-dessus.

Définition. D ≤ D' si et seulement si on peut prolonger D en une dérivation équivalente à D', i.e. s'il existe D" telle que D @ D" ≡ D'.

Soit Der(L) l'ensemble des dérivations issues d'un lambda L.

Théorème de Lévy. (Der(L)/≡ , ≤) a la structure d'un sup-demi-treillis.

Remarque : ce résultat justifie la généralité de la démonstration que nous avons donnée du théorème du losange, en passant par le lemme du cube. La structure des dérivations du λ-calcul est investigée très complètement dans la thèse de J.J. Lévy.

Exercices.

1. Donner une version catégorique du théorème exprimant que la catégorie ayant pour objets les λ-termes et pour flèches les dérivations quotientées par permutation admet des sommes amalgamées.

2. (Plotkin) Montrer qu'il n'existe pas de λ-terme se dérivant à la fois sur ([x]((y x) (y z)) z) et sur (([x](x x) (y z)). En déduire que la propriété duale d'inf-demi-treillis n'est pas vraie.

3. Montrer que la structure de sup-demi-treillis des dérivations n'induit pas une structure de sup-demi-treillis de l'ordre de β-réduction sur les termes issus d'un même terme.

Remarque. Les termes marqués peuvent être enrichis d'autres décorations. Par exemple, Barendregt dans son livre considère des termes dont les opérateurs d'abstraction sont marqués par des entiers naturels. J.J. Lévy dans sa thèse considère des termes marqués par des étiquettes, suites mémorisant l'histoire du calcul menant à ce terme. Dans la prochaine section nous allons marquer les abstractions par des entiers positifs appelés poids.

5. Récurrence sur les calculs

5.1 Le théorème des développements finis

Théorème des développements finis. Tout développement est fini, dans le sens où il est ultimement vide.

Autrement dit, toute suite de calculs qui ne réduit à chaque étape que des résidus de radicaux du terme d'origine, termine nécessairement.

La méthode de preuve utilise des termes marqués dont les variables dont pondérées par des entiers strictement positifs. Nous développons maintenant ce formalisme.

5.2 Termes pondérés

```
type pondéré =
      PRef of num
   |  PAbs of num * pondéré                    (* 1er arg = poids *)
   |  PApp of bool * pondéré * pondéré;;
```

On étend le poids à un terme pondéré par sommation des poids de ses variables :

```
(* plibres est un environnement de poids *)
let rec poids_env plibres = function
```

```
        PRef(n)       → nth plibres n
    |   PAbs(p,e)     → poids_env (p::plibres) e
    |   PApp(_,g,d)   → (poids_env plibres g) + (poids_env plibres d);;

let poids = poids_env [];;   (* poids d'un terme fermé *)
```

Comme au paragraphe précédent nous supposerons que les marques ne valent true que sur des occurrences de radicaux. On suppose qu'on étend de la manière naturelle aux termes pondérés les notions de substitution (algorithme psubst), et donc de calcul, résidu, etc...

On laisse au lecteur la preuve du lemme suivant, par récurrence contextuelle.

Lemme de substitution. Soit plibres un environnement de poids de longueur $n \geq 0$, soit $M \in \Lambda_{n+1}$, $N \in \Lambda_n$, et Q = psubst N M. Alors pour tout entier p≥poids_env plibres N, on a poids_env (p::plibres) M ≥ poids_env plibres Q.

Corollaire. Sous les mêmes conditions, avec P=PApp(b,PAbs(p,M),N), on a poids_env plibres P > poids_env plibres Q.

Il suffit de remarquer pour le corollaire que (poids_env plibres N)>0. On voit donc que la réduction d'un terme bien pondéré en fait décroître le poids, ce qui justifie la définition suivante.

Un terme pondéré est dit <u>décroissant</u> si et seulement si pour toute occurrence de radical marqué PApp(true,PAbs(p,M),N), le poids p est supérieur ou égal au poids de N. Plus précisément, on vérifie :

```
let rec decr_env plibres = function
      PRef(_)       →  true
    | PAbs(p,M)     →  decr_env (p::plibres) M
    | PApp(b,M,N)   →  decr_env plibres M & decr_env plibres N &
                       (not b or let (PAbs(p,_)) = M
                       in p≥(poids_env plibres N));;
let décroissant = decr_env [];;
```

Proposition. A tout terme marqué correspond un terme pondéré décroissant de même structure.

Démonstration. On associe au terme marqué M le terme pondéré pond(M) calculé comme suit :

```
let rec pondère plibres = pond_with 1
where rec pond_with p = function
    MRef(n)        → (PRef(n), nth plibres n)
  | MAbs(M)        → let (M',p') = pondère (p::plibres) M
                     in (PAbs(p,M'),p')
  | MApp(b,M,N)    → let (N',p') = pond_with 1 N in
                     let (M',p") = pond_with (if b then p' else 1) M
                     in (PApp(b,M',N'), p'+p");;

let pond x = fst(pondère [] x);;
```

On montre que pond(M) est décroissant par récurrence contextuelle sur M. En effet, on vérifie : pondère plibres M = (M' , poids_env plibres M').

On laisse au lecteur le soin de vérifier le lemme technique suivant, par récurrence sur M, et en utilisant le lemme de substitution ci-dessus.

Lemme de préservation de la décroissance.
Soit plibres un environnement de poids de longueur $n \geq 0$, soit $M \in \Lambda_{n+1}$, $N \in \Lambda_n$, $p \geq$ poids_env plibres N, tels que decr_env (p::plibres) M et decr_env plibres N. On a decr_env plibres (psubst N M).

Lemme de préservation des pondérations décroissantes.
Soit M un terme marqué pondéré décroissant, U un ensemble de radicaux marqués de M : $U \subseteq M$, et $N = M \backslash U$. N est un terme marqué pondéré décroissant, tel que poids(N)$\leq$poids(M), l'inégalité étant stricte si U est non vide.

Démonstration.
L'énoncé concerne un terme fermé. Plus généralement, pour $M \in \Lambda_n$, et plibres un environnement de poids quelconque de longueur $n \geq 0$, avec decr_env plibres M, on montre poids_env plibres N $\leq$ poids_env plibres M, et decr_env plibres N, par récurrence sur $M \backslash U$.

- Cas M=(M1 M2) $\twoheadrightarrow$ N=(N1 N2), top$\notin$U, avec M1 $\twoheadrightarrow$ N1 et M2 $\twoheadrightarrow$ N2. Par hypothèse de récurrence on a poids_env plibres N1 $\leq$ poids_env plibres M1, poids_env plibres N2 $\leq$ poids_env plibres M2, et donc poids_env plibres N $\leq$ poids_env plibres M. De même on a decr_env plibres M1, et decr_env plibres N. Si top $\in$N (N marqué), c'est que N1=[x:p]Q1, et alors forcément M1=[x:p]P1, avec top $\in$M. Par hypothèse decr_env plibres M, on doit avoir p $\geq$ poids_env plibres M2 $\geq$ poids_env plibres N2 par hypothèse de récurrence, et on a donc decr_env plibres N.

- Cas M=([x:p]M1 M2) $\twoheadrightarrow$ N=psubst N2 N1, top$\in$U, avec M1 $\twoheadrightarrow$ N1 et M2 $\twoheadrightarrow$ N2. Par hypothèse de récurrence on a poids_env plibres N1 $\leq$ poids_env plibres M1, poids_env plibres N2 $\leq$ poids_env plibres M2, et comme U $\subseteq$ M on a top$\in$M, d'où p $\geq$ poids_env plibres M2 par hypothèse decr_env plibres M. On a également decr_env (p::plibres) M1 et decr_env plibres M2. Par récurrence, on obtient donc aussi decr_env (p::plibres) N1 et decr_env plibres N2. Par le lemme de préservation de la décroissance on en déduit decr_env plibres N. Finalement, on a poids_env plibres M $\geq$ poids_env plibres ([x:p]N1 N2) > poids_env plibres N par le corollaire du lemme de substitution.

- les autres cas ne présentent pas de difficulté.

Remarque.
En général les radicaux créés ne peuvent être marqués sans violer la condition de décroissance. Considérez par exemple :

$$[x_2]([y_2](y\ x)\ [z_1]z) \Rightarrow [x_2]([z_1]z\ x).$$

5.3 Démonstration du théorème des développements finis.

Soit M un terme marqué. Tout développement de M sans étape vide est de longueur inférieure à poids(pond(M)).

Remarque.
Le théorème a d'abord été prouvé par Curry pour le λ-I-calcul, une restriction qui exige que toutes les variables liées apparaissent au moins une fois. Le théorème n'a été prouvé dans toute sa généralité qu'en 1965 par Schroer. La démonstration ci-dessus est inspirée d'une preuve due à Hyland et Barendregt. Ici nous donnons un poids uniforme à toutes les occurrences d'une variable liée, ce qui simplifie légèrement la

preuve, et permet d'assimiler la pondération d'une variable à un type (dans Barendregt, pond est obtenu en pondérant les occurrences de variables de la droite vers la gauche par des puissances de 2 successives).

5.4 Sous-dérivations strictes

Considérons deux dérivations parallèles co-initiales de même longueur :

$A = (M, [U_1; \ldots U_n])$ et $B = (M, [V_1; \ldots V_n])$. On dit que B est une sous-dérivation stricte de A, et on écrit $B \ll A$, ssi $B = A \backslash u$, avec $u \in R(M)$ tel que $\exists k \leq n$ tel que, avec $U = u \backslash [U_1; \ldots ; U_{k-1}]$, on ait $\varnothing \neq U \subseteq U_k$.

Lemme. La relation « est nœthérienne.

Démonstration.

Par l'absurde. On suppose qu'il existe une suite infinie :

$A_1 \gg A_2 \gg A_3 \gg \ldots$ Appelons k la colonne sélectionnée dans la définition ci-dessus. Soit r la colonne sélectionnée infiniment souvent maximale. Après un certain nombre d'étapes il n'y plus de colonne sélectionnée supérieure à r. On obtient alors un développement infini de U_r, ce qui est impossible.

On tire de ce lemme la justification de récurrences sur les dérivations. Nous en verrons une application dans la section suivante.

6. Le théorème de standardisation

Le théorème du losange nous a montré que la notion de calcul était déterministe. En particulier, tout terme M normalisable admet une forme normale Normal(M) unique, et il existe une dérivation $M \Rightarrow^* \mathrm{Normal}(M)$. Mais ceci ne nous suffit pas pour construire un interpréteur du λ-calcul. On aimerait que cet interpréteur puisse être défini comme une fonction récursive ϕ associant à tout terme M un ensemble $\phi(M)$ de ses radicaux, avec $\phi(M) = \varnothing$ si et seulement si M est en forme normale. L'interpréteur sera dit correct s'il mène à la forme normale de tout terme normalisable. C'est à dire, pour tout terme M normalisable il existe n, $M_1, \ldots, M_n$, tels que $M=M_1$, $M_n=\mathrm{Normal}(M)$, et pour tout i, avec $1 \leq i < n$, on a :

$$M_{i+1}=M_i\backslash\phi(M_i).$$

Par exemple, si on choisit pour $\phi(M)$ l'ensemble des radicaux les plus internes de M (stratégie dite d'appel par valeur), on obtient un interpréteur qui n'est pas correct, comme le montre l'exemple ([x]I Ω). Si on choisit pour $\phi(M)$ l'ensemble de tous les radicaux de M (stratégie dite complète ou de Gross-Knuth), on obtient un interpréteur correct (pourquoi?).

Nous allons maintenant voir qu'il existe des stratégies correctes qui calculent de l'extérieur vers l'intérieur, tout en ne réduisant qu'un radical à chaque étape.

Définition.
Soit u et v deux occurrences. On dit que u est à gauche de v, et on écrit $u \angle v$, si et seulement si soit $u<v$, soit il existe w, u' et v' tels que u=w@(A::u') et v=w@(B::v'). La relation $\angle$ est un ordre total strict.

Lemme de préservation des occurrences à gauche.
Soit u et v des occurrences de radicaux dans un terme quelconque, avec $u \angle v$. On a :

1. $u\backslash v=\{u\}$.
2. $u \angle w$ pour toute occurrence w d'un radical créé par v.
3. $u \angle w$ implique $u \angle w'$ pour tout w' dans $w\backslash v$.

Mêmes résultats avec V ensemble d'occurrences de radicaux, tel que $u \angle v$ pour tout v dans V.

Nous laissons la preuve de ce lemme, par cas sur les positions respectives des occurrences, au lecteur.

Pour tout terme M réductible, on définit gauche(M) comme le radical le plus à gauche de M. On peut le calculer par l'algorithme :

```
exception NormalForm;;
let gauche = searchleft top
  where rec searchleft u = function
      Ref(_)          → raise NormalForm
    | Abs(e)          → searchleft (C::u) e
    | App(Abs(_),_)   → rev(u)
    | App(g,d)        → try searchleft (A::u) g
                        with NormalForm → searchleft (B::u) d;;
```

Définition.

Considérons une dérivation réduisant un radical à chaque étape : (M,S), avec $S=[u_1; \ldots ;u_n]$. Soit $S_{ij} = [u_i; \ldots ;u_j]$, $M_i=M\backslash S_{1i}$. La dérivation est dite <u>standard</u> si et seulement si pour tous $1 \leq i<j \leq n$, si $u_j \in v\backslash S_{i,j-1}$, avec $v \in R(M_{i-1})$, alors $u_i \angle v$.

C'est à dire, pour toute étape de réduction i, il n'existe pas de radical v à gauche de u_i dans M_{i-1} dont l'un des résidus par la dérivation considérée soit contracté à une étape ultérieure j.

Exemples.

([x]<u>(I x)</u> (I y)) ⇨ ([x]x <u>(I y)</u>) ⇨ ([x]x y)	est standard
([x]<u>(I x)</u> (I y)) ⇨ ([x]x <u>(I y)</u>) ⇨ <u>([x]x y)</u> ⇨ y	ne l'est pas

Remarques.

1. Si D@D' est standard, D et D' le sont aussi. La réciproque n'est pas vraie en général.

2. Il n'y a pas d'extension évidente de la notion de dérivation standard aux dérivations parallèles. En effet, on a vu qu'une réduction en parallèle de n radicaux peut se séquentialiser en n réductions élémentaires, mais de l'intérieur vers l'extérieur, alors qu'une dérivation standard doit procéder de l'extérieur vers l'intérieur. Il est facile de montrer qu'une étape de réduction parallèle U peut se séquentialiser en réductions élémentaires procédant de l'extérieur vers l'intérieur : on réduit le radical u le plus à gauche de U, et on itère sur U\u; la construction s'arrête par le théorème des développements finis. Plus généralement :

Théorème de standardisation. Toute dérivation est équivalente à une dérivation standard unique.

Avant de montrer ce théorème, on définit récursivement une dérivation st(D) associée à une dérivation parallèle D.

Définition : R(D). Soit D=(M,S) une dérivation parallèle de longueur n issue d'un terme M, avec $S=[U_1; \dots ;U_n]$. On définit R(D) comme l'ensemble des occurrences de radicaux de M dont au moins un résidu est réduit dans D:

$$R(D) = \{v\in R(M) \mid \exists\, k\leq n \;\; u\backslash S_{1,k-1}\cap U_k \neq \varnothing\}.$$

Définition : st(D). Soit D=(M,S) une dérivation parallèle de longueur n issue d'un terme M. On définit récursivement une dérivation st(D), qui réduit un radical à chaque étape. Soit $S=[U_1; \dots ;U_n]$. Il y a deux cas.

- D est vide, c.a.d. $U_i= \varnothing$ pour tout i. Alors st(D) est la dérivation vide.

- sinon, soit u l'occurrence la plus à gauche de R(D). Par le lemme ci-dessus, on obtient que u est une occurrence de radical dans M, qui est préservée par la dérivation D, jusqu'à être réduite, disons à l'étape k :

$$u\in R(M),\; u\backslash S_{1,k-1} = \{u\},\; u\in U_k.$$

Soit D'= st(D\u), qui est bien définie, car D\u«D. On prend pour st(D) la dérivation issue de M qui réduit u, et se poursuit par D'. C'est-à-dire, avec D'=(M',S') : st(D)=(M,u::S').

La construction est illustrée par la figure 3 ci dessous.

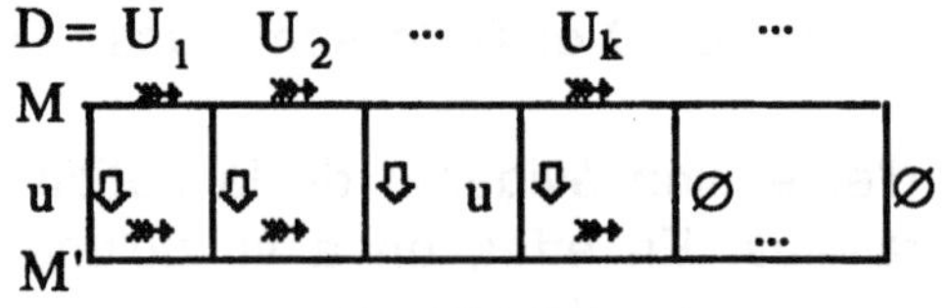

Figure 3

Démonstration du théorème de standardisation (Klop). Par construction st(D) est équivalente à D. Il suffit de montrer que st(D) est standard, et qu'une dérivation standard est unique dans sa classe d'équivalence. Nous laissons les détails de ces preuves au lecteur.

Remarque. Il est faux que R(D)=R(D') pour deux dérivations équivalentes D et D'. Par exemple, avec M=([x]**I** (**I I**)), D=[{top}] et D'=[{top,[B]}] sont deux dérivations équivalentes menant de M à sa forme normale **I**.

Définition. Une dérivation <u>normale</u> est une dérivation qui réduit à chaque étape le radical le plus à gauche.

Pour tout M, il existe au plus une dérivation normale de longueur donnée. Elle est calculée par l'interpréteur <u>normal</u>, défini par $\phi(M)$ = {gauche(M)} si M est réductible.

Théorème. L'interpréteur normal est correct.

Ceci est un corollaire immédiat du théorème de standardisation, car la standard d'une dérivation menant à la forme normale est normale (pourquoi?).

Voici la procédure qui itère le calcul de l'interpréteur normal.

```
let onestep e = let u=gauche(e) in réduit e u;;
let rec normal e = try normal (onestep e) with NormalForm → e;;
```

Remarques.

1. Cette spécification n'est pas satisfaisante, car cet interpréteur recalcule à chaque fois le radical à réduire à partir du sommet du terme. Il n'y a pas de moyen direct d'écrire un interpréteur normal calculant récursivement sur la structure d'un lambda, à cause du phénomène de création des radicaux “vers le haut”. Nous verrons comment résoudre cette difficulté dans la prochaine section.

2. Cet interpréteur n'est utile que pour les termes normalisables. Il boucle sur les termes non-normalisables, alors qu'intuitivement il est possible en général d'obtenir une information positive sur les approximations successives des termes obtenus le long d'un calcul, même ne se terminant pas sur une forme normale. Nous verrons plus loin comment calculer progressivement ces approximations.

3. En général, l'interpréteur normal n'est pas optimal (en nombre de réductions effectuées). Lorsqu'il y a des duplications, il vaut mieux souvent procéder de l'intérieur vers l'extérieur. Pourtant, l'interpréteur en appel par valeur n'est pas correct, car il peut faire des étapes non nécessaires; même lorsqu'il est correct, il n'est donc pas optimal non plus. Pour obtenir un interpréteur optimal, il faut compter pour une étape de calcul la réduction en parallèle de radicaux appartenant à la même famille. Intuitivement, deux radicaux sont dans la même famille s'ils sont résidus, ou créés de la même manière, par des radicaux de la même famille. Cette notion a été étudiée formellement par J.J. Lévy, qui a montré l'optimalité de cette stratégie. Remarquez toutefois qu'une telle stratégie généralise notre notion d'interpréteur, car il doit se rappeler de l'histoire de ses calculs. Cette mémorisation peut s'effectuer par des marques sur les λ-termes (étiquettes).

7. Approximations.

7.1. Forme normale de tête.

Les formes normales de tête sont les lambdas de la forme:

$$[x_1, \dots, x_n](x\ M_1 \dots M_p)$$

avec $x, x_1, \dots, x_n$ variables, $M_1 \dots M_p$ lambdas quelconques, $n,p \geq 0$.

Une forme normale de tête représente une approximation de forme normale, car elle est invariante (en n,x et p) par β-réduction.

```
type head = Head of num * num * lambda list;; (* n, x, [M1 ; ... ; Mp] *)
```

Voici l'algorithme de mise en forme normale de tête d'un lambda, par réduction normale :

```
let head = hnf 0 []            (* head : lambda → head *)
where rec hnf n args = function
        Ref(x)     → Head(n,x,args)
      | Abs(e)     → (match args with
```

```
                    []           → hnf (n+1) [] e
                  | first::rest  → hnf n rest (subst first e))
  | App(g,d)  →  hnf n (d::args) g;;
```

On peut maintenant organiser l'interpréteur normal en itérant la fonction head.

```
type normal = Normal of num * num * normal list;;

let rec norm e =           (* norm : lambda → normal *)
    let (Head(n,x,args)) = head(e) in Normal(n,x,map norm args);;
```

L'objection levée dans la remarque 1 ci-dessous est maintenant levée, mais l'objection 2 demeure. Voyons comment calculer progressivement une approximation de forme normale, pour des termes non forcémént normalisables.

7.2. Arbres de Böhm.

Il est possible de définir un interpréteur qui calcule, de l'extérieur vers l'intérieur mais non forcément de la gauche vers la droite, une suite d'approximations décrivant un arbre infini généralisant la notion de forme normale. Un tel arbre s'appelle l'arbre de Böhm du lambda.

```
type approximation = BT of num * num * (générateur list)
and générateur == (unit → approximation);;
```

L'approximation $BT(n,x,[f_1; f_2 ;...; f_p])$ représente l'information qu'on a sur un λ-terme lorsqu'on a reconnu qu'il se réduit sur la forme normale de tête :

$[x_1,x_2 ,..., x_n](x\ e_1\ e_2 \dots e_p)$. Le générateur f_i permet de calculer l'approximation de e_i par l'exécution de $f_i()$. La procédure approx ci-dessous calcule l'approximation d'un λ-terme par réduction normale :

```
(* approx : lambda → approximation *)
let rec approx = apx 0 []
where rec apx n args = function
```

```
      Ref(x)     →  BT(n,x,map (fun e () → approx(e)) args)
    | Abs(e)     →  (match args with
                          []              → apx (n+1) [] e
                        | first::rest     → apx n rest (subst first e))
    | App(g,d)   →  apx n (d::args) g;;
```

Exemple. Avec $\mathbf{D}$ = [x,y](y (x x y)), Θ = ($\mathbf{D}$ $\mathbf{D}$) et M = (Θ [x,y](y x)), on obtient approx(M) = BT(1,1,[f]), avec f() calculant récursivement la même approximation BT(1,1,[f]). L'arbre de Böhm de M est donc l'arbre infini :

```
[x](x | )
      |
      [x](x |
            |
            [x](x |
                  |
                  [x](x |
                        |
                       ...
```

Exercice. Redéfinir le type générateur en employant une structure de donnée paresseuse, plutôt que des fermetures.

Les lambdas normalisables ont un arbre de Böhm fini. Certains lambdas ont un arbre de Böhm vide (indiqué par le symbole $\perp$), car la procédure head (ou approx) peut ne pas terminer.

Définition. On dit que le lambda M est <u>défini</u> si et seulement si head(M) termine.

Par exemple, M ci-dessus est défini, bien qu'il n'ait pas de forme normale. Par contre, Ω n'est pas défini, il a pour arbre de Böhm $\perp$. Les termes non définis sont absurdes du point de vue calculatoire : ils n'apportent aucune information.

Proposition. Si (M N) est défini, alors M est défini.
Nous laissons en exercice la preuve de cette proposition, qui utilise le théorème de standardisation.

Nous donnons maintenant une autre caractérisation des termes définis. On rappelle que $\underline{M}$ désigne la fermeture $[u_1, u_2, \ldots u_k]M$ de $M \in \Lambda_k$.

Définition. On dit que le lambda M est <u>solvable</u> si et seulement s'il existe n et $N_1, \ldots, N_n$, tels que $(\underline{M}\ N_1 \ldots N_n) \cong \mathbf{I}$,

Théorème de Wadsworth. Un lambda est défini si et seulement s'il est solvable.

Démonstration. Soit M un lambda, qu'on peut supposer fermé. Si M est solvable, alors il est défini, par la proposition ci-dessus.

Réciproquement, si $M \Rightarrow^* [x_1,x_2,\ldots, x_n](m\ e_1\ e_2 \ldots e_p)$, alors, en définissant $\mathbf{K}^p = [x, y_1,y_2,\ldots, y_p]x$, on obtient :

$(M\ (\mathbf{K}^p\ \mathbf{I})_1 \ldots (\mathbf{K}^p\ \mathbf{I})_n) \Rightarrow^* (\mathbf{K}^p\ \mathbf{I}\ e'_1\ e'_2 \ldots e'_p) \Rightarrow^* \mathbf{I}$.

8. Variations sur la notion de calcul

La notion de calcul que nous avons étudiée jusqu'à présent s'appelle la β-réduction du λ-**K**-calcul pur. Il existe de nombreuses variations sur le thème du λ-calcul. Dans le chapitre 3 nous verrons diverses variantes typées. Donnons ici brièvement quelques variantes du calcul pur, c'est à dire sans types.

8.1. Eta-réduction

On considère quelquefois une règle de calcul supplémentaire, appelée η-réduction, qui correspond à une forme faible de l'extensionalité :

$$[x](M\ x) \rightarrow M \quad (x \text{ non libre dans } M) \qquad (\eta)$$

ou, en notation abstraite : $[](M^+\ 1) \rightarrow M$, avec $M^+ = \text{lift } 1\ M$.

Cette règle est justifiée par l'utilisation du λ-calcul pour dénoter des fonctions. En effet, pour tout lambda N, on a $([x](M\ x)\ N) \cong (M\ N)$ lorsque x n'apparait pas libre dans M, et la règle η est donc justifiée par extensionalité.

La théorie de la $\beta\eta$-réduction n'est pas aussi élégante que celle de la β-réduction seule, à cause de l'ambiguité des expressions de la forme ([x](M x) N), qui font se chevaucher des radicaux des deux sortes. Par une sorte de "miracle syntaxique", les deux sortes de réduction mènent au même résultat (M N), mais la notion de résidu n'est plus claire.

La propriété de confluence de la $\beta\eta$-réduction est conséquence des deux lemmes suivants, dont nous laissons la démonstration au lecteur. On utilise le symbole → pour la η-réduction (qui est étendue par congruence sur la structure des termes comme la β-réduction), et ➡ pour la $\beta\eta$-réduction.

Lemme de retard de la règle η.
Si M ➡* N, alors il existe P tel que M ⇨* P →* N.

Lemme de commutation des règles β et η.
a. Si M ⇨ N_1 et M → N_2 alors il existe P tel que N_1 →* P
et N_2 ⇨ P ou N_2 = P.
b. Si M → N_1 et M → N_2 alors il existe P tel que N_1 → P et N_2 → P.

8.2. Réduction faible

La réduction faible s'obtient en supprimant la règle ξ de congruence de la relation ⇨ par rapport à l'abstraction. Autrement dit, la réduction faible ne calcule pas à l'intérieur des abstractions. Ce calcul possède davantage de formes normales, puisque toutes les abstractions sont maintenant irréductibles.

La réduction faible se justifie par l'utilisation du λ-calcul pour décrire des calculs donnant ultimement des données concrètes. La règle ξ peut alors être interprétée comme une règle de manipulation de programmes plutôt qu'une règle de calcul proprement dite.

Nous verrons plus loin un autre formalisme de calcul, appelé logique combinatoire, et nous montrerons son équivalence avec la réduction faible dans le λ-calcul.

8.3. λ-I-calcul

Il s'agit ici d'une restriction à la syntaxe du λ-calcul : l'abstraction [x]M n'est autorisée que lorsque l'expression M a au moins une occurrence libre de la variable x. Les expressions autorisées sont appelées <u>strictes</u>, car elles représentent des fonctions strictes, qui ont toujours besoin de calculer leurs arguments.

Voici l'algorithme qui vérifie si un lambda est strict :

```
let rec strict = function
      Ref(_)    → true
    | Abs(e)    → (match locaux(e) with [] → false | _ → strict e)
    | App(g,d)  → (strict g) & (strict d);;
```

Nous verrons plus tard une interprétation des termes du λ-calcul typé comme les preuves d'un calcul propositionnel intuitioniste. La restriction du λ-I-calcul correspond alors à la logique dite <u>relevante</u>, qui ne reconnait la validité d'une preuve de $A \Rightarrow B$ que si la preuve de B utilise effectivement l'hypothèse A.

La théorie syntaxique du λ-I-calcul est plus simple que celle du λ-K-calcul étudiée précédemment. Les termes solvables sont exactement les termes normalisables. Un terme est fortement normalisable si et seulement s'il est normalisable. L'interpréteur par valeur est correct. La dérivation normale est de longueur maximale dans sa classe.

8.4. λ-calcul linéaire

Il s'agit ici d'une restriction supplémentaire : l'abstraction $[x]M$ n'est autorisée que lorsque l'expression M a exactement une occurrence de la variable x. Les expressions autorisées sont appelées <u>linéaires</u>, car elles représentent des fonctions qui utilisent leur argument de façon linéaire.

```
let rec linear = function
      Ref(_)    → true
    | Abs(e)    → (match locaux(e) with   [_] → linear e
                                        | _   → false)
    | App(g,d)  → (linear g) & (linear d);;
```

La théorie syntaxique devient presque triviale : les dérivations équivalentes ont toutes la même longueur.

9. Séparabilité.

9.1 Enonçé du théorème de Böhm

On rappelle les définitions : **T** = [x,y]x, **F** = [x,y]y.

Définition. On dit que les lambdas M et N sont <u>séparables</u> si et seulement s'il existe des termes fermés $P_1,\ldots,P_n$ tels que ($\underline{M}$ P_1 ... P_n) ⇨* **T** et ($\underline{N}$ P_1 ... P_n) ⇨* **F**.

On remarque que M et N sont séparables si et seulement si, pour tous termes fermés M' et N', il existe des termes fermés $P_1,\ldots,P_n$ tels que :
($\underline{M}$ P_1 ... P_n) ⇨* M' et ($\underline{N}$ P_1 ... P_n) ⇨* N'.

Théorème de Böhm.
Deux formes β-normales non η-convertibles sont séparables.

Corollaire. Tout modèle identifiant deux formes β-normales non η-convertibles est incohérent.

Ici incohérent veut dire identifiant tous les termes fermés. Ce corollaire est immédiat, en remarquant que par définition de **T** et **F** on peut facilement remplacer dans la définition de séparable **T** et **F** par des termes fermés P et Q arbitraires.

La preuve du théorème de Böhm nécessite l'introduction de certaines notions sur les arbres de Böhm.

9.2 Chaînes, accessibilité.

Définition. Une <u>chaîne</u> de longueur s est une suite :
$S = [(n_1, m_1, p_1, k_1); (n_2, m_2, p_2, k_2) \ldots ; (n_s, m_s, p_s, k_s)]$,
vérifiant, pour tout $1 \leq i \leq s$: $n_i \geq 0$, $1 \leq m_i \leq \Sigma[j=1,i]n_j$, $1 \leq k_i \leq p_i$.

Intuitivement, une telle chaîne dénote un chemin de longueur s dans un arbre de Böhm. Les n_i sont les arités des formes normales de tête successives, les m_i les variables de tête, les p_i leur nombre d'arguments, l'occurrence dans l'arbre de Böhm est la liste des k_i.

Définition. Soit M est un terme fermé, B une approximation, et Q une liste d'entiers de longueur s. On dit que la chaîne S de longueur s est accessible dans M vers B modulo Q ssi :

- soit $s=0$, $S=[]$, $Q=[]$, B = approx M
- soit $s=t+1$, et :

• $T = [(n_1, m_1, p_1, k_1); (n_2, m_2, p_2, k_2) ... ; (n_t, m_t, p_t, k_t)]$ est accessible dans M vers $B' = BT(n,m,[f_1; f_2; ... ; f_p])$ modulo $Q'=[q_1; q_2; ... ; q_t]$,

• $Q=Q'@[q_s]$, avec $q_s \geq 0$,

• $p_s=p+q_s$, $n_s=n+q_s$,

• $m_s=m+\Sigma[j=r,s]q_j$, avec r minimum tel que $m \geq \Sigma[j=r,s](n_j-q_j)$,

• $S=T@[(n_s, m_s, p_s, k_s)]$, avec $1 \leq k_s \leq p_s$.

• Si $k_s=k \leq p$ on a $B = f_k()$, si $k_s= p_s+1-j$, avec $1 \leq j \leq q_s$, on a $B = BT(0,j,[])$.

L'intuition est la suivante. S est un chemin dans l'arbre de Böhm d'un terme $\beta\eta$-convertible à M. A chaque étape i on approxime le sous-terme à l'occurrence indiquée par la liste des k_i, et on fait q_i η-expansions. L'entier r est la hauteur dans cet arbre à laquelle est liée la variable m, il est bien défini par hypothèse M fermé. On suppose que le long de ce chemin les sous-termes correspondants sont définis. En particulier, M est défini, et à chaque étape où $k \leq p$, $f_k()$ termine.

Remarque.
M et S étant donnés, si S est accessible dans M vers B modulo Q, alors B et Q sont calculables de manière unique. Autrement dit, B et Q sont des fonctions récursives partielles de (M,S). On ne peut pas espérer mieux, car nous verrons qu'il est indécidable en général si un terme est défini ou non.

Définition. Soit S une chaîne, M et M' deux termes fermés tels que S est accessible dans M vers B modulo Q, et accessible dans M' vers B' modulo Q', avec $B = BT(n,m,[f_1; f_2; ... ; f_p])$, $B' = BT(n',m',[f_1; f_2; ... ; f_{p'}])$. On dit que S distingue M et M' ssi $(m+n',p+n') \neq (m'+n,p'+n)$.

9.3 Technique de débõhmification.

Proposition 1. Soient M et N deux termes fermés, tels qu'il existe une chaîne S qui les distingue. Alors M et N sont séparables.

Donnons d'abord l'idée de la preuve ("Böhm out technique"). La chaîne S représente un chemin compatible dans les arbres de Böhm de M et N, modulo certaines η-expansions, qui mène à une différence irréconciliable, c'est à dire à des sous-termes définis avec des approximations incompatibles. La première étape consiste à projeter M et N le long de S, pour faire apparaître des instances de ces sous-termes par β-réduction de respectivement $(M\ P_1 \dots P_s)$ et $(N\ P_1 \dots P_s)$. On se ramène ainsi à l'un des 2 cas illustrés par les exemples suivants :

a) $[u,v](u\ M_1 \dots M_i)$ à différencier de $[u,v](v\ N_1 \dots N_j)$, ce qui est facile par application finale de $U = [u_1, u_2 \dots, u_i]\mathbf{T}$ et $V = [v_1, v_2 \dots, v_j]\mathbf{F}$.

b) $[u](u\ M_1 \dots M_i)$ à différencier de $[u](u\ N_1 \dots N_j)$, avec $i \neq j$. Dans ce cas, en supposant par exemple $i>j$, avec $k=i-j$, on applique $X = [u_1, u_2 \dots, u_i]\mathbf{I}$, puis $Y = [v_1, v_2 \dots, v_k]\mathbf{T}$, puis k fois $\mathbf{F}$.

La seule difficulté qui se présente est dans la première étape, lorsqu'une variable apparait plusieurs fois le long de la chaîne S (ou simultanément sur S et en tant que variable de tête de l'un des deux sous-termes accédés par S). Dans ce cas il faut d'abord substituer à cette variable une expression du genre $[u_1, u_2 \dots, u_n, w](w\ u_1 \dots u_n)$, avec n suffisamment grand, ce qui permet de se ramener au cas où les variables n'apparaissent qu'une fois.

Avant de détailler la démonstration, il convient de définir un certain nombre de notions. Tout d'abord, nommons les combinateurs utilisés dans la preuve.

Définitions.

1. Combinateurs de n-uplets.

$$\mathbf{P_n} = [u_1, u_2, \dots, u_{n+1}](u_{n+1}\ u_1\ u_2 \dots u_n) \quad (n>0)$$

Par exemple, le combinateur de paire est $\mathbf{P} = \mathbf{P_2}$.

2. Combinateurs de projections.

$$\Pi_{\mathbf{k,n}} = [u_1, u_2, \dots, u_n]u_k \qquad (n \geq k>0)$$

Par exemple, $\mathbf{I} = \Pi_{1,1}$, $\mathbf{T} = \Pi_{1,2}$, et $\mathbf{F} = \Pi_{2,2}$.
On écrit aussi $\mathbf{K}_n$ pour $\Pi_{1,n+1}$. Par exemple, $\mathbf{K} = \mathbf{K}_1$.

On vérifie : $(\mathbf{P}_n\ M_1\ M_2\ \ldots\ M_n\ \Pi_{k,n}) \Rightarrow^* M_k$.

Ensuite, soient M, N, S comme dans l'énoncé de la proposition, avec : $S = [(n_1, m_1, p_1, k_1); (n_2, m_2, p_2, k_2) \ldots ; (n_s, m_s, p_s, k_s)]$. On suppose donc que S est accessible dans M vers B modulo Q, et accessible dans N vers B' modulo Q', avec $B = BT(n,m,[f_1; f_2; \ldots ; f_p])$, $B' = BT(n',m',[f_1; f_2; \ldots ; f_{p'}])$, tels que $(m+n',p+n') \neq (m'+n,p'+n)$.

Supposons $s>0$, et considérons la variable liée la plus externe. On définit son ensemble d'utilisations U comme l'ensemble des niveaux, entre 1 et s+1, où cette variable apparaît comme variable de tête le long du chemin considéré, dans M ou N. Formellement, $i \in U$, avec $1 \leq i \leq s$, ssi $m_i = \Sigma[j=1,i]n_j$, et $s+1 \in U$ ssi $m=n+\Sigma[j=1,s]n_j$ ou $m'=n'+\Sigma[j=1,s]n_j$. On définit aussi son ensemble d'arités comme $\{p_i \mid i \in U, i \leq s\}$, auquel on adjoint p et/ou p' quand $s+1 \in U$. On dit que la variable est schizophrène ssi $|U|>0$. Définitions similaires pour les n_1 variables liées au premier niveau. On appelle schizophrénie le nombre ζ de telles variables schizophrènes.

Démonstration.
On suppose tout d'abord, sans perte de généralité, que M et N ont été convertis par β-réduction en forme normale de tête. Le résultat général suivra par le théorème de Church-Rosser.

On montre la proposition par récurrence lexicographique sur $<s, \zeta, n_1>$.

- $s>0$. Il y a trois cas, suivant le nombre d'utilisations $u=|U|$ de la variable la plus externe.
  - $u>1$: la variable est schizophrène. On prend $P=\mathbf{P}_\alpha$, avec α son arité maximum. On vérifie que (M P) et (N P) donnent lieu, après β-réduction, à une situation de même s, mais de moindre schizophrénie. Remarquons toutefois que la chaîne S doit être ajustée, car les n_i peuvent avoir augmenté.
  - $u=0$: On choisit pour P un terme clos arbitraire, disons $\mathbf{I}$, et on vérifie que (M P) et (N P) donnent lieu à une situation de même s et de même schizophrénie, mais avec un n_1 inférieur.
  - $u=1$: Avec $U=\{i\}$, on prend $P=\Pi_{k,p}$, où $k=k_i$, et $p=p_i$, et on vérifie

que (M P) et (N P) donnent lieu à une situation de plus petit s.

• s=0. Soit $M = [u_1, u_2, \dots, u_n](m\ M_1\ M_2 \dots M_p)$,

$N = [u_1, u_2, \dots, u_{n'}](m'\ N_1\ N_2 \dots N_{p'})$,

avec $(m+n', p+n') \neq (m'+n, p'+n)$.

Soit $i=n+1-m$, et $i'=n+1-m'$. Il y a deux cas :

• $i \neq i'$. Si $n \geq n'$, avec $k=p'+n-n'$, on prend $P_i = (\mathbf{K}_p\ \mathbf{T})$, $P_{i'} = (\mathbf{K}_k\ \mathbf{F})$, et tous les autres P_j, avec $1 \leq j \leq n$, égaux à $\mathbf{I}$. Si $n<n'$, avec $k=p+n'-n$, on prend $P_i = (\mathbf{K}_k\ \mathbf{T})$, $P_{i'} = (\mathbf{K}_{p'}\ \mathbf{F})$, et tous les autres P_j, avec $1 \leq j \leq n'$, égaux à $\mathbf{I}$.

• $i=i'$. On a donc $p+n' \neq p'+n$. De nouveau deux cas :

• $p+n'>p'+n$. Avec $k=p+n'-(p'+n)$, et $q=p$ si $n \geq n'$, $p+n'-n$ sinon, on prend $P_j = \mathbf{I}$, pour $1 \leq j<i$, $P_i = (\mathbf{K}_q\ \mathbf{I})$, $P_{i+1} = (\mathbf{K}_k\ \mathbf{T})$, et finalement $P_{i+j+1} = \mathbf{F}$, pour $1 \leq j \leq k$.

• $p+n'<p'+n$. Symétrique du précédent. Avec $k=p'+n-(p+n')$, et $q=p'$ si $n \leq n'$, $p'+n-n'$ sinon, on prend $P_j = \mathbf{I}$, pour $1 \leq j<i$, $P_i = (\mathbf{K}_q\ \mathbf{I})$, $P_{i+1} = (\mathbf{K}_k\ \mathbf{F})$, et finalement $P_{i+j+1} = \mathbf{T}$, pour $1 \leq j \leq k$.

Ceci termine la démonstration de la proposition 1.

Exemple.

Soit $M = [u](u\ \mathbf{I}\ (u\ \mathbf{I}\ \mathbf{I}))$ et $N = [u](u\ \mathbf{I}\ (u\ u\ \mathbf{I}))$. $S = [(1, 1, 2, 2); (0, 1, 2, 2)]$. La variable u étant schizophrène, d'arité maximale 2, on prend $P_1 = \mathbf{P}$. On projette ensuite par $P_2 = [u_1, u_2]u_2$, $P_3 = [u_1, u_2]u_1$, ce qui mène à $\mathbf{I}$ et $\mathbf{P}$ respectivement. On sépare finalement par $P_4 = [u_1, u_2]\mathbf{T}$, $P_5 = \mathbf{I}$, $P_6 = [u_1, u_2]\mathbf{F}$.

Proposition 2.
Soient M et N deux lambdas normalisables non distinguables. Alors M et N sont βη-convertibles.

Démonstration. Par récurrence sur la somme des tailles des approximations de M et N. Soit $B=T(n,m,[f_1; f_2; \dots ; f_p])$=approx M, $B'=BT(n',m',[g_1; g_2; \dots ; g_{p'}])$=approx N, et supposons par exemple $n'>n$. Soit $q=n'-n$. Par q η-expansions de la forme normale de tête de M, on obtient M', avec approx $M' = BT(n',m+q,[f_1; f_2; \dots ; f_p; h_1; \dots ; h_q])$. La chaîne vide est accessible dans M vers B et dans N vers B'. Comme elle ne distingue pas M et N, on doit avoir m+n'= m'+n, et p+n'=p'+n, d'où m'=m+q, p'=p+q. Par hypothèse de récurrence, avec N' la forme normale de tête de N, on obtient que M'/j est βη-convertible à N'/j, pour $1 \leq j \leq p'$, et donc que M et N sont βη-convertibles.

Corollaire. Soient M et N deux formes β-normales non η-convertibles. Il existe une chaîne S qui les distingue.

Le théorème de Böhm est maintenant conséquence directe de la proposition 1 et de ce corollaire.

On remarque qu'il n'est guère possible de faire mieux, dans la mesure où des formes normales η-convertibles ne sont pas séparables. Par exemple, considérez **I** et A=[u,v](u v).

L'importance du théorème de Böhm est qu'il donne une condition très forte sur la construction de modèles. Deux termes normalisables non β-convertibles ne sont identifiables dans un modèle non-trivial que s'ils sont βη-convertibles. Ceci fixe en gros les trois degrés de liberté d'un modèle par rapport au modèle purement syntaxique des arbres de Böhm :

- le modèle peut ou non vérifier η
- le modèle peut identifier plus ou moins les termes non solvables
- le modèle peut être plus ou moins riche en points non définissables.

10. Bibliographie sommaire.

L'ouvrage de base sur le λ-calcul non typé est le livre de Henk Barendregt : "The Lambda Calculus, Its Syntax and Semantics", North-Holland, revised edition, 1984. Une monographie plus courte, du même auteur, est disponible en tant que chapitre du "Handbook of Mathematical Logic", édité par J. Barwise, North-Holland, 1977.

Le meilleur livre général sur le λ-calcul (typé ou non) et la théorie des combinateurs est "Introduction to Combinators and λ-calculus", de Roger Hindley et Jonathan Seldin, Cambridge University Press, 1986.

Pour le λ-calcul typé, et ses connections avec la théorie des catégories, on lira "Introduction to higher order categorical logic", de J. Lambek et P. J. Scott, Cambridge University Press, 1986, ainsi que la monographie de Pierre-Louis Curien : Categorical Combinators, Sequential Algorithms and Functional Programming, Pitman & Wiley, 1986.

A titre historique, on pourra consulter la monographie d'origine d'Alonzo Church : "The calculi of lambda-conversion", Ann. of Math. Studies 6, Princeton University, 1941, ainsi que les œuvres d'Haskell Curry et de ses collaborateurs : Combinatory Logic, chez North-Holland (Vol 1, 1968, et Vol 2, 1972).

Un certain nombre de résultats importants ne sont disponibles que dans des thèses. On citera tout particulièrement la thèse de J.J. Lévy, Université Paris 7, 1977, celle de J.Y. Girard, Université Paris 7, 1972, et celle de J.W. Klop, Université d'Utrecht, 1980, disponible comme la monographie "Combinatory Reduction Systems" du Mathematisch Centrum d'Amsterdam.

Il n'existe pas de présentation synthétique des modèles du λ-calcul. Il faut lire les travaux de recherche de D. Scott, G. Plotkin, G. Berry, A. Meyer, J. Y. Girard, K. Koymans, P. L. Curien, R. Statman. On lira aussi avec profit les notes de cours de G. Longo (Carnegie-Mellon University, 1988).

Un recueil important de résultats de recherche est le volume anniversaire dédié à H. Curry : "Essays on Combinatory Logic, Lambda Calculus and Formalism", Academic Press, 1980.

Enfin, signalons que la théorie des combinateurs est présentée sous forme de récréations mathématiques dans un livre de R. Smullyan : "To mock a mockingbird", Knopf, 1985.

Normalization

Helmut Schwichtenberg
Mathematisches Institut, Universität München
Theresienstraße 39, D–8000 München 2, Germany

The aim of this paper is to present a central technique from proof theory, Gentzen's normalization for natural deduction systems, and to discuss some of its applications.

By normalization we mean a collection of algorithms transforming a given derivation into a certain normal form. A derivation is called *normal* if it does not contain any "detour" i.e. an application of an introduction rule immediately followed by an application of an elimination rule. Such normalization algorithms are useful because they allow to "straighten out" complex derivations and in this way extract hidden information.

We will treat many applications which demonstrate this, e.g. the subformula principle, Herbrand's theorem, the interpolation theorem, an exact characterization of the initial cases of transfinite induction provable in arithmetic and a proof that normalization in (the usual finitary) arithmetic is impossible.

From the computer science point of view, an even more interesting field of application for normalization algorithms is the possiblility to extract the constructive content of a maybe complex mathematical argument. Such algorithms can yield verified programs from derivations proving that certain specifications can be fulfilled. Of course, the feasability of programs obtained in this way will depend to a large extent on a good choice of the derivation, which should be done on the basis of a good idea for an algorithm. However, in this approach it is possible to use ordinary mathematical machinery for the development of programs.

Chapter 1 deals with normalization for minimal propositional logic, or more precisely for its implicational fragment. In Section 1.1 it is shown that — by adding stability axioms — classical logic can be embedded in it. In Sections 1.2–1.6 we then treat normalization for this calculus, with special emphasis on complexity questions. In Section 1.7 normalization (for a natural deduction system) is compared with cut–elimination (for a sequent calculus). Section 1.8 discusses a decision procedure for minimal implicational logic.

In Chapter 2 the method of collapsing types developed in (Troelstra and van Dalen 1988) is used to lift these results to minimal first order logic or more precisely to its $\to\forall$–fragment, which again suffices for classical logic. Section 2.4 contains some applications of normalization: the subformula principle, Herbrand's theorem and the interpolation theorem.

The final Chapter 3 treats normalization for arithmetic. Since normalization for finitary arithmetic with the induction rule is impossible (this is proved in Section 3.5), we extend in Section 3.3 the normalization technique to arithmetic with the ω–rule. This is used in Section 3.4 to give an exact characterization of the initial cases of transfinite induction provable in arithmetic as well as in some subsystems of arithmetic obtained by restricting the complexity of the induction formulas.

The expert will certainly note that most of the results and proofs elaborated here are well–known. The only novel points are the following.

NATO ASI Series, Vol. F 79
Logic, Algebra, and Computation
Edited by F. L. Bauer

- We have based our treatment of normalization in Chapters 1 and 2 on a slight generalization of β–conversion: we not only allow $(\lambda xr)s$ to be converted into $r_x[s]$, but more generally $(\lambda \vec{x}x.r)\vec{s}s$ to be converted into $(\lambda \vec{x}.r_x[s])\vec{s}$. This allows a particularly simple proof of the existence of the normal form (Theorem 1.2.2 below), which also provides an easy estimate of the number of conversion steps needed, and it also makes the results on strong normalization (Section 1.5) and the uniqueness of the normal form (Section 1.3) slightly stronger.
- In Section 1.7 we give an argument that normalization (for a natural deduction system) and cut–elimination (for a sequent calculus) are essentially different, using a recent result of (Hudelmaier 1989).
- In Section 1.8 we present a decision algorithm for implicational logic also due to (Hudelmaier 1989), together with a new proof of its correctness and completeness.
- In Theorem 3.5.1 it is stated that a certain weak form of a normalization theorem does not hold for arithmetic with the induction rule.

1. Normalization for propositional logic

1.1 Minimal implicational logic as a typed λ–calculus

Formulas are built up from propositional variables denoted by P, Q by means of $(\varphi \to \psi)$. We write $\varphi_1, \ldots, \varphi_m \to \psi$ for $(\varphi_1 \to (\varphi_2 \to \ldots(\varphi_m \to \psi)\ldots))$. Derivations are built up from assumption variables u^φ, v^ψ by means of the rule $\to^+$ of implication introduction (or λ–abstraction)

$$(\lambda u^\varphi r^\psi)^{\varphi\to\psi}$$

and the rule $\to^-$ of implication elimination (or application)

$$(t^{\varphi\to\psi}s^\varphi)^\psi.$$

A derivation r^ψ whose free assumption variables are among $u_1^{\varphi_1}, \ldots, u_m^{\varphi_m}$ is also called a derivation of ψ from $\varphi_1, \ldots, \varphi_m$. For readability we often leave out formula superscripts when they are obvious from the context or non–essential.

For obvious reasons we will also use the word term for derivations and type for formulas. The possibility to treat derivations as terms and formulas as types has been discovered by H. B. Curry and elaborated by W. A. Howard in (Howard 1980a). This correspondence can easily be shown to be an isomorphism; it is called the Curry–Howard–isomorphism.

More formally, it can be seen easily that a closed derivation (i. e. one without free assumption variables) is determined by

1. a type free λ–term describing the derivation and
2. the derived formula.

The formulas in the derivation can be left out since they can easily be reconstructed form the given derived formula.

As an example, a derivation of

$$\varphi \to (\psi \to \varphi)$$

is given by $\lambda u \lambda v u$, and a derivation of

$$(\varphi \to (\psi \to \chi)) \to ((\varphi \to \psi) \to (\varphi \to \chi))$$

is given by $\lambda u \lambda v \lambda w((uw)(vw))$. Both derivations can be easily written in the more usual tree form.

Note that our minimal implicational logic contains full classical propositional logic, as follows. Choose a particular propositional variable and denote it by $\bot$ (falsity). Associate with any formula φ in the language of classical propositional logic a finite list $\vec{\varphi}$ of formulas in our implicational language, by induction on φ:

$$\begin{aligned} P &\mapsto P \\ \neg\varphi &\mapsto \vec{\varphi} \to \bot \\ \varphi \to \psi &\mapsto \vec{\varphi} \to \psi_1, \ldots \vec{\varphi} \to \psi_m \\ \varphi \wedge \psi &\mapsto \vec{\varphi}, \vec{\psi} \\ \varphi \vee \psi &\mapsto (\vec{\varphi} \to \bot), (\vec{\psi} \to \bot) \to \bot \end{aligned}$$

Then, if φ is a formula in the language of full classical propositional logic and $\varphi_1, \ldots, \varphi_m$ is its associated list, φ is derivable in classical propositional logic iff each φ_i is derivable in minimal implicational logic from stability assumptions $\neg\neg P \to P$ (with $\neg\psi$ denoting $\psi \to \bot$) for all propositional variables P in φ. The essential step in the proof is to show that from the stability of ψ we can infer the stability of $\varphi \to \psi$: a derivation of

$$(\neg\neg\psi \to \psi) \to (\neg\neg(\varphi \to \psi) \to (\varphi \to \psi))$$

is given by

$$\lambda u_1 \lambda u_2 \lambda u_3 (u_1 \lambda u_4 (u_2 \lambda u_5 (u_4 (u_5 u_3)))).$$

1.2. Conversion

We are interested in the following process of simplification of terms:

$$(\lambda\vec{u}u.r)\vec{s}s \quad \text{converts into} \quad (\lambda\vec{u}.r_u[s])\vec{s}.$$

Here $\vec{u}$ and $\vec{s}$ denote finite lists $u_1 \dots u_m$ and $s_1 \dots s_m$, and $\lambda\vec{u}u.r$ denotes the term $\lambda u_1 \dots \lambda u_m \lambda ur$. Terms of the form $(\lambda\vec{u}u.r)\vec{s}s$ are called *convertible*.

Note that converting $(\lambda\vec{u}u.r)\vec{s}s$ into $(\lambda\vec{u}.r_u[s])\vec{s}$ may be viewed as first converting $(\lambda\vec{u}u.r)\vec{s}s$ "permutatively" into $(\lambda\vec{u}(\lambda ur)s)\vec{s}$ and then performing the inner conversion to obtain $(\lambda\vec{u}.r_u[s])\vec{s}$. One may ask why we take this conversion relation as our basis and not the more common $(\lambda ur)s \mapsto r_u[s]$. The reason is that our notion of level is defined with the clause $\text{lev}(\varphi \to \psi) = \max(\text{lev}(\varphi)+1, \text{lev}(\psi))$ and not $= \max(\text{lev}(\varphi), \text{lev}(\psi))+1$; this in turn seems reasonable since then the level of $P_1, \dots, P_m \to Q$ (i. e. of $(P_1 \to (P_2 \to \dots (P_m \to Q)\dots))$) is 1 and hence independent of m. But given this definition of level, and given the need in some arguments (e. g. in Theorem 1.2.1) to perform conversions of highest level first, we must be able to convert $(\lambda uv.r)st$ with u of a low and v of a high level into $(\lambda u.r_v[t])s$. —In addition, since we allow more conversions here, the results on strong normalization and upper bounds for the length of arbitrary reduction sequences get stronger.

We write $r \to r'$ if r' is obtained from r as follows. Mark some occurrences of convertible subterms in r. Then convert them all simultaneously. Hence new convertible subterms generated by such a conversion can not be converted. More precisely, $r \to r'$ is defined by the following rules

1. $u \to u$.
2. If $r \to r'$, then $\lambda ur \to \lambda ur'$,
3. If $r \to r'$ and $s \to s'$, then $rs \to r's'$.
4. If $r \to r'$, $\vec{s} \to \vec{s'}$ and $s \to s'$, then $(\lambda\vec{u}u.r)\vec{s}s \to (\lambda\vec{u}.r'_u[s'])\vec{s'}$.

As a special case, we take

$$r \to_1 r'$$

to mean that r' is obtained from r by converting exactly one convertible subterm in r. Finally

$$r \to^* r' \qquad (r \text{ reduces to } r')$$

denotes the transitive and reflexive closure of $\to$ (or equivalently of $\to_1$).

A term is said to be in *normal form* if it does not contain a convertible subterm.

We want to show now that any term reduces to a normal form. This can be seen easily if we follow a certain order in our conversions. To define this order we have to make use of the fact that all our terms (i.e. derivations) have types (i.e. formulas).

Define the *level* of a formula by

$$\text{lev}(P) = 0,$$
$$\text{lev}(\varphi \to \psi) = \max(\text{lev}(\varphi) + 1, \text{lev}(\psi))$$

A convertible derivation

$$(\lambda\vec{u}^{\vec{\varphi}}u^{\varphi}.r)\vec{s}s$$

is also called a *cut* with *cut-formula* φ. By the level of a cut we mean the level of its cut–formula. The *cut–rank* of a derivation r is the least number bigger than the levels of all cuts in r. Now let t be a derivation of cut–rank $k+1$. Pick a cut

$$(\lambda \vec{u} u.r)\vec{s}s$$

of the maximal level k in t, such that s does not contain another cut of level k. (E.g., pick the rightmost cut of level k.) Then it is easy to see that replacing the picked occurrence of $(\lambda \vec{u} u.r)\vec{s}s$ in t by $(\lambda \vec{u}.r_u[s])\vec{t}$ reduces the number of cuts of the maximal level k in t by 1. Hence

Theorem 1.2.1. *We have an algorithm which reduces any given term into a normal form.*

We now want to give an estimate of the number of conversion steps our algorithm takes until it reaches the normal form. The key observation for this estimate is the obvious fact that replacing one occurrence of

$$(\lambda \vec{u} u.r)\vec{s}s \quad \text{by} \quad (\lambda \vec{u}.r_u[s])\vec{s}$$

in a given term t at most squares the length of t; here the *length* of t is taken to be the number of variables in t (except those immediately following a λ–symbol).

A bound $s_k(l)$ for the number of steps our algorithm takes to reduce the rank of a given term of length l by k can be derived inductively, as follows. Let $s_0(l) := 0$. To obtain $s_{k+1}(l)$, first note that by induction hypothesis it takes $\leq s_k(l)$ steps to reduce the rank by k. The length of the resulting term is $\leq l^{2^s}$ where $s := s_k(l)$ since any step (i.e. conversion) at most squares the length. Now to reduce the rank by one more the number of additional steps is obviously bounded by that length. Hence the total number of steps to reduce the rank by $k+1$ is bounded by

$$s_k(l) + l^{2^{s_k(l)}} =: s_{k+1}(l).$$

Theorem 1.2.2. (Upper bound for the complexity of normalization) *The normalization algorithm given in the proof of Theorem 1.2.1 takes at most $s_k(l)$ steps to reduce a given term of cut–rank k and length l to normal form, where*

$$s_0(l) := 0 \quad \text{and} \quad s_{k+1}(l) := s_k(l) + l^{2^{s_k(l)}}.$$

1.3. Uniqueness

We shall show that the normal form of a term is uniquely determined; this will be done by an argument which also applies to type–free terms, i.e. terms without formula superscripts. The main idea of the proof (due to J. B. Rosser and W. W. Tait) is to use the relation $r \twoheadrightarrow r'$ defined in Section 1.2. Its crucial property is given by

Lemma 1.3.1. *If $r \twoheadrightarrow r'$ and $t \twoheadrightarrow t'$ then $r_v[t] \twoheadrightarrow r'_v[t']$.*

The proof is by induction on the definition of $r \twoheadrightarrow r'$. All cases are obvious except possibly Rule 4. So assume $r \twoheadrightarrow r'$, $\vec{s} \twoheadrightarrow \vec{s'}$ and $s \twoheadrightarrow s'$. Then

$$r_v[t] \twoheadrightarrow r'_v[t'], \quad \vec{s}_v[t] \twoheadrightarrow \vec{s'}_v[t'] \quad \text{and} \quad s_v[t] \twoheadrightarrow s'_v[t']$$

by induction hypothesis, and hence

$$\underbrace{(\lambda \vec{u}u.r_v[t])\vec{s}_v[t]s_v[t]}_{((\lambda \vec{u}u.r)\vec{s}s)_v[t]} \twoheadrightarrow \underbrace{(\lambda \vec{u}.(r'_v[t'])_u[s'_v[t']])\vec{s'}_v[t']}_{((\lambda \vec{u}.r'_u[s'])\vec{s'})_v[t']}.$$

by definition of $\twoheadrightarrow$. □

Lemma 1.3.2. *Assume $r \twoheadrightarrow r'$ and $r \twoheadrightarrow r''$. Then we can find a term r''' such that $r' \twoheadrightarrow r'''$ and $r'' \twoheadrightarrow r'''$.*

The proof is by induction on the definition of $r \twoheadrightarrow r'$. Again all cases are obvious except possibly the situation where either $r \twoheadrightarrow r'$ or $r \twoheadrightarrow r''$ is obtained via Rule 4. By symmetry we may assume the former. But then the claim follows from Lemma 1.3.1: If

$$(\lambda \vec{u}u.r)\vec{s}s \twoheadrightarrow (\lambda \vec{u}.r'_u[s'])\vec{s'}$$

and

$$(\lambda \vec{u}u.r)\vec{s}s \twoheadrightarrow (\lambda \vec{u}u.r'')\vec{s''}s'',$$

then

$$(\lambda \vec{u}u.r)\vec{s}s \twoheadrightarrow (\lambda \vec{u}.r'_u[s'])\vec{s'} \twoheadrightarrow (\lambda \vec{u}.r'''_u[s'''])\vec{s'''}$$

and

$$(\lambda \vec{u}u.r)\vec{s}s \twoheadrightarrow (\lambda \vec{u}u.r'')\vec{s''}s'' \twoheadrightarrow (\lambda \vec{u}.r'''_u[s'''])\vec{s'''},$$

and if

$$(\lambda \vec{u}u\vec{v}v.r)\vec{s}s\vec{t}t \twoheadrightarrow (\lambda \vec{u}\vec{v}v.r'_u[s'])\vec{s'}\vec{t'}t'$$

and

$$(\lambda \vec{u}u\vec{v}v.r)\vec{s}s\vec{t}t \twoheadrightarrow (\lambda \vec{u}u\vec{v}.r''_v[t''])\vec{s''}s''\vec{t''}$$

then

$$(\lambda \vec{u}u\vec{v}v.r)\vec{s}s\vec{t}t \twoheadrightarrow (\lambda \vec{u}\vec{v}v.r'_u[s'])\vec{s'}\vec{t'}t' \twoheadrightarrow (\lambda \vec{u}\vec{v}.r'''_{u,v}[s''',t'''])\vec{s'''}\vec{t'''}$$

and

$$(\lambda \vec{u}u\vec{v}v.r)\vec{s}s\vec{t}t \twoheadrightarrow (\lambda \vec{u}u\vec{v}.r''_v[t''])\vec{s''}s''\vec{t''} \twoheadrightarrow (\lambda \vec{u}\vec{v}.r'''_{u,v}[s''',t'''])\vec{s'''}\vec{t'''}.$$

□

Theorem 1.3.3. (Church–Rosser) *Assume $r \to^* r'$ and $r \to^* r''$. Then we can find a term r''' such that $r' \to^* r'''$ and $r'' \to^* r'''$.*

The proof is immediate from Lemma 1.3.2. □

Corollary 1.3.4. (Uniqueness of the normal form) *Assume $r \to^* r'$ and $r \to^* r''$, where both r' and r'' are in normal form. Then r' and r'' are identical.*

1.4. Complexity of normalization: a lower bound

In Theorem 1.2.2 we have obtained an upper bound on the number of conversion steps our particular normalization algorithm of Theorem 1.2.1 takes to reach the normal form. This upper bound was superexponential in the length of the given term. It is tempting to think that by choosing a clever normalization strategy one might be able to reduce that bound significantly. It is the purpose of the present section to show that this is impossible. More precisely, we will construct terms r_n of length $3n$ and show that any normalization algorithm needs at least $2_{n-2} - n$ conversions (with $2_0 := 1, 2_{n+1} := 2^{2_n}$) to reduce r_n to its normal form.

The fact that there is no elementary algorithm (i.e. whose time is exponentially bounded) to compute the normal form of terms also follows from (Statman 1979), where it is shown more generally that the problem whether two terms r_1 and r_2 have the same normal form is not elementary recursive. The simple example treated here is taken from (Schwichtenberg 1982, p. 455).

The pure types k are defined inductively by $0 := P$ (some fixed propositional variable) and $k+1 = k \to k$. We define iteration terms I_n of pure type $k+2$ by

$$I_n :\equiv \lambda f \lambda u(f(f(\ldots f(fu)\ldots))),$$

with n occurrences of f after $\lambda f \lambda u$; here f, u are variables of type $k+1, k$, respectively. Let $f \circ g$ be an abbreviation for $\lambda u(f(gu))$, and let $r = s$ mean that r and s have the same normal form. With this notation we can write

$$I_n = \lambda f(\underbrace{f \circ f \circ \ldots \circ f}_{n}).$$

The main point of our argument is the following simple lemma, which can be traced back to Rosser (cf. (Church 1941, p. 30))

Lemma 1.4.1.

$$\begin{aligned}(I_m f) \circ (I_n f) &= I_{m+n} f,\\ I_m \circ I_n &= I_{m \cdot n},\\ I_m I_n &= I_{n^m}. \square\end{aligned}$$

As an immediate consequence we have

$$r_n :\equiv \underbrace{I_2 I_2 \ldots I_2}_{n} = I_{2_n}.$$

Now consider any sequence of reduction steps transforming r_n into its normal form, and let s_n denote the total number of reduction steps in this sequence.

Theorem 1.4.2. $s_n \geq 2_{n-2} - n$.

Proof. The length of r_n is $3n$. Note that any conversion step can at most square the length of the original term. Hence we have

$$\begin{aligned} 2_n &< \mathrm{length}(I_{2_n}) && \text{(the normal form of } r_n\text{)} \\ &\leq \mathrm{length}(r_n)^{2^{s_n}} \\ &= (3n)^{2^{s_n}} \\ &\leq 2^{2^{n+s_n}} && (\text{since } 3n \leq 2^{2^n}), \end{aligned}$$

and the theorem is proved. □

1.5. Strong normalization

In Section 1.2 we have proved that any term can be reduced to a normal form, and in Section 1.3 we have seen that this normal form is uniquely determined. But it is still conceivable that there might be an odd reduction sequence which does not terminate at all. It is the aim of the present Section to show that this is impossible. This fact is called the strong normalization theorem.

For the proof we employ a powerful method due to W. W. Tait, which is based on so-called strong computability predicates. These are defined by induction on the types (i.e. formulas) as follows.

A term r^φ with φ of level 0 (i. e. a propositional variable) is strongly computable iff r is strongly normalizable, i.e. every reduction sequence starting from r terminates. A term $r^{\varphi \to \psi}$ is strongly computable iff for all strongly computable s^φ also $(rs)^\psi$ is strongly computable.

A term r is *strongly computable under substitution* iff for all strongly computable $\vec{s}$ the result of substituting $\vec{s}$ for all variables free in r is again strongly computable.

Lemma 1.5.1. *Let φ be a formula.*

1. *Any strongly computable term r^φ is strongly normalizable.*
2. *u^φ is strongly computable.*

We prove 1 and 2 simultaneously by induction on φ. For φ of level 0 both claims are obvious. Now consider $\varphi \to \psi$. For 1, assume that $r^{\varphi \to \psi}$ is strongly computable. By induction hypothesis 2 and the definition of strong computability we know that $(ru)^\psi$ is strongly computable and hence that any reduction sequence starting with ru terminates (by induction hypothesis 1). But this obviously implies that the same is true for r. For 2, assume that $\vec{r}$ are strongly computable. We have to show that $u\vec{r}$ (which is to be of level 0) is strongly computable, i. e. that any reduction sequence starting with $u\vec{r}$ terminates. But this follows from induction hypothesis 1, which says that any reduction sequence starting from r_i terminates. □

Lemma 1.5.2. *If $r \to_1 r'$ and r is strongly computable, then r' is strongly computable.*

Proof. Let $\vec{s}$ be strongly computable. We have to show that $r'\vec{s}$ is strongly computable, i. e. that any reduction sequence starting from $r'\vec{s}$ terminates. But this is obviously true, because otherwise we would also have an infinite reduction sequence for $r\vec{s}$. □

Lemma 1.5.3. *Any term r is strongly computable under substitution.*

The proof is by induction on the height of r.

Case u. Obvious.

Case rs. Let $\vec{t}$ be strongly computable. We have to show that $r[\vec{t}]s[\vec{t}]$ is strongly computable. But this holds, since by induction hypothesis we know that $r[\vec{t}]$ as well as $s[\vec{t}]$ are strongly computable.

Case $\lambda\vec{u}u.r$. Let $\vec{t}$ be strongly computable. We have to show that $\lambda\vec{u}u.r[\vec{t}]$ is strongly computable. So let $\vec{s}, s$ and $\vec{r}$ be strongly computable. We must show that $(\lambda\vec{u}u.r[\vec{t}])\vec{s}s\vec{r}$ is strongly computable, i.e. that any reduction sequence for it terminates. So assume we have an infinite reduction sequence. Since $r[\vec{t}], \vec{s}, s$ and $\vec{r}$ all are strongly normalizable, there must be a term $(\lambda\vec{u}u.r[\vec{t}]')\vec{s'}s'\vec{r'}$ with $r[\vec{t}] \to^* r[\vec{t}]', \vec{s} \to^* \vec{s'}, s \to^* s'$ and $\vec{r} \to^* \vec{r'}$ in that reduction sequence where a "head conversion" is applied, which we may assume to yield

$$(\lambda\vec{u}.(r[\vec{t}]')[s'])\vec{s'}\vec{r'}.$$

But $r[\vec{t}] \to^* r[\vec{t}]'$ implies $\lambda\vec{u}.r[s,\vec{t}] \to^* \lambda\vec{u}.(r[\vec{t}]')[s']$, and hence the fact that $\lambda\vec{u}.r$ is (by induction hypothesis) strongly computable under substitution together with Lemma 1.5.2 implies that $(\lambda\vec{u}.(r[\vec{t}]')[\vec{s'}]$ is strongly computable. But then, again by Lemma 1.5.2, also $(\lambda\vec{u}.(r[\vec{t}])'[s'])\vec{s'}\vec{r'}$ is strongly computable and therefore strongly normalizable. This contradicts our assumption above that we have an infinite reduction sequence. □

From Lemma 1.5.3 and both parts of Lemma 1.5.1 can conclude immediately

Theorem 1.5.4. *Any term r is strongly normalizable.* □

1.6. Complexity of normalization: an upper bound

By Section 1.5 we already know that the full reduction tree for a given term is finite; hence its height bounds the length of any reduction sequence. But it is not obvious how a reasonable estimate for that height might be obtained.

However, using a technique due to (Howard 1980b) (which in turn is based on (Sanchis 1967) and (Diller 1968)) it can be shown that we have the following superexponential universal bound.

Theorem 1.6.1. *Let r be a term of the typed λ–calculus of level 0. Let m be a bound for the levels of subterms of r and $k \geq 2$ be a bound for the arities of subterms of r. Then the length of an arbitrary reduction sequence for r with respect to $\to_1$ is bounded by*

$$k^{2_m(m+2\cdot\mathrm{height}(r)+2k+2)}.$$

For the proof see (Schwichtenberg 1990). □

1.7. Cut elimination versus normalization

Up to now we have considered pure implicational logic by means of Gentzen's rules of natural deduction. Now it is also common to use another type of logical calculus, the ***sequent calculus*** also introduced by Gentzen. Instead of formulas it treats ***sequents*** $\Gamma \Rightarrow \varphi$, where Γ is a finite set of formulas. The rules of the sequent calculus for pure implicational logic are the following; here we write Γ, φ for $\Gamma \cup \{\varphi\}$.

Axiom. $\vdash \Gamma, \varphi \Rightarrow \varphi$ for φ atomic.

→-right. If $\vdash \Gamma, \varphi \Rightarrow \psi$, then $\vdash \Gamma \Rightarrow \varphi \to \psi$.

→-left. If $\vdash \Gamma, \varphi \to \psi \Rightarrow \varphi$ and $\vdash \Gamma, \varphi \to \psi, \psi \Rightarrow \chi$, then $\vdash \Gamma, \varphi \to \psi \Rightarrow \chi$.

Cut. If $\vdash \Gamma \Rightarrow \chi$ and $\vdash \Gamma, \chi \Rightarrow \varphi$, then $\vdash \Gamma \Rightarrow \varphi$.

It is easy to see that the sequent calculus is equivalent to natural deduction, in the sense that $\vdash \Gamma \Rightarrow \varphi$ iff from Γ we can derive φ by means of the rules $\to^+$ and $\to^-$ and the assumption rule.

Now a normal derivation of φ from Γ has the property that all formulas occurring in this derivation are subformulas of either φ or a formula in Γ. The same property holds for derivations of $\Gamma \Rightarrow \varphi$ in the sequent calculus which do not use the cut rule. Hence it is of interest to know that the cut rule can always be eliminated from derivations in the sequent calculus.

Gentzen proved this Cut Elimination Theorem in his thesis. Here we prove it in such a way that we also obtain a good bound on the length of the resulting cut-free derivation, in the form $2^{j(d)} \cdot l(d)$, where $l(d)$ is the length of the original derivation and $j(d)$ is the maximum taken over all paths in d of the sum of the degrees of all cut formulas on the path. The notion of degree used here is rather peculiar. Its crucial property is

$$\deg(\varphi \to \psi) + \deg(\psi \to \chi) < \deg((\varphi \to \psi) \to \chi). \tag{1.1}$$

This can be achieved if we define

- $\deg(\varphi) = 2$ for φ atomic,
- $\deg(\varphi \to \psi) = 1 + \deg(\varphi) \cdot \deg(\psi)$.

For then we have, writing $a := \deg(\varphi)$, $b := \deg(\psi)$ and $c := \deg(C)$, $1 + ab + 1 + bc = 2 + (a + c)b \leq 2 + abc < 1 + c + abc = 1 + (1 + ab)c$ and hence 1.1.

More formally, we define the relation $\vdash^{\alpha}_{m} \Gamma \Rightarrow \varphi$ (to be read: $\Gamma \Rightarrow \varphi$ is derivable with height $\leq \alpha$ and cut-rank $\leq m$) with α, m natural numbers inductively by the following rules.

Axiom. $\vdash^{\alpha}_{m} \Gamma, \varphi \Rightarrow \varphi$ for φ atomic.

→-right. If $\vdash^{\alpha}_{m} \Gamma, \varphi \Rightarrow \psi$, then $\vdash^{\alpha+1}_{m} \Gamma \Rightarrow \varphi \to \psi$.

→-left. If $\vdash^{\alpha}_{m} \Gamma, \varphi \to \psi \Rightarrow \varphi$ and $\vdash^{\alpha}_{m} \Gamma, \varphi \to \psi, \psi \Rightarrow \chi$, then $\vdash^{\alpha+1}_{m} \Gamma, \varphi \to \psi \Rightarrow \chi$.

Cut. If $\vdash^{\alpha}_{m} \Gamma \Rightarrow \chi$ and $\vdash^{\alpha}_{m} \Gamma, \chi \Rightarrow \varphi$, then $\vdash^{\alpha+1}_{m+\deg(\chi)} \Gamma \Rightarrow \varphi$.

Then the bound mentioned above is a consequence of the following

Theorem 1.7.1. (Cut Elimination Theorem) *If* $\vdash^{\alpha}_{m+1} \Gamma \Rightarrow \varphi$, *then* $\vdash^{2\alpha}_{m} \Gamma \Rightarrow \varphi$.

This theorem is due to (Hudelmaier 1989); its present formulation and proof is the result of Buchholz' analysis (Buchholz 1989) of Hudelmaier's arguments.

We need some Lemmata before we can give the proof.

Lemma 1.7.2. (Weakening Lemma) *If* $\vdash^{\alpha}_{m} \Gamma \Rightarrow \varphi$, *then* $\vdash^{\alpha}_{m} \Gamma, \Delta \Rightarrow \varphi$. □

Lemma 1.7.3. (Inversion)

i. If $\vdash^{\alpha}_{m} \Gamma \Rightarrow \varphi \rightarrow \psi$, *then* $\vdash^{\alpha}_{m} \Gamma, \varphi \Rightarrow \psi$.

ii. If $\vdash^{\alpha}_{m} \Gamma, \varphi \rightarrow \psi \Rightarrow \chi$, *then* $\vdash^{\alpha}_{m} \Gamma, \psi \Rightarrow \chi$.

iii. If $\vdash^{\alpha}_{m} \Gamma, (\varphi \rightarrow \psi) \rightarrow \chi \Rightarrow \vartheta$, *then* $\vdash^{\alpha}_{m} \Gamma, \varphi, \psi \rightarrow \chi \Rightarrow \vartheta$.

Proof. By induction on α. We only treat one case of iii. Assume that

$$\vdash^{\alpha}_{m} \Gamma, (\varphi \rightarrow \psi) \rightarrow \chi \Rightarrow \vartheta$$

was inferred from

$$\vdash^{\alpha-1}_{m} \Gamma, (\varphi \rightarrow \psi) \rightarrow \chi \Rightarrow \varphi \rightarrow \psi \quad \text{and} \quad \vdash^{\alpha-1}_{m} \Gamma, (\varphi \rightarrow \psi) \rightarrow \chi, \chi \Rightarrow \vartheta.$$

By induction hypothesis we get

$$\vdash^{\alpha-1}_{m} \Gamma, \varphi, \psi \rightarrow \chi \Rightarrow \varphi \rightarrow \psi \quad \text{and} \quad \vdash^{\alpha-1}_{m} \Gamma, \varphi, \psi \rightarrow \chi, \chi \Rightarrow \vartheta.$$

Hence by i

$$\vdash^{\alpha-1}_{m} \Gamma, \varphi, \psi \rightarrow \chi \Rightarrow \psi.$$

Now $\rightarrow$-left yields $\vdash^{\alpha}_{m} \Gamma, \varphi, \psi \rightarrow \chi \Rightarrow \vartheta$. □

Lemma 1.7.4. (Cut Elimination Lemma)

i. If $\vdash^{\alpha}_{0} \Gamma \Rightarrow \varphi$ *and* $\vdash^{\beta}_{0} \Gamma, \varphi \Rightarrow \psi$ *and* φ *is atomic, then* $\vdash^{\alpha+\beta}_{0} \Gamma \Rightarrow \psi$.

ii. If $\vdash^{\alpha}_{0} \Gamma \Rightarrow \varphi \rightarrow \psi$ *and* $\vdash^{\beta}_{0} \Gamma, \varphi \rightarrow \psi \Rightarrow \chi$ *and* φ *is atomic and* $\beta \leq \alpha$, *then* $\vdash^{\alpha+\beta}_{m} \Gamma \Rightarrow \chi$ *with* $m = \deg(\psi)$.

iii. If $\vdash^{\alpha}_{0} \Gamma \Rightarrow (\varphi \rightarrow \psi) \rightarrow \chi$ *and* $\vdash^{\beta}_{0} \Gamma, (\varphi \rightarrow \psi) \rightarrow \chi \Rightarrow \vartheta$, *then* $\vdash^{\alpha+\beta+2}_{m} \Gamma \Rightarrow \vartheta$, *with* $m = \deg(\varphi \rightarrow \psi) + \deg(\psi \rightarrow \chi)$.

Proof. i. By induction on β. ii. Consider also

ii'. If $\vdash^{\alpha}_{0} \Gamma \Rightarrow \varphi \rightarrow \psi$ and $\vdash^{\alpha}_{0} \Gamma, \psi \Rightarrow \chi$ and $\vdash^{\beta}_{0} \Gamma, \varphi \rightarrow \psi \Rightarrow \varphi$ and φ is atomic and $\beta \leq \alpha$, then $\vdash^{\alpha+\beta+1}_{m} \Gamma \Rightarrow \chi$, with $m = \deg(\psi)$.

We prove ii and ii' simultaneously by induction on β.

ii. Assume

$$\vdash^{\beta-1}_{0} \Gamma, \varphi \rightarrow \psi \Rightarrow \varphi \text{ and } \vdash^{\beta-1}_{0} \Gamma, \varphi \rightarrow \psi, \psi \Rightarrow \chi.$$

Then we have $\vdash^{\alpha}_{0} \Gamma, \psi \Rightarrow \chi$ (since $\beta \leq \alpha$). Hence induction hypothesis ii' yields

$$\vdash^{\alpha+\beta}_{\deg(\psi)} \Gamma \Rightarrow \chi.$$

ii'. *Case* $\beta = 0$. Then $\varphi \in \Gamma$, hence

$$\vdash^{\alpha+1}_{\deg(\psi)} \Gamma \Rightarrow \chi.$$

Case $\vdash^{\beta-1}_{0} \Gamma, \varphi \rightarrow \psi \Rightarrow \varphi$ *and* $\vdash^{\beta-1}_{0} \Gamma, \varphi \rightarrow \psi, \psi \Rightarrow \varphi$. Then the claim follows immediately from the induction hypothesis.

Case $\vdash_0^{\beta-1} \Gamma, \varphi \to \psi \Rightarrow \vartheta$ *and* $\vdash_0^{\beta-1} \Gamma, \varphi \to \psi, \eta \Rightarrow \varphi$ *with* $\vartheta \to \eta \in \Gamma$ *and* $\vartheta \to \eta \neq \varphi \to \psi$. Then induction hypothesis ii yields

$$\vdash_{\deg(\psi)}^{\alpha+\beta-1} \Gamma \Rightarrow \vartheta,$$

and induction hypothesis ii' yields

$$\vdash_{\deg(\psi)}^{\alpha+\beta} \Gamma, \eta \Rightarrow \chi.$$

Now $\to$-left gives

$$\vdash_{\deg(\psi)}^{\alpha+\beta+1} \Gamma \Rightarrow \chi.$$

iii. By induction on β. Assume

$$\vdash_0^{\beta-1} \Gamma, (\varphi \to \psi) \to \chi \Rightarrow \varphi \to \psi \text{ and } \vdash_0^{\beta-1} \Gamma, (\varphi \to \psi) \to \chi, \chi \Rightarrow \vartheta.$$

From $\vdash_0^{\alpha} \Gamma \Rightarrow (\varphi \to \psi) \to \chi$ we get $\vdash_0^{\alpha} \Gamma, \varphi \to \psi \Rightarrow \chi$, hence $\vdash_0^{\alpha} \Gamma, \psi \Rightarrow \chi$, hence $\vdash_0^{\alpha+1} \Gamma \Rightarrow \psi \to \chi$. On the other hand, the Inversion Lemma, Parts i and iii, yields $\vdash_0^{\beta-1} \Gamma, \varphi, \psi \to \chi \Rightarrow \psi$, hence $\vdash_0^{\beta} \Gamma, \psi \to \chi \Rightarrow \varphi \to \psi$. So

$$\vdash_{\deg(\psi \to \chi)}^{\alpha+\beta+1} \Gamma \Rightarrow \varphi \to \psi.$$

Furthermore, from $\vdash_0^{\beta-1} \Gamma, (\varphi \to \psi) \to \chi, \chi \Rightarrow \vartheta$ we get $\vdash_0^{\beta-1} \Gamma, \chi \Rightarrow \vartheta$. Since $\vdash_0^{\alpha} \Gamma, \varphi \to \psi \Rightarrow \chi$, a cut yields

$$\vdash_{\deg(\chi)}^{\alpha+\beta} \Gamma, \varphi \to \psi \Rightarrow \vartheta.$$

Now one further cut gives $\vdash_m^{\alpha+\beta+2} \Gamma \Rightarrow \vartheta$. □

We now prove the Cut Elimination Theorem, by induction on α. Assume

$$\vdash_k^{\alpha-1} \Gamma \Rightarrow \chi \text{ and } \vdash_k^{\alpha-1} \Gamma, \chi \Rightarrow \varphi$$

with $m + 1 = k + \deg(\chi)$. If $k \neq 0$, then by induction hypothesis

$$\vdash_{k-1}^{2(\alpha-1)} \Gamma \Rightarrow \chi \text{ and } \vdash_{k-1}^{2(\alpha-1)} \Gamma, \chi \Rightarrow \varphi,$$

hence by cut $\vdash_m^{2\alpha} \Gamma \Rightarrow \varphi$, since $k - 1 + \deg(\chi) = m$. If $k = 0$, then the claim follows from the Cut Elimination Lemma together with (1). □

An interesting consequence of the fact that we have a (Kalmar) elementary bound on the length of the cut free derivation given by our algorithm in terms of the original derivation is the following: The cut elimination algorithm $d \mapsto d^{\mathrm{cf}}$ just described is essentially different from normalization $d \mapsto d^{\mathrm{nf}}$, in the sense that there cannot exist elementary translations $d \mapsto d^{\mathrm{seq}}$, $d \mapsto d^{\mathrm{nat}}$ from derivations in natural deductions to derivations in the sequent calculus and back, such that $d^{\mathrm{nf}} = ((d^{\mathrm{seq}})^{\mathrm{cf}})^{\mathrm{nat}}$. For then $d \mapsto d^{\mathrm{nf}}$ would be elementary, which it isn't by the counterexample in Section 1.4.

1.8. A decision algorithm for implicational logic

It is clearly decidable whether a given formula φ is derivable in minimal implicational logic: just search systematically for a normal derivation of φ. This search must terminate, since by the subformula property there are only finitely many such normal derivations. However it does not seem to be a good idea to try to implement this algorithm

Here we present another decision algorithm which is easy to implement and also seems to be rather efficient in cases of interest. It also amounts to searching for a "normal" proof, but now in a special calculus LH due to (Hudelmaier 1989), designed with the intention that most rules should be invertible. Again our formulation of LH and most proofs are taken from (Buchholz 1989).

The calculus LH is again a sequent calculus. To distinguish it from Gentzen's sequent calculus discussed in Section 1.7 we write

$$\vdash^{\alpha}_{LH} \Gamma \Rightarrow \varphi$$

if the sequent $\Gamma \Rightarrow \varphi$ is derivable with height $\leq \alpha$ in LH. The rules of LH are the following; again we write Γ, φ for $\Gamma \cup \{\varphi\}$

Axiom. $\vdash^{\alpha}_{LH} \Gamma, \varphi \Rightarrow \varphi$ for φ atomic.

H$\rightarrow$*-right.* If $\vdash^{\alpha}_{LH} \Gamma, \varphi \Rightarrow \varphi$, then $\vdash^{\alpha+1}_{LH} \Gamma \Rightarrow \varphi \rightarrow \psi$.

H$\rightarrow$*-left-atomic.* If $\vdash^{\alpha}_{LH} \Gamma, \varphi, \psi \Rightarrow \chi$ and φ is atomic, then $\vdash^{\alpha+1}_{LH} \Gamma, \varphi, \varphi \rightarrow \psi \Rightarrow \chi$.

H$\rightarrow$*-left-*$\rightarrow$. If $\vdash^{\alpha}_{LH} \Gamma, \varphi, \psi \rightarrow \chi \Rightarrow \psi$ and $\vdash^{\alpha}_{LH} \Gamma, \chi \Rightarrow \vartheta$, then $\vdash^{\alpha+1}_{LH} \Gamma, (\varphi \rightarrow \psi) \rightarrow \chi \Rightarrow \vartheta$.

Hudelmaier has observed — and we will prove it below — that this calculus is equivalent to minimal implicational logic. Now the point in these strange rules is that they are all invertible, with the sole exception of the last one which is only "half invertible":

Inversion Lemma 1.8.1.
i. If $\vdash^{\alpha}_{LH} \Gamma \Rightarrow \varphi \rightarrow \psi$, *then* $\vdash^{\alpha}_{LH} \Gamma, \varphi \Rightarrow \psi$.
ii. If $\vdash^{\alpha}_{LH} \Gamma, \varphi, \varphi \rightarrow \psi \Rightarrow \chi$ *and* φ *is atomic, then* $\vdash^{\alpha}_{LH} \Gamma, \varphi, \psi \Rightarrow \chi$.
iii. If $\vdash^{\alpha}_{LH} \Gamma, (\varphi \rightarrow \psi) \rightarrow \chi \Rightarrow \vartheta$, *then* $\vdash^{\alpha}_{LH} \Gamma, \chi \Rightarrow \vartheta$. □

Clearly the last rule H$\rightarrow$-left-$\rightarrow$ cannot be fully invertible. A counterexample is

$$P, ((Q \rightarrow \bot) \rightarrow \bot) \rightarrow Q \Rightarrow P,$$

which is clearly derivable, whereas

$$P, Q \rightarrow \bot, \bot \rightarrow Q \Rightarrow \bot$$

is not. Now the decision algorithm derived from the Inversion Lemma runs as follows. Given a sequent $\Gamma \Rightarrow \varphi$, first apply Parts i and ii of the Inversion Lemma as long as possible. If you end up with a sequent which does not contain left-iterated implications $(\varphi \rightarrow \psi) \rightarrow \chi$, then by the form of the LH-rules it is derivable if and only if it is an axiom. Now assume there are some left-iterated implications $(\varphi \rightarrow \psi) \rightarrow \chi$ among the premiss-formulas Γ. Choose one of them (this step may lead to backtracking!), form its premisses according to the rule H$\rightarrow$-left-$\rightarrow$, and continue with both sequents.

An example for the necessity of backtracking is the sequent

$$\neg\neg Q \to Q, \neg\neg(P \to Q), P \Rightarrow Q.$$

If the second occurence of a left–iterated implication is choosen, i.e. $\neg\neg(P \to Q)$, we obtain by H→–left–→

$$\neg\neg Q \to Q, P \to Q, \bot \to \bot, P \Rightarrow \bot$$

and

$$\neg\neg Q \to Q, \bot, P \Rightarrow Q.$$

Now the first of these sequents is clearly underivable, hence this path in the search–tree fails, and we have to backtrack and choose the other left–iterated implication $\neg\neg Q \to Q$ instead.

It remains to be shown that the calculus LH is equivalent to minimal implicational logic. One direction is easy, namely that any sequent derivable in LH is also derivable in minimal implicational logic. We only consider the rule H→–left–→, and argue informally. So assume Γ and $(\varphi \to \psi) \to \chi$. Then clearly $\psi \to \chi$ (for if we assume ψ, we certainly have $\varphi \to \psi$ hence χ). So by the first premiss $\varphi \to \psi$, hence χ by our assumption, hence ϑ by the second premiss.

For the other direction we need a Lemma.

Lemma 1.8.2. *If* $\vdash^{\alpha}_{LH} \Gamma, \varphi \to \psi \Rightarrow \varphi$ *and* $\vdash_{LH} \Gamma, \psi \Rightarrow \vartheta$ *and* φ *is atomic, then* $\vdash_{LH} \Gamma, \varphi \to \psi \Rightarrow \vartheta$.

The proof is by induction on α.

1. Assume $\varphi \in \Gamma$. Then from $\vdash_{LH} \Gamma, \psi \to \vartheta$ we get $\vdash_{LH} \Gamma, \varphi \to \psi \Rightarrow \vartheta$ by H→–left–atomic.

2. Let $\Gamma = \Delta, (\varphi_1 \to \psi_1) \to \chi_1$ and assume that

$$\vdash^{\alpha}_{LH} \Delta, (\varphi_1 \to \psi_1) \to \chi_1, \varphi \to \psi \Rightarrow \varphi$$

was inferred from

$$\vdash^{\alpha-1}_{LH} \Delta, \varphi_1, \psi_1 \to \chi_1, \varphi \to \psi \Rightarrow \psi_1 \tag{1.2}$$

and

$$\vdash^{\alpha-1}_{LH} \Delta, \chi_1, \varphi \to \psi \Rightarrow \varphi \tag{1.3}$$

by H→–left–→. First note that from the second premiss of the Lemma we get

$$\vdash_{LH} \Delta, \chi_1, \psi \Rightarrow \vartheta \tag{1.4}$$

by the Inversion Lemma, Part iii. Now from 1.3 and 1.4 we obtain by the induction hypothesis

$$\vdash_{LH} \Delta, \chi_1, \varphi \to \psi \Rightarrow \vartheta. \tag{1.5}$$

The rule H→–left–→ yields form 1.2 and 1.5

$$\vdash_{LH} \Delta, (\varphi_1 \to \psi_1) \to \chi_1, \varphi \to \psi \Rightarrow \vartheta.$$

3. Let $\Gamma = \Delta, \varphi_1, \varphi_1 \to \psi_1$ with φ_1 atomic and different from φ, and assume that

$$\vdash^{\alpha}_{LH} \Delta, \varphi_1, \varphi_1 \to \psi_1, \varphi \to \psi \Rightarrow \varphi$$

was inferred from

$$\vdash^{\alpha-1}_{LH} \Delta, \varphi_1, \psi_1, \varphi \to \psi \Rightarrow \varphi \tag{1.6}$$

by H→–left–atomic. First note that from the second premiss of the Lemma we get

$$\vdash_{LH} \Delta, \varphi_1, \psi_1, \psi \Rightarrow \vartheta \tag{1.7}$$

by the Inversion Lemma, Part ii. Now from 1.6 and 1.7 we obtain by the induction hypothesis

$$\vdash_{LH} \Delta, \varphi_1, \psi_1, \varphi \to \psi \Rightarrow \vartheta. \tag{1.8}$$

An application of H→–left–atomic to 1.8 yields

$$\vdash_{LH} \Delta, \varphi_1, \varphi_1 \to \psi_1, \varphi \to \psi \Rightarrow \vartheta.$$

□

Theorem 1.8.3. *If* $\vdash \Gamma \Rightarrow \vartheta$, *then* $\vdash_{LH} \Gamma \Rightarrow \vartheta$.

Proof. By the Cut Elimination Theorem in Section 1.7 it suffices to prove that $\vdash^{\alpha}_{0} \Gamma \Rightarrow \vartheta$ implies $\vdash_{LH} \Gamma \Rightarrow \vartheta$. This is done by induction on α. Since the claim is obvious for the rules *Axiom* and *→–right* and since *Cut* cannot occur, we only treat the rule *→–left*.

Case 1. Let $\Gamma = \Delta, \varphi \to \psi$ with φ atomic and assume that

$$\vdash^{\alpha}_{0} \Delta, \varphi \to \psi \Rightarrow \vartheta$$

was inferred from

$$\vdash^{\alpha-1}_{0} \Delta, \varphi \to \psi \Rightarrow \vartheta \tag{1.9}$$

and

$$\vdash^{\alpha-1}_{0} \Delta, \varphi \to \psi, \psi \Rightarrow \vartheta. \tag{1.10}$$

From 1.10 we get by the Inversion Lemma 1.7.3, Part ii

$$\vdash^{\alpha-1}_{0} \Delta, \psi \Rightarrow \vartheta. \tag{1.11}$$

By induction hypothesis we can replace $\vdash^{\alpha-1}_{0}$ in 1.9 and 1.11 by $\vdash_{LH}$. Now the Lemma yields

$$\vdash_{LH} \Delta, \varphi \to \psi \Rightarrow \vartheta.$$

Case 2. Let $\Gamma = \Delta, (\varphi \to \psi) \to \chi$ and assume that

$$\vdash^{\alpha}_{0} \Delta, (\varphi \to \psi) \to \chi \Rightarrow \vartheta$$

was inferred from

$$\vdash^{\alpha-1}_{0} \Delta, (\varphi \to \psi) \to \chi \Rightarrow \varphi \to \psi \tag{1.12}$$

and

$$\vdash^{\alpha-1}_{0} \Delta, (\varphi \to \psi) \to \chi, \chi \Rightarrow \vartheta. \tag{1.13}$$

From 1.12 we get by the Inversion Lemma 1.7.3, Part iii

$$\vdash_0^{\alpha-1} \Delta, \varphi, \psi \to \chi \Rightarrow \varphi \to \psi$$

and hence

$$\vdash_0^{\alpha-1} \Delta, \varphi, \psi \to \chi \Rightarrow \psi. \tag{1.14}$$

From 1.13 we get by the Inversion Lemma 1.7.3, Part ii

$$\vdash_0^{\alpha-1} \Delta, \chi \Rightarrow \vartheta. \tag{1.15}$$

By induction hypothesis we can replace $\vdash_0^{\alpha-1}$ in 1.14 and 1.15 by $\vdash_{LH}$. Now H→-left-→ yields

$$\vdash_{LH} \Delta, (\varphi \to \psi) \to \chi \Rightarrow \vartheta. \square$$

It is also possible to prove the Theorem directly for natural deductions in minimal implicational logic. We sketch the proof. So let a normal derivation of ϑ from assumptions Γ be given. We may assume that in any branch (see Section 2.4) of this normal derivation the minimal formula (see Section 2.4) is atomic, and use induction on the length of this derivation.

Case 1. $\Gamma = \Delta, \varphi \to \psi$ with φ atomic. In

$$\begin{array}{c}\dfrac{\varphi \to \psi \qquad \begin{array}{c}|\\ \varphi\end{array}}{\psi}\\ |\\ \vartheta\end{array}$$

we can apply the induction hypothesis to the subderivations of φ from $\Delta, \varphi \to \psi$ and of ϑ from Δ, ψ (any assumption $\varphi \to \psi$ here can be cancelled, since we already have assumed ψ). So we get

$$\vdash_{LH} \Delta, \varphi \to \psi \Rightarrow \varphi \quad \text{and} \quad \vdash_{LH} \Delta, \psi \to \vartheta,$$

and the claim follows by the Lemma.

Case 2. $\Gamma = \Delta, (\varphi \to \psi) \to \chi$. Replace an uppermost occurrence of the assumption $(\varphi \to \psi) \to \chi$

$$\begin{array}{c}\dfrac{(\varphi \to \psi) \to \chi \qquad \dfrac{\dfrac{\begin{array}{c}|\\ \varphi \to \psi\end{array} \qquad \varphi}{\psi}}{\varphi \to \psi}}{\chi}\\ |\\ \vartheta\end{array}$$

by

$$\dfrac{\varphi \to \psi \qquad \dfrac{\begin{matrix} | \\ \varphi \to \psi \end{matrix} \qquad \varphi}{\psi}}{\begin{matrix} \chi \\ | \\ \vartheta \end{matrix}}$$

Apply the induction hypothesis to the subderivation of ψ from $\Delta, \varphi, \psi \to \chi$, and of ϑ from Δ, χ (any assumption $(\varphi \to \psi) \to \chi$ here can be cancelled, since we already have assumed χ). So we get

$$\vdash_{LH} \Delta, \varphi, \psi \to \chi \Rightarrow \psi \quad \text{and} \quad \vdash_{LH} \Delta, \chi \Rightarrow \vartheta,$$

and the claim follows by an application of H$\to$-left-$\to$.

2. Normalization for first–order logic

We restrict our attention to the $\to\forall$–fragment of first–order logic with just introduction and elimination rules for both symbols, i.e. with minimal logic formulated in natural deduction style. This restriction does not mean a loss in generality, since it is well known that full classical first–order can be embedded in this system; the argument for that fact is sketched in Section 2.1. Equality is not treated as a logical symbol, but can be added via suitable equality axioms.

We extend our results and estimates on normalization to first–order logic by the method of collapsing types. Applications include the subformula property, Herbrand's theorem and the interpolation theorem.

2.1. The $\to\forall$–fragment as a typed λ–calculus

Assume that a fixed (at most countable) supply of function variables $f, g, h, \ldots$ and predicate variables $P, Q, \ldots$ is given, each with an arity ≥ 0. *Terms* are built up from object variables x, y, z by means of $fr_1 \ldots r_m$. *Formulas* are built up from prime formulas $Pr_1 \ldots r_m$ by means of $(\varphi \to \psi)$ and $\forall x\varphi$. *Derivations* are built up from assumption variables x^φ, y^φ by means of the rule $\to^+$ of implication introduction

$$(\lambda x^\varphi r^\psi)^{\varphi\to\psi},$$

the rule $\to^-$ of implication elimination

$$(t^{\varphi\to\psi} s^\varphi)^\psi,$$

the rule $\forall^+$ of $\forall$–introduction

$$(\lambda x r^\varphi)^{\forall x\varphi},$$

provided that no assumption variable y^ψ free in r^φ has x free in its type ψ, and finally the rule $\forall^-$ of $\forall$–elimination

$$(t^{\forall x\varphi} s)^{\varphi_x[s]}.$$

Each of the rules $\to^+, \forall^+$ and $\forall^-$ has a uniquely determined derivation as its *premiss*, whereas $\to^-$ has the two derivations $t^{\varphi\to\psi}$ and s^φ as premisses. Here $t^{\varphi\to\psi}$ is called the *main premiss* and s^φ is called the *side premiss*.

As an example we give a derivation of

$$\forall x(Px \to Qx) \to (\forall x Px \to \forall x Qx).$$

Such a derivation is

$$\lambda u^{\forall x(Px\to Qx)} \lambda v^{\forall x Px} \lambda x((ux)(vx)).$$

Derivations can be easily written in the more usual tree form. We will continue to use the word term for derivations (as long as this does not lead to confusion with the notion of (object) term inherent in first–order logic), and type for formula.

Note that our ($\to\forall$–fragment of) minimal logic contains full classical first–order logic. This can be seen as follows:

1. Choose a particular propositional variable and denote it by $\bot$ (falsity). Associate with any formula φ in the language of classical first–order logic a finite list $\vec{\varphi}$ of formulas in our $\to\forall$–fragment, by induction on φ:

$$\begin{aligned}
P\vec{r} &\mapsto P\vec{r}\\
\neg\varphi &\mapsto \vec{\varphi} \to \bot\\
\varphi \to \psi &\mapsto \vec{\varphi} \to \psi_1, \ldots, \vec{\varphi} \to \psi_n\\
\varphi \wedge \psi &\mapsto \vec{\varphi}, \vec{\psi}\\
\varphi \vee \psi &\mapsto (\vec{\varphi} \to \bot), (\vec{\psi} \to \bot) \to \bot\\
\forall x\varphi &\mapsto \forall x\varphi_1, \ldots, \forall x\varphi_m\\
\exists x\varphi &\mapsto \forall x(\vec{\varphi} \to \bot) \to \bot
\end{aligned}$$

2. In any model $\mathcal{M}$ where $\bot$ is interpreted by falsity, we clearly have that a formula φ in the language of full first–order logic holds under an assignment α iff all formulas in the assigned sequence $\vec{\varphi}$ hold under α.

3. Our derivation calculus for the $\to\forall$–fragment is complete in the following sense: A formula φ is derivable from stability assumptions

$$\forall\vec{x}(\neg\neg P\vec{x} \to P\vec{x})$$

for all predicate symbols P in φ iff φ is valid in any model under any assignment.

2.2. Strong normalization

Here we use the method of collapsing types (cf.(Troelstra and van Dalen 1988,p.560)) to transfer our results and estimates concerning strong normalization from implicational logic to first–order logic.

The notions concerning conversion introduced in Section 1.2 can be easily extended to first–order logic. In particular, we have

$$(\lambda\vec{x}x.r)\vec{s}s \quad \text{converts into} \quad (\lambda\vec{x}.r_x[s])\vec{s},$$

where the variables $\vec{x}, x$ now can be either assumption variables or else object variables. The rules generating the relation $r \to r'$ are extended by requiring $r \to r$ for object terms r of our first–order logic. Again a derivation is said to be in *normal form* if it does not contain a convertible subderivation.

For any formula φ of first–order logic we define its *collapse* φ^c by

$$(P\vec{r})^c \equiv P$$
$$(\varphi \to \psi)^c \equiv \varphi^c \to \psi^c$$
$$(\forall x\varphi)^c \equiv \top \to \varphi^c$$

where $\top :\equiv \bot \to \bot$ with $\bot$ a fixed propositional variable (i.e. $\top$ means truth). The *level* of a formula φ of first–order logic is defined to be the level of its collapse φ^c. For any derivation r^ψ in first–order logic we can now define its *collapse* $(r^\psi)^c$. It is plain from this definition that for any derivation r^ψ in first–order logic with free assumption variables $x_1^{\varphi_1}, \ldots, x_m^{\varphi_m}$ the collapse $(r^\psi)^c$ is a derivation $(r^c)^{\psi^c}$ in implicational logic with free assumption variables $x_1^{\varphi_1^c}, \ldots, x_m^{\varphi_m^c}$.

$$(x^\varphi)^c \equiv x^{\varphi^c}$$
$$(\lambda x^\varphi r)^c \equiv \lambda x^{\varphi^c} r^c$$
$$(t^{\varphi\to\psi}s)^c \equiv t^c s^c$$
$$(\lambda x r)^c \equiv \lambda x^\top r^c$$
$$(t^{\forall x\varphi}s)^c \equiv t^c(\lambda z^\bot z^\bot)^\top$$

Note that for any derivation r^ψ, assumption variable x^φ and derivation s^φ we have that $r^c[s^c]$ is a derivation in implicational logic (where the substitution of s^c is done for the assumption variable x^{φ^c}), which is the collapse of $r[s]$. Also for any derivation r^ψ, object variable x and term s we have that $r_x[s]$ is a derivation of $\psi_x[s]$ with collapse $(r_x[s])^c \equiv r^c$.

Lemma 2.2.1. *If $r \rightarrow_1 r'$ in first-order logic, then $r^c \rightarrow_1 (r')^c$ in implicational logic.*

The proof is by induction on the generation of $r \rightarrow_1 r'$. We only treat the case

$$(\lambda x r)s \rightarrow_1 r_x[s].$$

If x is an assumption variable, then

$$\begin{aligned}((\lambda x^{\varphi} r^{\psi})s^{\varphi})^c &\equiv (\lambda x^{\varphi^c} r^c)s^c \\ &\rightarrow_1 r^c[s^c] \\ &\equiv (r[s])^c),\end{aligned}$$

by the note above. If x is an object variable, then

$$\begin{aligned}((\lambda x^{\varphi})s)^c &\equiv (\lambda x^{\top} r^c)(\lambda z^{\perp} z^{\perp})^{\top} \\ &\rightarrow_1 r^c \\ &\equiv (r[s])^c),\end{aligned}$$

again by the note above. □

Hence from Theorem 1.5.1 we can conclude

Theorem 2.2.2. *Any derivation r in first-order logic is strongly normalizable.* □

Also we can apply Theorem 1.6.1 to obtain an upper bound for the length of arbitrary reduction sequences.

Theorem 2.2.3. *Let r be a derivation in first-order logic of a formula of level 0, i.e. a prime formula. Let r^c be the collapse of r into implicational logic. Let m be a bound for the levels of subterms of r^c and $k \geq 2$ be a bound for the arities of subterms of r^c. Then the length of an arbitrary reduction sequence for r with respect to $\rightarrow_1$ is bounded by*

$$k^{2_m(m+2\cdot\mathrm{height}(r^c)+2k+2)}$$

2.3. Uniqueness

The Church–Rosser Theorem and hence the uniqueness of the normal form for derivations in first-order logic can be proved exactly as in Section 1.3. We do not repeat this here.

2.4. Applications

Here we want to draw some conclusions from the fact that any derivation in first-order logic can be transformed into normal form. The arguments in this section are based on Prawitz' book (Prawitz 1965). We begin with an analysis of the form of normal derivations.

Let a derivation r be given. A sequence $r_1, \ldots, r_m$ of subderivations of r is a *branch* if

1. r_1 is an assumption variable,
2. r_i is the main premiss of r_{i+1}, and
3. r_m is either the whole derivation r or else the side premiss of an instance of the rule $\to^-$ within r.

It is obvious that any subderivation of r belongs to exactly one branch. The *order* of the branch ending with the whole derivation r is defined to be 0, and if the order of the branch through the main premise t of some instance $t^{\varphi\to\psi}s^{\varphi}$ of the rule $\to^-$ in r is k, then the order of the branch ending with that s^{φ} is defined to be $k+1$.

The relation "φ is a *subformula* of ψ" is defined to be the transitive and reflexive closure of the relation "immediate subformula", defined by

1. φ and ψ are immediate subformulas of $\varphi \to \psi$,
2. $\varphi_x[r]$ is an immediate subformula of $\forall x \varphi$.

We will also need the notion "φ is a *strictly positive subformula* of ψ", which is defined to be the transitive and reflexive closure of the relation "immediate strictly positive subformula", defined by

1. ψ is an immediate strictly positive subformula of $\varphi \to \psi$,
2. $\varphi_x[r]$ is an immediate strictly positive subformula of $\forall x \varphi$.

In a normal derivation r any branch $r_1^{\varphi_1}, \ldots, r_m^{\varphi_m}$ has a rather perspicious form: all elimination rules must come before all introduction rules. Hence, if i is maximal such that $r_i^{\varphi_i}$ ends with an elimination rule, then φ_i must be a strictly positive subformula of all φ_j for $j \neq i$. This φ_i is called the *minimal formula* of the branch. Also, any φ_j with $j \leq i$ is a strictly positive subformula of φ_1, and any φ_j with $j \geq i$ is a strictly positive subformula of φ_m.

Theorem 2.4.1. (Subformula property) *If r^{φ} is a normal derivation with free assumption variables among $x_1^{\varphi_1}, \ldots, x_m^{\varphi_m}$ and s^{ψ} is a subderivation of r^{φ}, then ψ is a subformula of φ or of some φ_i.*

The proof is by induction on the order of branches in r, using the property of branches in normal derivations mentioned above. □

We write $\varphi_1, \ldots, \varphi_m \vdash \varphi$ to mean that there is a derivation r^{φ} with free assumption variables among $x_1^{\varphi_1}, \ldots, x_m^{\varphi_m}$.

Theorem 2.4.2. (Herbrand) *Assume that $\forall \vec{x}_1 \varphi_1, \ldots, \forall \vec{x}_m \varphi_m \vdash \psi$ with quantifier-free $\varphi_1, \ldots, \varphi_m, \psi$. Then we can find $\vec{r}_{11}, \ldots, \vec{r}_{1n_1}, \ldots, \vec{r}_{m1}, \ldots, \vec{r}_{mn_m}$ such that*

$$\varphi_1[\vec{r}_{11}], \ldots, \varphi_1[\vec{r}_{1n_1}], \ldots, \varphi_m[\vec{r}_{m1}], \ldots, \varphi_m[\vec{r}_{mn_m}] \vdash \psi$$

Proof. To simplify notation let us assume $\forall x\varphi \vdash \psi$ with quantifier–free φ, ψ. By Section 2.2 we can construct from the given derivation a normal derivation r^{ψ} with free assumption variables among $x^{\forall x\varphi}$. By induction on the order of branches it is easy to see that any branch must end with the derivation of a quantifier–free formula and must begin with the rule $\forall^-$, i. e. with $x^{\forall x\varphi}r_i$. Now replace any such subderivation by $y_i^{\varphi[r_i]}$, with new assumption variables y_i. □

Our next application is the Craig interpolation theorem. We shall use the notation $\varphi_1, \ldots, \varphi_m \vdash^c \varphi$ (c for classical) to mean that there is a derivation r^{φ} with free assumption variables among $x_1^{\varphi_1}, \ldots, x_m^{\varphi_m}$ and some stability assumptions $y^{\forall \vec{x}(\neg\neg P\vec{x} \to P\vec{x})}$ for P predicate variable in $\vec{\varphi}, \varphi$, where again $\neg\psi$ denotes $\psi \to \bot$ with a fixed propositional variable $\bot$.

Theorem 2.4.3. (Interpolation) *Assume* $\Gamma, \Delta \vdash \varphi$. *Then we can find a finite list* $\vec{\gamma}$ *of formulas such that*

$$\Gamma \vdash^c \vec{\gamma} \text{ and } \vec{\gamma}, \Delta \vdash^c \varphi$$

(where $\Gamma \vdash^c \vec{\gamma}$ *means* $\Gamma \vdash^c \gamma_i$ *for each* γ_i *in* $\vec{\gamma}$*), and any object or predicate variable free in* $\vec{\gamma}$ *occurs free both in* Γ *and in* Δ, φ.

For the proof we shall use a somewhat more explicit formulation of the theorem: Let r^{φ} be a derivation with free assumption variables among $\vec{u}^{\Gamma}, \vec{v}^{\Delta}$. Then we can find a finite list $\vec{r_1}^{\vec{\gamma}}$ of derivations with free assumption variables among $\vec{u}^{\Gamma}$ and stability assumptions and a derivation r_2^{φ} with free assumption variables among $\vec{y}^{\vec{\gamma}}, \vec{v}^{\Delta}$ and stability assumptions, such that any object or predicate variable free in $\vec{\gamma}$ occurs free both in Γ and in Δ, φ.

For brevity we shall not mention stability assumptions any more (they will only be used in Case 2*b*(*ii*) below), and write "r^{φ} with $\vec{u}^{\Gamma}$", to mean the derivation r^{φ} with free assumption variables among $\vec{u}^{\Gamma}$.

The proof is by induction on the height of the given derivation, which by Section 2.2 we can assume to be normal. We distinguish two cases according to whether it ends with an introduction rule (i. e. $\to^+$ or $\forall^+$) or with an elimination rule.

Case 1a. $(\lambda x^{\varphi} r^{\psi})^{\varphi \to \psi}$ with $\vec{u}^{\Gamma}, \vec{v}^{\Delta}$. By induction hypothesis for r^{ψ} with $x^{\varphi}, \vec{u}^{\Gamma}, \vec{v}^{\Delta}$ we have $\vec{r}_1^{\vec{\gamma}}$ with $\vec{u}^{\Gamma}$ and r_2^{ψ} with $\vec{y}^{\vec{\gamma}}, x^{\varphi}, \vec{v}^{\Delta}$. An application of $\to^+$ to the latter derivation yields $(\lambda x^{\varphi} r_2^{\psi})^{\varphi \to \psi}$ with $\vec{y}^{\vec{\gamma}}, \vec{v}^{\Delta}$.

Case 1b. $(\lambda x r^A)^{\forall x\varphi}$ with $\vec{u}^{\Gamma}, \vec{v}^{\Delta}$, where x is not free in Γ, Δ. By induction hypothesis for r^{φ} with $\vec{u}^{\Gamma}, \vec{v}^{\Delta}$ we have $\vec{r}_1^{\vec{\gamma}}$ with $\vec{u}^{\Gamma}$ and r_2^{φ} with $\vec{y}^{\vec{\gamma}}, \vec{v}^{\Delta}$. Since x is not free in Γ, we know that x is not free in $\vec{\gamma}$. An application of $\forall^+$ to the latter derivation yields $(\lambda x r_2^{\varphi})^{\forall x\varphi}$ with $\vec{y}^{\vec{\gamma}}, \vec{v}^{\Delta}$.

Case 2a. $(w^{\chi \to \vartheta} s^{\chi} \vec{t})^A$ with $\vec{u}^{\Gamma}, \vec{v}^{\Delta}$.

Subcase i. $w^{\chi \to \vartheta}$ is among $\vec{u}^{\Gamma}$. By induction hypothesis for s^{χ} with $\vec{u}^{\Gamma}, \vec{v}^{\Delta}$ we have $\vec{s}_1^{\vec{\gamma}}$ with $\vec{v}^{\Delta}$ and s_2^{χ} with $\vec{y}^{\vec{\gamma}}, \vec{u}^{\Gamma}$. By induction hypothesis for $(u^{\vartheta}\vec{t})^{\varphi}$ with $u^{\vartheta}, \vec{u}^{\Gamma}, \vec{v}^{\Delta}$ we have $\vec{t}^{\vec{\delta}}$ with $u^{\vartheta}, \vec{u}^{\Gamma}$ and t_2^{φ} with $\vec{z}^{\vec{\delta}}, \vec{v}^{\Delta}$. From these derivations we obtain

$$(\lambda \vec{y}^{\vec{\gamma}} (\vec{t}_1^{\vec{\delta}})_u [w^{\chi \to \vartheta} s_2^{\chi}])^{\vec{\gamma} \to \vec{\delta}} \quad \text{with} \quad \vec{u}^{\Gamma}$$

and

$$(t_2^{\varphi})_z [\vec{x}^{\vec{\gamma} \to \vec{\delta}} \vec{s}_1^{\vec{\gamma}}] \quad \text{with} \quad x^{\vec{\gamma} \to \vec{\delta}}, \vec{v}^{\Delta},$$

where $\vec{\gamma} \to \vec{\delta}$ means $\vec{\gamma} \to \delta_1, \ldots, \vec{\gamma} \to \delta_n$.

Subcase ii. $w^{\chi\to\vartheta}$ is among $\vec{v}^{\Delta}$. By induction hypothesis for s^{χ} with $\vec{u}^{\Gamma}, \vec{v}^{\Delta}$ we have $s_1^{\vec{\gamma}}$ with $\vec{u}^{\Gamma}$ and s_2^{χ} with $\vec{y}^{\vec{\gamma}}, \vec{v}^{\Delta}$. By induction hypothesis for $(u^{\vartheta}\vec{t})^{\varphi}$ with $u^{\vartheta}, \vec{u}^{\Gamma}, \vec{v}^{\Delta}$ we have $t_1^{\vec{\delta}}$ with $\vec{u}^{\Gamma}$ and t_2^{φ} with $\vec{z}^{\vec{\delta}}, u^{\vartheta}, \vec{v}^{\Delta}$. From these derivations we obtain

$$(\vec{s}_1, \vec{t}_1)^{\vec{\gamma},\vec{\delta}} \quad \text{with} \quad \vec{u}^{\Gamma}$$

and

$$(t_2^{\varphi})_u[w^{\chi\to\vartheta} s_2^{\chi}] \quad \text{with} \quad \vec{y}^{\vec{\gamma}}, \vec{z}^{\vec{\delta}}, \vec{v}^{\Delta}.$$

Case 2b. $w^{\forall x\chi} s\vec{t}$ with $\vec{u}^{\Gamma}, \vec{v}^{\Delta}$.

Subcase i. $w^{\forall x\chi}$ is among $\vec{u}^{\Gamma}$. By induction hypothesis for $(u^{\chi[s]}\vec{t})^{\varphi}$ with $u^{\chi[s]}$, $\vec{u}^{\Gamma}$, $\vec{v}^{\Delta}$ we have $t_1^{\vec{\gamma}}$ with $u^{\chi[s]}, \vec{u}^{\Gamma}$ and t_2^{φ} with $\vec{y}^{\vec{\gamma}}, \vec{v}^{\Delta}$. Let $\vec{z}$ be all variables free in $\vec{\gamma}$ that are in s, but not free in Γ. We now construct derivations

$$(\lambda \vec{z}(\vec{t}_1^{\vec{\gamma}})_u[w^{\forall x\chi}s])^{\forall\vec{z}\vec{\gamma}} \quad \text{with} \quad \vec{u}^{\Gamma}$$

and

$$(t_2^{\varphi})_{\vec{y}}[\vec{x}^{\forall\vec{z}\vec{\gamma}}\vec{z}] \quad \text{with} \quad \vec{x}^{\forall\vec{z}\vec{\gamma}}, \vec{v}^{\Delta},$$

where $\forall\vec{z}\vec{\gamma}$ means $\forall\vec{z}\gamma_1, \ldots, \forall\vec{z}\gamma_m$. Note that any object or predicate variable free in $\forall\vec{z}\vec{\gamma}$ is both free in Δ, φ and free in Γ.

Subcase ii. $w^{\forall x\chi}$ is among $\vec{v}^{\Delta}$. By induction hypothesis for $(u^{\chi[s]}\vec{t})^{\varphi}$ with $u^{\chi[s]}$, $\vec{u}^{\Gamma}$, $\vec{v}^{\Delta}$ we have $\vec{t}_1^{\vec{\delta}}$ with $\vec{u}^{\Gamma}$ and t_2^{φ} with $\vec{y}^{\vec{\delta}}, u^{C[s]}, \vec{v}^{\Delta}$. Let $\vec{z}$ be all variables free in $\vec{\delta}$ that are in s, but not free in Δ, φ. We now construct derivations

$$(\lambda v^{\forall\vec{z}\neg\vec{\delta}}(v\vec{z}\vec{t}_1))^{\neg\forall\vec{z}\neg\vec{\delta}} \quad \text{with} \quad \vec{u}^{\Gamma}$$

and

$$(t^{\neg\neg\varphi\to\varphi}\lambda v^{\neg\varphi}.x^{\neg\forall\vec{z}\neg\vec{\gamma}}\lambda\vec{z}\lambda\vec{y}(v(t_2^{\varphi})_u[w^{\forall x\chi}s]))^{\varphi} \quad \text{with} \quad x^{\neg\forall\vec{z}\neg\vec{\delta}}, \vec{v}^{\Delta}$$

and stability assumptions (which are used to build $t^{\neg\neg\varphi\to\varphi}$). Note again that any object or predicate variable free in $\neg\forall\vec{z}\neg\vec{\delta}$ is both free in Γ and free in Δ, φ. □

3. Normalization for arithmetic

3.1. Ordinal notations

We want to discuss the derivability and underivability of initial cases of transfinite induction in arithmetical systems. In order to do that we shall need some knowledge and notations for ordinals. Now we do not want to assume set theory here; hence we introduce a certain initial segment of the ordinal (the ordinals $< \varepsilon_0$) in a formal, combinatorial way, i.e. via ordinal notations. Our treatment is based on the Cantor normal form for ordinals; cf. (Bachmann 1955). We also introduce some elementary relations and operations for such ordinal notations, which will be used later.

We define the two notions

- α is an ordinal notation
- $\alpha < \beta$ for ordinal notations α, β

simultaneously by induction:

1. If $\alpha_m, \dots, \alpha_0$ are ordinal notations and $\alpha_m \geq \dots \geq \alpha_0$ (where $\alpha \geq \beta$ means $\alpha > \beta$ or $\alpha = \beta$), then
$$\omega^{\alpha_m} + \dots + \omega^{\alpha_0}$$
is an ordinal notation. Note that the empty sum denoted by 0 is allowed here.
2. If $\omega^{\alpha_m} + \dots + \omega^{\alpha_0}$ and $\omega^{\beta_n} + \dots + \omega^{\beta_0}$ are ordinal notations, then
$$\omega^{\alpha_m} + \dots + \omega^{\alpha_0} < \omega^{\beta_n} + \dots + \omega^{\beta_0}$$
iff there is an $i \geq 0$ such that $\alpha_{m-i} < \beta_{n-i}$, $\alpha_{m-i+1} = \beta_{n-i+1}, \dots, \alpha_m = \beta_n$, or else $m < n$ and $\alpha_m = \beta_n, \dots, \alpha_0 = \beta_{n-m}$

It is easy to see (by induction on the levels in the inductive definition) that $<$ is a linear order with 0 being the smallest element.

We shall use the notation 1 for ω^0, a for $\omega^0 + \dots + \omega^0$ with a copies of ω^0 and $\omega^\alpha a$ for $\omega^\alpha + \dots + \omega^\alpha$ again with a copies of ω^α.

We now define addition for ordinal notations:

$$\omega^{\alpha_m} + \dots + \omega^{\alpha_0} + \omega^{\beta_n} + \dots + \omega^{\beta_0} := \omega^{\alpha_m} + \dots + \omega^{\alpha_i} + \omega^{\beta_n} + \dots + \omega^{\beta_0}$$

where i is minimal such that $\alpha_i \geq \beta_n$.

It is easy to see that $+$ is an associative operation which is strictly monotonic in the second argument and weakly monotonic in the first argument. Note that $+$ is not commutative: $1 + \omega = \omega \neq \omega + 1$.

The *natural* (or Hessenberg) sum of two ordinal notations is defined by

$$(\omega^{\alpha_m} + \dots + \omega^{\alpha_0}) \# (\omega^{\beta_n} + \dots + \omega^{\beta_0}) := \omega^{\gamma_{m+n}} + \dots + \omega^{\gamma_0},$$

where $\gamma_{m+n}, \dots, \gamma_0$ is a decreasing permutation of $\alpha_m, \dots, \alpha_0, \beta_n, \dots, \beta_0$.

Again it is easy to see that # is associative, commutative and strictly monotonic in both arguments.

We will also need to know how ordinal notations of the form $\beta + \omega^\alpha$ can be approximated from below. First note that

$$\delta < \alpha \rightarrow \beta + \omega^\delta a < \beta + \omega^\alpha.$$

Furthermore, for any $\gamma < \beta + \omega^\alpha$ we can find a $\delta < \alpha$ and an a such that

$$\gamma < \beta + \omega^\delta a.$$

We now define 2^α for ordinal notations α. Let $\alpha_m \geq \cdots \alpha_0 \geq \omega > k_n \geq \cdots \geq k_1 > 0$. Then

$$2^{\omega^{\alpha_m}+\cdots+\omega^{\alpha_0}+\omega^{k_n}+\cdots+\omega^{k_1}+\omega^0 a} := \omega^{\omega^{\alpha_m}+\cdots+\omega^{\alpha_0}+\omega^{k_n-1}+\cdots+\omega^{k_1-1}} 2^a.$$

It is easy to see that $2^{\alpha+1} = 2^\alpha + 2^\alpha$ and that 2^α is strictly monotonic in α.

In order to work with ordinal notations in a purely arithmetical system we set up a bijection between ordinal notations and nonnegative integers (i.e., a Gödel numbering). For its definition it is useful to refer to ordinal notations in the form

$$\omega^{\alpha_m} a_m + \cdots + \omega^{\alpha_0} a_0 \quad \text{with} \quad \alpha_m > \cdots > \alpha_0.$$

For any ordinal notation α we define its Gödel number $|\alpha|$ inductively by

$$|0| := 0,$$

$$|\omega^{\alpha_m} a_m + \cdots + \omega^{\alpha_0} a_0| := \Big(\prod_{i \leq m} p_{|\alpha_i|}^{a_i}\Big) - 1.$$

For any nonnegative integer x we define its corresponding ordinal notation $o(x)$ inductively by

$$o(0) = 0$$

$$o\Big(\Big(\prod_{i \leq m} p_i^{a_i}\Big) - 1\Big) = \sum_{i \leq m} \omega^{o(i)} a_i$$

where the sum is to be understood as the natural sum.

Lemma 3.1.1.
1. $o(|\alpha|) = \alpha$,
2. $|o(x)| = x$.

This can be proved easily by induction. □

Hence we have a bijection between ordinal notations and nonnegative integers. Using this bijection we can transfer our relations and operations on ordinal notations to computable relations and operations on nonnegative integers. We will use the notations

$$x \prec y \quad \text{for} \quad o(x) < o(y),$$
$$\omega^x \quad \text{for} \quad |\omega^{o(x)}|,$$
$$x \oplus y \quad \text{for} \quad |o(x) + o(y)|.$$

3.2. Provability of initial cases of transfinite induction

We now set up some formal systems of arithmetic and derive initial cases of the principle of transfinite induction in them, i.e. of

$$\forall x(\forall y \prec x : Py \rightarrow Px) \rightarrow \forall x \prec a : Px$$

for some numeral a and a predicate variable P. In Section 3.4 we will see that our results here are optimal in the sense that for larger segments of the ordinals transfinite induction is underivable. All these results are due to (Gentzen 1943).

Our arithmetical systems are based on a fixed (possibly countably infinite) supply of function constants and predicate constants which are assumed to denote fixed functions and predicates on the nonnegative integers for which a computation procedure is known. Among the function constants there must be a constant S for the successor function and 0 for (the 0–place function) zero. Among the predicate constants there must be a constant $=$ for equality and $\bot$ for (the 0–place predicate) falsity. In order to formulate the general principle of transfinite induction we also assume that predicate variables $P, Q, \ldots$ are present.

Terms are built up from object variables x, y, z by means of $fr_1 \ldots r_m$, where f is a function constant. We identify closed terms which have the same value; this is a convenient way to express in our formal systems the assumption that for each function constant a computation procedure is known. Terms of the form $SS \ldots S0$ are called *numerals*. We use the notation S^i0 or even i for them. *Formulas* are built up from prime formulas $Pr_1 \ldots r_m$ with P a predicate constant or a predicate variable by means of $(\varphi \rightarrow \psi)$ and $\forall x\varphi$. As usual we abbreviate $\varphi \rightarrow \bot$ by $\neg\varphi$.

The *axioms* of our arithmetical systems will always include the Peano–axioms

$$\forall xy(Sx = Sy \rightarrow x = y),$$

$$\forall x(Sx \neq 0).$$

Any instance of the induction scheme

$$\varphi[0], \forall x(\varphi[x] \rightarrow \varphi[Sx]) \rightarrow \forall x\varphi[x]$$

with φ an arbitrary formula is an axiom of full arithmetic Z. We will also consider subsystems Z_k of Z where the formulas φ in the induction scheme are restricted to Π^0_k–formulas; the latter notion is defined inductively, as follows.

1. Any prime formula $P\vec{r}$ is a Π^0_k–formula, for any $k \geq 1$.
2. If φ is quantifier–free and ψ is a Π^0_k–formula, then $\varphi \rightarrow \psi$ is a Π^0_k–formula.
3. If φ is a Π^0_k–formula and ψ is a Π^0_l–formula, then $\varphi \rightarrow \psi$ is a Π^0_p–formula with $p = \max(k+1, l)$.
4. If φ is a Π^0_k–formula, then so is $\forall x\varphi$.

Note that a formula is a Π^0_k–formula iff it is logically equivalent to a formula with a prefix of k alternating quantifiers beginning with $\forall$ and a quantifier–free kernel. For example, $\forall x\exists y\forall zPxyz$ is a Π^0_3–formula. In addition, in any arithmetical system we have the equality axioms

$$\forall x(x = x),$$

$$\forall \vec{x}\vec{y}(x_1 = y_1, \ldots, x_m = y_m \to f\vec{x} = f\vec{y}),$$

$$\forall \vec{x}\vec{y}(x_1 = y_1, \ldots, x_m = y_m, P\vec{x} \to P\vec{y})$$

for any function constant f and predicate constant or predicate variable P. We also require for any such P the stability axioms

$$\forall \vec{x}(\neg\neg P\vec{x} \to P\vec{x}).$$

We express our assumption that for any predicate constant a decision procedure is known by adding the axiom

$$P(S^{i_1}0)\ldots(S^{i_m}0)$$

whenever $P\vec{i}$ is true, and

$$\neg P(S^{i_1}0)\ldots(S^{i_m}0)$$

whenever $P\vec{i}$ is false.

We finally allow in any of our arithmetical systems an arbitrary supply of true Π^0_1–formulas as axioms. Our (positive and negative) results concerning initial cases of transfinite recursion will not depend on which of those axioms we have chosen, except that for the positive results we always assume

$$\forall x(x \not\prec 0) \tag{3.1}$$

$$\forall yz(z \prec y \oplus \omega^0, z \not\prec y, z \neq y \to \bot) \tag{3.2}$$

$$\forall x(x \oplus 0 = x) \tag{3.3}$$

$$\forall xyz(x \oplus (y \oplus z) = (x \oplus y) \oplus z) \tag{3.4}$$

$$\forall x(0 \oplus x = x) \tag{3.5}$$

$$\forall x(\omega^x 0 = 0) \tag{3.6}$$

$$\forall xy(\omega^x(Sy) = \omega^x y \oplus \omega^x) \tag{3.7}$$

$$\forall xyz(z \prec y \oplus \omega^x, x \neq 0 \to z \prec y \oplus \omega^{fxyz}(gxyz)) \tag{3.8}$$

$$\forall xyz(z \prec y \oplus \omega^x, x \neq 0 \to fxyz \prec x) \tag{3.9}$$

where in 3.9 f and g are function constants.

Theorem 3.2.1. (Gentzen) *Transfinite induction up to ω_n (with $\omega_1 := \omega, \omega_{n+1} := \omega^{\omega_n}$) i.e. the formula*

$$\forall x(\forall y \prec x : \varphi[y] \to \varphi[x]) \to \forall x \prec \omega_n : \varphi[x]$$

is derivable in Z.

Proof: To any formula φ we assign a formula φ^+ (with respect to a fixed variable x) by

$$\varphi^+ :\equiv \forall y(\forall z \prec y : \varphi_x[z] \to \forall z \prec y \oplus \omega^x : \varphi_x[z]).$$

We first show

$$\varphi \text{ is progressive } \to \varphi^+ \text{ is progressive,}$$

where "ψ is *progressive*" means $\forall x(\forall y \prec x : \psi[y] \to \psi[x])$. So assume that φ is progressive and

$$\forall y \prec x : \varphi^+[y]. \tag{3.10}$$

We have to show $\varphi^+[x]$. So assume further

$$\forall z \prec y : \varphi[z] \tag{3.11}$$

and $z \prec y \oplus \omega^x$. We have to show $\varphi[z]$. Case $x = 0$. From $z \prec y \oplus \omega^0$ we have by 3.2 $z \prec y \vee z = y$. If $z \prec y$, then $\varphi[z]$ follows from 3.11, and if $z = y$, then $\varphi[z]$ follows from 3.11 and the progressiveness of φ. Case $x \neq 0$. From $z \prec y \oplus \omega^x$ we obtain $z \prec y \oplus \omega^{fxyz} gxyz$ by 3.8 and $fxyz \prec x$ by 3.9. From 3.10 we obtain $\varphi^+[fxyz]$. By the definition of φ^+ we get

$$\forall u \prec y \oplus \omega^{fxyz} v : \varphi[u] \to \forall u \prec (y \oplus \omega^{fxyz} v) \oplus \omega^{fxyz} : \varphi[u]$$

and hence, using 3.4 and 3.7

$$\forall u \prec y \oplus \omega^{fxyz} v : \varphi[u] \to \forall u \prec y \oplus \omega^{fxyz}(Sv) : \varphi[u].$$

Also from 3.11 and 3.6, 3.3 we obtain

$$\forall u \prec y \oplus \omega^{fxyz} 0 : \varphi[u].$$

Using an appropriate instance of the induction scheme we can conclude

$$\forall u \prec y \oplus \omega^{fxyz} gxyz : \varphi[u]$$

and hence $\varphi[z]$.

We now show, by induction on n, how to obtain a derivation of

$$\forall x(\forall y \prec x : \varphi[y] \to \varphi[x]) \to \forall x \prec \omega_n : \varphi[x].$$

So assume the left–hand side, i.e. assume that φ is progressive. Case 0. From $x \prec \omega_0$ we get $x = 0$ by 3.5, 3.2 and 3.1, and $\varphi[0]$ follows from the progressiveness of φ by 3.1. Case $n+1$. Since φ is progressive, by what we have shown above also φ^+ is progressive. Applying the induction hypothesis to φ^+ yields $\forall x \prec \omega_n : \varphi^+[x]$, and hence $\varphi^+[\omega_n]$ by the progressiveness of $\varphi^+[x]$. Now the definition of φ^+ (together with 3.1 and 3.5) yields $\forall z \prec \omega^{\omega_n} : \varphi[z]$. □

Note that in these derivations the induction scheme was used for formulas of unbounded complexity.

We now want to refine Theorem 3.2.1 to a corresponding result for the subsystems Z_k of Z. Note first that if φ is a Π^0_k–formula, then the formula φ^+ constructed in the proof of Theorem 3.2.1 is a Π^0_{k+1}–formula, and for the proof of

$$\varphi \text{ is progressive } \to \varphi^+ \text{ is progressive}$$

we have used induction with a Π^0_k induction formula.

Now let φ be a Π^0_1–formula, and let $\varphi^0 :\equiv \varphi, \varphi^{i+1} :\equiv (\varphi^i)^+$. Then φ^k is a Π^0_{k+1}–formula, and hence in Z_k we can derive that if φ is progressive, then also $\varphi^1, \varphi^2, \ldots \varphi^k$ are all progressive. Let $\omega_1[m] := m, \omega_{i+1}[m] = \omega^{\omega_i[m]}$. Since in Z_k we can derive that φ^k is progressive, we can also derive $\varphi^k[0], \varphi^k[1], \varphi^k[2]$ and generally $\varphi^k[m]$ for any m, i.e. $\varphi^k[\omega_1[m]]$. But since

$$\varphi^k \equiv (\varphi^{k-1})^+ \equiv \forall y(\forall z \prec y : \varphi^{k-1}[z] \to \forall z \prec y \oplus \omega^x : \varphi^{k-1}[z]),$$

we first get (with $y = 0$) $\forall z \prec \omega_2[m] : \varphi^{k-1}[z]$ and then $\varphi^{k-1}[\omega_2[m]]$ by the progressiveness of φ^{k-1}. Repeating this argument we finally obtain $\varphi^0[\omega_{k+1}[m]]$. Hence we have

Theorem 3.2.2. *Let φ be a Π^0_1–formula. Then in Z_k we can derive transfinite induction for φ up to $\omega_{k+1}[m]$ for any m, i.e.*

$$Z_k \vdash \forall x(\forall y \prec x : \varphi[y] \to \varphi[x]) \to \forall x \prec \omega_{k+1}[m] : \varphi[x].\square$$

If more generally we start out with a Π^0_l–formula φ instead, where $1 \leq l \leq k$, then a similar argument yields the following result of (Parsons 1973)

Theorem 3.2.3. *Let φ be a Π^0_l–formula, $1 \leq l \leq k$. Then in Z_k we can derive transfinite induction for φ up to $\omega_{k+2-l}[m]$ for any m, i.e.*

$$Z_k \vdash \forall x(\forall y \prec x : \varphi[y] \to \varphi[x]) \to \forall x \prec \omega_{k+2-l}[m] : \varphi[x].\square$$

Our next aim is to prove that these bounds are sharp. More precisely, we will show that in Z (no matter how many true Π^0_1–formulas we have added as axioms) one cannot derive transfinite induction up to ε_0, i.e. the formula

$$\forall x(\forall y \prec x : Py \to Px) \to \forall x Px$$

with a free predicate variable P, and that in Z_k one cannot derive transfinite induction up to ω_{k+1}, i.e. the formula

$$\forall x(\forall y \prec x : Py \to Px) \to \forall x \prec \omega_{k+1} : Px.$$

This will follow from the method of normalization applied to arithmetical systems, which we have to develop first.

3.3. Normalization for arithmetic with the ω–rule

We will show in Section 3.5 that a normalization theorem does not hold for a system of arithmetic like Z in Section 3.2, in the sense that for any formula φ derivable in Z there is a derivation of the same formula φ in Z which only uses formulas of a level bounded by the level of φ. The reason for this failure is the presence of the induction axioms, which can be of arbitrary level.

Here we remove that obstacle against normalization and replace the induction axioms by a rule with infinitely many premisses, the so–called ω–rule (suggested by Hilbert and studied by Lorenzen, Novikov and Schütte), which allows to conclude $\forall x\varphi[x]$ from $\varphi[0]$, $\varphi[1]$, $\varphi[2]$, ...

Clearly this ω–rule can also be used to replace the rule $\forall^+$. As a consequence we do not need to consider free object variables.

So we introduce the system Z^∞ of ω–arithmetic as follows. Z^∞ has the same language and — apart from the induction axioms — the same axioms as Z. Derivations in Z^∞ are infinite objects; they are built up from assumption variables x^φ, y^ψ and constants ax^φ for any axiom φ of Z other than an induction axiom by means of the rules

$$(\lambda x^\varphi r^\psi)^{\varphi\to\psi}$$

$$(t^{\varphi\to\psi} s^\varphi)^\psi$$

$$\langle r_i^{\varphi[i]} \rangle_{i<\omega}^{\forall x\varphi}$$

$$(t^{\forall x\varphi} i)^{\varphi[i]}$$

denoted by $\to^+, \to^-, \omega$ and $\forall^-$, respectively.

More precisely, we define the notion of an $\vec{x}$–derivation (i. e. a derivation in Z^∞ with free assumption variables among $\vec{x}$) of height $\leq \alpha$ and degree $\leq k$ inductively, as below.

Note that derivations are infinite objects now. They may be viewed as mappings from finite sequences of natural numbers (= nodes in the derivation tree) to lists of data including the formula appearing at that node, the rule applied last, a list of assumption variables including all those free in the subderivation (starting at that node), a bound on the height of the subderivation, and a bound on the degree of the subderivation.

Intuitively, the degree of a derivation is the least number $\geq$ the level of any subderivation λxr in a context $(\lambda xr)s$ or $\langle r_i \rangle_{i<\omega}$ in a context $\langle r_i \rangle_{i<\omega} j$, where the level of a derivation is the level of its type, i.e. the formula it derives. This notion of a degree is needed for the normalization proof we give below.

* Any assumption variable x^φ and any axiom ax^φ is an $\vec{x}$–derivation of height $\leq \alpha$ and degree $\leq k$, for any list $\vec{x}$ of assumption variables (containing x in the first case), ordinal α and number k.

$\to^+$ If r^ψ is an $\vec{x}, x, \vec{y}$–derivation of height $\leq \alpha_0 < \alpha$ and degree $\leq k$, then $(\lambda x^\varphi r^\psi)^{\varphi\to\psi}$ is an $\vec{x}, \vec{y}$–derivation of height $\leq \alpha$ and degree $\leq k$.

$\to^-$ If $t^{\varphi\to\psi}$ and s^φ are $\vec{x}$–derivations of heights $\leq \alpha_i < \alpha$ and degrees $\leq k_i \leq k$ $(i = 1, 2)$, then $(t^{\varphi\to\psi} s^\varphi)^\psi$ is an $\vec{x}$–derivation of height $\leq \alpha$ and degree $\leq m$ with $m = \max(k, \mathrm{lev}(\varphi \to \psi))$, if $t^{\varphi\to\psi}$ is generated by the rule $\to^+$, or of degree $\leq k$ otherwise.

ω If $r_i^{\varphi[i]}$ are $\vec{x}$–derivations of heights $\leq \alpha_i < \alpha$ and degrees $\leq k_i \leq k$ $(i < \omega)$, then $\langle r_i^{\varphi[i]} \rangle_{i<\omega}^{\forall x\varphi}$ is an $\vec{x}$–derivation of height $\leq \alpha$ and degree $\leq k$.

$\forall^-$ If $t^{\forall x\varphi}$ is an $\vec{x}$–derivation of height $\leq \alpha_0 < \alpha$ and degree $\leq k$, then $(t^{\forall x\varphi} i)^{\varphi[i]}$ is an $\vec{x}$–derivation of height $\leq \alpha$ and degree $\leq m$ with $m = \max(k, \mathrm{level}\ \forall x\varphi)$, if $t^{\forall x\varphi}$ is generated by the rule ω, or of degree $\leq k$ otherwise.

We now embed our systems Z_k (i. e. arithmetic with induction restricted to Π_k^0–formulas) and hence Z into Z^∞.

Lemma 3.3.1. *Let r^ψ be a derivation in Z_k with free assumption variables among $\vec{x}^{\vec{\varphi}}$ which contains $\leq m$ instances of the induction scheme all with induction formulas of level $\leq k$. Let σ be a substitution of numerals for object variables such that $\vec{\varphi}\sigma, \psi\sigma$ do not contain free object variables. Then we can find an $\vec{x}^{\vec{\varphi}\sigma}$–derivation $(r_\infty)^{\psi\sigma}$ in Z^∞ of height $\leq \omega^m + h$ for some $h < \omega$ and degree $\leq k$.*

Proof. First note that from any normal derivation in first–order logic we can construct a normal derivation r_0^ψ with the same free assumption variable $\vec{x}^{\vec{\varphi}}$, such that in r_0^ψ any branch has a prime formula as its minimal formula (cf. Section 2.4). For if φ is a minimal formula which is not prime we can first apply elimination rules until a prime formula is reached and later build α up again by the corresponding introduction rules.

The lemma is proved by induction on the height of the given derivation r. By the Normalization Theorem 2.2.3 and the note above we can assume that r is normal with prime minimal formulas. The only case which requires some argument is when r consists of two applications of $\to^-$ to an instance of the induction scheme. Then r must have the form

$$\mathrm{ax}^{\varphi[0],\forall x(\varphi[x]\to\varphi[Sx])\to\forall x\varphi[x]}s^{\varphi[0]}(\lambda x\lambda y^{\varphi[x]}t^{\varphi[Sx]}).$$

By induction hypothesis we obtain derivations

$$\begin{array}{rll} s_\infty^{\varphi[0]} & \text{of height} & \leq \omega^{m-1}+h_0 \\ t_\infty^{\varphi[1]}[s_\infty^{\varphi[0]}] & \text{of height} & \leq \omega^{m-1}\cdot 2+h_1, \\ t_\infty^{\varphi[2]}[t_\infty^{\varphi[1]}[s_\infty^{\varphi[0]}]] & \text{of height} & \leq \omega^{m-1}\cdot 3+h_2 \end{array}$$

and so on, all of degree $\leq k$. Combining all these derivations of $\varphi[i]$ as premisses of the ω–rule yields a derivation t_∞ of $\forall x\varphi[x]$ of height $\leq \omega^m$ and degree $\leq k$. □

A derivation is called *convertible* if it is of the form $(\lambda x r)s$ or else $\langle r_i\rangle_{i<\omega}j$, which can be converted into $r_x[s]$ or r_j, respectively. Here $r_x[s]$ is obtained from r by substituting s for all free occurences of x in r. A derivation is called *normal* if it does not contain a convertible subderivation. Note that a derivation of degree 0 must be normal.

We want to define an operation which by repeated conversions transforms a given derivation into a normal one with the same end formula and no more assumption variables. The methods employed in Sections 1 and 2 to achieve such a task have to be adapted properly in order to deal with the new situation of infinitary derivations. Here we give a particularly simple argument due to (Tait 1965).

Lemma 3.3.2. *If r is an $\vec{x}, x^\varphi, \vec{y}$–derivation of height $\leq \alpha$ and degree $\leq k$ and s^φ is an $\vec{x}, \vec{y}$–derivation of height $\leq \beta$ and degree $\leq l$, then $r_x[s]$ is an $\vec{x}, \vec{y}$–derivation of height $\leq \beta+\alpha$ and degree $\leq \max(k, l, \text{level } s)$.*

This is proved by a straightforward induction on the height of r. □

Lemma 3.3.3. *For any $\vec{x}$–derivation r^φ of height $\leq \alpha$ and degree $\leq k+1$ we can find an $\vec{x}$–derivation $(r^k)^\varphi$ of height $\leq 2^\alpha$ and degree $\leq k$.*

The proof is by induction on α. The only case which requires some argument is when r is of the form ts with t of height $\leq \alpha_1 < \alpha$ and s of height $\leq \alpha_2 < \alpha$. We first consider the subcase where $t^k = \lambda x t_1$ and $\text{lev}(t) = k+1$. Then $\text{lev}(s) \leq k$ by the definition of level, and hence $(t_1)_x[s^k]$ has degree $\leq k$ by Lemma 3.3.2. Furthermore, also by Lemma 3.3.2, $(t_1)_x[s^k]$ has height $\leq 2^{\alpha_2}+2^{\alpha_1} \leq 2^{\max(\alpha_2,\alpha_1)+1} \leq 2^\alpha$. Hence we can take $(ts)^k$ to be $(t_1)_x[s^k]$. If we are not in the above subcase, we can simply take $(ts)^k$ to be $t^k s^k$. This derivation clearly has height $\leq 2^\alpha$. Also it has degree $\leq k$, which can be seen as follows. If $\text{lev}(t) \leq k$ we are done. If however $\text{lev}(t) \geq k+2$, then t must be of the form $t_0t_1\dots t_m$ for some assumption variable or axiom t_0 (since r has degree $\leq k+1$). But then t^k has the form $t_0t_1^k\dots t_m^k$ and we are done again. (To be completely precise, this last statement has to be added to the formulation of the Lemma above and proved simultaneously with it). □

As an immediate consequence we obtain

Theorem 3.3.4. (Normalization for Z^∞) *For any $\vec{x}$–derivation r^φ of height $\leq \alpha$ and degree $\leq k$ we can find a normal $\vec{x}$–derivation $(r^*)^\varphi$ of height $\leq 2_k\alpha$ (where $2_0\alpha = \alpha, 2_{m+1}\alpha = 2^{2_m^\alpha}$).*

3.4. Unprovable initial cases of transfinite induction

We now apply the technique of normalization for arithmetic with the ω–rule for a proof that transfinite induction up to ε_0 is underivable in Z, i.e. of

$$Z \nvdash \forall x(\forall y \prec x : Py \to Px) \to \forall x Px$$

with a predicate variable P, and that transfinite induction up to ω_{k+1} is underivable in Z_k, i.e. of

$$Z_k \nvdash \forall x(\forall y \prec x : Py \to Px) \to \forall x \prec \omega_{k+1} : Px.$$

Our proof is based on an idea of Schütte, which consists in adding a so–called ***progression rule*** to the infinitary systems. This rule allows to conclude Pj (where j is any numeral) from all Pi for $i \prec j$.

More precisely, we define the notion of an $\vec{x}$–derivation in $Z^\infty +$ Prog(P) of height $\leq \alpha$ and degree $\leq k$ by the inductive clauses of Section 3.2 and the additional clause Prog(P):

> If r_i^{Pi} are $\vec{x}$–derivations of heights $\leq \alpha_i < \alpha$ and degrees $\leq k_i \leq k$ $(i \prec j)$, then $\langle r_i^{Pi}\rangle_{i \prec j}^{Pj}$ is an $\vec{x}$–derivation of height $\leq \alpha$ and degree $\leq k$.

Since this progression rule only deals with derivations of prime formulas it does not affect the degrees of derivations. Hence the proof of normalization for Z^∞ carries over unchanged to $Z^\infty +$ Prog(P). In particular we have

Lemma 3.4.1. *For any $\vec{x}$–derivation r^φ in $Z^\infty +$ Prog(P) of height $\leq \alpha$ and degree $\leq k+1$ we can find an $\vec{x}$–derivation $(r^k)^\varphi$ in $Z^\infty +$ Prog(P) of height $\leq 2^\alpha$ and degree $\leq k$.*

We now show that from the progression rule for P we can easily derive the progressiveness of P.

Lemma 3.4.2. *We have a normal derivation of $\forall x(\forall y \prec x : Py \to Px)$ in $Z^\infty +$ Prog(P) with height ≤ 5.*

Proof. By the ω–rule it suffices to derive $\forall y \prec j : Py \to Pj$ for any j with height ≤ 4. We argue informally. Assume $\forall y \prec j : Py$. By $\forall^-$ we have $i \prec j \to Pi$ for any i. Now for any $i \prec j$ we have $i \prec j$ as an axiom; hence Pi for any such i. An application of the progression rule yields Pj, with a derivation of height ≤ 3. Now by $\to^+$ and ω the claim follows. □

The crucial observation now is that a normal derivation of $P|\beta|$ must essentially have a height of at least β. However, to obtain the right estimates for our subsystems Z_k we cannot apply Lemma 3.4.1 down to degree 0 (i.e. to the normal form) but must stop already at degree 1. Such derivations, i.e. those of degree ≤ 1, will be called ***almost normal***; they can also be analyzed easily. An almost normal derivation r in $Z^\infty +$ Prog(P) is called a ***$P|\vec{\alpha}|, \neg P|\vec{\beta}|$–refutation*** if r derives a formula $\vec{\varphi} \to \psi$ with $\vec{\varphi}$ and the free assumptions in r among $P|\vec{\alpha}| := P|\alpha_1|, \ldots, P|\alpha_m|$ and $\neg P|\vec{\beta}| := \neg P|\beta_1|, \ldots, \neg P|\beta_n|$ and true prime formulas, and ψ a false prime formula or else among $\neg P|\vec{\beta}|$.

Lemma 3.4.3. *Let r be an almost normal $P|\vec{\alpha}|, \neg P|\vec{\beta}|$-refutation of height $\leq |r|$ with $\vec{\alpha}$ and $\vec{\beta}$ disjoint. Then*

$$\min \vec{\beta} \leq |r| + \#\vec{\alpha},$$

where $\#\vec{\alpha}$ denotes the number of ordinals in $\vec{\alpha}$.

Proof. By induction on $|r|$. Note that we may assume that r does not contain either ω or else $\forall^-$. Note also that r cannot be an equality axiom $\mathrm{ax}^{P|\gamma|,|\gamma|=|\delta| \to P|\delta|}$ with $\gamma = \delta$ true, since we have assumed that $\vec{\alpha}$ and $\vec{\beta}$ are disjoint. We distinguish cases according to the last rule in r.

Case $\to^+$. By our definition of refutations the claim follows immediately from the induction hypothesis.

Case $\to^-$. Then $r \equiv t^{\varphi \to (\vec{\vartheta} \to \psi)} s^{\varphi}$. If φ is a true prime formula, the claim follows from the induction hypothesis for t. If φ is a false prime formula, the claim follows from the induction hypothesis for s. If φ is $\neg\neg P|\gamma|$ (and hence $t \equiv \mathrm{ax}^{\forall x(\neg\neg Px \to Px)}|\gamma|$), then since the level of $\neg\neg P|\gamma|$ is 2 the derivation $s^{\neg\neg P|\gamma|}$ must end with an introduction rule, i. e. $s \equiv \lambda x^{\neg P|\gamma|} s_0^{\perp}$ (for otherwise, since no axiom contains some $\neg\neg Pr_0$ as a strictly positive subformula, we would get a contradiction against the assumption that r has degree ≤ 1). The claim now follows from the induction hypothesis for s_0. The only remaining case is when φ is $P|\gamma|$. Then t is an almost normal $P|\gamma|, P|\vec{\alpha}|, \neg P|\vec{\beta}|$ -refutation and s is an almost normal $P|\vec{\alpha}|, \neg P|\vec{\beta}|, \neg P|\gamma|$ -refutation. We may assume that γ is not among $\vec{\alpha}$, since otherwise the claim follows immediately from the induction hypothesis for t. Hence we have by the induction hypothesis for t

$$\min \vec{\beta} \leq |t| + \#\vec{\alpha} + 1 \leq |r| + \#\vec{\alpha}.$$

Case Prog(P). Then $r \equiv \langle r_\delta^{P|\delta|} \rangle_{\delta<\gamma}^{P|\gamma|}$. By induction hypothesis, since r_δ is a $P|\vec{\alpha}|$, $\neg P|\vec{\beta}|$, $\neg P|\delta|$ -refutation, we have for all $\delta < \gamma$

$$\min(\vec{\beta}, \delta) \leq |r_\delta| + \#\vec{\alpha} < |r| + \#\vec{\alpha}$$

and hence

$$\min(\vec{\beta}, \gamma) \leq |r| + \#\vec{\alpha}.$$

□

Now we can show the following result of (Mints 1971) and (Parsons 1973)

Theorem 3.4.4. *Transfinite induction up to ε_0 is underivable in Z, i.e.*

$$Z \nvdash \forall x(\forall y \prec x : Py \to Px) \to \forall x Px$$

with a predicate variable P, and transfinite induction up to ω_{k+1} is underivable in Z_k, i.e.

$$Z_k \nvdash \forall x(\forall x \prec x : Py \to Px) \to \forall x \prec \omega_{k+1} : Px.$$

Proof. We restrict ourselves to the second part. So assume that transfinite induction up to ω_{k+1} is derivable in Z_k. Then by the embedding of Z_k into Z^∞ (Lemma 3.3.1) and the normal derivability of the progressiveness of P in $Z^\infty + \mathrm{Prog}(P)$ with finite height (Lemma 3.4.2) we can conclude that $\forall x \prec \omega_{k+1} : Px$ is derivable in $Z^\infty + \mathrm{Prog}(P)$ with height $< \omega^m + h$ for some $m, h < \omega$ and degree $\leq k$. Now $k-1$ applications of Lemma 3.4.1 yield a derivation of the same formula $\forall x \prec \omega_{k+1} : Px$ in $Z^\infty + \mathrm{Prog}(P)$ with height $\leq \gamma \leq 2_{k-1}(\omega^m + h) < \omega_{k+1}$ and degree ≤ 1, hence also a derivation of $P|\gamma + 1|$ in $Z^\infty + \mathrm{Prog}(P)$ with height $\leq \gamma$ and degree ≤ 1. But this contradicts Lemma 3.4.3.

□

3.5. Normalization for arithmetic is impossible

The normalization theorem for first-order logic applied to arithmetic Z is not particularly useful since we may have used in our derivation induction axioms of arbitrary complexity. Hence it is tempting to first eliminate the induction scheme in favour of an induction rule allowing to conclude $\forall x\varphi[x]$ from a derivation of $\varphi[0]$ and a derivation of $\varphi[Sx]$ with an additional assumption $\varphi[x]$ to be cancelled at this point (note that this rule is equivalent to the induction scheme), and then to try to normalize the resulting derivation in the new system Z with the induction rule. We will apply our results from Section 3.4 to show that even a very weak form of the normalization theorem cannot hold in Z with the induction rule.

Theorem 3.5.1. *The following weak form of a normalization theorem for Z with the induction rule is false: For any $\vec{x}^{\vec{\varphi}}$-derivation r^{ψ} with $\vec{\varphi}, \psi$ Π^0_l-formulas there is an $\vec{x}^{\vec{\varphi}}$-derivation $(r^*)^{\psi}$ containing only Π^0_k-formulas, with k depending only on l.*

Proof. Assume that such a normalization theorem would hold. Consider the Π^0_3-formula

$$\forall x(\forall y \prec x : Py \to Px) \to \forall x \prec \omega_{n+1} : Px$$

expressing transfinite induction up to ω_{n+1}. By Theorem 3.2.1 it is derivable in Z. Hence there exists a derivation of the same formula containing only Π^0_k-formulas, for some k independent of n. Hence Z_k derives transfinite induction up to ω_{n+1} for any n. But this clearly contradicts Theorem 3.4.1. □

Bibliography

Bachmann, H.: Transfinite Zahlen. Berlin: Springer 1955

Barendregt, H.P.: The lambda calculus. Amsterdam: North-Holland 1984

Buchholz, W.: Bemerkungen zur Dissertation Hudelmaier. Unveröffentlichtes Manuskript, München: 1989

Church, A.: The calculi of lambda-conversion. Annals of Math. Studies No.6, Princeton: 1941

Diller, J.: Zur Berechenbarkeit primitiv-rekursiver Funktionale endlicher Typen. In: Contributions to mathematical logic (K. Schütte, ed.), pp. 109–120. Amsterdam: North-Holland 1968

Gandy, R.O.: An early proof of normalization. In: To H.B. Curry: Essays on combinatory logic, lambda calculus and formalism (J.P. Seldin and J.R. Hindley, eds.), pp. 453–455. Academic Press 1980a

Gandy, R.O.: Proofs of strong of normalization. In: To H.B. Curry: Essays on combinatory logic, lambda calculus and formalism (J.P. Seldin, J.R. Hindley, eds.) pp. 457–477. Academic Press 1980b

Gentzen, G.: Beweisbarkeit und Unbeweisbarkeit von Anfangsfällen der transfiniten Induktion in der reinen Zahlentheorie. Mathematische Annalen *119*, 140–161 (1943)

Girard, J.Y.: Proof theory and logical complexity. Napoli: Bibliopolis 1987

Howard, W.A.: The formulae–as–types notion of construction. In: To H.B. Curry: Essays on combinatory logic, lambda calculus and formalism (J.P. Seldin and J.R. Hindley, eds.), pp. 479–490. Academic Press 1980a

Howard, W.A.: Ordinal analysis of terms of finite type. J. Symbolic Logic *45*, 493–504 (1980b)

Hudelmaier, J.: Bounds for cut–elimination in intuitionistic propositional logic. Dissertation, Tübingen: 1989

Mints, G.E.: Exact estimates of the provability of transfinite induction in the initial segmants of arithmetic. Zapiski Nauch. Sem Leningrad *20*, 134–144 (1971). Translated in: J. Soviet Math. *1*, 85–91 (1973)

Parsons, C.: Transfinite induction in subsystems of number theory (abstract) J. Symbolic Logic *36*, 544–545 (1973)

Prawitz, D.: Natural deduction. Stockholm: Almqvist & Wiksell 1965

Sanchis, L.E.: Functionals defined by recursion. Notre Dame J. Formal Logic *8*, 161–174 (1967)

Schütte, K.: Proof theory. Berlin: Springer 1977

Schwichtenberg, H.: Proof theory: Some applications of cut–elimination. In: Handbook of mathematical logic (J. Barwise, ed.), pp. 867–895. Amsterdam: North–Holland 1977

Schwichtenberg, H.: Complexity of normalization in the pure typed λ–calculus. In: The L.E.J. Brouwer centenary symposium (A.S. Troelstra and D. van Dalen, eds.), pp. 453–458. Amsterdam: North–Holland 1982

Schwichtenberg, H.: A normal form for natural deductions in a type theory with realizing terms. In: Atti del congresso logica e filosofia della scienza, oggi. Vol.1–Logica. (V.M. Abrusci and E.Casari, eds.), pp. 95–138. Bologna: CLUEB 1986

Schwichtenberg, H.: An upper bound for reduction sequences in the typed λ–calculus. Archive Math. Logic, to appear 1990

Statman, R.: The typed λ–calculus is not elementary recursive. Theoretical Computer Science *9*, 73–81 (1979)

Tait, W.W.: Infinitely long terms of transfinite type. In: Formal systems and recursive functions (J. Crossley and M. Dummett, eds.), pp. 176–185. Amsterdam: North–Holland 1965

Takeuti, G.: Proof theory. Second edition. Amsterdam: North–Holland 1987

Troelstra, A., editor. Metamathematical investigations of intuitionistic arithmetic and analysis. Lecture Notes in Mathematics, Vol. 344, Berlin: Springer 1973

Troelstra, A. and van Dalen, D.: Constructivism in mathematics. An introduction. Studies in Logic and the Foundations of Mathematics, Vol. 121, 123 Amsterdam: North–Holland 1988

Computability - Logical and Recursive Complexity

STANLEY S. WAINER

School of Math. and CTCS, University of Leeds, U.K.

Introduction

The basis of this short course is the strong analogy between programs and proofs (of their specifications). The main theme is the classification of computable number-theoretic functions according to the logical complexity of their formal specification or termination proofs. A significant sub-branch of mathematical logic has grown around this theme since the 1950's and new ideas are presently giving rise to further developments. The methods employed are chiefly those from proof theory, particularly "normalization" as presented in the accompanying lectures of Helmut Schwichtenberg, and "ordinal assignments". Since program-termination corresponds to well-foundedness of computation trees, it is hardly surprising that transfinite ordinals and their constructive representations play a crucial role, measuring the logical complexity of programs and of the functions which they compute.

Note that we are concerned here solely with computable (i.e. recursive) number-theoretic functions

$$f : N^k \to N$$

where $N = \{0, 1, 2, 3, \dots\}$. Of course computation in general involves many different sorts of data-structures, but among these N is of basic foundational importance, its theory is rich and well-developed, and it therefore best illustrates the underlying ideas linking computation theory with proof theory.

When asked for the main outstanding problem in the theory of recursion, Gödel is reported to have answered: *classify the recursive functions!*

We are still trying; but what follows is a report on some *partial* success toward the classification problem.

The only way we know of constructing "new" recursive functions is by "diagonalizing" through the class already at hand. Thus in order to define a recursive function which is not provably specifiable in a given formal theory, we need to diagonalize through those specifications which *are* provable. We also need to know they are *true* and this presupposes the *consistency* of the theory. The lesson is that we can successfully (usefully) classify those recursive functions which are provably specifiable in formal theories for which "ordinal analyses of consistency strength" are established. This represents a very large class of recursive functions! - but by no means all!

During the preparation of this work the author was supported by Research Grant GR/D/90345 and Senior Fellowship B/ITF/216 from the UK Science and Engineering Research Council.

NATO ASI Series, Vol. F 79
Logic, Algebra, and Computation
Edited by F. L. Bauer

§1. Programs and Proofs

A number-theoretic ***program-specification*** is a so-called Π^0_2 - or $\forall E$ - statement of the form

$$\forall \ \textit{input } x \ \ \exists \ \textit{output } y \ \ R(x,y)$$

where $R(x,y)$ expresses the fact that the output y is "correctly related" to the input x.

A number-theoretic program $p(x) \Leftarrow (......)$ ***satisfies*** the above specification if for every input $x \in N$, the program computes an output $y = p(x)$ such that $R(x, p(x))$ is true.

Note that we are specifying ***totally-defined*** functions.

QUESTION 1. How does a given formal theory of arithmetic restrict the class of programs it can provably specify?

QUESTION 2. How can we synthesise programs from their specification proofs in a given formal theory?

Illustration. (Kreisel, Parsons, Mints, Takeuti...). *From specification-proofs in the theory of Σ^0_1-Induction, we can synthesise primitive recursive programs.*

Rather than giving a complete proof of this old established result (see e.g. Copestake-Wainer [1988]), we shall merely indicate the main proof-theoretic ideas underlying it in order to show up the clear and natural correspondence which exists, between proof-rules on one hand and recursive programming constructs on the other. A complete proof can be read off from the results in §4.

First we must describe the theory of Σ^0_1-*Induction*. It is a certain sub-theory of full first-order classical *Peano-Arithmetic* PA, which can be formalized in the following way (convenient for proof-theoretical analysis):

We derive not single formulas A, B, C standing alone, but in general ***finite sets*** of formulas $\Gamma = \{A_0, A_1, \ldots, A_m\}$ or $\Delta = \{B_o, B_1, \ldots, B_k\}$, the intuitive meaning being (A_0 **or** A_1 **or** ... A_m) etcetera. The single formula B is identified with the singleton set $\{B\}$ but we always omit $\{\}$ in such a case and simply write B. Thus Γ, A stands for $\Gamma \cup \{A\}$ and Δ, B, C stands for $\Delta \cup \{B\} \cup \{C\}$ etc.

The only ***terms*** of PA are the constant 0, the variables $x, y, z, \ldots$ and those which can be constructed from these by repeated applications of ***given*** function symbols such as S (for successor).

The ***atomic*** or ***prime formulas*** are constructed from terms by applying relation symbols which come in dual pairs: R and its complement $\bar{R}$.

In this way ***negation*** becomes a ***defined symbol*** using De Morgan's laws thus: $\neg R \equiv \bar{R}, \neg\bar{R} \equiv R, \neg(A \wedge B) \equiv \neg A \vee \neg B, \neg(A \vee B) \equiv \neg A \wedge \neg B, \neg\exists \equiv \forall x \neg, \neg\forall x \equiv \exists x \neg$. ***Implies*** is then also defined as $(A \rightarrow B) \equiv \{\neg A, B\}$.

Full Peano Arithmetic has

Logical axioms

$\Gamma, \neg A, A$ where A is atomic.

Arithmetical axioms

Γ, Δ where Δ is a set of prime formulas defining one of the *given* elementary relations. For example, to define the relation $R(x,y,z) \equiv$ "$x + y = z$" we need axioms:

$$\begin{array}{llll} \Gamma, & R(x,0,x) & & \\ \Gamma, & \bar{R}(x,y,z), & R(x,Sy,Sz) & \\ \Gamma, & \bar{R}(x,y,z), & \bar{R}(x,y,z'), & z = z' \end{array}$$

and all substitution instances of them.

Logical rules

$$\begin{array}{lllll} (\wedge) & \vdash \Gamma, (A \wedge B) & \text{if} & \vdash \Gamma, A \ \text{ and } \ \vdash \Gamma, B & \\ (\vee) & \vdash \Gamma, (A \vee B) & \text{if} & \vdash \Gamma, A \ \text{ or } \ \vdash \Gamma, B & \\ (\forall) & \vdash \Gamma, \forall x A(x) & \text{if} & \vdash \Gamma, A(y) & - y \text{ not free in } \Gamma \\ (\exists) & \vdash \Gamma, \exists x A(x) & \text{if} & \vdash \Gamma, A(t) & - \text{ some term } t \\ (Cut) & \vdash \Gamma & \text{if} & \vdash \Gamma, C \ \text{ and } \ \vdash \Gamma, \neg C. & \end{array}$$

The Induction rule

(Ind) $\vdash \Gamma, A(x)$ if $\vdash \Gamma, A(0)$ and $\vdash \Gamma, \neg A(x), A(Sx)$ where x is not a free variable in Γ.

The Theory of **Σ_1^0-Induction** has the same axioms and logical rules as PA above, but the rule of induction is restricted to apply not to *all* formulas $A(x)$ but only those in Σ_1^0- or $\exists$-form; for example $A(x) \equiv \exists y R(x,y)$.

Now we have set up the formal system, let us return to the Illustration:

Suppose we have a proof in Σ_1^0-Induction, of a specification:

$$\forall x \exists y R(x,y)$$

This is not a Σ_1^0-formula because of the $\forall x$ at the front. However it is a simple matter to check through all the rules and axioms and see that from any proof of $\Gamma, \forall x A(x)$ we can extract a proof of $\Gamma, A(z)$. This is called $\forall$-inversion.

Therefore from the original specification proof we can extract a proof of

$$\exists y R(x, y)$$

with x now a free variable.

Now what was the final rule applied in proving

$$\exists y R(x, y) \quad ?$$

It could only have been an $\exists$-rule, or an Induction-rule, or a Cut-rule. We therefore have just 3 cases to consider, though in fact the Cut-case is best treated as being split into 2 subcases.

In each case we exhibit a typical (goal-directed) proof of

$$\exists y R(x, y)$$

and from the proof we can immediately see how to extract/synthesize a program

$$p(x) \Leftarrow (.....)$$

which satisfies the specification.

The $\exists$-case

For example, the simplest application of $\exists$-rule would be

$$\begin{array}{ll} \exists y R(x, y) & \\ \quad | & \\ R(x, t(x)) & \text{some term } t. \end{array}$$

Here, the witnessing term already satisfies R for input x, so the extracted program is

$$\boxed{\textit{Assignment} \quad p(x) \Leftarrow t(x)}$$

In classical logic there would in general be a finite set of possible witnessing expressions $\{R(x, t_1(x)), R(x, t_2(x)), ..., R(x, t_k(x))\}$ in the final goal, and we can only be sure that for each input x, **one** of the terms $t_i(x)$ will succeed. Assuming R is computable, we can decide which by a case analysis. This comment also applies (but is not made explicit) in the Induction and Cut cases below.

The Induction-case

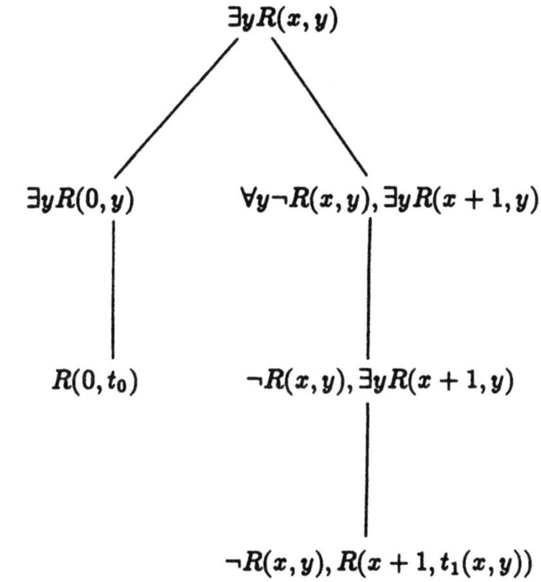

From this we read off witnessing terms t_0 for the base case, and $t_1(x,y)$ for the successor case - provided that y is already assumed to satisfy R at x! The synthesized program is therefore

$$\boxed{\textit{PrimitiveRecursion} \quad \begin{array}{rcl} p(0) & \Leftarrow & t_0 \\ p(x+1) & \Leftarrow & t_1(x,p(x)) \end{array}}$$

The Cut-case

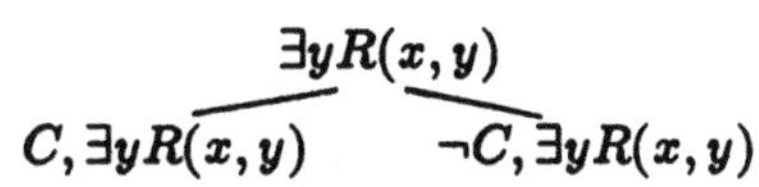

The problem here is that the cut-formula C may be arbitrarily complex and we are stuck unless we can somehow restrict its complexity.
Proof theory at this point comes to our rescue!

CUT ELIMINATION THEOREM. (Gentzen,....) *Though we cannot eliminate cuts completely from proofs in Σ_1^0-Induction, we can always effectively transform any such proof into another wherein all cut-formulas are Σ_1^0.*

So assume this done at the outset. Then we know that the cut-formula C has the form

$$C \equiv \exists z Q(x,z)$$

There are now essentially two possible ways in which the left-hand branch of our Cut-proof of $\exists y R(x,y)$ could come about:

Sub-case 1

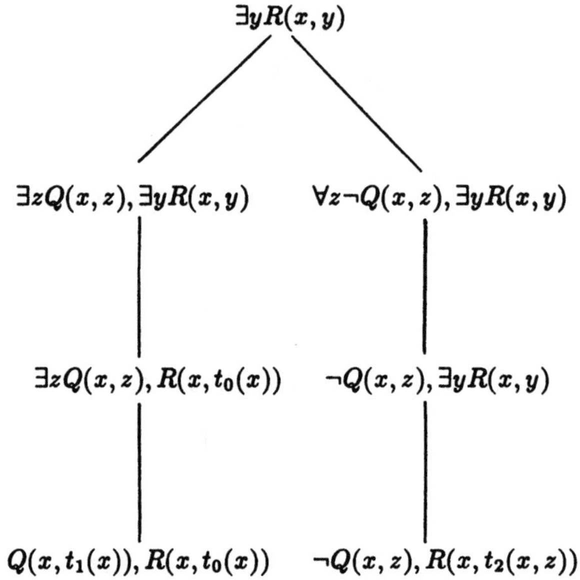

Thus *either* the term $t_1(x)$ witnesses Q at x, in which case (putting $z = t_1(x)$) the term $t_2(x,z)$ witnesses R, *or else* $t_1(x)$ fails to witness Q in which case $t_0(x)$ witnesses R. Therefore we can extract the following program satisfying R :

Conditional	$p(x) \Leftarrow$ **if** $Q(x,t_1(x))$	**then**	$t_2(x,t_1(x))$
		else	$t_0(x)$

Sub-case 2

It is possible that $\exists y R(x,y)$ might be a "side-formula" in the left-hand branch and that that branch actually contains a *proof* of

$$C \equiv \exists z Q(x,z).$$

In this case we can assume inductively that a program $q(x) \Leftarrow (\dots)$ satisfying the sub-specification C has already been synthesized!

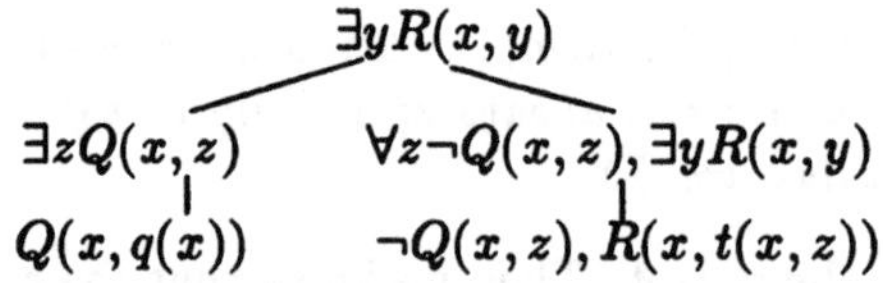

From this situation we extract the program

Composition	$p(x) \Leftarrow t(x,q(x))$

□

COROLLARY. *The non-primitive recursive Ackermann function is not provably specifiable in Σ_1^0-Induction.*

Actually it requires Π_2^0-Induction, the "next" sub-theory of PA, in which the Induction rule is restricted to Π_2^0-formulae.

How can we make some computational use of this next higher level of induction?

Illustration.

Suppose f is some *given* function and let

$$It(x,y,z) \quad \text{mean} \quad z = f^x(y) = f \circ \ldots \circ f(y).$$

We can derive various programs satisfying

$$\forall x, y \exists z \; It(x,y,z)$$

from the following Π_2^0-inductive proof:

$$\forall x \forall y \exists z \; It(x,y,z)$$

$$\forall y \exists z \; It(x,y,z)$$

$$\forall y \exists z \; It(0,y,z) \qquad \exists y \forall z \; \neg It(x,y,z), \forall y \exists z \; It(x+1,y,z)$$

$$\exists z \; It(0,y,z) \qquad \forall z \; \neg It(x,t_1,z), \exists z \; It(x+1,y,z)$$

$$It(0,y,t_0) \qquad \neg It(x,t_1,z), It(x+1,y,t_2)$$

To complete this proof the theorem-prover has to produce terms t_0, t_1, t_2 so that the following hold:

$$It(0,y,t_0)$$
$$It(x,t_1,z) \rightarrow It(x+1,y,t_2).$$

Obviously t_0 has to be the term "y" since $f^0(y) = y$. But there are two possible ways of choosing (t_1, t_2), giving rise to two different synthesized programs.

I Choose $t_1 \equiv y$ and hence $t_2 \equiv f(z)$.

The extracted program is then the standard recursive definition of iteration:

$$\boxed{\begin{array}{lcl} p(0,y) & \Leftarrow & y \\ p(x+1,y) & \Leftarrow & f(p(x,y)) \end{array}}$$

As this is a simple primitive recursion it could already have been proved in Σ_1^0-Induction.

II Choose $t_2 \equiv z$ and hence $t_1 \equiv f(y)$.

The extracted program is now tail-recursive:

$$\boxed{\begin{array}{lcl} p(0,y) & \Leftarrow & y \\ p(x+1,y) & \Leftarrow & p(x,f(y)) \end{array}}$$

so it could be reformulated as a while-loop:

$$\boxed{\textbf{while } x \neq 0 \textbf{ do } x := x-1 \ ; \ y := f(y) \textbf{ od}}$$

The reason for program II being tail-recursive is that we insisted on choosing $t_2 \equiv z$. This meant that we had to change t_1 from y to $f(y)$. Thus program II involves substituting $f(y)$ for the original parameter y. A moment's thought will convince the reader that in order to prove that program II is defined one needs to show, by induction, that for each x the following Π_2^0-formula holds

$$\forall y \ \exists z(p(x,y) = z)$$

Thus Π_2^0-Induction was needed in deriving the tail-recursive/while-program, because without it the term t_1 would not have arisen and we would then not have had the freedom to replace y by $f(y)$.

Π_2^0-Induction therefore seems to provide a useful framework for program transformations from primitive recursions into while-loops. We have the ratios

$$\frac{\text{I}}{\text{II}} = \frac{\text{Prim.Rec.}}{\text{While-loop}} = \frac{\Sigma_1^0\text{-Ind}}{\Pi_2^0\text{-Ind}}$$

The cost of the transformation is an increase in induction-complexity of the specification proof, from Σ_1^0 to Π_2^0!

Can we use proof theory to further analyse, and quantify the complexity of this transformation? See later.

§2. Terminating Recursions and Well Foundedness

"Recursion on the variable a" is the definition of a function value $f(a,x)$ in terms of other values $f(p(a,x),s(x))$ where $p(a,x)$ is in some sense a predecessor of a. For the recursion to terminate the sequences

$$a, p(a,x), p(p(a,x),s(x)), p(p(p(a,x),s(x)),s^2(x))...$$

must "hit bottom" after finitely many steps. I.e. predecessors should form a *well-founded* tree. Furthermore we may as well assume that if a and a' have identical trees of predecessors, then they are themselves identical, i.e. $a = a'$.

For the predecessors to be linearly (hence well-)ordered, each $p(a,x)$ should be a predecessor of $p(a,y)$ whenever $x < y$. The ordinal height of the well ordering should then provide a useful measure of complexity of the recursion.

These informal ideas motivate the definitions below. Note that we may consider $p(a,x)$ as a *sequence* of predecessors $P_0(a)$, $P_1(a)$, $P_2(a), \dots$ one for each value of x where $P_x(a) = p(a,x)$.

2.1 DEFINITION. A *well-founded recursion structure* (WFRS) is a structure $A = \langle A, 0_A, P_0, P_1, \dots \rangle$ where for each $x \in N$, $P_x : A \to A$ with $P_x(0_A) = 0_A$ and

$$(*) \qquad \forall a \in A. \forall x \in N. \forall s : N \to N. \exists n (P_{s^n(x)} \cdots P_{s^2(x)} P_{s(x)} P_x(a) = 0_A)$$

$$(**) \qquad \forall a, b \in A (\forall x (P_x(a) = P_x(b)) \to a = b).$$

2.2 DEFINITION. A WFRS *with successor* is one for which there is a function $a \mapsto a +_A 1$ such that for each $x \in N$ and each $a \in A$, $P_x(a +_A 1) = a$.

2.3 DEFINITION. A *well-ordered recursion structure* (WORS) is a WFRS with successor in which, for every $a \in A$ and all $x, y \in N$,

$$x < y \to \exists m (P_x(a) +_A 1 = P_y^m(a)).$$

2.4 DEFINITION. *Recursion* over a WFRS is any scheme

$$\begin{cases} f(0_A; x_1 \dots x_k) & = g_0(x_1 \dots x_k) \\ f(a; x_1 \dots x_k) & = Term(g_1 \dots g_m, f_{P(a)}; x_1 \dots x_k) \end{cases}$$

where $g_0, g_1 \dots g_m$ are *given* functions and for each non-zero $a \in A$, and all $z, y_1 \dots y_k \in N$,

$$f_{P(a)}(z, y_1 \dots y_k) = f(P_z(a); y_1 \dots y_k).$$

A number-theoretic function $h : N^k \to N$ is said to be defined (from given functions $g_0, g_1 \dots g_m$) by *A-recursion* as above, if there is some fixed element $a_h \in A$ such that for all $x_1 \dots x_k \in N$,

$$h(x_1 \dots x_k) = f(a_h; x_1 \dots x_k).$$

Remark. a_h measures the level of complexity of h, as defined by the given A-recursion. Note that by condition (*), *every A-recursion terminates.*

2.5 **Definition.** REC(A) denotes the smallest class of number-theoretic functions containing all "elementary" ones and closed under explicit definitions and all A-recursions. (The elementary functions are those definable explicitly from $0,\ 1,\ +,\ .$ using bounded sums and products).

What we would like to have now, is a convenient uniform way of representing well-founded recursion structures (with successor), so that we can hopefully compare them, and develop ways of effectively computing one from another. It is at this point that we need to introduce a constructively-motivated version of transfinite ordinals.

2.6 **Definition.** The set Ω of countable *tree-ordinals* $\alpha, \beta, \gamma, \ldots$ is inductively generated by the rule:

$$\begin{aligned} \alpha \in \Omega \quad & \text{if } \alpha = 0 \\ & \text{or } \alpha = \beta + 1 = \beta \cup \{\beta\} \text{for some } \beta \in \Omega \\ & \text{or } \alpha \text{ is a function from } N \text{into } \Omega \end{aligned}$$

If α is a function from N into Ω we denote its value at $n \in N$ by α_n and write

$$\alpha = \sup \alpha_n.$$

2.7 **Definition.** The predecessor functions $P_0, P_1, \ldots$ on Ω are defined as follows, (following Cichon)

$$P_x(0) = 0, \quad P_x(\alpha + 1) = \alpha, \quad P_x(\sup \alpha_n) = P_x(\alpha_x).$$

Thus if $A \subset \Omega$ is non-empty, closed under predecessors and $+1$, then $\langle A, 0, P_0, P_1, \ldots \rangle$ is a WFRS with successor provided (**) holds.

2.8 **Representation Theorem.** *Every* WFRS *A with successor is isomorphic to a subset $\Omega(A) = \{o(a) : a \in A\}$ of Ω where the map $o : A \to \Omega$ is given by:*

$$o(0_A) = 0, \quad o(a +_A 1) = o(a) + 1, \quad o(a) = \sup_x (o(P_x^A a) + 1) \quad \textit{otherwise.}$$

Proof: Given any WFRS A, the well-foundedness condition (*) allows us to prove properties of $a \in A$ by induction up the tree of predecessors .

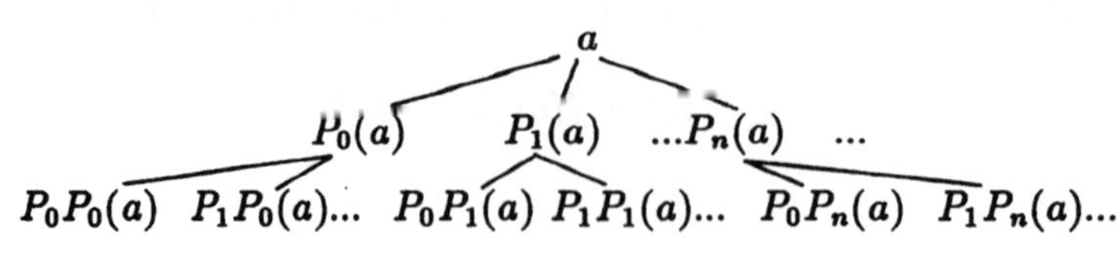

For example we can easily show that the map $o : A \to \Omega$ is injective, i.e. $o(a) = o(b) \to a = b$.

Case $a = 0_A$: then $o(a) = 0$ so $o(b) = 0$ and hence $b = 0_A$.

Case $a = a' +_A 1$: then $o(a) = o(a') + 1 = o(b)$, so $b = b' +_A 1$ and therefore $o(a') + 1 = o(b') + 1$, hence $o(a') = o(b')$. By the induction hypothesis applied to a' we therefore have $a' = b'$ and consequently $a = b$.

Case otherwise: then $o(a) = \sup(o(P_x^A(a)) + 1) = o(b)$ so b is neither 0_A nor a successor $b' +_a 1$, and hence $o(b) = \sup(oP_x^A(b) + 1)$ by definition. Consequently for every x, $oP_x^A(a)+1 = oP_x^A(b)+1$ and hence $P_x^A(a) = P_x^A(b)$ by the induction hypothesis applied to $P_x^A(a)$. Then $a = b$ follows immediately from (**).

It only remains to check that predecessors P_x are preserved by the map o , i.e. that

$$P_x(o(a)) = o(P_x^A(a)).$$

This is immediate however. If $a = 0_A$ or $a = a' +_A 1$ then it follows directly from the definition of o. Otherwise we have $o(a) = \sup(o(P_x^A(a)) + 1)$, so by 2.7,

$$P_x(o(a)) = P_x(o(P_x^A(a)) + 1) = o(P_x^A(a))$$

□

Note: The Representation Theorem tells us that Ω should provide all the recursion structures we are likely to need, so henceforth we will work inside Ω.

If we start with a *well-ordered* recursion structure A as given by Definition 2.3, and if $a \in A$ is neither 0_A nor a successor $a' +_A 1$, then $o(a) = \sup \alpha_x$ where $\alpha_x = o(P_x^A(a) +_A 1)$ and

$$x < y \to \exists m(P_x(a) +_A 1 = P_y^m(a)).$$

Therefore

$$x < y \to \exists m(\alpha_x = P_y^m(\sup \alpha_x)).$$

This motivates the following definition.

2.9 DEFINITON. A tree-ordinal $\alpha \in \Omega$ is said to be *structured* if for every "sub-tree" of α of the form $\lambda = \sup \lambda_n$ we have for all $x, y \in N$,

$$x < y \to \exists m(\lambda_x = P_y^m(\lambda)).$$

The set of all structured tree-ordinals is denoted Ω^S.

2.10 DEFINITION. For each $\alpha \in \Omega$ and every $x \in N$, set

$$\alpha[x] = \{P_x(\alpha), P_xP_x(\alpha), \ldots, P_x^n(\alpha) = 0\}.$$

Thus, by the definition of the predecessor functions on Ω we have for each $x \in N$,

$$0[x] = \emptyset, \quad \alpha + 1[x] = \alpha[x] \cup \{\alpha\}, \quad \sup \alpha_n[x] = \alpha_x[x].$$

2.11 DEFINITION. Let $\prec$ denote the *sub-tree partial ordering* on Ω where $0 \preceq \beta$, $\beta \prec \alpha+1$ if and only if $\beta \preceq \alpha$ and $\beta \prec \sup \alpha_n$ if and only if $\beta \preceq \alpha_k$ for some k.

For each fixed $\alpha \in \Omega$ let $\prec_\alpha$ denote the restriction of $\prec$ to $\{\beta : \beta \prec \alpha\}$.

2.12 LEMMA. *For each* $\alpha \in \Omega^S$ *we have*

(1) $$x < y \rightarrow \alpha[x] \subset \alpha[y]$$

(2) $$\beta \prec \alpha \rightarrow \beta \in \bigcup_{x \in N} \alpha[x]$$

Hence $\prec_\alpha$ *is a well-ordering with bottom* 0 *and* $\beta + 1 \preceq \alpha$ *whenever* $\beta \prec \alpha$.

PROOF: We prove 1) and 2) simultaneously, by induction over the generation of α in Ω^S

If $\alpha = 0$ there is nothing to do. If $\alpha = \alpha' + 1$ then for each x, $\alpha[x] = \alpha'[x] \cup \{\alpha'\}$ and the induction hypothesis holds for α', so we immediately have 1); and for 2) suppose $\beta \prec \alpha$. Then either $\beta \prec \alpha'$ in which case $\beta \in \alpha'[x]$ for some x, or $\beta = \alpha'$. Whichever is the case, we have $\beta \in \alpha[x]$ some x. If $\alpha = \sup \alpha_n$ then for each x, $\alpha[x] = \alpha_x[x]$, and by structuredness, $\alpha_x \in \alpha[y]$ when $x < y$, and therefore by the induction hypothesis applied to α_x, $\alpha[x] = \alpha_x[x] \subset \alpha_x[y] \subset \alpha[y]$. For 2) suppose $\beta \prec \alpha$. Then $\beta \preceq \alpha_k \prec \alpha_{k+1}$ some k, so by the induction hypothesis applied to α_{k+1} we have $\beta \in \alpha_{k+1}[x]$ for some $x > k$, hence $\beta \in \alpha[x]$.

To show that $\prec_\alpha$ is a well-order, little remains to be done. We know already that $\prec$ is well-founded because of the inductive nature of the generation of each $\alpha \in \Omega$. So we only need check that $\prec_\alpha$ is a linear ordering, i.e. that whenever $\beta, \gamma \prec \alpha$ we have either $\beta \prec \gamma$ or $\beta = \gamma$ or $\gamma \prec \beta$. But if $\beta, \gamma \prec \alpha$ then by 1) and 2) there is some $x \in N$ such that $\beta, \gamma \in \alpha[x]$. Clearly $\alpha[x]$ is a finite linear sub-ordering of $\prec_\alpha$:

$$0 = P_x^n(\alpha) \prec P_x^{n-1}(\alpha) \prec \ldots \prec P_x^2(\alpha) \prec P_x(\alpha)$$

so if $\beta \neq \gamma$ then one of them is $\prec$ the other.

Therefore $\prec_\alpha$ is a well-order whenever $\alpha \in \Omega^S$. Obviously 0 is its bottom element and if $\beta \prec \alpha$ then $\beta \in$ some $\alpha[x]$ and from this it follows, by a trivial induction on α, that $\beta + 1 \preceq \alpha$. □

REMARK. The above lemma is important because it shows that Ω^S singles out those tree-ordinals α whose orderings $\prec_\alpha$ are exact copies of ordinary set-theoretic ordinals $0 \prec 1 \prec 2 \prec \ldots \prec \beta \prec \beta + 1 \prec \ldots \prec \alpha$. In addition however the ordering has a further structural property which is crucial for the definition of number-theoretic functions by recursion. Namely, each *limit point* $\lambda = \sup \lambda_n \preceq \alpha$ comes equipped with a *fixed choice* of so-called *fundamental sequence* approximating to it: i.e.

$$\lambda_0 \prec \lambda_1 \prec \lambda_2 \prec \ldots \prec \lambda_x \prec \ldots \prec \lambda$$

EXAMPLE. let $\lambda = \sup \lambda_x$ where $\lambda_x = x+1$ and $\lambda' = \sup \lambda'_x$ where $\lambda'_x = x+2$. Then λ, λ' are "order-isomorphic" but *incomparable* under $\prec$.

2.13 DEFINITIONS. For each $\alpha \in \Omega^S$, $x \leq y \in N$, $s : N \to N$,

1) $G_\alpha(x) =$ size of $\alpha[x] =$ least $n.P^n_x(\alpha) = 0$.

2) i_{xy} = identity embedding of the numbers less than x into the numbers less than y.

3) $G_\alpha(i_{xy})$ = the finite map from numbers less than $G_\alpha(x)$ into numbers less than $G_\alpha(y)$, given by the rule:

$$G_\alpha(i_{xy})(m) = G_\gamma(y)$$

where $\gamma \in \alpha[x]$ is such that $m = G_\gamma(x)$.

4) $H_0(s)(x) = x$ and for all $\alpha \neq 0$, $H_\alpha(s)(x) = s^{n+1}(x)$ where n is the least such that $P_{s^n(x)}P_{s^{n-1}(x)} \dots P_{s^2(x)}P_{s(x)}P_x(\alpha) = 0$.

2.14 THE DIRECT LIMIT REPRESENTATION OF $\alpha \in \Omega^s$.

1) $\prec_\alpha$ is the direct limit of the functor $G_\alpha : N \to N$;

$$\prec_\alpha \; = \varinjlim G_\alpha.$$

2) The functional H_α witnesses the well-foundedness of $\prec_\alpha$;

$$\prec_\alpha \textit{ is wellfounded} \equiv \forall s \forall x (H_\alpha(s)(x)\downarrow).$$

2.15 LEMMA. *The function G_α and functional H_α satisfy the following recursive definitions.*

$$G_0(x) = 0, \quad G_{\alpha+1}(x) = G_\alpha(x) + 1, \; G_\lambda(x) = G_{\lambda_x}(x).$$
$$H_0(s)(x) = x, \quad H_{\alpha+1}(s)(x) = H_\alpha(s)(s(x)), \; H_\lambda(s)(x) = H_{\lambda_x}(s)(x).$$

PROOF: immediate from the definitions above.

2.16 ARITHMETIC ON Ω^S .

$$\textit{Addition} \qquad \begin{aligned} &\alpha + 0 = \alpha \\ &\alpha + (\beta + 1) = (\alpha + \beta) + 1 \\ &\alpha + \lambda = sup(\alpha + \lambda_x) \end{aligned}$$

Then $G_{\alpha+\beta}(x) = G_\alpha(x) + G_\beta(x)$ and $H_{\alpha+\beta}(s) = H_\alpha(s) \circ H_\beta(s)$.

$$\textit{Multiplication}(\alpha \neq 0) \qquad \begin{aligned} &\alpha.0 = 0 \\ &\alpha.(\beta + 1) = \alpha.\beta + \alpha \\ &\alpha.\lambda = sup(\alpha.\lambda_x) \end{aligned}$$

Then $G_{\alpha.\beta}(x) = G_\alpha(x).G_\beta(x)$ and $H_{\alpha.\beta}(s) = H_\beta(H_\alpha(s))$.

$$\textit{Exponentiation}(\alpha \neq 0, 1) \qquad \begin{aligned} &\alpha^0 = 1 \\ &\alpha^{\beta+1} = \alpha^\beta.\alpha \\ &\alpha^\lambda = sup(\alpha^{\lambda_x}) \end{aligned}$$

Then $G_{\alpha^\beta}(x) = G_\alpha(x)^{G_\beta(x)}$ and $H_{\alpha^\beta}(s) = H_\alpha^{(\beta)}(s)$, the β-times iterate of H_α on s.

The "first" limit $\omega = sup(x + 1)$ *where* $x + 1 = 0 + 1 \ldots + 1$ ($x + 1$*times*).
The "first" epsilon $\varepsilon_o = sup(\omega, \omega^\omega, \omega^{\omega^\omega}, \ldots)$.

We end this chapter with some basic observations concerning recursion on WFRS's $A \subset \Omega$.

2.17 DEFINITION. If $\alpha \in \Omega$ is such that $A = \{\beta : \beta \prec \alpha\}$ forms a WFRS (with the standard predecessors P_x) then we often denote $REC(A)$ instead by $REC(\alpha)$.

2.18 LEMMA. *Suppose $\alpha \in \Omega$ is such that every "limit" $\lambda = \sup \lambda_n \preceq \alpha$ has the property $\forall n(\lambda_n \neq 0)$. Then for every $\beta \prec \alpha$*

$$(1) \qquad G_\beta \in \mathrm{REC}(\alpha)$$

$$(2) \qquad s \in \mathrm{REC}(\alpha) \to H_\beta(s) \in \mathrm{REC}(\alpha)$$

PROOF: G_β and $H_\beta(s)$ are definable by the following $\alpha - recursions$:

$$G_0(x) = 0$$
$$G_\beta(x) = G_{P_x(\beta)}(x) + 1$$

$$H_0(s)(x) = x$$
$$H_\beta(s)(x) = H_{P_x(\beta)}(s)(s(x))$$

□

Note: that H is given by a tail-recursion, so it could alternatively be defined by the while-loop

$$\textbf{while } \beta \neq 0 \textbf{ do } \beta := P_x(\beta)\ ;\ x := s(x) \textbf{ od}$$

Note also the difference between G_β and $H_\beta = H_\beta(\text{succ})$

$$G_\beta(x) = \text{succ} \circ G_{P_x(\beta)}(x), \quad H_\beta(x) = H_{P_x(\beta)} \circ \text{succ}\ (x).$$

The difference may at first seem slight, but in fact it is far from it! Much interesting work has been generated from analysis of the difference, e.g. Girard's Π^1_2-Logic.

2.19 LEMMA. *If $A \subset \Omega$ forms a* WFRS *with $\omega \in A$ then $REC(A)$ is closed under primitive recursion.*

PROOF: Suppose h is defined from $g_0, g_1 \in REC(A)$ by a *primitive* recursion:

$$\begin{cases} h(0,x) & = g_0(x) \\ h(z+1,x) & = g_1(z,x,h(z,x)) \end{cases}$$

Define $f : A \times N \times N \to N$ by recursion over A:

$$f(0;z,x) = g_0(x)$$
$$f(\alpha;z,x) = \textbf{if} z = 0 \textbf{ then} g_0(x)$$
$$\textbf{else} g_1(z-1,x,f(P_{z-1}(\alpha);z-1,x)).$$

Putting $\alpha = z$ and noting that $P_{z-1}(z) = z - 1$, it is now easy to see by induction on z that

$$f(z;z,x) = h(z,x).$$

Putting $\alpha = \omega$ and noting that $P_{z-1}(\omega) = z - 1$, we then have, for all $z, x \in N$,

$$f(\omega;z,x) = h(z,x).$$

Therefore $h \in REC(A)$ since $\omega \in A$. □

2.20 Remark. For every totally-defined number theoretic function f there is an $\alpha_f \in \Omega$ such that

$$f = G_{\alpha_f}.$$

For choose $\alpha_f = \sup k_x$ where $k_x = 0 + 1 + 1 \ldots + 1(f(x) times)$. Then $G_{\alpha_f}(x) = G_{k_x}(x) = f(x)$.

That we can write down the above definition of α_f corresponds only to the fact that we are told - in advance - that f is indeed totally defined. The aim of subrecursive hierarchy classifications is to find, for *recursive* functions f, more delicate ordinal assignments $\alpha_f \in \Omega$ whose size and structure reflect the complexity of the *proofs* that f is totally defined!

§3. Unravelling Inductive Proofs and Recursions

To assess the computational complexity of a recursion one needs to 'unravel' it somehow, or in other words, 'simulate' its intended sequence of computation steps. There should therefore be an analogous procedure for measuring the complexity of inductive proofs. There is! - it goes back to Gentzen, was developed by Schütte, then by many others since; and it constitutes one of the principal tools in the proof-theory trade.

Suppose we have a proof of an arithmetical formula $A(x)$, by induction:

$$A(0) \quad A(x) \rightarrow A(x+1)$$

We can unravel it into an infinite sequence of 'sub-proofs':

$$A(0)\ ,\ A(1)\ ,\ A(2)\ ,\ \ldots\ ,\ A(n)\ ,\ \ldots$$

where each $A(n)$ requires n nested Cuts:

$$A(0)\ ,\ A(0) \rightarrow A(1)\ ,\ A(1) \rightarrow A(2)\ ,\ \ldots\ ,\ A(n-1) \rightarrow A(n)$$

Therefore we could eliminate the induction entirely, provided we are prepared to allow an *infinite* proof-rule:

Derive $\forall x A(x)$ from $A(0), A(1), A(2), \ldots, A(n), \ldots$.

How can we then measure the size and complexity of the resulting infinite proofs? By using tree-ordinals to represent proof-trees!

The above rule is called the *ω-rule* and the "semi-formal" system we are about to define is called *ω-Arithmetic*:

3.1 Definition of ω-Arithmetic (Buchholz)

We are going to define inductively, the notion of an α-bounded infinitary proof $\vdash^\alpha \Gamma$ where $\alpha \in \Omega^S$. The finite sets of formulas Γ, and most of the logical rules, will be just like those of PA in §1, except that a new unary relation "$x \in N$" - meaning x is a natural number - is added to the language, together with a new axiom "$0 \in N$" and a new rule saying N is closed under sucessor. The new ω-rule is formulated in such a way that it incorporates both Induction and $\forall$-rules. This means that we will no longer need to consider formulas containing free variables. All formulas will be "closed" ones.

The tree-ordinals control proof-structure in a strict way using a neat but delicate mechanism due to Buchholz - we call it the Accumulation Rule (Acc).

Axioms. (any $\alpha \neq 0$) $\vdash^\alpha \Gamma, A$ where A is any *true* closed atomic formula of PA.

$\vdash^\alpha \Gamma, \text{“}0 \in N\text{”}$

$\vdash^\alpha \Gamma, \text{“}m \notin N\text{”}, \text{“}m \in N\text{”}$ for any numeral m.

Rules.

$(\wedge)$	$\vdash^{\alpha+1} \Gamma, (A \wedge B)$	if	$\vdash^\alpha \Gamma, A$ and $\vdash^\alpha \Gamma, B$.
$(\vee)$	$\vdash^{\alpha+1} \Gamma, (A \vee B)$	if	$\vdash^\alpha \Gamma, A$ or $\vdash^\alpha \Gamma, B$.
$(\exists)$	$\vdash^{\alpha+1} \Gamma, \exists x A(x)$	if	$\vdash^\alpha \Gamma, A(n)$ for some numeral n.
(Cut)	$\vdash^{\alpha+1} \Gamma$	if	$\vdash^\alpha \Gamma, \neg C$ and $\vdash^\alpha \Gamma, C$.
(ω)	$\vdash^{\alpha+1} \Gamma, \forall x A(x)$	if	$\vdash^\alpha \Gamma, A(n)$ for every n.
(N)	$\vdash^{\alpha+1} \Gamma, \text{“}n+1 \in N\text{”}$	if	$\vdash^\alpha \Gamma, \text{“}n \in N\text{”}$.
(Acc)	$\vdash^\alpha \Gamma$	if	$\vdash^\beta \Gamma$ where $\beta \in \alpha[k], k =$ max $\{2\} \cup \{3m : \text{“}m \notin N\text{”}$ in $\Gamma\}$.

Convention: the occurrence of a formula “$m \notin N$” in a set Γ is a *declaration* of the integer parameter m. In the following we shall often write

$$m_1 \in N,\ m_2 \in N, ..., m_r \in N \quad \vdash^\alpha \quad \Gamma_o$$

instead of

$$\vdash^\alpha m_1 \notin N,\ m_2 \notin N, ..., m_r \notin N, \Gamma_o.$$

3.2 **Definition.** For any formula A of arithmetic, let A^N denote the fomula constructed in the same way, but with all quantifiers relativized to N, i.e. $\forall x B(x)$ is replaced by $\forall x(x \notin N \vee B^N(x))$ and $\exists x B(x)$ is replaced by $\exists x(x \in N \wedge B^N(x))$.

3.3 **Completeness Theorem.** *For every true statement A of arithmetic there is an $\alpha \in \Omega^S$ such that*

$$\vdash^\alpha A^N.$$

Proof: by induction on the logical structure of A. *A atomic*: then $A \equiv A^N$ is an axiom. $A \equiv B_o \vee B_1$: then one of B_o or B_1 is true, so by the induction hypothesis we have $\beta \in \Omega^S$ such that either $\vdash^\beta B_0^N$ or $\vdash^\beta B_1^N$. Therefore by the $\vee$-rule, $\vdash^{\beta+1} A^N$. $A \equiv B_o \wedge B_1$: then both B_o and B_1 are true, so by the induction hypothesis we have $\beta_0, \beta_1 \in \Omega^S$ such that $\vdash^{\beta_o} B_o^N$ and $\vdash^{\beta_1} B_1^N$.
Note that for any $\gamma, \delta \in \Omega^S, \vdash^\delta \Gamma$ implies $\vdash^{\gamma+\delta} \Gamma$. Define $\alpha = sup\alpha_x \in \Omega^S$ where $\alpha_o = \beta_o + 1, \alpha_1 = \alpha_o + \beta_1 + 1$ and $\alpha_{n+2} = \alpha_{n+1} + 1$. Then $\beta_o \in \alpha[2]$ and $\alpha_o + \beta_1 \in \alpha[2]$ so by the accumulation rule and the above note, $\vdash^\alpha B_o^N$ and $\vdash^\alpha B_1^N$. Therefore by the $\wedge$-rule, $\vdash^{\alpha+1} A^N$. $A \equiv \exists x B(x)$: then for some n, $B(n)$ is true, so by the induction hypothesis we have a $\beta \in \Omega^S$ such that $\vdash^\beta B^N(n)$. By the N-rule, $\vdash^{\beta+n} n \in N$, and by the accumulation rule, $\vdash^{\beta+n} B^N(n)$. Thus by the $\wedge$-rule,

$$\vdash^{\beta+n+1} n \in N \wedge B^N(n)$$

and by the $\exists$-rule, with n as witnessing term,

$$\vdash^{\beta+n+2} \exists x(x \in N \wedge B^N(x))$$

i.e. $\vdash^\alpha A^N$ where $\alpha = \beta + n + 2$. $A \equiv \forall x B(x)$: then for every n, $B(n)$ is true, so by the induction hypothesis we have $\beta_o, \beta_1, \beta_2, ... \in \Omega^S$ such that $\vdash^{\beta_n+1} n \notin N \vee B^N(n)$.

Now define $\alpha = sup\alpha_x \in \Omega^S$ where $\alpha_o = \beta_o + 1$ and $\alpha_{x+1} = \alpha_x + \beta_{x+1} + 1$. Then using the "note" above, we have, for every n,

$$\vdash^{\alpha_n} n \notin N \vee B^N(n) \quad where \quad \alpha_n \in \alpha[\max(2,3n)].$$

Therefore by the accumulation rule, for every n,

$$\vdash^{\alpha} n \notin N \vee B^N(n)$$

and then by the ω-rule,

$$\vdash^{\alpha+1} \forall x(x \notin N \vee B^N(x))$$

i.e. $\vdash^{\alpha+1} A^N$ as required. □

REMARK. The correspondence A *true* $\mapsto \alpha \in \Omega^S$ is constructed according to the evidence we have for believing in the "truth-definition" of A. The stronger the evidence, the more structure we ought to infer about α. For example, what *more* do we know about α if A is a theorem of PA?

3.4 EMBEDDING THEOREM. *If* Γ, *with free variables* $x_1, ..., x_r$, *is a theorem of* PA *then for some integer* ℓ *and all* $n_1, n_2, ..., n_r \in N$,

$$n_1 \in N, ..., n_r \in N \vdash^{\omega.\ell} \Gamma^N(n_1, ..., n_r).$$

(See Buchholz-Wainer [1987])

3.5 DEFINITION. A Σ_1^N-formula is one of the form

$$\exists x_1 \exists x_2, ..., \exists x_r(x_1 \in N \wedge ... \wedge x_r \in N \wedge D)$$

where D is built up from the variables $x_1, \ldots x_r$, using the given elementary relations $R, \bar{R}$ of PA and the propositional connectives $\vee, \wedge$.

3.6 DEFINITION. A closed Σ_1^N-formula is said to be *true at* m where m is some natural number, if it is true when N is interpreted as the finite set $\{n : 3n < m\}$. A set Γ of closed Σ_1^N-formulas is said to be *true at* m if any one of its members is true at m.

Note $m < m'$ and Γ true at $m \rightarrow \Gamma$ true at m'.

3.7 BOUNDING THEOREM (BUCHHOLZ). *Suppose that* Γ *is a set of closed* Σ_1^N*-formulas and*

$$n_1 \in N, \ldots n_r \in N \ \vdash^{\alpha} \Gamma$$

where all of the cut-formulas C *involved in the proof are* Σ_1^N*!*

Then putting $k = \max(2, 3n_1, \ldots, 3n_r)$ *we have*

$$\Gamma \textit{ is true at } m = H_{2\alpha}(\text{succ})(k).$$

PROOF: For shorthand we will denote the function $H_{2\alpha}(\text{succ})$ by h_α so that by 2.15, 2.16 we have

$$(1) \qquad h_0(x) = x + 1, \quad h_{\alpha+1}(x) = h_\alpha \circ h_\alpha(x), \quad h_\lambda(x) = h_{\lambda_x}(x).$$

Some properties relating to rate-of-growth of these functions will be needed in the course of the proof. These are numbered (2)-(3) and will be verified separately after the proof is complete.

Proceed by induction over α, according to the inductive generation of the proof

$$n_1 \in N, \ldots n_r \in N \vdash^{\alpha} \Gamma.$$

Axiom case. if Γ contains an axiom then Γ is automatically true in $h_\alpha(k)$ whatever α is.

$\wedge$-case. if $\Gamma = \Gamma_0, (D_0 \wedge D_1)$ with $\alpha = \beta + 1$, where

$$n_1 \ldots n_r \in N \vdash^\beta \Gamma_0, D_0 \quad \text{and } n_1 \ldots n_r \in N \vdash^\beta \Gamma_0, D_1$$

then by the induction hypothesis both Γ_0, D_0 and Γ_0, D_1 are true at $h_\beta(k)$. Hence Γ itself is true at $h_\beta(k)$ and therefore also at $h_\alpha(k)$ because $h_\beta(k) < h_\beta \circ h_\beta(k) = h_\alpha(k)$ by property

(2) $\qquad h_\beta$ is strictly increasing.

$\vee$-case. similar to the $\wedge$-case.

$\exists$-case. if $\Gamma = \Gamma_0$, $\exists x(x \in N \wedge B(x))$ and $\alpha = \beta + 1$ where for some numeral m,

$$n_1 \ldots n_r \in N \vdash^\beta \Gamma_0, \quad m \in N \wedge B(m)$$

then by the induction hypothesis, Γ_0, $m \in N \wedge B(m)$ is true at $h_\beta(k)$ and hence also at $h_\alpha(k)$ as above. This automatically implies that Γ_0, $\exists x(x \in N \wedge B(x)) = \Gamma$ is true at $h_\alpha(k)$.

Cut-case. suppose $\alpha = \beta + 1$ and there is a Σ_1^N cut-formula $C \equiv \exists z_1 \ldots \exists z_s (z_1 \in N \wedge \ldots z_s \in N \wedge D)$ such that

$$n_1 \ldots n_r \in N \vdash^\beta \Gamma, \neg C \quad \text{and } n_1 \ldots n_r \in N \vdash^\beta \Gamma, C.$$

Now $\neg C \equiv \forall z_1 \ldots \forall z_s (z_1 \notin N \vee \ldots \vee z_s \notin N \vee \neg D)$ and

$$\vdash^\beta \Gamma, \forall z B(z) \text{ implies } \vdash^\beta \Gamma, B(m) \quad \text{for every } m.$$

Therefore for all $m_1, \ldots, m_s \in N$ we have

$$n_1 \ldots n_r \in N \vdash^\beta \Gamma, m_1 \notin N \vee \ldots \vee m_s \notin N \vee \neg D$$

But $\vdash^\beta B_1 \vee B_2$ implies $\vdash^\beta B_1, B_2$ so therefore

$$n_1 \ldots n_r \in N, m_1 \ldots m_s \in N \vdash^\beta \Gamma, \neg D.$$

Hence by the induction hypothesis,

(*) $\qquad \Gamma, \neg D$ is true at $h_\beta(\max(k, 3m_1 \ldots, 3m_s))$.

Applying the induction hypothesis also to the second premise of the cut we obtain

$$\Gamma, C \text{ is true at } h_\beta(k)$$

i.e. *there are* numbers $m_1 \ldots m_s$ such that the max of $3m_1, \ldots, 3m_s$ is $< h_\beta(k)$ and

(**) $\qquad \Gamma, D$ is true at $h_\beta(k)$.

Therefore for these particular values of $m_1 \ldots m_s$ we have, since $h_\beta(k) < h_\beta \circ h_\beta(k) = h_\alpha(k)$,

$$\Gamma, D \text{ is true at } h_\alpha(k)$$

and from (*) since $\max(k, 3m_1, \ldots, 3m_s) < h_\beta(k)$,

$$\Gamma, \neg D \text{ is true at } h_\beta \circ h_\beta(k) = h_\alpha(k).$$

But the only way that both Γ, D and $\Gamma, \neg D$ can be true at $h_\alpha(k)$ is when Γ itself is true at $h_\alpha(k)$, as required.

ω-CASE. the ω-rule cannot be applied because it introduces the $\forall$-quantifier, and there aren't any in a Σ_1^N-formula.

N-CASE. if $\Gamma = \Gamma_0$, "$m+1 \in N$" and $\alpha = \beta + 1$ where

$$n_1 \ldots n_r \in N \vdash^\beta \Gamma_0, \text{“} m \in N \text{”}$$

then by the induction hypothesis, Γ_0, "$m \in N$" is true at $h_\beta(k)$. So either Γ_0 is true at $h_\beta(k)$, in which case Γ is true at $h_\beta(k)$ and hence at $h_\alpha(k)$; or else "$m \in N$" is true at $h_\beta(k)$ in which case we have $3m < h_\beta(k)$, hence $3(m+1) < h_\alpha(k)$ and hence Γ is true at $h_\alpha(k)$ as required. The proof that $3m < h_\beta(k)$ implies $3m+3 < h_{\beta+1}(k)$ in this case runs as follows. Remember that $\beta \neq 0$. First suppose $\beta = 1$, so that $3m < h_\beta(k) = k+2$ and $(n_1 \notin N, \ldots, n_r \notin N, \Gamma_0, m \in N)$ is an axiom. If $k = 2$ then, since we are dealing with the case where Γ_0 is not true at $h_\beta(k)$, the only possibility is that $m = 0$. If $k \neq 2$ then k must be a multiple of 3. Either way, since $3m < k+2$ we must have $3m \leq k$, and hence $3m+3 < k+4 = h_2(k)$. Now suppose $\beta \neq 1$. Then an easy induction on β gives $h_\beta(k) \geq h_2(k) = k+4$ and therefore

$$3m < h_\beta(k) \text{ implies } 3m+3 < h_\beta \circ h_\beta(k) = h_{\beta+1}(k).$$

ACCUMULATION CASE. Suppose $n_1 \ldots, n_r \in N \vdash^\beta \Gamma$ where $\beta \in \alpha[k]$. Then by the induction hypothesis, Γ is true at $h_\beta(k)$. Therefore Γ is true at $h_\alpha(k)$ because of property

(3) $$\beta \in \alpha[k] \to h_\beta(k) < h_\alpha(k).$$

The majorisation properties (2), (3) required in the above, follow from the following

3.8 LEMMA. *Let $s : N \to N$ be strictly increasing, > 0. Then for every $\alpha \in \Omega^S$ we have*

(2) $H_\alpha(s)$ *is strictly increasing*

(3) $$\beta \in \alpha[x] \to H_\beta(s)(x) < H_\alpha(s)(x)$$

PROOF: We check (2) and (3) simultaneously, by an induction on α. If $\alpha = 0$ there is nothing to do, since $H_0(s) =$ identity and $0[x] = \emptyset$. Next suppose $\alpha = \gamma + 1$. Then $H_\alpha(s)(z) = H_\gamma(s)(s(z))$ for all $z \in N$ so (2) $H_\alpha(s)$ is strictly increasing since both $H_\gamma(s)$ and s are, and (3) if $\beta \in \alpha[x] = \gamma[x] \cup \{\gamma\}$ then $H_\beta(s)(x) \leq H_\gamma(s)(x) < H_\gamma(s)(s(x)) = H_\alpha(s)(x)$.

Finally suppose $\alpha = \sup \alpha_z \in \Omega^S$. To prove (2) suppose $y < z$. Then $H_\alpha(s)(y) = H_{\alpha_y}(s)(y)$ and by the induction hypothesis applied to α_y we have $H_{\alpha_y}(s)(y) < H_{\alpha_y}(s)(z)$. But $\alpha_y \in \alpha[z] = \alpha_z[z]$ so by the induction hypothesis applied to α_z we have $H_{\alpha_y}(s)(z) < H_{\alpha_z}(s)(z) = H_\alpha(s)(z)$. Therefore $H_\alpha(s)(y) < H_\alpha(s)(z)$ and $H_\alpha(s)$ is strictly increasing. For part (3) notice that if $\beta \in \alpha[x] = \alpha_x[x]$ we can apply the induction hypothesis immediately to α_x in order to give $H_\beta(s)(x) < H_{\alpha_x}(s)(x) = H_\alpha(s)(x)$.

Note: The Bounding Theorem uses the functions

$$h_\beta = H_{2^\beta}(\text{succ})$$

so in order to apply property (3) to the h_β's we need the additional sub-lemma:

$$\beta \in \alpha[k] \to 2^\beta \in 2^\alpha[k]$$

This is easily checked by induction on α. □

3.9 COROLLARY TO BOUNDING THEOREM. *If a specification $\forall x \in N.\exists y \in N.R(x,y)$ is provable in ω-Arithmetic by an α-bounded proof in which all cut-formulas are Σ_1^N, then it is satisfiable by a function bounded by:*

$$H_{2^\alpha}(\text{succ}) \circ K \text{ where } K(n) = \max(2, 3n).$$

PROOF: From $\vdash^\alpha \forall x(x \notin N \vee \exists y(y \in N \wedge R(x,y)))$ we obtain, by removing "$\forall x$" and "$\vee$",

$$n \in N \vdash^\alpha \exists y(y \in N \wedge R(n,y)) \text{ for every n.}$$

So by the Bounding Theorem, for every n there is a $y_n < H_{2^\alpha}(\text{succ})(K(n))$ such that $R(n, y_n)$. □

Question What do we do if the specification proof uses more complex cut-formulas than Σ_1^N?

Answer Use a Cut-Elimination Theorem to transform the proof into one whose cuts are all Σ_1^N. There is a price to pay however, namely an (iterated) exponential increase in the ordinal bound.

3.10 CUT-ELIMINATION THEOREM. (Gentzen,...) *Define the cut-rank of a proof to be the l.u.b. of the lengths of all cut-formulas involved; where $length(atomic) = 0, length(A \vee B) = length(A \wedge B) = max(length(A), length(B)) + 1$ and $length(\forall x B) = length(\exists x B) = length(B) + 1$.*

Then we have

$$\vdash^\alpha \Gamma \ \textit{with cut} - \textit{rank} \ c+1 \ \rightarrow \vdash^{\alpha'} \Gamma \ \textit{with cut} - \textit{rank} \ c$$

where α' is any exponential 2^α or 3^α or ... or ω^α.

PROOF: See the lectures of Helmut Schwichtenberg, or Buchholz-Wainer [1987]. □

3.11 COROLLARY. *If $\vdash^\alpha \forall x \in N.\exists y \in N.R(x,y)$ with finite cut-rank c, then it is satisfiable by a function bounded by*

$$H_{\alpha^*}(\text{succ}) \circ K \textit{ where } \alpha^* = \exp_2^c(\alpha) \textit{ or } \exp_\omega^c(\alpha)$$

PROOF: Apply the Cut-Elimination Theorem to the α-bounded proof, repeatedly until all the cuts are reduced to being Σ_1^N. Then apply 3.9, noting if necessary that $H_{2^\beta} \leq H_{\omega^\beta}$. □

§4. Classifying Provably Recursive Functions

A **recursive** function is one which is computable by (say) a Turing machine M which terminates everywhere. By coding up Turing machine **states** in a natural way we obtain an "elementary" termination relation:

$$T_M(x,y) \equiv \text{ computation of } M \text{ on input } x$$
$$\text{terminates within } y \text{ tape-cells.}$$

and an "elementary" value-extractor:

$$U_M(x,y) = \text{output left on the tape if } T_M(x,y).$$

Thus for every recursive function f there is a T.M. such that the following hold:

(1) $$\forall x \exists y T_M(x,y)$$

(2) $$T_M(x,y) \rightarrow f(x) = U_M(x,y).$$

In other words, M gives a "computational specification" of f :

$$\forall x \exists y \exists z (T_M(x,y) \wedge z = U_M(x,y)).$$

4.1 DEFINITION. $f \in SPACE(h)$ if there is a Turing machine M which computes f, and a linear function k such that the following holds:

$$\forall x \exists y \leq h \circ k(x).T_M(x,y).$$

4.2 THEOREM. *For every recursive function f there is an $\alpha \in \Omega^S$ such that $\|\alpha\| = \omega$ and*

$$f \in \text{ SPACE } (H_\alpha(\text{succ})).$$

PROOF: Suppose f is computed by a Turing machine which uses no more than $g(x)$ tape-cells for each input x. We can assume that g is strictly increasing. Define $\alpha = \sup \alpha_x \in \Omega^S$ where, for each x, $\alpha_x = 0 + 1 + 1 + \ldots + 1$ ($g(x)$ times). Then every predecessor of α is just an integer, so the set-theoretic ordinal "height" of α is just ω, i.e. $\|\alpha\| = \omega$. Furthermore

$$H_\alpha(\text{succ})(x) = H_{\alpha_x}(\text{succ})(x) = \text{succ}^{g(x)}(x) = x + g(x)$$

so $g \leq H_\alpha(\text{succ})$ and $f \in$ SPACE$(H_\alpha(\text{succ}))$. □

REMARK. Although from 4.2 we have

$$\text{RECURSIVE} = \bigcup \{\text{SPACE} H_\alpha(\text{succ}) : \alpha \in \Omega^S, \|\alpha\| = \omega\}$$

this is in no way a *useful* classification of all recursive functions, because in general there is no useful comparison which can be made between one $\alpha \in \Omega^S$ and another, even when they both have $\|\alpha\| = \omega$! There are infinitely-many (in fact continuum-many) different α 's in Ω^S, all with $\|\alpha\| = \omega$, but no two of which are comparable under $\prec$ or any other "reasonable" order. The Classification Problem is to find a $\sigma \in \Omega^S$ for which

$$\text{RECURSIVE} = \bigcup_{\alpha \prec \sigma} \text{ SPACE } (H_\alpha(\text{succ})).$$

No satisfactory way of locating such a σ is known. Certainly there could be no way of *effectively constructing* such a σ, because if there were, $H_\sigma(\text{succ})$ would be a recursive function which dominates *all* recursive functions (itself included).

We therefore adopt a more modest aim: given a formal theory $\mathbf{T}$ of arithmetic, construct a tree-ordinal $\tau \in \Omega^S$ such that

$$\bigcup_{\alpha \prec \tau} \text{SPACE}(H_\alpha(\text{succ}))$$

is precisely the class of all recursive functions which are "provably specifiable" or "provably terminating" in the theory $\mathbf{T}$.

4.3 DEFINITION. Call a function *provably recursive* in a theory $\mathbf{T}$ if it is computed by a Turing machine M in such a way that

$$\mathbf{T} \vdash \forall x \exists y T_M(x, y).$$

Let PROV.REC($\mathbf{T}$) denote the class of all functions provably recursive in $\mathbf{T}$

The most obvious theory to look at is PA.

4.4 THEOREM. (Kreisel, Schwichtenberg, Wainer ...)

$$\text{PROV.REC(PA)} = \text{REC}(\varepsilon_0) = \bigcup_{\alpha \prec \varepsilon_0} \text{SPACE}(H_\alpha(\text{succ})).$$

PROOF: Henceforth we write H_α for $H_\alpha(\text{succ})$. Recall that $\varepsilon_0 = \sup(\omega, \omega^\omega, \omega^{\omega^\omega}, \ldots)$.

(1) To show $\text{PROV.REC(PA)} \subset \bigcup_{\alpha \prec \varepsilon_0} \text{SPACE}(H_\alpha)$:

Suppose f is provably recursive in PA, so for some Turing machine M which computes f, $\forall x \exists y T_M(x, y)$ is a theorem of PA. Then by the Embedding Theorem 3.4 there is a number ℓ such that in ω-Arithmetic,

$$\vdash^{\omega.\ell} \forall x \in N . \exists y \in N . T_M(x, y).$$

The proof has bounded cut-rank c , so by Corollary 3.11 the termination statement $\forall x \exists y T_M(x, y)$ is satisfiable by a function bounded by

$$H_{\alpha^*} \circ K \text{ where } \alpha^* = exp_\omega^c(\omega.\ell) \prec \varepsilon_0.$$

But this means that $f \in \text{SPACE}(H_{\alpha^*})$.

(2) To show $\bigcup_{\alpha \prec \varepsilon_0} \text{SPACE}(H_\alpha) \subset \text{REC}(\varepsilon_0)$:

If $f \in \text{SPACE}(H_\alpha)$ then for some linear function k, f is definable explicitly by

$$f(x) = U_M(x, H_\alpha(k(x))).$$

But U_M and k both belong to $\text{REC}(\varepsilon_0)$ because they are "elementary" and $H_\alpha \in \text{REC}(\varepsilon_0)$ by Lemma 2.18. Therefore $f \in \text{REC}(\varepsilon_0)$.

(3) To show $\text{REC}(\varepsilon_0) \subset \text{PROV.REC(PA)}$:

By the correspondence set up in §1, between primitive recursive definitions and proof-rules in the theory of Σ_1^0-Induction, it follows that if f is "elementary" or if f is defined explicitly from functions already known to be provably recursive in PA, then f also is provably recursive in PA.

Suppose then that $f = f(\alpha; \ldots)$ for some fixed $\alpha \prec \varepsilon_0$, where $f(\alpha; \ldots)$ is defined from given functions $g_0, g_1, \ldots g_m \in \text{PROV.REC(PA)}$, by a recursion:

$$f(0; \underline{x}) = g_0(\underline{x})$$
$$f(\alpha; \underline{x}) = \text{Term}(g_1 \ldots g_m, f_{P(\alpha)}; \underline{x}).$$

Note that any $\alpha \prec \varepsilon_0$ is generated from smaller tree ordinals $\alpha_1 \ldots \alpha_n$ by $\alpha = \omega^{\alpha_1} + \omega^{\alpha_2} + \ldots + \omega^{\alpha_n}$. This means that we can represent each $\alpha \prec \varepsilon_0$ by a *number* computed in a way reflecting the generation of α. Furthermore the functions $(\alpha, x) \rightarrow P_x(\alpha)$ will be computable (primitive recursive in fact). Therefore we can implement the above recursion on a Turing machine M taking input $(\alpha; x)$. Then proving termination $\forall x \exists y T_M(\alpha, x, y)$ clearly amounts to proving $\forall x$ ($f(\alpha; x)$ is defined). Obviously the way to prove this is by ***transfinite induction up to*** α :

$$TI(\alpha, A) \equiv (A(0) \wedge \forall \beta \prec \alpha(\forall x A(P_x(\beta)) \rightarrow A(\beta))) \rightarrow \forall \beta \prec \alpha A(\beta),$$

with $A(\beta) \equiv \forall x \exists y T_M(\beta, x, y)$. Thus if we can show that for each $\alpha \prec \varepsilon_0$ and any formula A, $TI(\alpha, A)$ is a theorem of PA, then we will have $f = f(\alpha; \ldots) \in$ PROV.REC(PA) as required.

First note that $TI(\omega, A)$ is just a reformulation of the ordinary principle of induction, which is already built into PA. Second, note that if we can prove, inside PA,

$$(*) \qquad TI(\alpha, A') \wedge TI(\gamma, A) \rightarrow TI(\gamma + \omega^\alpha, A)$$

where A' is some appropriate formula depending on A, then we can successively generate, inside PA, the principle of ***transfinite induction up to any given predecessor of*** ε_0. Then we are finished.

To show why (*) holds we work very informally "inside" PA:

We will use $TI(\alpha, A')$ for an appropriate A', to prove

$$\forall \beta \prec \gamma + \omega^\alpha . A(\beta)$$

from the assumptions:

(i) $TI(\gamma, A)$

(ii) $A(0)$

(iii) $\forall \beta \prec \gamma + \omega^\alpha . (\forall x . A(P_x(\beta)) \rightarrow A(\beta))$.

Case $\alpha = 0$: From (i), (ii) and (iii) we have $A(\beta)$ for every $\beta \prec \gamma$. Hence by (iii) we have $A(\gamma)$ also. Thus $\forall \beta \prec \gamma + 1 . A(\beta)$ as required.

Case $\alpha \rightarrow \alpha{+}1$: Assume true for α. Note that if $\beta \prec \gamma{+}\omega^{\alpha+1}$ then $\beta \prec \gamma{+}\omega^\alpha.(n{+}1)$ for some n. Hence the required result will follow if we can prove that for each $n \in N$,

$$\forall \beta \prec \gamma + \omega^\alpha (n+1) . A(\beta).$$

But this is done by ordinary induction on n. For $n = 0$ it amounts to the induction hypothesis on α. For $n + 1$ it amounts to $\forall \beta \prec \gamma' + \omega^\alpha . A(\beta)$ where $\gamma' = \gamma + \omega^\alpha n$. This follows from the induction hypothesis on α together with the sub-induction hypothesis on n to change γ into γ'.

Case $\alpha = \sup \alpha_z$: Assume true for each $\alpha_z, z \in N$. Note that if $\beta \prec \gamma + \omega^\alpha$ then $\beta \prec \gamma + \omega^{\alpha_z}$ for some z, so by the induction hypothesis applied to α_z we already have $A(\beta)$. Hence we have $\forall \beta \prec \gamma + \omega^\alpha . A(\beta)$ automatically in this case. This completes the proof of Theorem 4.4. □

Theorem 4.4 is just one in a whole hierarchy of similar results

$$\text{PROV.REC}(\mathbf{T}) = \text{REC}(\tau) = \bigcup_{\alpha \prec \tau} \text{SPACE}(H_\alpha)$$

the crucial element in each case being a careful proof-theoretic analysis of the upper-bound τ on the "provable ordinals" of the given theory.

4.5 EXAMPLES.

(i) $\mathsf{T} = (\Sigma^0_1 - IND)$, $\tau = \omega^\omega$.

(ii) $\mathsf{T} = (\Pi^0_2 - IND)$, $\tau = \omega^{\omega^\omega}$.

(iii) $\mathsf{T} = PA$, $\tau = \varepsilon_0$.

(iv) $\mathsf{T} = (ID_1)$, $\tau =$ Howard ordinal.

(v) $\mathsf{T} = (\Pi^1_1 - CA)_0$, $\tau =$ a much bigger ordinal (see e.g. Wainer [1989]).

In each case the termination statement for H_τ is *true* (by transfinite induction up to τ) but *not provable* in T because H_τ dominates all the provably recursive functions of T. For interesting independence results related to H_τ see Paris-Harrington [1977], Ketonen-Solovay [1981], Kirby-Paris [1982], Cichon [1983].

REMARK. For the "small" theories displayed in 4.5, i.e. the sub systems of PA, there is an irregularity in the relationship with the corresponding REC() classes, which turns out to be of some interest. In fact we have

$$\text{PROV.REC.}(\Sigma^0_1 - \text{IND}) = \bigcup_{\alpha \prec \omega^\omega} \text{SPACE}(H_\alpha) = \text{REC}(\omega.\omega)$$

$$\text{PROV.REC.}(\Pi^0_2 - \text{IND}) = \bigcup_{\alpha \prec \omega^{\omega^\omega}} \text{SPACE}(H_\alpha) = \text{REC}(\omega.\omega^\omega)$$

The *correct* way to formulate the relationship between REC-definitions and SPACE-complexity is in general as follows (for appropriate β):

$$\text{REC}(\omega.\beta) = \bigcup_{\alpha \prec \omega^\beta} \text{SPACE}(H_\alpha).$$

This brings us back to the "recursive program" - "while-program" transformation considered in §1.

4.6 DEFINITION. WHILE(α) denotes the class of number-theoretic functions definable by programs built up from the following constructs:

Assignment	$x :=$ given function(x)
Conditional	**if** $R(x)$**then** S_0**else** S_1
Composition	S_0; S_1
Whileloop	$\beta := \gamma$; **while** $\beta \neq 0$ **do** $\beta := P_x(\beta); S$ **od** (for any $\gamma \prec \alpha$).

4.7 THEOREM (TAIT 1961, ... , FAIRTLOUGH 1989). *Provided that* $\{\gamma : \gamma \prec \beta\}$ *satisfies certain basic conditions such as "closure under addition" and "uniform Turing machine representability", we have*

$$\text{REC}(\omega.\beta) = \bigcup_{\alpha \prec \omega^\beta} \text{SPACE}(H_\alpha) = \text{WHILE}(\omega^\beta).$$

PROOF: (sketch only). 1) First suppose $f \in \text{REC}(\omega.\beta)$, say $f = f(\alpha; \ldots)$ where $f(\alpha; \ldots)$ is given by recursion as in 2.4 with $\alpha \prec \omega.\beta$. Implement the recursion on a Turing machine M. Then consider the proof of

$$\forall x \exists y T_M(\alpha, x, y)$$

by transfinite induction up to α. This proof can be embedded in ω-Arithmetic and the ordinal bound will be (roughly) equal to α. In fact there will be a fixed k such that for all n,

$$n \in N \vdash^{k.\alpha} \exists y \in N.T_M(\alpha, n, y)$$

and this proof will only use Σ_1^N-cuts. By Corollary 3.9 we therefore have

$$f \in \text{SPACE}(H_{2^{k.\alpha}}).$$

But $\alpha \preceq \omega.\gamma$ for some $\gamma \prec \beta$, so therefore

$$H_{2^{k.\alpha}} \leq H_{2^{k.\omega.\gamma}} \leq H_{\omega^\gamma}(h)$$

for some "elementary" function h. But then as long as $3 \leq \gamma$ we'll have $h \leq H_{\omega^\gamma}$, so by 2.16

$$H_{2^{k.\alpha}} \leq H_{\omega^\gamma}(H_{\omega^\gamma}) = H_{\omega^{\gamma.2}}$$

Hence $f \in \text{SPACE}(H_{\omega^{\gamma.2}})$ and $\omega^{\gamma.2} \prec \omega^\beta$.

2) If $f \in \text{SPACE}(H_\alpha)$ where $\alpha \prec \omega^\beta$ then f is definable explicitly from H_α itself and certain given elementary functions. But $H_\alpha \in \text{WHILE}(\omega^\beta)$ because it is definable by the loop:

$$\gamma := \alpha \text{ ; } \textbf{while } \gamma \neq 0 \textbf{ do } \gamma := P_x(\gamma) \text{ ; } x := succ(x) \textbf{ od} .$$

Therefore $f \in \text{WHILE}(\omega^\beta)$.

3) Finally suppose $f \in \text{WHILE}(\omega^\beta)$. Say it is definable by a while-loop starting on $\alpha = \omega^\gamma \prec \omega^\beta$, with sub-program S as in 4.6. Assume that $s \in \text{REC}(\omega.\beta)$ is a function which bounds the amount of Turing machine space needed to compute S on input x. Then provided the representation of $P_x(\alpha)$ does not take up "too much" space, we can bound the space needed to compute the while-loop by (guess what?)

$$H_\alpha(s)(x) = H_{P_x(\alpha)}(s)(s(x))$$

Therefore, since $\alpha = \omega^\gamma$,

$$f \in \text{SPACE}(H_{\omega^\gamma}(s))$$

and hence f is elementary-definable from $H_{\omega^\gamma}(s)$. But using 2.16, $H_{\omega^\gamma}(s) = H_\omega^{(\gamma)}(s)$ and H_ω is just the iteration operator:

$$H_\omega(g)(x) = g^{x+1}(x).$$

Therefore $H_\omega^{(\gamma)}(s)$ is definable from s by $\omega.\gamma$-recursion, i.e. $H_{\omega^\gamma}(s) = H_\omega^{(\gamma)}(s) \in \text{REC}(\omega.\beta)$ since $\gamma \prec \beta$. Hence we have shown $f \in \text{REC}(\omega.\beta)$ and this completes the sketch of the proof. □

REMARK. Thus (in general "pure mathematical" terms) the cost of transformation from recursive to while-programs is (at worst) an exponential increase in the well-ordering needed to prove termination.

References

[1952] G. Kreisel, *On the interpretation of non-finitist proofs II,* Journ. of Symbolic Logic 17, 43-58.

[1953] A. Grzegorczyk, *Some classes of recursive functions,* Rozprawy Matem. IV, Warsaw.

[1961] W.W. Tait, *Nested Recursion,* Math. Annalen 143, 236-250.

[1962] S. Feferman, *Classification of recursive functions by means of hierarchies,* Trans. Amer. Math. Soc. 104, 101-122.

[1965] J.W. Robbin, *Subrecursive hierarchies,* Ph.D. Princeton.

[1968] W.W. Tait, *Normal derivability in classical logic,* in J. Barwise Ed., Springer Lecture Notes in Math. 72, 204-236.

[1968] R.L. Constable, *Extending and refining hierarchies of computable functions,* T.R.25, Computer Science Dept., Univ. of Wisconsin.

[1970] S.S. Wainer, *A classification of the ordinal recursive functions,* Archiv. f. math. Logik 13, 136-153.

[1971] H. Schwichtenberg, *Eine klassifikation der ε_0 - rekursiven funktionen*, Zeitschrift f. math. Logik 17, 61-74.

[1972] C. Parsons, *On n-quantifier induction,* Journ. of Symbolic Logic 37, 466-482.

[1972] S.S. Wainer, *Ordinal recursion, and a refinement of the extended Grzegorczyk hierarchy,* Journ. of Symbolic Logic 37, 281-292.

[1973] G.E. Mints, *Quantifier-free and one-quantifier systems,* Jour. of Soviet Math. 1, 71-84 (Translation of original paper in Russian which appeared in 1971).

[1977] K. Schütte, *Proof theory,* Springer Verlag, Berlin.

[1977] H. Schwichtenberg, *Proof theory : some applications of cut-elimimation* in J. Barwise Ed., "Handbook of Mathematical Logic", North Holland, Amsterdam, 867-896.

[1977] J. Paris and L. Harrington, *A mathematical incompleteness in Peano arithmetic,* ibid 1133-1142.

[1981] J. Ketonen and R. Solovay, *Rapidly growing Ramsey functions,* Annals of Math. 113, 267-314.

[1982] L. Kirby and J. Paris, *Accessible independence results for Peano arithmetic,* Bull. London Math. Soc. 14, 285-293.

[1983] E.A. Cichon, *A short proof of two recently discovered independence results using recursion theoretic methods,* Proc. Amer. Math. Soc. 87 (4), 704-706.

[1984] H.E. Rose, *Subrecursion : functions and hierarchies,* Oxford Logic Guides 9, Oxford Univ. Press.

[1985] W. Sieg, *Fragments of arithmetic,* Annals of Pure and Applied Logic 28, 33-71.

[1987] G. Takeuti, *Proof theory,* (2nd. edition) Studies in Logic 81, North Holland, Amsterdam.

[1987] W. Buchholz and S.S. Wainer, *Provably computable functions and the fast growing hierarchy,* in S. G. Simpson Ed., "Logic and Combinatorics", AMS Contemporary Math. 65, 179-198.

[1987] J.-Y. Girard, *Proof theory and logical complexity,* Bibliopolis, Naples.

[1988] C.S. Copestake and S.S. Wainer, *A proof theoretic approach to the termination of computer programs,* Report 26.88, Centre for Theoretical Computer Science, Leeds University.

[1989] S.S. Wainer, *Slow growing versus fast growing,* Journ. of Symbolic Logic 54, 608-614.

[1989] M.V. Fairtlough and S.S. Wainer, *Ordinal complexity of recursive definitions,* Report 28.89, Centre for Theoretical Computer Science, Leeds University.

[Address] School of Mathematics and Centre for Theoretical Computer Science, University of Leeds, Leeds LS2 9JT, UK

Reflecting the Open-Ended Computation System of Constructive Type Theory *

Robert L. Constable
Stuart F. Allen
Douglas J. Howe

Cornell University
Ithaca, NY 14853

Abstract

The computation system of constructive type theory is open-ended so that theorems about computation will hold for a broad class of extensions to the system. We show that despite this openness it is possible to completely reflect the computation system into the language in a clear way by adding simple primitive concepts that anticipate the reflection. This work provides a hook for developing methods to modify the built-in evaluator and to treat the issues of intensionality and computational complexity in programming logics and provides a basis for reflecting the deductive apparatus of type theory.

1 Introduction

In this paper we use the term "reflection" to refer to constructions which allow a language to talk about its syntax. This capability is important in natural language and is also an important mechanism in formal languages. In Lisp it is used to provide an extensible syntax. Formal logical calculi also use it to provide an extensible inference system [4], by allowing users to state new rules of inference and prove that they are sound. Reflection need not always be explicitly provided for when a language is designed because it can sometimes be achieved through *gödelization*, a technique used by Gödel to prove his incompleteness theorem by reflecting the relation "p is a proof of P" inside the pure language of arithmetic.

Reflection provides not only a basis for reasoning about computation, but also a means of modifying evaluation, say to make it more efficient. For instance it is possible to provide other function evaluation procedures such as "call by value" in addition to the basic lazy

*Supported in part by NSF grant CCR-8616552 and ONR grant N00014-88-K-0409.

NATO ASI Series, Vol. F 79
Logic, Algebra, and Computation
Edited by F. L. Bauer

evaluation. Reflection also provides a basis for reasoning about syntax. We can define basic operators on terms at the reflected level, e.g. substitution, renaming, pattern matching, unification, etc. These can be given just as at the metalevel, providing an internal account of the basic system operators. Also, because the term structure is so general, its reflection provides a way to study syntax and metamathematics rigorously inside the system in such a way that the results are applicable to the reflected system. Moreover, because the theory is constructive, the metatheorems are also applicable.

We believe that reflection will be a good mechanism for treating at least two basic concepts that have proved troublesome in formal programming logics, namely the notion of the *structure* of algorithms, formulas, proofs, etc., and a notion of *computational complexity.* In some ways reflection seems like an obvious way to treat these concepts, but formal reflection mechanisms are subtle and sometimes opaque. So there is a bias away from them (see for example [13, 10]), and they have not been used in this way. One of the results of our work is a clean reflection mechanism which will support an account of structural and computational complexity in programming logics.

One reason that we can make the reflection mechanism especially clear is that we describe it in constructive type theory, an exceptionally rich language. One reason it is so useful is that the theory it reflects is so rich, unlike the situation with a quantifier free theory based on Lisp syntax. However, one of the characteristics of this theory is its *open-ended computation* system. The "openness" notion is that the theory should remain sound when new computational forms are added as primitives as long as they satisfy certain simple conditions. This feature of type theory is very important to its role in the foundation of constructive mathematics because in mathematics the notion of construction is never finally closed off. We believe that this feature is also interesting in programming languages and logics. For example, building a programming logic on the principle guarantees that all rules and theorems will remain valid if a new operator is added, for example a new form of iteration, or a non-deterministic choice operator, or a form to represent a more efficient access mechanism to a data structure, or forms to reflect the theory in itself. (For example, the underlying computational system of Nuprl has been extended several times, once to add induction for recursive types [9], again to allow generators for infinite objects [9], then a fixed point operator [12]. Similarly Martin-Löf type theory has been extended twice since 1973 [7], to add well-founded trees [8] and lazy numbers. Because of the open-endedness principle, no previously established theorems became false in these extensions. Other programming languages, notably Lisp, have undergone similar evolution. Open-endedness will be discussed further in section 3.

In a sense, open-endedness is assumed by programming language designers who freely add new constructs in the evolution of a language, e.g. new procedure calling mechanisms. Difficulties do not show up until one tries to axiomatize the language, then rules must often be amended to take 1account of new kinds of execution. Constructive type theory is designed from the start to accomodate these extensions without changing the rules already established. Our results show that it is possible to preserve this important design principle and nevertheless allow a reflection mechanism. It is not clear *a priori* that this can be done.

This is acheived by providing primitives that anticipate reflection, as opposed to taking a fixed language and finding a gödelization; we cannot use a fixed encoding of the terms into a fixed class such as the integers because the class of terms can be extended after any such encoding is attempted.

One of the by-products of our analysis is a new method for explaining the syntax of type theory itself. This can be done because the type theoretic notation for inductive definitions and the basic types and sets used to define the theory itself are isomorphic to standard informal accounts of these concepts. So the definition of the syntax and computation rules is presented first in this natural high level notation which then turns out to be exactly the notation being reflected into the internal definition of a formal language. This capability is not so strikingly useful for less rich theories.

We put off the more complex issue of how to reflect the deductive apparatus of type theory [2], but the practical value of such a reflection is well illustrated by Howe's development of a reflected theory of term rewriting for a fragment of type theory [5, 6]. The work described here is a necessary prerequisite to "deductive reflection", and is sufficiently involved that it bears this separate treatment. In particular, this is a good basis for exploring the notion of complete reflection in programming languages.

2 Preliminaries

2.1 Terms

Constructive type theory can be defined starting with a class of terms which will denote the types and their inhabitants. For example, there may be a term for the list induction combinator; following [3] it has the form *list_ind*(l; b; $h, t, v.g$). In this term, l, b and g are *subterms* and $h, t,$ and v are *bound variables* whose scope is g i.e., all occurrences of these variables in g are *bound*. Typically there will be many term constructors such as pairing of terms a and b, often denoted $< a, b >$, and injecting a term a into a binary dijoint union, often denoted $inl(a)$ or $inr(a)$. In Martin-Löf '82 there are 30 such operators, in Nuprl there are 38. It simplifies matters to regularize term formation. Here we will stipulate that all terms, however they may be written, may be uniquely put into the form $op(\overline{v}_1.t_1; \ldots; \overline{v}_n.t_n)$, where $\overline{v}_i$ is a vector of *bound variables* and t_i is a *subterm*. We consider the *operator*, *op*, as a component; let us call a term constructed in this way an *instance* of *op*. If there are n subterms of an instance of an operator, we say that the operator has *arity* n; each operator has a unique arity. And for each argument place i, there is a unique number of binding variables, the length of $\overline{v}_i$, required for binding into t_i. When the arity is zero, we omitt the parentheses and write only *op*.

Terms are built inductively from variables using operators, constants being instances of operators of arity zero. We treat variables as a special category of term, and indeed, we shall assume they are gotten by injecting into the class of terms, some discrete class of objects, which it will be convenient for us to call *variables* as well. When we must take special care in distinguishing these latter objects we call variables from the terms we call variables, we shall call the terms *variable injections*.

The inductive character of terms justifies primitive recursion and proof by induction over the structure of terms. For example, we might define substitution by *term-induction* of the following kind. The result of substituting a for variable x in term t is denoted $subst(a;x;t)$.

To define this operation, let arity (t) denote the arity of the operator and let *bindings* and *subterms* be functions which list the bound variable for each subterm position and the subterm of that position, e.g. for $op(\overline{v}_1.t_1;\ldots;\overline{v}_n.t_n)$

$$\begin{aligned}&arity(op) = n\\&bindings(i) = \overline{v}_i\\&subterm(i) = t_i \quad for\ 1 \leq i \leq n.\end{aligned}$$

If we also use a special form for term induction, substitution is defined as

$$\begin{aligned}&subst(a;x;t) =\\&term\_ind(t;\ if\ t = x\ then\ a\ else\ t;\\&\quad op, bvd, subst, v.(make\_term(op, bvd,\\&\qquad\qquad \lambda i.\ if\ x \text{ in } bvd(i) \text{ then } subst(i)\\&\qquad\qquad\qquad else\ v))\ ).\end{aligned}$$

Here is a list of the operators of Nuprl which we will mention in this article prior to anticipating the reflection of the theory.

$void \qquad any(a)$

$int \qquad 0\ 1\ 2 \ldots$
$ind(n;x,y.s;b;u,v.t)$
$add(n;m)\ sub(n;m)\ mult(n;m)\ div(n;m)\ rm(n;m)$
$int\_eq(a;b;e;ne) \qquad less(a;b;e;ge)$

$list(A) \qquad nil \qquad cons(a;b)$
$list\_ind(e;b;h,t.v.g)$

$fun(A;x.B) \qquad \lambda(x.b)$
$ap(f;a)$

$prod(A;x.B) \qquad pair(a;b)$
$spread(p:u,v.t)$

$union(A;B) \qquad inl(a) \qquad inr(b)$
$decide(d;u.s;v.t)$

$equal(A;a;b) \qquad axiom$

$rec(t.A)$
$rec\_ind(r; h, z.t)$

$fix(f.b)$.

2.2 Abbreviations

To aid in readability of terms, Nuprl has built-in abbreviations or so-called *display forms* for them. Here are those for the above terms. The term $list(A)$ is written A *list*. Elements of a list are built with *cons*, but we sometimes write $cons(a; b)$ as $a.b$. The dependent function space constructor, $fun(A; x.B)$, is abbreviated $x : A \rightarrow B$; in the case of no dependency, it is displayed as $A \rightarrow B$.

The dependent product space constructor, $prod(A; x.B)$, is displayed as $x : A \times B$, and the disjoint union $union(A; B)$ is displayed as $A|B$ or as $A + B$.

2.3 Evaluation

Computational meaning in type theory arises from rules for evaluating terms, e.g. saying that $1 + 1$ evaluates to 2. Evaluation is a partial procedure from closed terms to closed terms. A more involved example than $1 + 1$ is needed to see the key points. We consider the evaluation of the list induction form. Let us abbreviate $list\_ind(\,s\,;\, b\,;\, h, t, v.\,g\,)$ by $F(s)$. Then $F(s)$ evaluates to a term e iff either s evaluates to *nil* and b evaluates to e, or else, for some a and l, s evaluates to $a.l$ and $g[a, l, F(l)/h, t, v]$ evaluates to e.

In general the evaluation relation will be of the form $op(\overline{v}_1.t_1; \ldots; \overline{v}_n.t_n)$ evaluates to t. For a fixed closed system, such as the rules in Martin-Löf '82 or Nuprl as of 1986 [3], there are a finite number of such rules, and they determine the *computation system*.

Here are the evaluation rules for the forms introduced above. The term *void* evaluates to itself, and $any(a)$ does not evaluate to anything (the evaluation procedure does not give a value).

The term *int* as well as 0 1 ... all evaluate to themselves. If we let $F(n)$ denote the induction term $ind(n; x, y.s; b; u, v.t)$, then $F(0)$ evaluates to the value of b; and for $n + 1$ positive, $F(n+1)$ evaluates to $t[(n+1)/u, F(n)/v]$, and for $n-1$ negative, $F(n-1)$ evaluates to $s[(n-1)/x, F(n)/y]$. So in general to evaluate the term $F(n)$ one must first evaluate the subterm n and obtain zero or a positive or negative number. The term $int\_eq(a; b; e; ne)$ evaluates to the value of e if both a and b evaluate to equal integers and to the value of ne if they evaluate to unequal integers.

The terms $list(A)$, nil, and $cons(a; b)$ evaluate to themselves. The procedure for evaluating $list\_ind(l; b; h, t, v.s)$ was given above.

The terms $fun(A; x.B)$ and $\lambda(x.b)$ evaluate to themselves. To evaluate the term $ap(f; a)$, first evaluate the subterm f, and if its value is $lambda(x.b)$ for some variable x and subterm b, then continue by evaluating $b[a/x]$.

The terms $prod(A; x.B)$ and $pair(a; b)$ evaluate to themselves. To evaluate the term $spread(p; u, v.t)$, first evaluate p, and if the result is a term of the form $pair(a; b)$, then continue by evaluating $t[a/u, b/v]$.

The terms $union(A; B)$, $inl(a)$, and $inr(b)$ evaluate to themselves. To evaluate the term $decide(d; u.s; v.t)$, first evaluate d, and if the result is a term of the form, $inl(a)$, then continue by evaluating $s[a/u]$; and if the result of evaluating d is a term of the form $inr(b)$, then continue by evaluating $t[b/v]$.

The terms $equal(A; a; b)$ and *axiom* evaluate to themselves.

The term $rec(t.A)$ evaluates to itself. To evaluate the term $rec\_ind(r; h, z.t)$, evaluate $t[r/z, \lambda w.rec\_ind(r; h, z.t)/h]$.

To evaluate the term $fix(f.b)$, evaluate $b[fix(f.b)/f]$.

Here are some examples of evaluation. The term $spread(p; u, v.u)$ is the first projection, selecting the first member of a pair, e.g. $spread(pair(1; 2); u, v.u)$ evaluates to 1. Although $spread(2; u, v.u)$ is a well-formed term, it does not evaluate to anything since the first argument, called also the *principal argument*, does not evaluate to a pair. The term $spread(ap(\lambda(x.pair(1; 2)); x); u, v.v)$ evaluates to 2 since the application of the function results in a pair and the spread selects its second argument.

The induction form requires some practice before it seems natural. It is a little easier to play with a simpler form which omits the negative case. It is $ind(n; b; x, y.h)$. Letting $F(n)$ denote the form where h is x+y and b is 0, here is how it evalututates for n equals 2. $F(2)$ is the value of $2 + F(1)$, and $F(1)$ is the value of $1 + F(0)$, and $F(0)$ evaluates to 0. So $F(2)$ evaluates to the value of $2 + (1 + 0)$. The form $ind(n; 0; x, v.f(v))$ can be seen to produce the values 0,$f(0)$, $f(f(0))$, $f(f(f(0)))$ for n equal 0, 1,2,3 respectively. The form $ind(n; f; x, v.\lambda(x.f(v(x))))$ can be seen to produce the functions f, $\lambda(x.f(f(x)))$, $\lambda(x.f(f(f(x))))$ for n equals 0,1,2 respectivley. The form $ind(n; int; x, v.prod(int; v))$ can be seen to produce the sequence of types *int*, $prod(int; int)$, and $prod(int; (prod(int; int))$ for n equals 0,1 and 2 respectively.

The term $fix(f.f)$ does not evaluate to anything. An attempt to evaluate it indefinitely will produce an unbounded computation since after any finite number of reductions, the result is the term itself.

2.4 Open-Endedness

For reasons discussed in the introduction, we are interested in *open-ended computation systems*. This means that we must be able to provide for the introduction of new terms and their incorporation into the evaluation relation. It is not clear how this can be specified in such a way that the system is truly open-ended and yet can be completely described to the point where the mechanism can be reflected into the theory itself. We consider the issue further before introducing our solution.

Let us look more closely at what might be meant by open-ended computation systems. We are aware of two radically different interpretations. One, which we may call *profound* open-endedness, construes open-endedness to be a property of computation systems. On this view, one might also consider other open-ended sorts of systems or classes.

The need for open-ended computation systems in the *intuitionistic theory of types* (see ML82) arises because the intuitionist does not believe that all methods of computation can be effectively described at once, or at least not now. Hence, the rejection of Church's thesis. Yet, intuitionistic type theory needs a type of all possible representations of, for example, (effectively computable) functions from numbers to numbers. As a result, one must be able to choose new representations of number-theoretic functions and introduce them at will.

The other interpretation of open-endedness of (computation) systems is independent of constructivism. Here, the open-endedness is not a property of individual systems, but rather, consists in pragmatic constraints on our *use* of systems. Whatever theorems and procedures we have developed *within those constraints* for one system, will apply to any *different* system which is a suitable extension of the first. The language of constructive type theory can be given a natural re-interpretation using conventional (non-open-ended) classes, using ZF, say, which preserves the suitability of the language as a programming language and logic [1]. It happens that when one carefully states the method for introducing new terms under the "profound" interpretation, the results are readily re-interpreted as conditions for suitable extension of fixed evaluation systems. Our opinion is that this particular approach even to merely pragmatic open-endedness is quite elegant. Also, the method for reflecting the profoundly open-ended computation system will work just as well for closed systems.

An examination of [8] would reveal that to introduce a new form of expression into the language, one must specify the number of constituent subexpressions required to form expressions of that form, and one must specify which variables become bound in which constituent subexpressions, and one must explain how to evaluate terms of that form. We have refined this procedure, as part of an effort, upon which we hope to report later, to clarify for ourselves a number of issues in the design of type theoretic languages. Here is an approximation adequate for explaining our reflection method. We assume a discrete class of objects we shall call the *variables*. We assume an open-ended class of discrete objects to which we may always, at will, introduce a new member; we shall call the members of this class *operators*. But, whenever one introduces a new operator, one is required to associate with it several other entities that meet several conditions. Introducing a new operator and carrying out these related tasks constitute the procedure for introducing a new form of expression:

1. Introduce a new operator.
2. Indicate a number as the *arity*.
3. Indicate, for each i from 1 to the arity, called the *number of binding variables at place* i.
4. By a *term* we shall mean a (finite) tree each of whose nodes is labelled either with a variable or else with
   - an operator and

- for each i from 1 to the arity, a list of variables, the length of which is the number of binding variables at place i.

Give a partial term-valued procedure on terms; this procedure may apply the evaluation procedure. Further, this procedure must meet several conditions which we will not describe here. We shall call this procedure the *evaluation fragment.*

To evaluate a term t, see if the root is labeled with an operator rather than a variable; if so, then apply the evaluation fragment for that operator to t; if execution results in a term s, then return s.

There is much about this procedure for introducing forms of expression that demands explanation and justification. For example, the availablity of the full evaluation procedure in the evaluation fragments requires a finer analysis. The conditions imposed on the evaluation fragment when introducing it are quite important as well; they are needed to guarantee some global features of the open-ended system, such as idempotence of evaluation, and the harmlessness of changing bound variables. And, of course, the specification of open-ended classes suitable for use as the operator class is not unproblematic.

But, our topic is reflection, and the features of the computation system to be reflected here are adequately represented for the purpose. These are the variables, the open-ended class of operators, the tree structure of terms, and the evaluation procedure. The main thing to notice is that the whole process of computation-system extension is *operator driven.* The open-endedness of the class of terms arises solely from the open-endedness of the class of operators, and the extension of evaluation to new terms is incremental in the operators.

The merely pragmatic open-endedness is realized by interpreting the class of operators in the above procedure as a parameter ranging over fixed discrete classes. To use a fixed computation system in an open-ended fashion, one proves only theorems true for every computation system reachable from it by a series of extensions through the above procedure.

2.5 Informal Semantics

We will not be concerned in this article with the semantics of the theory nor with its proof rules and proof theory. The reflection of the notion of proof is the concern of [2] and the semantics is presented in [1, 3]. But a brief reminder of the informal meaning of the terms listed above might be helpful.

The term *int* is of course meant to denote the integers, and the associated familiar terms have their obvious meaning. The unfamiliar induction term has already been explained to some extent. Not only is it a primitive-recursion combinator used for calculation, but via the so-called *propositions-as-types* principle, it also serves as the computational content of the induction principle. This point is by now explained in a number of articles on constructive type theory, in [3, 8] for instance.

The term $list(A)$ donotes the type of lists over the type A, provided of course that A is a type. Lists are built from the empty list, *nil* by the constructor *cons*. The list of the first three positive integers is built as $cons(1; cons(2; cons(3; nil)))$. Sometimes we write these lists as (1,2,3) or as 1.2.3.*nil*. The usual head and tail functions for lists can be built from the list induction combinator, *list_ind*.

The term $fun(A; B)$ denotes the type of functions from A to B when these are types. The form $fun(A; x.B)$ allows for the type of the range to depend on the argument to the function. It is called a *dependent function space*. The $\lambda(x.b)$ term is the format used to write functions as *lambda terms*. We will also use the form $\lambda x.b$ as our informal way of writing functions. So $\lambda x.x + x$ is the informal doubling function which is written inside the theory as $\lambda(x.add(x; x))$. The term $ap(f; a)$ denotes the *application* of the function f to the argument a which is informally written $f(a)$.

The term $prod(A; B)$ denotes the cartesian product of the types A and B. The elements of this type are pairs, written $pair(a; b)$ where a is from A and b is from B. In the form $prod(A; x.B)$ it denotes the *dependent product type* in which the type of the second component can depend on the value of the first. That is, b must come from the type $B[a/x]$. The term $spread(p; u, v.h)$ is used to break apart a pair as the evaluation rules show.

The term $union(A; B)$ represents the disjoint union of the types A and B. The canonical members of this type are either of the form $inl(a)$ or $inr(b)$ when a is form A and b is from B. The term $decide(d; u.g; v.h)$ is used to decide which disjunct the element d is in, as shown by the evaluation rules.

The term $rec(T.B)$ is used to denote a recursively defined type when B is a monotone operator on types. We often see these definitions written as $T := B$ where B contains instances of T. The term $rec\_ind(t; h, z.b)$ denotes the recursion combinator on a recursive type. For a full account see [3, 9].

3 Defining Reflection

Our aim is to reflect the tree structure of terms as described above, to reflect the evaluation procedure as a whole, and to reflect the reflection apparatus itself. Before proceeding with that, let us contrast this aim with a variety of other related goals which one might attempt. As was mentioned earlier, we will not carry out the reflection of the deductive apparatus, although it could be done rather directly since the possible formal proofs are inductively definable from the terms, see [2].

The more detailed structure of evaluation and evaluation fragments, as well as the constraints on evaluation fragments (which were mentioned but not described), will not be reflected. In particular, although we could reflect the fixed-point construction of evaluation from (further parameterized) evaluation fragments, we found it to be very complex; we do not know of a practical benefit great enough to justify reflecting in such detail. However, we do want to reflect, but not here, a principle of induction over complete evaluations, which is critical to proving important global, i.e., not operator-specific, facts about terms that have values (see [12] for relevant discussion).

We will not try to reflect our knowledge about the reflection mechanism. Although we will be able to use our reflective devices to state propositions about the syntax, even the design criteria to be worked out below for the reflective apparatus itself, we will not concern ourselves here with providing formal deductive apparatus sufficient for proving these propositions (which we know to be true). In practice, it is, of course, of prime importance that sufficient deductive apparatus be designed (by appropriate choice of new axioms or inference rules). Finally, we do not even partially reflect the semantics of type theory, which must begin with the reflection of type membership.

3.1 Types and Representation

A typical use of types within the language of intutitionistic type theory, is as a collection of expressions referring to constructive objects of a a given sort. For example, one may define a type N for the purpose of collecting represention of some kind for natural numbers, or N $\rightarrow$ N to collect some manner of representations for (effectively computable) numeric functions. We may use a type T to refer not only to itself but also to the sort of objects its members are designed to represent. Two terms t and s which are members of a type T, or $t, s \in T$, are said to be equal in type T, or $t = s \in T$, when they represent the same object of sort T. The intended use of type theoretic language requires that the representation associated with any given type be preserved under evaluation, and that only closed terms can be used as representations.

Here is the representation of natural numbers used for the type N. (Although we use *int* in Nuprl, N is a simpler type for which to make this point.) Zero is represented by any term that evaluates to the term 0. The successor of a number n is represented by any term that evaluates to a term succ(t), where n is represented by t. This interleaving of evaluation with canonical forms is typical of the method of lazy evaluation. Note that there will be infinitely many representations of each number, and that whether an arbitrary term represents a number is undecidable.

To make use of this representation, the type N is introduced and inhabited by the representations of numbers, equal members of N being terms that represent the same number. We may take the design of the type N as our paradigm for designing types for representing terms.

3.2 Anticipation

We now turn to the question of whether we can completely reflect the open-ended computation system described above. Our hope is that by using type theory (rather than gödelizing) we can offer an especially clear analysis on account of the ease of designing representations and types.

However, looking ahead to designing the representation of terms, we may expect to use a representation for operators, since they are constituents of terms. How can we guarantee that whenever a new operator is introduced we will also have a name for it. After considering a few alternatives, we decided to *extend the structure* of terms; for each operator, not only will the instances of that operator be among the terms, but so will a new degenerate form of term gotten simply by injecting the operators into the *term* class. We shall call such degenerate terms *operator injections*. So, from now on we shall admit three kinds of terms: variable injections, operator injections, and instances of operators. As we shall see, with this single anticipation of the reflection scheme to come, the design will develop quite smoothly. Just as we have reduced the notion of open-endedness to one specific type, the operators, we reduce reflection to the notion of *anticipating it in the type system*. Given the type system before adding these new types, the extension illustrates the power of open-endedness to provide useful new forms of computation.

3.3 Reflecting Terms

We shall only outline the design and implementation of the reflection scheme in this article. We can determine what new type and members are needed by describing precisely the computation system as set out in section 2. One can view this effort as an explantion of the basics of computation in the theory itself.

There must be types for *operators, variables* and *terms*. These will be denoted Op, Var, $Term$. (In general we notate the internal reflective notions by capitalizing the first letter of the informal type used in the metalanguage.) Var should be a discrete type handy enough to use in typical syntactic operations on variables, such as finding an unused one. The handiest would be lists over some finite alphabet, perhaps *Letterlist*, but for purpose of demonstration, we shall not require this here.

$Term$ can be easily defined as a recursive type given the types Var and Op. From the form of an operator in a term, it is easy to compute its arity (Arity) and number of bound variables (Bvnum). Here is a plausible defintion of Term:

$$Term = rec(t.union(Op; union(Var; prod(Op; list(prod(list(Var); t))))))$$

Using the abbreviations mentioned above, this can be displayed as

$$rec(t.OP + (Var + Op \times (Varlist \times t)list)),$$

Arity: $Op \rightarrow N$,

Bvnum: $o : Op \rightarrow i : [1, Arity(o)] \rightarrow N$.

(Here we employ the dependent function space constructor to give an exact typing of these functions.)

So, defining $Term$ boils down to defining Op; but this is just what we anticipated. Define Op by stipulating that $a = b \in Op$ iff, for some operator op, a and b evaluate to the operator injection of op.

Once it is clear how terms are to be represented, then we can define (at the metalevel) a function *rep* from term to *term* which carries out the representation, and an inverse function, *unrep*. The details are quite tedious, so we only sketch the ideas. First in order to define *rep* we need some representation of the elements of Op; let us say that there is some function *repop* and $repop(op) = \widehat{op}$, but we won't say just what $\widehat{op}$ is in all cases. Just to be concrete for an example, suppose that $repop(spread) = SPREAD$ and $repop(inr) = INR$. We also need some representation of elements of Var, given by a function $repvar(v) = \hat{v}$. (Since we have not said just what Var is, we cannot say what its representation is.) With these notions in place we can say that

$$rep(op(\overline{v}_1.t_1;\ldots;\overline{v}_n.t_n)) = pair(\widehat{op};(pair(\hat{\overline{v}}_1;rep(t_1)),\ldots,pair(\hat{\overline{v}}_n;rep(t_n))))$$

where $\hat{\overline{v}}_i$ is a list of representations of the variables in the list $\overline{v}_i$.

To see how this works on an example, consider the term $spread(x;u,v.u)$. Let $varinj(x) = inr(inl(x))$, and $complx(x) = inr(inr(x))$.

$$\begin{aligned}
&rep(spread(x;u,v.u) = complx(body)\\
&\text{where } body = pair(SPREAD;list) \text{ and}\\
&\text{where } list = cons(first;rest) \text{ and}\\
&\text{where } first = pair(nil;principalarg) \text{ and}\\
&\text{where } principalarg = varinj(\hat{x}) \text{ and}\\
&\text{where } rest = cons(second;nil) \text{ and}\\
&\text{where } second = pair(varlist;varinj(\hat{u})) \text{ and}\\
&\text{where } varlist = cons(\hat{u};cons(\hat{v};nil)).
\end{aligned}$$

There is also a formal analogue of term induction, given by the new operator $Term\_ind$. It has arity 2 and is written generally as $Term\_ind(t;b;o,s,b,v.g)$ where intuitively o picks out the operator of t, s the subterm, b the bound variables and v the value of $Term\_ind$ on subterms. This operator is defined using rec_ind which is the built-in recursion combinator for recursive types discussed above.

By induction on $Term$ we can define inside the theory versions of rep and $unrep$. They are functions on $Term$. Again we do not have enough detail to define them completely, but the idea might be clear from an example. Consider the internal version of rep, called according to our conventions, Rep. On the above example,

$$Rep(complx(body)) = complx(pair(INR;cons(pair(nil;Rep(inr(body)));nil))).$$

3.4 Reflecting Evaluation

The evaluation rules given in section 2.2 were expressed as fragments of an overall evaluation procedure on terms. To be more explicit, let us call this procedure val. Here is the way its definition would begin:

$val(t) =$

if t is $\lambda x.b$, then the value is t

if t is $ap(f;a)$ then if $val(f)$ is $\lambda x.b$ then the value is $val(b[a/x])$

if t has o as outer operator, then use the evaluation fragment for o with val to evaluate subtertms.

We see that *val* is a recursive procedure, $val(t) = F(val, t)$ where F is the right-hand side of the above definition as a function of *val* and t. One approach to defining an internal val is to write the above procedure as an internal partial function on Terms. Doing this would require new primitive forms for the evaluation fragments, say such as $Evalfrag(o; t; e)$ where e is the evaluation procedure on subterms. But typing this function directly would require a more elaborate type theory, the one in [12] would suffice. There is a simple alternative to using the more general types, namely, to introduce a new primitive predicate defining the domain of the above procedure and a new operator which evaluates terms.

The evaluator is represented as a partial function by two components. There is a relation, i.e., a 2-place type-valued function, to represent evaluation, and a function, Val, to compute the value of a term when it exists. The type s *Evaluates_to* t should be inhabited when s represents a term s', and t represents a term t', and s' evaluates to t'.

The term $Val(s)$ is to be defined as follows, assuming that we have a procedure for choosing a *standard* representative of a term: to evaluate $Val(s)$, see if s represents a terms s'; if so, then see if s' evaluates to some t'; if so, then return the standard representative of term t'. Thus, the implementation of Val only awaits the design and implementation of $Term$ (term representation) and a standard-representation function.

All that remains to design is the reflection of the rest of the reflection apparatus, namely, standard term-representation. The specification for $s\,Reps\,t$ and for $s\,Evaluates\_to\,t$, are analogous; and $Rep(s)$ should be designed to represent the standard representation function, much like $Val()$ represents the evaluation function (only, $Rep(s)$ is total on terms).

If we restrict our attention to a specific closed subclass of terms, say all those in the theory as of 1990, called it Term90 (it consists simply of enumerating the operators that are already defined), then we can define an internal evaluation function, say $Val90$ which directly mimics the recursive procedure for *val* on all of the terms now known. It would have the form $Val90 = \lambda t.fix(v.F(v,t))$ where F dispatches on the form of the outer operator; for example, if t is $ap(f; a)$ then the procedure is to apply v to f and if the result of that application is $\lambda x.b$ then to return v applied to $b[a/x]$. We can also define the functions $Comp(n,t)$ and $Steps(n,t)$ which respectively reduce t for n steps and return the number of steps it takes to reduce t to canonical form if that happens in less then n steps else n.

In order to prove that $Val90$ computes the same result as Val, we need to argue by induction on the terminating computations, $Val(t)$. A good principle can be stated by letting $Term\dagger = \{x : Term | \exists y : Term.x\,Evaluates\_to\,y\}$ (likewise for Term90).

Computation Induction: $\forall t : Term\dagger .P(t)$ provided that for all t in $Term\dagger$ we know that if $P(s)$ holds for all terms s evaluated in the computation of t then $P(t)$.

Using this principle we can prove

Computation Details: $\forall t : Term90\dagger .Val(t) = Val90(t)$.

The proof proceeds by Computation Induction.

Using this connection to a concrete evaluator, we can build alternative evaluators inside the system which are provably equivalent to the original. For example, we can introduce notions such as graph-reduction or environment based evaluation and show that they are equivalent to the built-in evaluator. In some cases we can show that an eager evaluation strategy for a class of terms produces the same results as the built-in lazy evaluator. It is also possible to incorporate provable optimizations.

3.5 The Denotation of Terms

From our stance outside the theory, we can see that Term represents the class of terms. For example we know that $inr(inr(pair(0; nil)))$ is the member of Term representing 0, and that $inl(0)$ represents itself. But there is no way inside the theory to say these things, that is to express the meaning or the denotation or the *reference* of a term. This is a very different notion from that of the value of a term. For example, the term $Val(inr(inr(pair(0; nil))))$ evaluates to $inr(inr(pair(0; nil)))$. To express this concept of reference we need a new notion and a new primitive operator, Ref. It appears in terms of the form $Ref(t; v; l)$ (when t is closed we use $Ref(t)$ as well). These terms are well-formed when t is a member of Term, v is a list of variables including all the free variables of t, and l is a list of values at least as long as the list v of variables.

The details of evalutating $Ref(t; v; l)$ depend on the details of rep and $unrep$ which we have not been concerned with in this article. Essentially Ref behaves like $unrep$ except that variables are mapped to their values. Thus if x belongs to Var, then as a term it is a variable injection, $inr(inl(x))$, call it ix, then $Ref(ix; cons(ix; nil); cons(5; nil))$ computes to 5 whereas the $unrep$ of ix is just x.

The meaning which the theory must axiomatize is that $Ref(t; v; l)$ is the denotation of the term t when its free variables are given the corresponding values in the list l, i.e. when listing the pairing of the elements of v with the corresponding ones of l produces an environment in which t is defined.

Ref allows us to define this type, $[A] = \{x : Term | Ref(x) in A\}$. This suggests a way of viewing types as predicates on $Term$.

3.6 Completeness

We claim without proof, since the point seems sufficiently clear, that with the above representations of terms and evaluation, there is a complete representation of evaluation inside the theory. So for example, given any function term $\lambda(x.b)$,call it f, the function from $term$ to $term$ given by $\lambda Y.val(ap(f; Y))$ for Y a metavariable over terms is represented by an internal function from $Term$ to $Term$ given by $\lambda(y.Val(rep(ap(f; Y))[y/Y])$. The term $rep(ap(f; Y))[y/Y]$ is a way of saying that we want the representation function to be abstract with respect to y, that is y is not to be represented as a variable but *is* a variable.

In this article we have been very brief in discussing the representation details. We want to mention that there are a number of ways to meet the design criteria listed above, given the sort of type theory we have in mind. It should be an open-ended extensional system with at least the primitives of Martin-Löf '82 or Nuprl. Depending on certain other details, there are various ways to proceed. For example, if there are recursive types, then $Term$ can be defined naturally using them as sketched above, otherwise it must be encoded somewhat, say using the W-types of Martin-Lőf '82, or even using quotient or set types of [3], along with the other usual constructors. If there are partial types, as in extensions of Nuprl [3, 12], then *Val* can be interpreted as a partial function. Otherwise we can use *Evaluates_to* to specify its domain exactly and make it a total function.

Among the details omitted is a definition of the formal variables. They could be simply natural numbers, that is we could implement *Var* as N.

There is no significant latitude for implementing *s Evaluates_to t* and *Val(s)*. They are simply introduced as primitives in the most direct possible way.

It turns out that *s Reps t* and *Rep(t)* need not be introduced as primitive since given the other primitives, they are expressible in terms of the other conventional operators by recursion on *Term*.

To sum up, the only primitives we introduce are the type *Op*, exploiting operator injection, the operator *Val(t)*, and the type constructor *s Evaluates_to t*. ¿From these, together with conventional operators and type constructors, all the rest of our reflection apparatus can be expressed. Although in future work we will show that the implementation meets the design and that the reflection mechanism itself can be reflected, we feel that it is the design, described in some detail above, which conveys most of the novel ideas.

3.7 Stance

We can stand, as we now are, outside the theory and talk of *term* and *rep* and so forth. From this vantage point we can speak about the *real integers* and *actual functions* of our meta theory, saying things like *int* represents the integers.

Once we start to use the theory, we can *get inside it* and then *int* are the integers as far as we are concerned, and $\lambda(x.b)$ is an actual function since the val operation is somehow *animated* by the implementation, and when we invoke

$$val(pair(ap(\lambda(x.mult(x;x));17);ap(\lambda(x.mult(x;mult(x;x)));7)))$$

we really see the pair 289 and 343. (In Nuprl *val* is invoked by entering E-mode and presenting the system a term.)

This is not be the ideal *stance* however, because we cannot do things like find the representative of an object or analyze it as a term, syntactically. Perhaps the right place to stand and the right place to provide the system's automated help is at the level of represented terms, i.e. for *Term*. (Using color could allow us to present *Term* in just the same way we present *term* except or color.) Given t in *Term* we can speak of its structure, its intensional or syntactic properties; we can evaluate it, using *Val* and get some idea of the resources required using $Step(n,t)$. We can speak of the *real values* that a term t represents using $Ref(t)$, and we can reflect t into the metatheory using $Rep(t)$. From this vantage point we can *see in both directions* so to speak, and we can use Nuprl as its own metatheory. (In this stance, to execute t we enter $Val(t)$ in E mode.)

The programming language Lisp was originally presented in terms of M-expressions and S-expressions. The M-expressions included *eval* which was used to *animate* the language.

But the S-expression level came to be the one really used. This corresponds to adopting for Nuprl the stance that we are working with *Term*. In the setting of Lisp it is worth mentioning that this reflected theory provides many, perhaps all, of the capabilities of Smith's 3-Lisp [11]. The tower of reflective evaluators he describes seems to correspond to a sequence of ever deeper *Val* functions working on ever deeper representations of terms made possible by the internal representation function, *Rep*.

References

[1] S. F. Allen. *A non-type-theoretic semantics for type-theoretic language*. PhD thesis, Cornell University, 1987.

[2] S. F. Allen, R. L. Constable, D. J. Howe, and W. E. Aitken. The Semantics of Reflected Proof. *Proceedings of the Fifth Annual IEEE Symposium on Logic in Computer Science*. IEEE Computer Society,pp. 95-107, June 1990.

[3] R. L. Constable, S. Allen, H. Bromely, W. Cleaveland, and et al. *Implementing Mathematics with the Nuprl Development System*. NJ:Prentice-Hall, 1986.

[4] M. Davis and J. Schwartz. Metamathematical extensibility for theorem verifiers and proof checkers. In *Comp. Math. with Applications, v. 5*, pages 217–230, 1979.

[5] D. Howe. Computational metetheory in nuprl. *CADE-9*, 230:404–415, 1988.

[6] D. Howe. *Automating Reasoning in an Implementation of Constructive Type Theory*. PhD thesis, Cornell University, Ithaca, NY, 1989.

[7] P. Martin-Lof. An intuitionistic theory of types: predicative part. In *Logic Colloquium '73.*, pages 73–118. Amsterdam:North-Holland, 1973.

[8] P. Martin-Lof. Constructive mathematics and computer programming. In *Sixth International Congress for Logic, Methodology, and Philosophy of Science*, pages 153–75. Amsterdam:North Holland, 1982.

[9] P. Mendler. *Inductive Definition in Type Theory*. PhD thesis, Cornell University, Ithaca, NY, 1988.

[10] N. Shanker. Towrds mechanical metamathematics. Technical report, Institute for Computing Science, University of Texas at Austin, 1984. Tech. report 43.

[11] B. C. Smith. Reflection and semantics in lisp. *Conference Record of the Eleventh Annual ACM Symposium on Principles of Programming Languages*, pages 23-35, 1984.

[12] S. Smith. *Partial Objects in Type Theory*. PhD thesis, Cornell University, 1989.

[13] R. Weyhrauch. Prolegomena to a theory of formal reasoning. *Artificial Intelligence*, 13:133–170, 1980.

Some Lectures on Modal Logic

Anil Nerode* **
Mathematical Sciences Institute
Cornell University
Ithaca, New York 14853
USA

KEYWORDS / ABSTRACT : modal / frames / tableaux / dynamic / concurrency / logic / nonmonotonic / autoepistemic / belief / intuitionistic / constructive

An exposition of some modal logics useful for teaching and research in computer science. §1. Preface §2. Propositional modal logic §3. Modal frames §4. Propositional tableaux §5. Modal axioms and their frame semantics §6. Modal predicate tableaux with constant domains §7. Autoepistemic logic §8. Nonmonotonic reasoning §9. Classical concurrent dynamic logic §10. Intuitionistic concurrent dynamic logic.

§1. Preface

Formal modal logics were developed to make precise the mathematical properties of differing conceptions of such notions as possibility, necessity, belief, and knowledge, notions which arise in philosophy and natural languages. But modal logics have emerged in the last twenty five years as useful tools to express essential ideas throughout computer science and artificial intelligence. What is the best way to teach modal logic to computer science and mathematics students who want to learn and do research in the areas in which modal logic is used? Classical modal logic texts are oriented for philosophers and are not oriented toward computer scientists or toward mathematicians. Computer science models have to be used as examples and the computational content of modal proof procedures has to be brought out. At a more advanced level, see Goldblatt [1982, 1987] for an exposition of some applications of modal logic to computer science, and see Wallen [1990] for an exposition of automated deduction for nonclassical logics. In the present notes we outline, without proofs, some modal logics with computer science examples using an easy–to–teach and easy–to–remember tableaux method suitable for hand computation and automated deduction. This exposition is consistent with that in Nerode [1990] for intuitionistic logic in computer science. An undergraduate or graduate semester applied logic course can easily cover classical, intuitionistic, and modal logics for computer science based on the first fifty pages of Smullyan [1967] plus the intuitionistic notes referred to plus §1–§6 of these notes. In §7–§10 we include very brief expositions of some AI and computer science examples and incorporate some of our own work as examples in §8 and §10. Others teaching such a course will have their own favorite examples and should, in any case, include some of the many examples alluded to in the next few paragraphs. An extensive set of references has been given to encourage this. What are the computer science applications of modal logics?

I. One source is **theories of program behaviour**. Modalities are implicit in the works of Turing [1949], Von Neumann [1963], Floyd [1967], Hoare [1973], and Burstall [1972] on program correctness. Underlying systems of modal logic were brought to the surface by Engeler

NATO ASI Series, Vol. F 79
Logic, Algebra, and Computation
Edited by F. L. Bauer

[1967,1968, 1975], Salwicki [1970, 1977], Mirkowska [1979, 1980], Banachowski et al. [1977], Pratt [1976, 1979], Harel [1978, 1979], Fischer and Ladner [1977], Goldblatt [1979, 1981], Kozen and Parikh [1980], Parikh [1980], Manna and Pnueli [1979], Pnueli [1977, 1981], Manna and Waldinger [1978], and others. The logics of programs developed include algorithmic logic, dynamic logic, process logic, and temporal logic. Here are the primitive modalities of the dynamic logic of sequential programs.

Let α be a sequential program, let s be a state of the machine executing α. Let $\varphi(s)$ be a predicate of states. Then there are modal operators constructing new predicates $([\alpha]\varphi)(s)$, $(<\alpha>\varphi)(s)$ about states s with the following characterization.

– $([\alpha]\varphi)(s)$ asserts that $\varphi(s')$ holds for any state s' such that there exists a legal execution sequence for α which starts execution in state s, terminates execution in state s'.

– $(<\alpha>)\varphi(s)$ asserts that $\varphi(s')$ holds for at least one state s' such that there exists a legal execution sequence for α which starts execution in state s, terminates execution in state s'.

Then $[\alpha]$ is a modal necessity operator and $<\alpha>$ is a modal possibility operator. There are separate operators $[\alpha]$, $<\alpha>$ for each program α. A modal Kripke semantics is evident to those acquainted with Kripke semantics. The program α induces a transition relation between states s in which α begins execution and states s' in which α terminates at the end of execution starting in state s. With the set S of all states of the machine as the set of "possible worlds", this transition relation is a modal accessibility relation between states with associated necessity operator $[\alpha]$ and possibility operator $<\alpha>$ as described above. See §9 for the classical dynamic logic of concurrent programs and see §10 for intuitionistic use of these same modalities for making conclusions about behaviour based on partial information about states.

Other logics have been designed for non–terminating and perpetual processes such as operating systems, and also for concurrent programs. In temporal logic, the program is fixed and considered part of the structure over which the logic is interpreted. Such a logic is sometimes called an "endogenous" logic The current location in the program during execution is stored in a special variable for that purpose, called the program counter, and is part of the state along with the values of the program variables. Instead of program operators, there are temporal operators that describe how the program variables, including the program counter, change with time. Temporal logic originally lacked the ability of dynamic logic to combine programs and deal with several programs in the same model, but because it deals with execution sequences, temporal logic (and another subject, process logic), can deal with correctness of perpetual programs and programs that sometimes halt, such as operating systems and communication networks. Pnueli [1977] suggested that temporal logics could be used to reason about concurrent programs, when the issue of termination ought to be suppressed from the discussion. Temporal logic began as a formal axiomatic subject (tense logic) in Prior [1955]. Temporal logic also has a natural "possible worlds" Kripke frame theory. Syntax and semantics of various temporal logics from a philosophical point of view and without computer science may be found in the excellent texts of Rescher and Urquhart [1971], McArthur [1976], van Benthem [1983]. The 1960's introduced the problem of program specification (what a program is supposed to do), program development (find a program which is supposed to satisfy the specification), and program verification (verify that a program satisfies its specification). Floyd [1967] developed the "inductive assertion method" for verifying that a flowchart program (built up from conditional branching, join of control, and assignment) for computing such a function satisfies "partial correctness" (if the program terminates on an input, the resulting output satisfies the specification). Hoare [1969] turned this into a calculus, much investigated since, based on the construct $\{P\}S\{Q\}$ representing "if the assertion P is true when the program S is initiated, then assertion Q is true if and when the program S terminates." Burstall [1974] developed a method for showing "total correctness" (partial correctness plus the program always terminates). He follows the execution of the program with symbolic data, using mathematical induction to prove general assertions about what happens at loops. Burstall himself makes the point that modalities are involved. In Burstall's proofs of total correctness, assertions to be proved have the form

"($\exists$time t)(at time t, program line 1 is executed and $P(t)$)".

In Floyd's proofs of partial correctness, assertions to be proved have the form "($\forall$ times t)(at time t, program line l is executed implies $P(t)$)". Pnueli [1977] systematized the modal logic suggested by Burstall as a classical logic augmented by $\Box$, $\Diamond$ corresponding to Burstall's suggestion.

$\Box P(t)$ is read "always" and means "now and in the future",

$\Diamond P(t)$ is read "eventually" and means "now or sometime in the future".

He assumed that time is the non–negative integers with the usual order and introduced a third operator $\circ$.

$\circ P(t)$ is read "next P" and means "$P(t+1)$".

These modal logics clarified <u>program correctness proofs</u>, and are equally suitable for concurrent or perpetual programs such as operating systems. An important topic is fairness. This takes many forms. A weak one is that a continuously active process will eventually be scheduled. A stronger requirement is that a process active infinitely often will be scheduled. Another is that a process which is active at least once will be scheduled. All these can be formulated in the Pnueli calculus mentioned, and treated as program specifications. But stronger notions such as "Of two processes, the one that is active first will be scheduled sooner" exceed the capacity of this calculus.

Gabbay et al. [1980] introduced a binary connective U, vUw, read " v until w", such that (vUw) is true at time t if v is true at times t' for $t' \geq t$ until a time $s \geq t'$ arrives when w is true. There are many other contributors. Computer science applications of temporal logic are a thriving specialty in their own right. There are many contributions to the specification and verification of sequential and concurrent systems. We omit extended discussion of temporal logics for computation due to a lack of space. But such examples should be included and can be found in, for example, Goldblatt [1987] and Galton [1987].

II. A second source of modalities in computer science is **theories of knowledge and belief for AI.** These have their roots in philosophy (Hintikka [1962], Stalnaker [1980]) and in Artificial Intelligence (Halpern and Moses [1984], Moore [1983, 1984, 1985, 1988], Marek [1986,], Marek and Truszczynski [1988,1989], Konolige [1986, 1988], and others). Modal operators commonly studied are

– "agent α knows φ", $[\alpha]_K\varphi$

– "agent α believes φ", $[\alpha]_B\varphi$.

This subject is introduced briefly in §7. Now think of "agent 1 believes", "agent 1 knows", "agent 2 believes", "agent 2 knows", etc. We may want all these operators present at once in the same logic. After all, what one agent believes or knows does not necesssarily coincide with what another believes or knows, or with the common beliefs or knowledge of several agents. One objective of such studies is to analyze, model, and machine simulate rational behavior based on knowledge and belief. These applications generally start by putting down reasonable axioms for belief or knowledge, and continue by trying to develop methods of determining whether a given proposition is believed or known on the basis of other propositions. The agents themselves may be machines, and we may be trying to reconcile their databases (beliefs, knowledge). Hintikka [1962] gave a Kripke model of beliefs of multiple agents. In its simplest form, there is one set of possible worlds, but a different accessibility relation for each agent, and an agent believes P if P is true at the worlds accessible to the agent. See Halpern and Moses [1985] for a survey of logics of knowledge and belief. Also see Konolige [1986, 1988]. For lack of space, we omit an exposition of common knowledge and information implicit in distributed systems.

III. A third source of modalities in computer science is **nonmomotonic reasoning.** Nonmonotonic logics deal with reasoning which draws conclusions based on the absence as well as the presence of information. The potential usefulness of developing such a subject for commonsense reasoning was pointed out by Minsky [1975]. There are proposals by McCarthy [1980, 1986] (circumscription), Reiter [1980] (default logic), McDermott and Doyle [1980] (truth maintenance systems), Clark [1978] (negation as failure in logic programming), Moore [1084] (autoepistemic logic). In §7 we introduce the noton of a "point of view", or "belief set" informally and develop autoepistemic logic in three styles as an example. In §8 we give a description of several forms of nonmonotonic reasoning in the context of nonmonotonic rule systems (Marek–Nerode–Remmel [1990]).

§2. Propositional modal logic

Propositional modal logic is based on connectives which construct new propositions from old. We treat propositional logic first. The modal propositional connectives are

"and"	$\wedge$
"or"	$\vee$
"implies"	$\rightarrow$
"not"	$\neg$
"box"	$\Box$
"diamond"	$\Diamond$

The primitive symbols will be:

An infinite list of propositional constants.
The list of logical connectives $\wedge$, $\vee$, $\rightarrow$, $\neg$, $\Box$, $\Diamond$.
parentheses (,) and the comma.

The inductive definition of (modal) proposition is:

1) Propositional constants are propositions,
2) If α, β are propositions, then $(\alpha \wedge \beta)$, $(\alpha \vee \beta)$, $(\alpha \rightarrow \beta)$, $(\neg\alpha)$ are propositions.
3) If α is a proposition, then $(\Box\alpha)$ is a proposition.
4) If α is a proposition, then $(\Diamond\alpha)$ is a proposition.

Sometimes we omit parentheses. We are indiscriminate in using both upper case Roman letters and lower case Greek letters for propositions.
Propositions constructed by rules 1), 2) alone are called classical propositions and constitute the language L. Propositions constructed using rules 1), 2), 3) are called modal propositions and constitute the language $L_{\Box}$. Propositions constructed using 1), 2), 3), 4) are also called modal propositions and constitute the language $L_{\Box,\Diamond}$.

The classical propositions of L are intended as truth functional modes of statement composition, that is the truth or falsity of a compound statement is determined by the truth or falsity of the parts. This is the import of the truth tables of propositions. Conversely, any truth table is the truth table of a proposition built from $\wedge$, $\vee$, $\neg$. Classical propositional logic was defined to deal with exactly all truth functional connectives.

As for the modal connectives,

"$\Box P$" is read "box P", or sometimes "necessarily P",
"$\Diamond P$" is read "diamond P", or sometimes "possibly P".

We prefer the readings "box" and "diamond", simply because the interpretations of the connectives $\Box$ and $\Diamond$ symbols in applications are often quite different from those associated with "necessary" and "possible". For example, "I know that", "I believe that", "John knows that", "John believes that" are often axiomatized using box with appropriate axioms. A further reason for neutral terminology is that the question as to what are the properties of necessity and possibility has been debated since the golden age of Greece.

Remark. Modal propositional connectives, unlike the classical connectives, never entered into the foundations of classical mathematics. These foundations rest only on the classical "truth functional" propositional connectives. The new connectives of modal logic are not intended to be truth–functional. "It is necessary that P" should not depend solely for its truth or falsity on the truth or falsity of P, otherwise it might as well be one of P, $\neg P$.

§3. Frames

In the early part of the century C. I. Lewis introduced modal logic as a deductive subject and gave a notion of theorem based on axioms and rules of inference. Kanger [1957] and Kripke [1959, 1963] gave a semantics based on the notions of frame and model.

First let us review truth valuations of classical logic L. An L–assignment is a mapping A with domain the set of propositional constants' and range $\{T, F\}$. Let $\underline{A}$ be the set of all propositional constants mapped into T by A. Each assignment A has a unique extension to a classical L–valuation v of L, such that

0) $v(P) = A(P)$ for all propositional constants P.
1) $v(A \wedge B) = T$ iff $v(A) = T$ and $v(B) = T$.
2) $v(A \vee B) = T$ iff $v(A) = T$ or $v(B) = T$.
3) $v(A \rightarrow B)$ T iff $v(A) \neq T$ or $v(B) = T$.
4) $v(\neg A) = T$ iff $v(A) \neq T$.

Equivalently,

0) For propositional constants P, P is true iff $P \in \underline{A}$.
1) $A \wedge B$ is true iff A is true and B is true.

2) $A \vee B$ is true iff A is true or B is true.
3) $A \rightarrow B$ is true iff A is not true or B is true.
4) $\neg A$ is true iff A is not true.

Frame semantics. For modal propositional logic Kripke introduced the notion of a frame $(\mathcal{I}, R)$ consisting of a non-empty set $\mathcal{I}$ of "possible worlds" and a binary relation $R \subseteq \mathcal{I} \times \mathcal{I}$. Then wRx is read "x is accessible from w". A model M is a triple $(\mathcal{I}, R, v)$, with $(\mathcal{I}, R)$ a frame and v(w) a "valuation function" with domain the set of "possible worlds" $\mathcal{I}$ and range contained in the set of L–valuations, which assigns to each w in $\mathcal{I}$ an L–valuation v(w). So the notation for the truth value of the valuation v(w) assigned to world w at proposition A is v(w)(A). Here is a definition of "φ is true at w in M". Reference to M is omitted when understood.

0) A propositional constant A is true at w iff $v(w)(A) = T$.
1) $A \wedge B$ is true at w iff A is true at w and B is true at w.
2) $A \vee B$ is true at w iff A is true at w or B is true at w.
3) $A \rightarrow B$ is true at w iff A is not true at w or B is true at w.
4) $\neg A$ is true at w iff A is not true at w.
5) $\Box A$ is true at w iff for every x accessible from w, A is true at x.
6) $\Diamond A$ is true at w iff for some x accessible from w, A is true at x.

Fix M. We indiscriminately write "A is true at w" as "$w \Vdash A$, or as "w forces φ". At times this avoids incorrect connotations of classical truth, and is a notation borrowed from set theory.

In the forcing notation, the inductive definition of $\Vdash$ is: for all w in $\mathcal{I}$,

0) For a propositional constant A, $w \Vdash A$ iff $v(w)(A) = T$
1) $w \Vdash A \wedge B$ iff $w \Vdash A$ and $w \Vdash B$.
2) $w \Vdash A \vee B$ iff $w \Vdash A$ or $w \Vdash B$.
3) $w \Vdash A \rightarrow B$ iff not($w \Vdash A$) or $w \Vdash B$.
4) $w \Vdash \neg A$ iff not ($w \Vdash A$).
5) $w \Vdash \Box A$ iff for all x in $\mathcal{I}$ such that $w R x$, $x \Vdash A$.
6) $w \Vdash \Diamond A$ iff for some x in $\mathcal{I}$ such that $w R x$, $x \Vdash A$.

Remark. Box "$\Box$" and diamond "$\Diamond$" are written as "propositional" connectives. But in model $(\mathcal{I}, R, v)$ the "$\Box$" in $w \Vdash \Box A$ is a universal quantifier over possible worlds accessible from A (if any), while the "$\Diamond$" in $w \Vdash \Diamond A$ is an existential quantifier over worlds accessible from w. So to construct models for modal propositional calculus, the appropriate method comes from classical predicate logic, not from classical propositional logic.

We assume as given an infinite sequence of constants to name worlds, "world constants". In the tableaux below these will be used ambiguously as names for worlds w and names for classical valuations w at worlds. That is, in the models associated with tableaux branches, $\mathcal{F}$ will be a set of world constants, and the accessibility relation R will be a relation between world constants. This means that for the models constructed by tableaux, the valuation map v will be such that v(w) = w. There we do not distinguish between names of worlds and names of valuations in the tableaux. The tableaux rules below are chosen to reflect exactly the definition of forcing in a frame.

§4. Propositional tableaux

A tableaux is a finite labelled tree, with apex at the top. Each node is labelled by a "signed formula" $Tv \Vdash A$ or $Fv \Vdash A$, with A a formula, v a world constant. These are read respectively "at world v, A is true", or "at world v, A is false". In addition, at the base of some branches is an "×", and these are called closed branches, the rest open. Tableaux are developed (extended to larger tableaux) by the rules below.

Here is the dynamic idea behind constructing a tableaux proof of A. To verify that A is valid in all models, we suppose not, and search for a counterexample by developing a tableaux with apex $Fw \Vdash A$, with w a world constant. If we develop such a tableaux according to the tableaux rules, all possible ways to falsify $w \Vdash A$ are taken into account. If an immediate contradiction occurs on every branch at some point of the tableaux construction, all ways of falsifying A have been exhausted. A is valid in all models. The resulting tableaux with contradictions on every branch is a proof of A. A closed branch is one with a contradiction "×" at the base. Open branches are those that are not closed. We develop the tableaux by <u>using</u> entries on <u>open</u> branches. An entry is <u>used</u> by placing an appropriate atomic tableaux, omitting its apex, at the base of some (or every) open branch through that entry. A branch is declared <u>closed</u> as soon as for some branch and some proposition B and some world constant w, that branch has entries of the form $Tw \Vdash B$, $Fw \Vdash B$. We place a cross "×" at the base of each branch so closed. A tableaux proof is a tableaux with all branches closed.

The tableaux proof system is based on the atomic tableaux for classical and modal connectives below, $\wedge$, $\vee$, $\rightarrow$, $\neg$, $\Box$, $\Diamond$. It is the equivalent of the system K traditionally studied in modal logic. See Fitting [1983] for closely related systems of prefixed tableaux, from which these tableaux stem. We also will extend this system with additional tableaux development rules to deal with validity in special classes of frames.

<u>Atomic tableaux. The classical connectives.</u>

and

$Tw\Vdash\varphi \wedge \psi$ → $Tw\Vdash\varphi$ → $Tw\Vdash\psi$

$Fw\Vdash\varphi \wedge \psi$ branches to $Fw\Vdash\varphi$ | $Fw\Vdash\psi$

or

$Tw\Vdash\varphi \vee \psi$ branches to $Tw\Vdash\varphi$ | $Tw\Vdash\psi$

$Fw\Vdash\varphi \vee \psi$ → $Fw\Vdash\varphi$ → $Fw\Vdash\psi$

implies

$Tw\Vdash\varphi \rightarrow \psi$ branches to $Fw\Vdash\varphi$ | $Tw\Vdash\psi$

$Fw\Vdash\varphi \rightarrow \psi$ → $Tw\Vdash\varphi$ → $Fw\Vdash\psi$

not

$Tw\Vdash\neg\varphi$ → $Fw\Vdash\varphi$

$Fw\Vdash\neg\varphi$ → $Tw\Vdash\varphi$

Example. Here is a tableaux proof of $\neg B \wedge (A \vee B) \rightarrow A$.

1	$Fw\Vdash\neg B \wedge (A \vee B) \rightarrow A$		
2	$Tw\Vdash\neg B \wedge (A \vee B)$	by 1	
3	$Fw\Vdash A$	by 1	
4	$Tw\Vdash(\neg B)$	by 2	
5	$Tw\Vdash(A \vee B)$	by 2	
6	$Tw\Vdash A$ | $Tw\Vdash B$	by 5	
7	× | $Fw\Vdash B$	by 3, 6	by 4
8	×	by 6, 7	

The number annotations on the left and the reason annotations on the right are not part of the formal tableaux proof, but are useful for reading a finished proof. Since "$w\Vdash$" plays no role in tableaux for propositions in the classical propositional calculus L, it can be omitted, getting the tableaux below.

$F\neg B \wedge (A \vee B) \rightarrow A$

$T\neg B \wedge (A \vee B)$

```
      FA
      |
     T¬B
      |
    T(A∨B)
    /    \
  TA      TB
  |       |
  ×       FB
          |
         ×
```

This is a classical tableaux in the sense of Smullyan [1968].

Example. (Peirce's Law) Here is a classical tableaux proof of another classical proposition.

```
1   F(((A→B)→A)→A)
            |
2          FA           by 1
            |
3     T((A→B)→A)         by 1
        /       \
4   F(A→B)      TA       by 3
      |          |
5    FB          ×     by 4    by 2, 4
      |
6    TA                by 4
7     ×               by 2, 6
```

Example. Here is another classical tableaux.

```
1        T((A∧(¬A))∨(B∨(C∧D)))
          /               \
2   T(A∧¬A)          T(B∨(C∧D))      by 1
       |              /      \
3     TA           TB     T(C∧D)   by 2
       |                   /
4    T¬A                 TC        by 3
       |                  |
5     FA                 TD        by 4    by 3
6      ×
```

This is not a proof. Here we get a contradiction on the left branch. Each of the other branches exhibit valuations making the topmost signed statement true. That is, any valuation making B true makes the topmost signed statement true, any valuation making C, D both true makes the topmost signed statement true. This exhibits the fact that counterexamples can be read off tableaux.

Atomic tableaux. The modal connectives.

Box.

$T w \Vdash \Box P$ | $T v \Vdash P$

provided $TwRv$ occurs on the branch already

$F w \Vdash \Box P$ | $F v \Vdash P$ | $TwRv$

for a new v not yet occurring on the branch

Diamond.

$T w \Vdash \Diamond P$ | $T v \Vdash P$ | $TwRv$

for a new v not yet occurring on the branch

$F w \Vdash \Diamond P$ | $F v \Vdash P$

provided $TwRv$ occurs on the branch already

Explanation. Recall the definition of forcing at a world in clauses 5), 6) above. For any tableaux entry $Tw \Vdash \Box P$ on an open branch, if v is a valuation constant already occurring in a signed formula on that branch, we wish to be able to adjoin $Tv \Vdash P$ to the end of that branch. For any tableaux entry $Fw \Vdash \Box P$ on an open branch, and any valuation constant v not occurring on that branch, we wish to be able to adjoin to the end of that branch $Fv \Vdash P$ followed by $TwRv$. These are the last of the rules of proof for modal propositional logic $L_{\Box,\Diamond}$.

Remark. In classical logic $\Diamond$ is equivalent to $\neg\Box\neg$. Diamond $\Diamond$ will not be mentioned again until §10. The latter uses intuitionistic frames. The modalities $\Diamond$, $\Box$ are not interdefinable in an intuionistic context.

The definition of semantic validity must be expressed with care. A proposition P is valid in a model $(\mathcal{F}, R, v)$ if P is forced by every w in $\mathcal{F}$. A proposition P is valid in a frame $(\mathcal{F}, R)$ if for every possible valuation function v for that frame, P is valid in the model $(\mathcal{F}, R, v)$. A proposition is valid if valid in every frame. So P is valid if for every frame $\mathcal{F}$, every valuation v, every world w of $\mathcal{F}$, w forces P.

It is useful to have the notion of a deduction of proposition B from premises $A_1, \ldots, A_n$. The notion of deduction is supposed to be a syntactical equivalent of the semantical assertion that for all frames $(\mathcal{F}, R)$ for which $A_1, \ldots, A_n$ are valid in $(\mathcal{F}, R)$, B is also valid for $(\mathcal{F}, R)$. The hypothesis is that for all i, all w in $\mathcal{F}$, $w \Vdash A_i$. This is not reflected in the tableaux proof rules above. We need an additonal

Atomic tableaux for deductions

For any premise A_i, any world constant v,

the tableaux below may be appended to any open branch

$$\begin{array}{c} | \\ Tv \Vdash A_i \end{array}$$

Then a deduction of conclusion B from premises $A_1, \ldots, A_n$ is a tableaux with all branches closed in which

1) the apex is $Fw \Vdash B$
2) The atomic tableaux for proofs are allowed.
3) Application of the tableaux deduction rule for premises $A_1, \ldots, A_n$ is allowed– i.e., the atomic tableaux for deduction indicated can be appended to the base of any open branch at will for any premise A_i and any valuation constant v.

Theorem. (Correctness). Every proposition with a tableaux proof (by rules 1–6) is valid. If proposition B has a tableaux deduction from $A_1, \ldots, A_n$, then B is valid in any frame in which $A_1, \ldots, A_n$ are valid.

Theorem. (Completeness) Every valid proposition has a tableaux proof. If B is valid in every frame in which $A_1, \ldots, A_n$ are valid, then there is a tableaux deduction of B from $A_1, \ldots, A_n$.

The completeness and correctness proofs mimic the classical case. They are straightforward by a "complete systematic tableaux procedure" like that of Smullyan for classical tableaux and of Fitting [1983] for prefixed tableaux. They will be supplied in a more complete version of these notes.

Example. All classical tautologies have tableaux proofs. For simply substitute "$Tw \Vdash \varphi$" for "TP", "$Fw \Vdash P$" for "FP" throughout the classical tableaux proof of the tautology.

Example. Here is a tableaux proof of $\Box(A \to B) \to (\Box A \to \Box B)$. It is the axiom used to prove that $\{P : \Box P \text{ is a theorem}\}$ is closed under modus ponens in Hilbert–style systems of propositional modal logic based on axioms and modus ponens as the sole rule of inference.

1	$F w \Vdash \Box(A \to B) \to (\Box A \to \Box B)$	
2	$T w \Vdash \Box(A \to B)$	by 1
3	$F w \Vdash \Box A \to \Box B$	by 1
4	$T w \Vdash \Box A$	by 3
5	$F w \Vdash \Box B$	by 3
6	$T wRv$ new v	by 5
7	$F v \Vdash B$	by 5
8	$T v \Vdash A$	by 4, 6
9	$T v \Vdash A \to B$	by 2, 6
10	$F v \Vdash A$ $T v \Vdash B$	by 9
11	× ×	by 8, 7

Example. (Modus Ponens) From premises A, $A \to B$, deduce B.

1	$F w \Vdash B$	
2	$T w \Vdash A \to B$	Premise
3	$T w \Vdash A$	Premise
4	$F w \Vdash A$ $T w \Vdash B$	By 2
	× ×	By 1, by 3

The semantical equivalent is that if A and $A \to B$ are valid in a frame, then so is B.

Example. From premise A, deduce $\Box A$. This is called the rule of necessitation. (Line 4 of the deduction below uses the deduction atomic tableaux.)

1	$F w \Vdash \Box A$	
2	$T wRv$	by 1
3	$F v \Vdash A$	by 1
4	$T v \Vdash A$	premise
	×	

Example. In contrast, $A \to \Box A$ is not valid.

1 $Fw \Vdash A \to \Box A$

2 $Tw \Vdash A$ by 1

3 $Fw \Vdash \Box A$ by 1

4 $TwRv$ by 3

5 $Fv \Vdash A$ by 3

This produces a frame $\mathcal{F} = \{w, v\}$, $R = \{(w, v)\}$, and a valuation v in which A is true at w but not at v.

$$\overset{A}{w} \longrightarrow v$$

In this frame, w does not force $A \to \Box A$, so $A \to \Box A$ is not valid in all frames. But $A \to \Box A$ is valid in those <u>models</u> $(\mathcal{F}, R, v)$ such that for all $w, x \in \mathcal{F}$, if wRx and under the valuation, if A is true at w, then A is true at x. So one has to be very careful in formulating any sort of "deduction theorem" saying that under certain circumstances, if B can be deduced from A, then $A \to B$ is provable. One would have to decode the forcing meaning, which unwinds the modal operators as quantifiers, and look at their scopes.

Compare

– "From A as premise, deduce conclusion B." This says the following. Suppose that in a frame $(\mathcal{F}, R)$, all possible valuations v that can be assigned to worlds in $\mathcal{F}$ give models $(\mathcal{F}, R, v)$ such that for all w in $\mathcal{F}$, w forces A. Then conclude that for all possible valuations v of that frame $(\mathcal{F}, R)$, in the model $(\mathcal{F}, R, v)$, every world w in $\mathcal{F}$ forces B.

– "$A \to B$ is a theorem." This says the following. In any model $(\mathcal{F}, R, v)$, for any w in $\mathcal{F}$, if w forces A, then w forces B.

The quantifier structure of the two statements is quite different.

§5. Some modal axioms

<u>Example</u>. $\Box A \to A$ is not valid. It is traditionally called "T". If $\Box$ is interpreted as "I know", then T says "knowledge is truth", so it is called the "knowledge axiom". If $\Box$ is interpreted as "I believe", then T says "What I believe is true". One <u>can</u> have false beliefs.

1 $Fw \Vdash \Box A \to A$

2 $Tw \Vdash \Box A$ by 1

3 $Fw \Vdash A$ by 1

There is no contradiction. Reading off this tableaux the worlds and the forced atomic statements at those worlds, a one world frame $(\mathcal{I}, R)$. $\mathcal{I} = \{w\}$ with empty accessibility relation R and A false at w makes $\Box A \rightarrow A$ false. A reflexive frame is one in which wRw for every world w. Looking at the tableaux line 2, we would get $Tw \Vdash A$, contradicting line 3. So $\Box A \rightarrow A$ is valid in all reflexive frames. Conversely, any proposition valid in all reflexive frames can be deduced from $\Box A \rightarrow A$.

<u>Reflexive tableaux development rule</u>.

If w is any world occurring in an entry, at the base of any open branch through that entry we may append the tableaux

|
TwRw

A proposition is valid in all reflexive frames if and only if it is provable by the standard modal tableaux plus the reflexive tableaux development rule.

<u>Example</u>. $\Box A \rightarrow \Box\Box A$ is not valid. Traditionally, this proposition is called "4". In newer papers, this is called the "positive introspection axiom", "What I believe, I believe I believe".

1	$Fw \Vdash \Box A \rightarrow \Box\Box A$	
2	$Tw \Vdash \Box A$	by 1
3	$Fw \Vdash \Box\Box A$	by 1
4	TwRv	new v by 3
5	$Fv \Vdash \Box A$	by 3
6	TvRu	new u by 5
7	$Fu \Vdash A$	by 5
8	$Tv \Vdash A$	by 2, 4

There is no contradiction. But reading off the true atomic statements from the tableaux, we get a three world frame $\mathcal{I} = \{w, v, u\}$, with wRv, vRu, and in the model with A true at v, but A not true at w or u. This is a counterexample to the validity of $\Box A \rightarrow \Box\Box A$, which is not true at w.

A
$w \longrightarrow v \longrightarrow u$

The labelled graph above has branches representing all accessibility relations and nodes representing all worlds. Labels of nodes are atomic propositions true at that world.

A transitive frame $(\mathcal{S}, R)$ is one such that for all w, v, u, if wRv and vRu, then wRu. Then from the tableaux we get TwRv, TvRu, so we get TwRu. Then we could apply line 2 and get $Tu \Vdash A$, contradicting line 7. So $\Box A \to \Box\Box A$ is valid in transitive frames. Conversely, any proposition valid in all transitive frames is deducible from $\Box A \to \Box\Box A$. If we wish to deal only with transitive accessibility relations, we can add the following rule of tableaux development directly to those already given.

<u>Transitivity tableaux development rule</u>.

If TwRu and TuRv occur on a branch, we may append to all (some) open branches through that pair, the tableaux

$$\begin{array}{c} | \\ TwRv \end{array}$$

Then a proposition is valid in all transitive frames if and only if it has a tableaux proof using the standard modal tableaux rules plus the transitive tableaux development rule.

<u>Example</u>. $\neg\Box\varphi \to \Box\neg\Box\varphi$ is not valid. In older papers this is often abbreviated E for the Euclidean axiom, or "5". In newer papers, this is called the "negative introspection axiom". "What I don't believe, I believe I don't believe"

1	$Fw \Vdash \neg\Box\varphi \to \Box\neg\Box\varphi$		
2	$Tw \Vdash \neg\Box\varphi$		by 1
3	$Fw \Vdash \Box\neg\Box\varphi$		by 1
4	$Fw \Vdash \Box\varphi$		by 2
5	TwRv		by 4
6	$Fv \Vdash \varphi$	new v	by 4
7	TwRu	new u	by 3
8	$Fu \Vdash \neg\Box\varphi$		by 3
9	$Tu \Vdash \Box\varphi$		by 8

If we read off the true atomic sentences, they are wRv, wRu. With φ declared false in all three valuations, we get a model

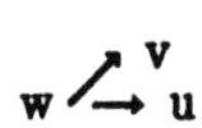

in which w does not force $\neg\Box\varphi \to \Box\neg\Box\varphi$. An Euclidean frame is one such that for all w, v, u in $\mathcal{S}$, wRu and wRv imply uRv. Looking at the tableaux, we had TwRu, TwRv. With the

Euclidean property, we get also TuRv, by line 9 this gives $Tv \Vdash \varphi$, contradicting line 6. So $\neg\Box\varphi \rightarrow \Box\neg\Box\varphi$ is valid in all Euclidean frames. Conversely, any proposition true in all Euclidean frames is deducible from $\neg\Box\varphi \rightarrow \Box\neg\Box\varphi$. For a Euclidean R, for any world w in $\mathcal{F}$, the restriction of R to $\{v \in \mathcal{F}: w\,R\,v\}$ is an equivalence relation, but this set does not necessarily contain w itself.

<u>Euclidean tableaux development rule.</u>

If a branch contains entries TwRu and TwRv, then we may append to every open branch through these two entries the tableaux

|
TuRv

Then a proposition is true in all Euclidean frames if and only if it has a tableaux proof using the standard modal tableaux plus the Euclidean Tableaux development rule.

<u>Example</u>. $\Box P \rightarrow \neg\Box\neg P$ is not valid. In the older literature, this is called axiom "D". In newer papers, this is called the serial axiom. "What I believe, I don't believe the negation of"

	$Fw \Vdash \Box P \rightarrow \neg\Box\neg P$	
2	$Tw \Vdash \Box P$	by 1
3	$Fw \Vdash \neg\Box\neg P$	by 1
4	$Tw \Vdash \Box\neg P$	by 3

There is no contradiction. The model with a single world w and empty accessibility R and P false at w will do to falsify $\Box P \rightarrow \neg\Box\neg P$. A serial frame is one such that for every world w, there is a world v such that wRv. In this case from $Tw \Vdash \Box P$, $Tw \Vdash \Box\neg P$, we get $Tv \Vdash P$, $Tv \Vdash \neg P$, a contradiction, so $\Box P \rightarrow \neg\Box\neg P$ is valid in serial frames. In fact, any proposition valid in all serial frames is deducible from $\Box P \rightarrow \neg\Box\neg P$.

<u>Serial tableaux deduction rule.</u>

For any world constant v occurring in an entry on an open branch, and any valuation constant u not on that branch, we may append to that open branch the tableaux below.

|
TvRu

Thus a proposition is valid in every serial frame if and only if it has a tabeaux proof using the standard modal tableaux plus the serial tableaux deduction rule.

<u>Hilbert systems for modal logic</u>. A standard set of axioms and rules of inference for a Hilbert

style modal logic called K defines the theorems (of K) as the smallest set of propositions such the following hold.

1. All classical tautologies with modal propositions substituted for variable are theorems. (These tautologies are the "axioms".)
2. All propositions $\Box(A \rightarrow B) \rightarrow (\Box A \rightarrow \Box B)$ are theorems.
3. If $A, A \rightarrow B$ are theorems, then B is a theorem.
4. If A is a theorem, then $\Box A$ is a theorem.

The notion of deductive closure of a set of premises S would replace "are theorems" by "are consequences of S " in 1, 2, 3, 4, and add
5) Premises in S are consequences of S.

We have already shown that each axiom and rule of inference holds for tableaux provability using the standard modal tableaux. Propositions proven by the tableaux method are valid in all frames. Any proof of completeness for the system based on 1)–4) shows that tableaux provability coincides with provability in this system.
Here is a list of commonly occurring systems.

– K is the proof system using the classical and modal atomic tableaux.
This tends to be a substructure of modal systems used for computer science.
– T is K plus the schema $\Box A \rightarrow A$ as premises for deductions. T tends to be regarded as the logic of knowledge (true beliefs).

• A proposition is provable in T iff valid in all reflexive frames iff provable by the tableaux of K plus the reflexive tableaux development rule.

– S4 is T plus the additional schema $\Box A \rightarrow A$ and $\Box A \rightarrow \Box\Box A$ added as premises for deductions.

• A proposition is provable in S4 iff valid in all reflexive, transitive frames iff provable by the tableaux of K plus the reflexive and transitive tableaux development rules.

– S5 is S4 plus the additional schema $\neg\Box A \rightarrow \Box\neg\Box A$ added as premises for deductions. A relation R on $\mathcal{F}$ is transitive, Euclidean and reflexive iff R is an equivalence relation.

proposition is provable in S5 iff valid in all frames with an equivalence relation on $\mathcal{F}$ as accesibility iff provable by the tableaux rules of K plus the reflexive, transitive, and Euclidean tableaux development rules.

There is more to say for S5.

Lemma. Suppose $(\mathcal{F}, R, v)$ is a model and $w \in \mathcal{F}$. Define a model $(\mathcal{G}, R', v')$ by setting $\mathcal{G} = \{w' \in \mathcal{F}: wRw'\}$, $R' = R \cap \mathcal{G} \times \mathcal{G}$, $v' = v$ restricted to $\mathcal{G}$. Then w forces φ in $(\mathcal{F}, R, v)$ iff w forces φ in $(\mathcal{G}, R', v')$.
The proof of the lemma is by induction on the definition of forcing.

According to this lemma, φ is forced by all v in all models $\mathcal{F}$ with R an equivalence relation on $\mathcal{F}$ iff forced by all v in all models with R an equivalence relation on $\mathcal{F}$ which has a single equivalence class, that is $R = \mathcal{F} \times \mathcal{F}$. A complete frame is one where the accessibility on $\mathcal{F}$ is $R = \mathcal{F} \times \mathcal{F}$.

The system S5 was used by Moore [1985] for autoepistemic logic (see below). The system S5 is suitable for reasoning about knowledge in distributed systems, provided that there are many S5 modal connectives $\Box_A$, one for each agent or machine A. This takes one beyond complete frames, the lemma no longer works for multiple agents, one is stuck with many equivalence relations, one for each agent, see Halpern Moses [1984, 198?], and also Lehmann [1984].

Complete tableaux development rule.

If world constants u, v occur in entries, then we may append to the base of any open branch through those entries the tableaux below.

$$\begin{array}{c} | \\ TuRv \end{array}$$

• A proposition is provable in S5 iff valid in all complete frames iff provable by the tableaux of K plus the complete tableaux development rule.

– K45 is K plus the additional schema $\neg\Box A \to \Box\neg\Box A$, $\Box A \to \Box\Box A$.

• A proposition is provable in K45 iff true in all transitive Euclidean frames iff provable by the tableaux of K plus the transitive and Euclidean development rules.

K45 is a candidate (Halpern and Moses [1986], Moore [1988]) for a logic of belief for a "logically omniscient completely introspective rational agent", see §7.

§6. Modal predicate tableaux with constant domain

We introduce very briefly a modal predicate logic within classical logic intended to describe a single "constant" domain, with different true atomic statements at different worlds. This logic can be extended to have several modalities $\Box_i$, we do not do this here. It is the predicate modal logic with the most direct current applications in computer science. Dynamic logic uses this model, there the constant domain is the set of all states of a machine, the $\Box_i$ are induced by programs or commands. This formulation covers theories of beliefs or knowledge for several agents at once about a fixed domain of individuals as well, one $\Box_i$ for each agent. The earliest example of such a theory is Hintikka [1962], see also Konolige [1986] or Halpern and Moses [1985] for further references. Here is the list of primitive symbols.

Predicate letters of degree n,
An infinite list of variables
an infinite list of (individual) constants
Logical connectives $\wedge$, $\vee$, $\rightarrow$, $\neg$, , $\exists$, $\forall$
Modal connectives $\Box$, $\Diamond$,
parentheses (,) and a comma.

The inductive definition of formula and free occcurence of variables is:

1) If R is a predicate letter of degree n and $\alpha_1, \ldots \alpha_n$ are variables or constants, then $R(\alpha_1, \ldots, \alpha_n)$ is a formula. (These are called the atomic formulas.) In atomic formulas all occurrences of all variables are free.

2) If α, β are formulas, then $(\alpha \wedge \beta)$, $(\alpha \vee \beta)$, $(\alpha \rightarrow \beta)$, $(\neg \alpha)$ are formulas. Occurrences of variables in these formulas are free or bound as they were in α, β.

3) If α is a formula, x is a variable, then $((\exists x)\alpha)$, $((\forall x)\alpha)$ are formulas. Occurrences of variables other than x are free or bound in these formulas as they were in α. variable x is bound in all its occurences in these formulas.

3) If α is a formula, then $(\Box\alpha)$ and $(\Diamond\alpha)$ are formulas. Variables are free or bound in these statements as they are in α.

A statement is a formula in which all occurrences of all variables are bound.
This determines a language $L_{\Box,\Diamond}$. It has a purely classical sublanguage L obtained by omitting all reference to clause 3).
We need the notion of substitution. If we write a formula φ as $\varphi(x)$ for x a variable, and c is a constant, then $\varphi(c)$ is the result of substituting c for all free occurrences of x throughout φ.

For the sake of defining the usual notion of "relational system" in a form exactly appropriate for tableaux, assume that L has no constants itself. Let C be a set of individual constants and extend L to a language L(C) by adding in C to L. An <u>assignment</u> A for L(C) is a map of the atomic statements of L(C) to {T, F}. Each assignment A is extended uniquely to a <u>valuation</u> V mapping the statements of L(C) to {T, F}, by the inductive definition below.

0) $V(P) = A(P)$ for all atomic statements P.
1) $V(A \wedge B) = T$ iff $V(A) = T$ and $V(B)$ T.
2) $V(A \vee B) = T$ iff $V(A) = T$ or $V(B) = T$.
3) $V(A \rightarrow B)$ T iff $V(A) \neq T$) or $V(B) = T$.
4) $V(\neg A) = T$ iff $V(A) \neq T$.
5) $V((\exists x)\varphi(x)) = T$ iff for some constant c of C, $V(\varphi(c)) = T$.
6) $V((\forall x)\varphi(x)) = T$ iff for all constants c of C, $V(\varphi(c)) = T$.

In the notation common in predicate logic, an assigment defines a relational system for L with <u>domain</u> C. This relational system has each relation symbol R of degree n of $L(C)_{\square,\diamond}$ denote

$$[(c, ..., c_n) \in \times^n C : V(R(c_1, ..., c_n)) = T].$$

The definition of <u>model</u> $(\mathcal{F}, R)$ for modal predicate logic with constant domains goes as follows. Frames are the same as in propositional calculus, a pair $(\mathcal{F}, R)$ consisting of a non–empty set $\mathcal{F}$ of "possible worlds" and an "accessibility relation" $R \subseteq \mathcal{F} \times \mathcal{F}$. A <u>model</u> is given by a set C of individual constants (the "constant domain") and a map v (the valuation function) assigning to each $w \in \mathcal{F}$ a valuation v(w) of L(C). The definition of "$w \Vdash \varphi$" for statements φ of $L(C)_{\square,\diamond}$ is as follows.

0) $w \Vdash P$ for atomic statements P iff $v(w)(P) = T$
1) $w \Vdash A \wedge B$ iff $w \Vdash A$ and $w \Vdash B$.
2) $w \Vdash A \vee B$ iff $w \Vdash A$ or $w \Vdash B$.
3) $w \Vdash A \rightarrow B$ iff $w \Vdash A$ implies $w \Vdash B$
4) $w \Vdash \neg A$ iff not $w \Vdash A$
5) $w \Vdash \square A$ iff for all w' in $\mathcal{F}$ such that $w R w'$, $w' \Vdash A$.
6) $w \Vdash \diamond A$ iff for some w' in $\mathcal{F}$ such that $w R w'$, w, $w \Vdash A$.
7) $w \Vdash ((\exists x)\varphi(x))$ iff $w \Vdash \varphi(c)$ for some c in C.
8) $w \Vdash ((\forall x)\varphi(x))$ iff $w \Vdash \varphi(c)$ for all c in C.

The reason these are called "constant domain" models is that the domain C of the relational system assigned to each world is precisely the same. In constant domain models we do not have to worry about any change in denotation of a constant from world to world. The constants are the same in every world and can be thought of as having the same denotation, and even may be thought of as denoting themselves. The big difference between worlds is that the atomic statements $R(c_1, ..., c_n)$ forced in one world may not be forced in another world.

The definition of "valid in a frame" and "valid in a model" and "valid" are as for propositional logic. Using the tableaux before, correctness and completeness are routine.

Constant domain tableaux. We need a countable list of world constants just as in modal propositional logic.. We also need a countable list of new individual constants, to be used in the tableaux to name elements of an intended constant domain. These individual constants are used, as in tableaux for classical predicate logic (Smullyan [1968]) as witnesses for existential quantifiers. Here is the motivation, similar to that for classical logic, but for frames. Each branch b of a tableaux is viewed an attempt to build a model in which each forcing statement on the branch holds as stated. So the frame would consist of the set $\mathcal{T}$ of world constants w mentioned on b; the constant domain C would be the set of all constants occurring on b; the model based on this frame has the valuation at world w with atomic statement $R(c_1, ..., c_n)$ true iff $Tw \Vdash R(c_1, ..., c_n)$ occurs on b. When a branch b is contradictory in such a tableaux development, this attempt to build a model has failed. When all such attempts to build a model have failed on all branches, we have a tableaux proof.

We add the usual atomic tableaux rules for predicate logic quantifiers (Smullyan [1968]) to those of modal propositional logic. The rules for $T(\forall x)\varphi(x)$, $F(\exists x)\varphi(x)$ are set up to handle the constant domain situation only, since we are allowed to instantiate using any constant already on the branch. We have assumed for simplicity that $L_{\Box,\Diamond}$ has no constants itself.

Quantifier atomic tableaux for constant domains.

Universal

$Tw \Vdash (\forall x)\varphi(x)$	$Fw \Vdash (\forall x)\varphi(x)$
$\mid$	$\mid$
$Tw \Vdash \varphi(c)$	$Fw \Vdash \varphi(c)$
For any c	For a new c not occurring on any entry above on the branch

Existential

$Tw \Vdash (\exists x)\varphi(x)$	$Fw \Vdash (\exists x)\varphi(x)$
$\mid$	$\mid$
$Tw \Vdash \varphi(c)$	$Fw \Vdash \varphi(c)$
For a new c not occurring on any entry above on the branch	For any c

Example.

1	$Fw\Vdash(\forall x)\Box A(x) \rightarrow \Box(\forall x)A(x)$	
2	$Tw\Vdash(\forall x)\Box A(x)$	by 1
3	$Fw\Vdash\Box(\forall x)A(x)$	by 1
4	$TwRv$	by 3
5	$Fv\Vdash(\forall x)A(x)$	by 3
6	$Fv\Vdash A(c)$ new c	by 5
7	$Tw\Vdash\Box A(c)$	by 2
8	$Tv\Vdash A(c)$	by 7
	×	

So $(\forall x)\Box A(x) \rightarrow \Box(\forall x)A(x)$ is provable.

<u>Example.</u> $(\forall x)\neg\Box\varphi \rightarrow \neg\Box(\exists x)\varphi$.

1	$Fw\Vdash(\forall x)\neg\Box\varphi \rightarrow \neg\Box(\exists x)\varphi$	
2	$Tw\Vdash(\forall x)\neg\Box\varphi$	by 1
3	$Fw\Vdash\neg\Box(\exists x)\varphi$	by 1
4	$Tw\Vdash\Box(\exists x)\varphi$	by 3
5	$Tw\Vdash\neg\Box\varphi(c)$	by 2
6	$Fw\Vdash\Box\varphi(c)$	by 5
7	$TwRv$	by 6
8	$Fv\Vdash\varphi(c)$	by 6
9	$Tv\Vdash(\exists x)\varphi$	by 4
10	$Tv\Vdash\varphi(d)$ new d	by 9

This is not a proof. With domain $C = \{c, d\}$, and two worlds w, v, with v accessible from w and no atomic proposition holding in w and $\varphi(d)$ holding in v, we get a counterexample.

$$w \longrightarrow \overset{\varphi(d)}{v}$$

Just as in classical predicate tableaux, the constructions are as helpful for finding counterexamples as for finding proofs. We remark that the same tableaux method applies in case more general situations than constant domains are allowed. But the semantics intended for individual constants then has to be very precisely specified before it becomes obvious what the appropriate tableaux rules for quantifiers are.

§7. Autoepistemic logic

Belief

Commonsense reasoning is often "nonmonotonic". I, a rational introspective agent, have beliefs in my current complete set of beliefs B based on incomplete information. Later I have to change to another belief set B' in which we may no longer have some of the previous beliefs in B. We assume my belief set is closed under classical logical consequence, that is, I believe the logical consequences of what I believe (principle of logical omniscience). We assume that my belief set contains all of my beliefs. We suppose that the facts about the external world (objective facts) and rules I know for sure (our knowledge base) are in all my belief sets.

Example. In my current belief set B might be the propositions

R: "If x is a bird, and I do not believe that x cannot fly, then x can fly"
F: "Tweety is a bird."

Suppose

"Tweety cannot fly."

is not derivable from my belief set B. My beliefs are assumed closed under classical deduction so we conclude that

"I do not believe that Tweety cannot fly."

is in B. So applying rule R of B, we deduce that "Tweety can fly" and thus also we deduce that

"I believe that Tweety can fly."

But B is closed under classical deduction, so this proposition is in B. I now visit New Zealand and see a Kiwi, and realize that Tweety is a Kiwi, and conclude

N: "Tweety cannot fly"

In my new belief set B' I retain rule R and fact F and put new fact N into B'.

Since

"Tweety can not fly"

is in B', and is therfore a belief,

"I believe that Tweety can not fly"

is in B', and the hypothesis of rule R is not satisfied for B', and we cannot conclude, using rule R, that "Tweety can fly" is in B'. We have withdrawn a conclusion of B. This is the nonmonotonicity of the reasoning. If indeed

"Tweety can fly"

is not derivable from B', since B' consists of all beliefs, we can conclude that

"I don't believe that Tweety can fly"

is in B' as well.

Moore's view (autoepistemic valuations)

The complete set B of all beliefs of an agent is the subject of Moore's autoepistemic logic [1984]. His is an account of how an agent reasons about the agent's own beliefs. This is the origin of the use of "autoepistemic", the notion of self knowledge. Let $L_\Box$ be the set of all modal propositions based on classical connectives and $\Box$. Let L be the subset of classical propositions. Moore reads $\Box P$ as "P is in the agent's complete current belief set B". In his exposition he begins with $L_\Box$ regarded as a classical propositional logic with every proposition of the form $\Box\varphi$ as an additional propositional letter along with the usual ones. Thus a classical deductive closed set of this classical $L_\Box$ is merely one closed under classical tableaux deductions, or one closed under tautologies and modus ponens. A classical $L_\Box$–assignment maps all propositional letters and all propositions $\Box\varphi$ into {T, F}, and each of these is extendible to a classical $L_\Box$–valuation with domain $L_\Box$ and values in {T, F}.

Definition. An autoepistemic theory is a set B of $L_\Box$–propositions for which there is a classical $L_\Box$–valuation v such that B consists of all P in $L_\Box$ such that $v(\Box P) = T$. Also v is said to be an autoepistemic interpretation of B.

Since v can be $L_\Box$–valued arbitrarily on any atomic proposition P and any modal $\Box Q$, there is no necessary connection between the truth values of these propositions.

Example. For propositional letters A, B, we can define an $L_\square$–valuation with $\square A$ true, $\square B$ true, $\square(A \wedge B)$ false. So A, B are in the corresponding autoepistemic theory, but $A \wedge B$ is not. This is simply an instance of the fact that we can $L_\square$–value propositions of the form $\square P$ arbitrarily and independently. So Moore allows in his definition of an autoepistemic theory B that an agent may be incapable of any reasoning from beliefs to beliefs. This makes it possible in this framework to study adding in reasoning abilities of limited strength by suitable axioms restricting the allowed $L_\square$–valuations. So the notion of autoepistemic theory allows the study of agents with varied reasoning abilities by introducing additional modal axioms reflecting these abilities.

Definition. A model of autoepistemic theory B is an autoepistemic interpretation of B such that all propositions in B are true.

Definition. An autoepistemic theory B is semantically complete if B contains every proposition true in all autoepistemic models of B.

Theorem (Moore [1985]). B is semantically complete iff
1) B is closed under classical $L_\square$–consequence.
2) If $P \in B$, then $\square P \in B$.
3) If $\neg(P \in B)$, then $\neg\square P \in B$.

These three properties were the definition of a stable set B of modal propositions given by Stalnaker [1980, 1981].

Example. We informally used the stability of B and B' in the Tweety example.
– We applied 3) to verify that rule R could be applied to yield that "Tweety can fly" is in B, with P the proposition "Tweety cannot fly".

– We applied 2) to verify that "I believe that Tweety cannot fly" is in B'. with P the proposition "Tweety cannot fly".

Is stability a reasonable condition for the complete belief set B of a rational agent?

Requirement 1) is that the agent should be "logically omniscient", that is, any classical logical consequence of the agent's belief set B should also be in B. This is a simplyfying idealization, since to recognize that a given proposition is a classical logical consequence of known axioms for a given B is at least an NP–complete problem (Halpern and Moses [1985]). Verifying conditions 2) and 3) for specific propositions both involve this NP–hard problem.

Reformulating, the condition that B is closed under $L_{\Box}$–consequence means exactly that a tautology with beliefs substituted for variables is a belief, and that beliefs be closed under modus ponens. For a modal point of view this commits us exactly to the closure conditions on B imposed by deductive closure in the Hilbert style version of system K described in section §5. Equivalently, this commits us exactly to system K, of modal atomic tableaux for $\Box$, together with the deduction rule for tableaux.

Definition. An autoepistemic theory B is sound with respect to a set of premises A iff every autoepistemic interpretation of B in which all the propositions of A are true is an autoepistemic model of B.

Definition. An autoepistemic theory B is grounded in a set of premises A iff B is contained in

$$Cn[A \cup \{\Box P : P \in B\} \cup \{\neg\Box P : \neg(P \in B)\}],$$

where Cn is the classical $L_{\Box}$–consequence relation.

Theorem (Moore [1985]). An autoepistemic theory B is grounded in A iff sound with respect to A.

Theorem (Moore [1985]). If A is a set of premises, then an autoepistemic theory T extending A is sound and semantically complete with respect to A iff

$$T=Cn[A \cup \{\Box P : P \in T\} \cup \{\neg\Box P : P \notin T\}]$$

Definition. An autoepistemic theory B is a stable expansion of a set of premises A if B contains A and is grounded in A.

Theorem. (Moore–Truszczynski [1989]) B is semantically complete with respect to A if and only if B is a stable expansion of A.

Moore identifies the possible complete sets of beliefs that a rational agent might hold after accepting A as the stable expansions of A. The problem in dealing with stable expansions is that there can be none, one, two, or many, and the are not so easy to identify.

Example. $\{\neg\Box P \rightarrow Q, \neg\Box Q \rightarrow P\}$ has at least two stable expansions, one containing P but not Q, one containing Q but not P.

Example. $\{\neg\Box P \rightarrow P\}$ has no stable expansions. Let B be a purported stable expansion. If P is in B , then B is not grounded and therefore not a stable expansion. Any stable If P is not in B, then $\neg\Box P$ is in B (B is stable), so P would be in B (B is grounded), a contradiction.

Marek's view (nonmonotone operators and S–entailment)

I am indebted to W. Marek for the outline of this section. We will repeat from scratch some of the same ground as in the previous section, but from a different viewpoint. This viewpoint starts out with a "rational agent", it does not lend itself as immediately to partially rational agents with limited reasoning powers as did the Moore exposition of the previous section. We begin with "list semantics" for $\Box$. We work again in $L_\Box$. We let L be the corresponding classical language without $\Box$. This exposition emphasizes the role of classical L–valuations v. These are valuations of the classical propositions only, obtained from assignments to the classical propositional letters (not the $\Box P$ propositions). If S is a set of modal propositions (called the "list"), a "list" consequence relation "$\vdash_{v,S}$" is defined from S.

1. For propositional constants P, $\vdash_{v,S} P$ iff $v(P) = T$.
2. $\vdash_{v,S} \neg\varphi$ iff not $\vdash_{v,S} \varphi$.
3) $\vdash_{v,S} (\varphi \vee \psi)$ iff $\vdash_{v,S} \varphi$ or $\vdash_{v,S} \psi$.
4) $\vdash_{v,S} \Box\varphi$ iff $\varphi \in S$.

Remark. We can interpret $\Box\varphi$ as "the agent believes φ", we can interpret S as the list of the agent's beliefs, we can interpret 4) as expressing that if the agent is asked if the agent believes φ , the agent consults the "list", and answers yes in case φ is on the list.

Definition (S–entailment). Let I be a set of modal propositions. Then

$I \vdash_S \varphi$ iff for all valuations v, $\vdash_{v,S} I$ implies $\vdash_{v,S} \varphi$.

Definition. An expansion of I is a collection S of modal propositions such that the fixed point condition $S = \{\varphi : I \vdash_S \varphi\}$.

Let I be given, suppose that S is being guessed by the agent. What does it mean for the guess to be correct?

1) Whatever is S–entailed should be in S (an adaquacy requirement).
2) Whatever is in S should be S–entailed (a completeness requirement).

Theorem (Moore). The following are equivalent.
a) S is an expansion of I.
b) $S = Cn(I \cup \{\Box\varphi : \varphi \in S\} \cup \{\neg\Box\varphi : \varphi \notin S\})$.
(Here Cn is classical consequence).

A set S of modal propositions is called <u>stable</u> (Stalnaker, 1980) if

1) closed under classical deduction,
2) $\varphi \in S$ implies $\Box\varphi \in S$,
3) $\varphi \notin S$ implies $\neg\Box\varphi \in S$.

Condition 3) makes the reasoning nonmonotonic. Stable theories are supposed to represent the set of all beliefs of a completely rational introspective agent.

The objective part of a set of modal propositions is its subset of classical propositions without $\Box$.

<u>Theorem</u> (Moore [1984]).
(i) If S is an expansion of I, then S is stable.
(ii) If S is stable, then S is an expansion (and in fact the unique expansion) of its objective part.

<u>Theorem</u> (Marek [1986], Konolige [1988]). Every collection of L–propositions closed under classical L–consequence is the objective part of a stable $L_\Box$–theory.

So stable $L_\Box$–theories are in a 1–1 correspondence with classical objective L–theories.

We now discuss how to generate expansions.
Let $L_{\Box,n}$ be the propositions of $L_\Box$ with $\Box$'s nested to at most depth n.

<u>Operation E</u>
Given a set A of L–propositions,
let E(0, A) be the set of classical L–consequences of A,
let E(n+1, A) be the set of classical $L_\Box$–consequences in $L_{\Box,n+1}$ of

$$E(n, T) \cup \{\Box\varphi : \varphi \in E(n, T)\} \cup \{\{\neg\Box\varphi : \varphi \in L_{\Box,n} - E(n, T)\}.$$

Let E(T) be the union of all E(n, T).

<u>Theorem</u> (Marek [1986]). If A is a set of propositions in L, then E(A) is the unique expansion of A.

Thus to find the expansion one has to find the objective part.

<u>Example</u>. Let I consist of $\neg\Box P \rightarrow Q$ and $\neg\Box Q \rightarrow P$. This I has two expansions, E(P) and E({Q}). (There are two more candidates $E(\emptyset)$ and E({P, Q}), but they are discarded by the same reasoning as above.) Why is E({P}) an expansion? Since Q is not in E({P}), we get

that $\neg\Box Q$ is in $\{\neg\Box\varphi : \varphi \notin E(\{P\})$. From this

$$E(\{P\}) = Cn[I \cup \{\Box\varphi : \varphi \in E(\{P\})\} \cup \{\neg\Box\varphi : \varphi \notin E(\{P\})\}]$$

can be proved. The non–trivial inclusion is from left to right, proved by induction using E(n, {P}).

Example. Let I consist of P and $\Box P \rightarrow Q$. This has the unique expansion $E(P \wedge Q)$.

Theorem (Moore [1984]). If S_1 and S_2 are two different stable theories, one cannot be contained in the other.

Thus stable theories act a little like classical complete theories.
Clearly if $Cn(I) = Cn(J)$, then I, J have exactly the same expansions.

Autoepistemic normal form

An autoepistemic (ae) clause is a modal statement of the form $A \rightarrow \sigma$, where

σ is in L, and
A is of the form $\Box\varphi_1 \wedge \ldots \wedge \Box\varphi_r \wedge \ldots \wedge \neg\Box\psi_1 \wedge \ldots \wedge \neg\Box\psi_s$,

where $\varphi_1, \ldots \varphi_r, \psi_1, \ldots \psi_s$ are in $L_\Box$.
We can take the theories we are concerned with to be generated by ae clauses. Here, imitating logic programming, we call "σ" the head and "A" the body of the ae clause.

Theorem (Marek and Truszczynski [1988, 1989]).
1) For every $I \subseteq L_\Box$ there exists an $I' \subseteq L_\Box$ with the same expansions such that the ae–clauses $A \rightarrow \sigma$ of I' all have A of $\Box$–nesting depth 1.
2) Such a I' can be computed in polynomial time.

This tells us that the problem of Byzantine generals does not exist in autoepistemic logic.

Normal form for expansions.

Theorem (Marek and Truszczynski [1988, 1989]).
Let $I = \{O_i = A_i \rightarrow \sigma_i : 1 < i < k\}$. Then
1) Every expansion of I is of form

$$E(\{O_i : i \in J\}) \text{ for suitably chosen } J \subset \{1,\ldots,k\}$$

2) A theory $S \subseteq L$ has the property that E(S) is an expansion of I if and only if there exists a

representation of S in the form $S = Cn(\{O_i: i \in J\})$ such that

(i) $I \subseteq E(S)$

(ii) For all $i \in J$, we have that $A_i \in E(S)$.

The problem is that a theory S may have numerous representations as $Cn(\{O_i: i \in J\})$ for various J. It is enough that <u>one</u> of these representations has the property 2(ii)..

<u>Example</u>. Let I consist of $\neg\Box P \rightarrow P$ alone. This I has no expansion. The only candidates are: 1) $E(\emptyset)$ and 2) $E(\{P\})$.

Re 1: P is not in $E(\emptyset)$, so $\neg\Box P \in E(\emptyset)$, and if it is an expansion, then $P \in E(\emptyset)$ by modus ponens.

Re 2: One can check that P is not an $L_\Box$ classical consequence of

$I \cup \{\Box\varphi : \varphi \in E(P)\} \cup \{\Box\varphi : \text{not } (\varphi \in E(P)\}$.

<u>Example</u>. $I = \{\neg\Box P \rightarrow (P \wedge Q),\ \neg\Box R \rightarrow P,\ \neg\Box R \rightarrow Q\}$. Then $E(\{P \wedge Q\})$ is an expansion of I for the following reason. The second and third clause give "epistemic support". If we select our representation of $E(\{P \wedge Q\})$ $(=E(\{P, Q\})$ from first clause, then we would not have the necessary epistemic support. Fortunately only one representation is required so it is an expansion.

<u>Theorem</u> (Marek [1986], Moore [1988]). There is an algorithm which, given $T \subseteq L$ and $\varphi \in L_\Box$, tests whether or not $\varphi \in E(T)$.

This algorithm cannot be polynomial time, but it is polynomial time in the characteristic function of Cn(T). The last two theorems imply that we can effectively compute all expansions in the propositional logic case.

<u>Definition</u>. φ is ae–consequence of I iff φ belongs to all expansions of I.

The previous two theorems imply that this notion is decidable.

<u>Example</u>. $I = \{\neg\Box P \rightarrow Q, \neg\Box Q \rightarrow P\}$. Proposition $P \vee Q$ is an ae consequence of I. Proposition $\Box P \vee \Box Q$ is ae–consequence of I. Formula $(\Box P \wedge \neg\Box Q) \vee (\neg\Box Q \wedge \Box P)$ is ae–consequence of I. Neither $\Box P$ nor $\Box Q$ are ae–consequence of I.

<u>Example</u>. (a) $I = \{\neg\Box P \rightarrow P\}$. I is consistent but it is ae–inconsistent (since there is no expression, intersection of expansions is $L_\Box$).

(b) $I = \{\neg\Box P \rightarrow P, \Box P \rightarrow P\}$ has a unique expansion $E(P)$, thus it is ae–consistent. Its subtheory $\{\neg\Box P \rightarrow P\}$ has no extensions. Thus there are ae–consistent theories with inconsistent subtheories. The subtheory $\{\Box P \rightarrow P\}$ has two expansions: E(TAUT), E(P). Thus we have a situation in which the smaller theory has a smaller set of consequences (previously had bigger...). The fact that I can have many or no expansions is disturbing. Are there conditions that imply uniqueness of expansions?

Gelfond stratification

A G–clause is a propositon of the form

$$(P_1 \wedge \ldots \wedge P_K \wedge \Box Q_1 \ldots \wedge \Box Q_r \wedge \neg\Box S_1 \wedge \ldots \wedge \neg\Box S_m) \rightarrow (T_1 \vee \ldots \vee T_u),$$

where all P_i's, Q_i's, S_k's, and T_t's are atoms. A theory I consisting of G–clauses is G–stratified if there exists a representation.

as a disjoint union $I = I_0 \cup \ldots \cup I_n$ such that

a) I_0 consists of the classical propositions in I.

b) Whenever clause

$$P_1 \wedge \ldots \wedge P_k \wedge \Box Q_1 \wedge \ldots \wedge \Box Q_1 \wedge \neg\Box S_1 \wedge \ldots \wedge \neg\Box S_m \rightarrow T_1 \vee \ldots \vee T_u$$

belongs to I_j, then

(i) $Q_1, \ldots, Q_1, S_1, \ldots, S_n$ do not appear on the right hand side of implication in any I_m, $m \geq j$ (that is, they are "defined" in $I_0 \ldots I_{j-1}$).

(ii) $P_1 \ldots P_k$ do not appear on the right hand side of implications in any I_m, $m > j$.

Theorem (Gelfond [1987]). If a theory I consisting of G–clauses is G–stratified, then it possesses a unique expansion.

Stratification

There is another notion of stratification. Theory I consisting of ae–clauses is stratified if there is a representation

$$I = I_0 \cup \ldots \cup I_k$$

such that for all $A \rightarrow \sigma \in I_j$,

(i) If an atom appears in σ then it does not appear in any formula in any I_k, $k < j$.

(ii) If an atom appears in A then it does not appear in the "head" of any formula in I_k, $k > j$

Theorem (Marek and Truszczynski [1988]). If I is stratified and $I = I_0 \cup \ldots \cup I_n$, then

(a) I has at most one expansion

(b) If $S = E(T)$ is an expansion of I and T is closed under Cn, and if T_i is an intersection of T with the language whose atoms are those appearing in I_i, then

(i) $E(T_i)$ is an expansion of I_i

(ii) S is an expansion of $T_i \cup I_{i+1} \cup \ldots \cup I_n$.

This theorem tells us how to compute expansions recursively: Compute an expansion of I_0, S_0. Then compute an expansion of $S_0 \cup I_1$, say S_1. Then compute an expansion of of $S_1 \cup I_2$, say S_2. At each step we are guaranteed at most one expansion. If we do not get one at any stage, there is no expansion for I.

Fixed Points

Let $\mathcal{S}$ be a modal logic such as K, S_4, S_5, etc. S is called an $\mathcal{S}$-fixed point over I iff $S = Cn_{\mathcal{S}}(I \cup \{\neg\Box\varphi : \varphi \notin S\})$. This definition is due to McDermott.

Theorem (Svarts [1989]). Expansions of I are precisely the KD45 fixed points over I.

However, one can prove the

Theorem (Marek–Truszczynski [1989]). Expansions of A are precisely the 5–fix points containing I. (In this theorem we refer to the Hilbert–style presentation of classical propositional logic with modality where axiom 5 is the only extra modal axiom assumed.

Pleonasm (Marek, Svarts, Truszczynski [1989]). Let S_1, S_2, be two modal logics. Let S_3 be an intermediate modal logic (that is, based on the axioms, $S_1 \subseteq S_3 \subseteq S_2$). If the set of S_1–fixed points coincides with the set of S_2–fixed points, then both of these sets coincide with the set of S_3 fixed points.

These results point attention to an as yet not completely explained phenomenon; different monotone logics can give rise to precisely the same class of extensions, and in this sense to the same non–monotone logic if extension is regarded as the central notion.

Example. Extensions for 5 and for KD45 coincide. There are also other such examples.

Example. $I = \{\Box P \to Q, \Box Q \to P\}$ has two expansions, $E_1 = E(\phi)$ and $E_2 = E(P, Q)$.

But P is in the second expansion because $\Box Q$ is there, that is, because Q is there, that is because $\Box P$ is there, that is because p is there. Hence the evidence for p being in E_2 is that "p is there", and there is definitely a circularity.

Let us eliminate this circularity. Define an operator A as follows. For $S \subseteq L_{\Box}$ put $A(S) = Cn(S \cup \{\Box\varphi : \varphi \in S\})$. and define

$$A_0(S) = S$$
$$A_{n+1}(S) = A(A_n(S))$$
$$A_{\infty}(S) = \cup_n A_n(S)$$

Call T an <u>iterative expansion</u> over I if $T = A_{\infty}(I \cup \{\neg\Box\varphi : \varphi \notin T\})$

<u>Theorem</u> (Marek and Truszczynski [1988]). If T is iterative expansion over I then T is an expansion of I.

It is easily verified from the definition that iterative expansions are precisely the N–fixedpoints , where N is the "poorest" modal logic, based on modal axiom N (necessitation) plus modus ponens alone. We remark that this poorest modal logic also possesses a tableaux system. However, one has to introduce a separate totally independent accessibility relation for <u>every</u> formula to get tableaux completeness.
Iterative expansions are fixed points with respect to the simplest modal logic in which there is classical tautologies, modus ponens, and necessitation, but no specific modal axiom such as K, T, 4 or 5.

<u>Connection with Logic Programming</u>
Given a logic program P, let $\Pi = \text{ground}(P)$ be the set of all ground instances of P. Then Π consists of expressions of the form

$$C: P \leftarrow Q_1,\dots,Q_r,\neg S_1,\dots,\neg S_1$$

To such clause assign its Gelfond translation

$$G(C) = Q_1 \wedge\dots\wedge Q_r \wedge \neg\Box S_1 \wedge \dots \wedge \neg\Box S_t \rightarrow P$$
$$G(\Pi) = \{G(C) : C \in \Pi\}$$

If P is stratified in the sense of Apt–Blair–Walker then $G(\Pi)$ is G–stratified.

<u>Theorem</u>.
(a) (Gelfond [1987]). Let P be stratified. Let M_p be its "perfect" model in sense of Apt–Blair–Walker. Then $E(Cn(M_p))$ is the only expansion of $G(\Pi)$.

(b) (Marek and Truszczynski [1988]) $E(Cn(M_p))$ is an iterative expansion of $G(\Pi)$.

The Kripke Frame view

The two semantics above, using $L_\Box$ and L valuations respectively are the classical logic way of doing things. It is natural that there is an equivalent in frame semantics.

Theorem (Moore [1984]). T is a stable autoepistemic theory if and only if T is the set of all valid modal propositions of a complete frame.

This was also proven by Halpern and Moses and Levesque.

Since the complete graphs (in which directed branches extend from every node to every node) are determined up to isomorphism by the cardinality of the nodes alone, one can restrict the complete frames for this theorem to those of the form $K = (\mathcal{I}, R)$, where the set of worlds $\mathcal{I}$ is a set of classical valuations (of the propositional letters) and R is $\mathcal{I} \times \mathcal{I}$. Introduce for each classical valuation V a copy (V, 0), to be used as a new world distinguished from world V if the latter is present in $\mathcal{I}$. Call it the distinguished V. Each pair consisting of K and a distinguished V gives rise to an ordinary Euclidean frame $K_V = (\mathcal{I}, R')$, where $\mathcal{I} = \mathcal{I} \cup \{(V, 0)\}$ and $R' = R \cup (\{(V,0)\} \times \mathcal{I})$. That, is, every W in K is accessible from (V,0) (including V if the latter is in $\mathcal{I}$), but (V, 0) is not accessible from any world in K. There is a natural extension of K_V to a model, where each world V in K is assigned valuation V, and (V, 0) is assigned valuation V.

Now let B be an autoepistemic theory. Such a model K_V with the assignment above, arising from a complete model, is called a

– "possible worlds" interpretation of B iff B consists of all modal propositions P such that P is valid in K.

– "possible worlds" model of B iff B consists of all modal propositions P such that $\Box P$ is valid in the model K_V.

Theorem (Moore [1988]). The "possible worlds" models K_V of B are precisely those "possible worlds" interpretations in which V is a member of $\mathcal{I}$.

This affords a back and forth translation of autoepistemic interpretations and autoepistemic models of stable theories to "possible worlds" interpretations and "possible worlds" models as defined above. Moore [1984, 1988] uses these semantic "possible world" characterizations to investigate stable expansions, and decision methods for semantic entailment. His method amounts to the use of the "list semantics" of the last section. There is a nice tableaux–based approach which can be redone in the style of the present lectures due to Niemelă [1986]. We omit these applications, which will be in an expanded version of these lectures, for lack of space.

§8. Nonmonotonic reasoning

In mathematics, a consequence drawn by a deduction from a set of premises can also drawn by the same deduction from any larger set of premises. The deduction remains a deduction no matter how the axioms are increased. This is monotonic reasoning, much imitated in other, less certain, disciplines. The very nature of monotonic reasoning makes mathematical proofs permanent, independent of new information. Thus it has been since Euclid and Aristotle. Theorems with complete proofs are never withdrawn due to later knowledge. It is little exaggeration to say that mathematicians never reject the completed proofs of their predecessors, except to complain about their constructivity. Mathematicians build directly on the works of their forebearers stretching back two and a half millenia to Euclid. Our current mathematical reasoning is merely a fleshed out version of Euclid's. Monotonic reasoning marks theoretical mathematics as a discipline. The traditional systems of mathematical logic are monotonic since they simply reflect mathematical usage. In the 1930's Tarski described a calculus of deductive systems and captured in a simple way the general concept of a monotonic formal system. His formulation includes all logics traditionally studied, intuitionistic, modal, and classical. He did not qualify his definition, as we do, with the adjective "monotone", because there were no other systems studied at that time.

Minsky [1975] suggested that there is another sort of reasoning which is not monotonic. This is reasoning in which we deduce a statement based on the absence of any evidence against the statement. Such a statement is in the category of beliefs rather than in the category of truths. Modern science offers as a tool for establishing provisional beliefs statistics. but in many instances we have no basis for applying statistics, due to a lack of governing distributions or a lack of samples for the problem at hand.

What role does belief play in our affairs? Often we must make sharp "yes or no" decisons between alternative actions. There may be no deductive or statistical base which justifies our choice, but we may not be able to wait for missing information, it may never materialize anyway. Often all we have as a basis for decision is surmise; that is, deductions from beliefs as well as truths and statistically derived statements. These beliefs are often accepted and used as premises for deduction and choice of action due to an unquantified lack of evidence against them.

A philosopher's much–quoted example is about Tweety. We observe only birds that can fly, and accept the belief that all birds can fly from the absence of evidence for the existence of non–flying birds. We are told that Tweety is a bird, and conclude that Tweety can't fly using our belief as premise. Later, we observe that Tweety is a pet Ostrich and clearly can't fly. We reject our previous belief set and conclusions as a basis for decision making, and are forced to choose a new belief. The new set of beliefs may also include equally uncertain statements, accepted due to a lack of evidence against. But we blithly draw consequences from the new belief set and make decisions on that basis till contrary evidence on some accepted belief is garnered, at which time we again have to acquire a new set of beliefs.

This has happened in the history of practically every subject except mathematics. The principles of physics, or biology, have been changed with every scientific revolution, even though unreflective scientists of each age think that final principles have been found. Even for mathematics, the Dutch mathematician and philosopher L.E.J. Brouwer would have argued that the belief in theorems established by "non–constructive methods" was unjustified, and that a new belief set based on constructive principles should be adopted in its place. Other mainstream mathematicians such as Hilbert did not agree with this position. Some philosophers of mathematics living now would argue that even within classical mathematics the idependence proofs for propositions of set theory such as the continuum hypothesis or the axiom of choice,indicate there are several incompatible axiomatic systems which, as belief sets, could be the foundation of mathematics.

One can envisage making up non–monotone logics describing the mathematical nature of belief. The exact result depends on the definition chosen for "lack of evidence against". McCarthy [1980] initiated the mathematical study of nonmonotonicity with his notion of circumscription. Its simplest form is as follows. With all relation symbols $\overline{R}$ but one of a model (the world we are discussing) held fixed, and given axioms $\varphi(\overline{R})$ relating that $\overline{R}$ to the other (fixed) relations of the model, the belief should be that, lacking further evidence to the contrary, we should believe that $\overline{R}$ denotes the least relation R, if any, satisfying $\varphi(\overline{R})$. If further evidence in the form of an axiom $\psi(\overline{R})$ becomes available, then we should believe that $\overline{R}$ denotes the least R satisfying $(\varphi \wedge \psi)(\overline{R})$, if any, instead, in a changed belief set.

There are now many different nonmonotonic system, abstracted from different questions in computer science and AI. Among the other systems that have been studied are:

– The theory of multiple believers of Hintikka,
– The truth maintenance systems of McDermott and Doyle,
– The default logics of Reiter,
– The autoepistemic logic of Moore,

– The theory of individual and common knowledge and belief of Halpern and Moses,
– The theory of logic programming with "negation as failure" of Clark.

This by no means exhausts the list. What issues in AI or computer science motivates these systems?

a) Suppose that we build a robot in a "blocks world" to navigate in a room and avoid obstacles and perform simple tasks, such as crossing the room with variable obstacles. We want the robot to learn principles from experience as to how to cross the room. At any given point, one may imagine that the robot should have a consistent deductively closed set of beliefs which are the current basis for its actions, including such provisional beliefs as "I can always traverse the left edge of the room since there has never been anything in the way there". But when such a principle is contradicted by new obstacles, the robot has to choose another belief set. So an important problem is to define what a belief set is and how to compute them and how to update them based on new evidence. Moore's autoepistemic logic is really a first try at this problem, mostly for propositional logic.

b) In computers, the operating system and program obey rules which compute how to change state. In the absence of exceptional behaviour, such as error conditions or failures to access resources, there is a system of decision rules (beliefs) computing how to change the state of the machine in this "normal behavior", or "default" case. But when an exceptional behavior happens, we are thrown to a different set of decision rules for change of state, a different set of "beliefs". One wants to be able to deduce what is true of the machine in states when it is a particular such "belief set". A logic for dealing with one such belief set at a time is Reiter's default logic.

c) In databases, facts and rules are stored as entries (the PROLOG model). Often also the database computes and stores conclusions, such as summary statistics or rules or tables computed from the database. These act as a deductive base for the set of current beliefs. When we query the database, we are asking for consequences of this belief set. When we update the database, all old entries that have changed have to be replaced, every consequence that uses these entires has to be recomputed and changed too. This is the process of replacing an old belief set by a new one. One often makes decisions on the basis of the absence of information in the database as well. A logic appropriate for describing a single such belief set is McDermott and Doyle's truth–maintenance systems. Also stable models for logic programming with negation as failure arise in this way.

d) There is an interesting point of connection with the philosophy of mathematics.
Non–monotone reasoning takes place during the process of discovery of mathematical theorems, when one posits temporarily some propositions on the basis of no evidence against them, and

explores the consequences of these beliefs until new mathematical facts force the abandonment of the beliefs. These non–monotone belief sets have their traces eradicated when demonstrative proofs are finished and published. The only hint left in the papers is in the motivational remarks explaining what obstacles were overcome and by what changes in viewpoint the proof was achieved.

Definition (Marek–Nerode–Remmel [1990]). A non–monotone formal system consists of a set U and a set of triples (α, β, γ), called rules, where $\alpha = (\alpha_1, \ldots, \alpha_k)$ is a finite sequence of elements of U, called premises, $\beta = (\beta_1, \ldots, \beta_l)$ is a finite sequence of elements from U, called guards, and γ is an element of U.

This is written, generalizing a notation of default logic, as

$$\frac{\alpha_1, \ \ldots, \ \alpha_k; \ \beta_1, \ \ldots, \ \beta_n}{\gamma}$$

The informal reading is : From $\alpha_1, \ldots, \alpha_k$ being established, and $\beta_1, \ldots, \beta_k$ not being established now or ever, conclude γ. You may substitute "computed" for "established for an informal reading in many applications. A subset S of U is called deductively closed if for every rule of the system, whenever $\alpha_1, \ldots, \alpha_k$ are in S and $\beta_1, \ldots, \beta_n$ are not in S, then γ is in S. There are no variables here, these are not schema. The intersection of all deductively closed sets containing a set I is generally not deductively closed, but the intersection of a descending chain of deductively closed sets is deductively closed. So any I is contained by Zorn's lemma in a minimal deductively closed set. But I may be contained in a lot of mutually incomparable minimal deductively closed sets over I. In AI the intersection of all deductively closed sets containing I is a (generally non–deductively closed) set called the set of secure consequences of I. The secured consequences of I are those that a "skeptical reasoner" would take as beliefs based on I.

The most important notion thus far introduced for nonmonotonic logics, the one that gives it its characteristic flavor, is the notion of an extension. Given a subset D of U, the D–projection of a non–monotone system (U, N) is the monotone formal system (U, M_N) obtained by omitting all rules from (U, N) with a guard conflicting with D and keeping the rest of the rules but without their guards. This D–projection is always a "monotone" system, that is, one with no guards. In a monotone system every set I is contained in a unique least deductively closed set.

Definition. An extension of I is a deductively closed set D in (U, N) containing D such that D is also the deductive closure of I in the D–projection (U, M_N).

Extensions play the role of "points of view" or "belief sets" D which are deductively closed with respect to the non–monotone rules, but whose elements also have "monotone deductions from I" when D is taken for granted. Such definitions capture much of the common content of the several theories of nonmonotonicity listed above.

There is some prejudice in the computer science community against nonmonotonic reasoning, based in many cases on the informality of the intended interpretations. At the level of abstraction above, however, one can see that non–monotone systems pervade ordinary mathematical practice, and are worthy of study on that basis alone. For example, the set of all marriages solving a "marriage problem" can be formulated as exactly the set of all extensions in an associated non–monotone system; similarly for the set of all colorings of a graph, the set of "chain covers" for a partial order, or the set of all maximal ideals of a boolean algebra. These mathematical examples have suggested to us a whole new set of techniques for finding extensions because of the availability of algorithms already investigated for some of these problems, not previously known to be relevant in the AI community. These do not arise in logic, but rather in operations research. This makes available known mathematical techniques from other areas of conventional mathematics for constructing and classifying belief sets (extensions), and simultaneously finds a common combinatorial thread among disparate parts of mathematics and disparate nonmonotonic systems. Finally, in recursion theory, prioric constructions can be construed as non–monotone systems. Sets constructed by the prioric argument are r.e. extensions. See Marek–Nerode–Remmel [1990] for development of nonmonotonic systems in this framework, including truth maintenance systems, default logic, autoepistemic logic, and logic programming with "negation as failure", with uniform semantics and with algorithms for computing extensions. Using this calculus, one can also develop uniformly non–monotone classical predicate logic, non–monotone intuitionistic predicate logic, etc.

§9. Classical Concurrrent Dynamic Logic

Hoare designed a logic for expressing program specifications and for proving "partial correctness" of programs. A basic construct of his logic was $A\{P\}B$, meaning that if A holds before the execution of program P, then B holds afterwards. Pratt (1976) was motivated by this to develop a modal logic of sequential non–deterministic programs in which each command α in a computer language implemented on the machine is associated with two distinct modal connectives $[\alpha]$ and $<\alpha>$, using also ideas from the algorithmic logic of Engeler and Salwicki. Dynamic logic will be well covered in a forthcoming article by Kozen and Tiuryn in the new *Handbook of Theoretical Computer Science.* Dynamic logic extends ordinary classical propositional or predicate logic as a language for expressing propositions φ about states of machines or processes by introducing formulas of the form $[\alpha]\varphi$, where α is a program. The intended meaning of $[\alpha]\varphi$ is: if a machine is in state s and then program α is executed in any way, at the termination of execution of α, φ holds at the new state s' of the machine. In dynamic logic, $\psi \rightarrow [\alpha]\varphi$ expresses Hoare's partial correctness assertion $\psi\{\alpha\}\varphi$, but unlike Hoare logic dynamic logic can also express termination of programs by $\neg[\alpha]\neg\varphi$, abbreviated $<\alpha>\varphi$, all within classical logic.

Peleg introduced concurrent dynamic logic within classical logic with an intended semantics of concurrent state transition system and proved a completeness theorem. Here is a definition of a Peleg concurrent transition system. Fix a non–empty set S, whose members are called "states". Suppose we are given a collection of "programs" as well. We are given that each program α denotes an "accessibility relation" $R_\alpha \subseteq S \times P(S)$, where $P(S)$ is the power set of the set of states. The idea behind $sR_\alpha T$ is that, starting in state s to execute α, at the end of the execution of α we are allowed to be in any state in T (but retain no knowledge of intermediate states). Through its accessibility relation R_α each program α induces a modal possibility (diamond) operator $<\alpha>$ and a modal necessity (box) operator $[\alpha]$ applying to propositions about states to produce propositions about states. These modal operators are defined without reference to any operations on programs to obtain programs, and as a result can be applied flexibly in many other situations than those in dynamic logic.

Concurrent diamond (classical case)
$<\alpha>\varphi$ is true at state s iff there is a set $V \subseteq S$ such that $(s, V) \in R_\alpha$, and for each $v \in V$, φ is true at state v.

Concurrent box (classical case)
$[\alpha]\varphi$ is true at state s iff $(s, V) \in R_\alpha$ implies that for all $v \in V$, φ is true at state v.

In the intuitionistic treatment below we will replace the notion of "complete state of knowledge about the machine" by a notion of "partial state of knowledge about the machine", which we call a "world state" because it represents knowledge available in a Kripke world in an intuitionistic frame approximating to the relational system underlying the dynamic logic. . The definition of state and diamond will remain formally the same as in classical dynamic logic in intuitionistic dynamic logic. But intuitionistic dynamic logic will require a change in the definition of box, since the context is changed from one classical relational system to a Kripke frame of relational systems and the box has a universal quantifier in it in Kripke frames which intuitionistically accomodates increasing knowledge.

To get a dynamic logic corresponding to a machine model and for the programs that run on it, one has to describe what are the intended atomic programs and how to build up complex programs from simple programs by program constructs. For dynamic logic as it has been developed thus far, one envisages the machine described by memory locations named by program variables. The state s of the machine is supposed to be determined by the simultaneous contents of all memory locations. A "program variable" x_i is associated with each memory location i. The state s of the machine is identified with a map s with domain the set X of all program variables x_i, $s(x_i)$ being interpreted as the content of the ith memory location. The class of atomic programs is the class of sequential assignments $(x_i \leftarrow t)$, where t is a "program term" built up from program variables and function symbols and individual constants from the language of the relational system. This assignment, applied as a program when the state is s_1, yields new state s_2, where s_2 coincides with s_1 except for argument x_i, where $s_2(x_i)$ is defined as $s_1(t)$. This s_2 is conventionally denoted as $s_1[s_1(t)/x_i]$. Then the accessibility relation corresponding to this assignment is:

Sequential Assignment. $(x_i \leftarrow t)$ denotes $R(x_i \leftarrow t)$, the set of all $(s_1, \{s_2\})$ such that $s_2 = s_1[s_1(t)/x_i]$, where t is a program term and x_i is a program variable.

Here are accessibilities for other standard constructs for building new programs from old.

Composition. $(\alpha; \beta)$ denotes $R(\alpha; \beta)$, the set of all (s, U) such that
$\exists V$ with $(s, V) \in R(\alpha)$, and for each $v \in V$, $\exists U_v \subset U$ with $(v, U_v) \in R(\beta)$, and
$U = \cup \{U_v : v \in V\}\}$.

Define $R(\alpha^{n+1})$ as $R(\alpha;\alpha^n)$.

Sequential iteration. (α^*) denotes $R(\alpha^*) = \cup R(\alpha^i)$

Non–deterministic "or". $(\alpha \cup \beta)$ denotes $R(\alpha \cup \beta) = R(\alpha) \cup R(\beta)$

Here is the single construct used to build a program and its accessibility from a formula φ.

Deterministic test. $(\varphi?)$ denotes $R(\varphi?) = \{(s, \{s\}) : \varphi \text{ holds at } s\}$

For concurrency, one has to add an extra program construct, "parallel and", denoted as "$\cap$". This $\alpha \cap \beta$ is supposed to mirror the fact that α and β are executed simultaneously, starting from a common state w, and is the dual of $\cup$. The intention is that α, β, executed in a state s, lead to a set of states reached by executing α and β in parallel with no interleaving.

Parallel "and". $(\alpha \cap \beta)$ denotes $R(\alpha \cap \beta) = \{(s, U \cup V) : (s, U) \in R(\alpha) \text{ and } (s, V) \in R(\beta)\}$.

§10. Intuitionistic concurrent dynamic logic

In classical logic the relational system for the values of states in dynamic logic is completely given in advance. We know for any atomic statement about the relational system whether it is true or false in advance. We are allowed to use the values of arbitrary terms and the truth of arbitrary statements instantly in all programs.

a) We assume that as soon as the state s_1 is given as a function on program variables for any program term t, the value $s_1(t)$ is available instantly for use in an assignment program resulting in next state $s_2 = s_1[s_1(t)/x_i]$. But new state s_2 depends on $s_1(t)$, which must be computed by following the inductive definition of program term t. At any stage of an actual computation, we have computed values $s_1(t)$ for some terms only. In effect, we are prepared to make some assignments only.

b) We assume that as soon as the state s_1 is given, we can instantly test whether φ holds (φ?), and can use this answer for determining the next state of the computation. But at a finite stage of the computation, we only have computed the validity of some φ.

We give an intuitionistic approach to partial knowledge via Kripke frames; see Nerode [1990] for extended explanations as to why Kripke frames of intuitionistic reasoning are a good way of reflecting increasing partial knowledge. Here each intuitionistic world of the Kripke frame represents a state of partial information about what constants exist, what atomic relations hold, and what (partial) functions entering into terms can be evaluated in that world. Thus in that world only certain tests and certain assignments can actually be carried out, those to which we have access. at that time. We also wish to be intuitionistic to be as constructive as possible for another reason. We believe that much more intuitionistic systems have to be developed with term extraction for many of these logics to make them tools for automated reasoning. The definition of state and the definition of diamond will remain formally the same as in classical dynamic logic in intuitionistic dynamic logic. But intuitionistic dynamic logic will require a change in the definition of box, since the context is changed from one classical relational system to a Kripke frame of relational systems and the box has a universal quantifier in it in Kripke frames which intuitionistically accomodates increasing knowledge. Wijesekera's thesis [1990] under Nerode developed a very general form of intuitionistic modal logic and semantics so as to cover all known computer science applications as theories. It has a deduction theorem, a completeneness theorem, independent definitions of [ ], < >, and < > does not distribute over box. We use this as the underlying logic for intuitionistic concurrent dynamic logic for understanding the intuitionistic consequences of partial information. See Nerode–Wijesekera [1990] for a complete discussion. Here we explain only the language and the semantics. Our intent is that a mapping of program variables into a relational system from the Kripke frame represents a "world state" of partial knowledge; a formula $\varphi(x_1, ..., x_n)$, with program variables $x_1, ..., x_n$ as its only possibly free variables, represents a proposition about world states, these world states assign values in the relational system to all program variables. Substitutions of program terms for program variables are used solely to change world states, and what is substituted is always a program term. The formula φ will have within it quantifications over individual variables, but these quantifications are to be regarded as solely used to build enough propositions about world states, and have nothing to do with the program variables. This is a shift in viewpoint from previous literature in dynamic logic. The standard viewpoint concentrates implicitly on statements, not formulas with free variables. Our viewpoint concentrates on the algebra of all formulas of any number of free variables with all program variable occurrences free and all other (dummy) variable occurrences bound.

Language

SYMBOLS

We use a first order language for intuitionistic logic with

– $y_0, y_1, ...$, an infinite sequence of individual dummy variables

– $c_0, c_1, ...$, an infinite sequence of individual constants

– $x_0, x_1, ...$, an infinite sequence of program variables.

– $A_0, A_1, ...$, an infinite sequence of predicate letters of various arities. Also equality $=$ and unary existence predicate E.

– $g_0, g_1, ...$, an infinite sequence of function symbols of various arities.

Let C be the set of all individyual constants.

PROGRAM TERMS

– $t_0, t_1, ...$, all program terms built up from constants, program variables and function symbols.

LOGIC TERMS

– $s_0, s_1, ...$, all logic terms built from constants, dummy variables and function symbols.

MIXED TERMS

– $u_0, u_1, ...$, all mixed terms built from constants, logic and program variables, and function symbols.

Let $X = \{x_0, x_1, ...\}$ be the set of all program variables.

ATOMIC PROGRAMS

"Assignment statements" $(x_i \leftarrow t_j)$ such that $i, j \in \omega$, where t_j is a program term.

LOGICAL CONNECTIVES (for building logical formulas)

T (truth), $\bot$ (falsum), $\wedge$ (and), $\vee$ (or), $\longrightarrow$ (implies), $\neg$ (not), $\longleftrightarrow$ (iff), $\exists$ (there exists), $\forall$ (for all).

PROGRAM OPERATORS (for building up programs)

$\cup$ (parallel or), $\cap$ (parallel and), ; (composition) , * (sequental iteration).

The mixed operators are ? (test), < > (possibility), [ ] (necessity). Here < > and [ ] are the modal operators.

PROGRAMS AND FORMULAS (Programs and formulas are defined by a simultaneous induction.)

1. Atomic programs are programs.

   If P is a predicate letter of arity n and $t_1, t_2, \ldots, t_n$ are mixed terms, then $P(t_1, t_2, \ldots, t_n)$ is a formula.

2. If α, β are programs and φ, ψ are formulas and y is an individual variable, then (α^*), $(\alpha ; \beta)$, $(\alpha \cap \beta)$, $(\alpha \cup \beta)$ are programs.

   $(\varphi \wedge \psi)$, $(\varphi \vee \psi)$, $(\varphi \to \psi)$, $(\neg\varphi)$, $((\exists y)\varphi(y))$, $((\forall y)\varphi(y))$, $(<\alpha>\varphi)$, $([\alpha]\varphi)$ are formulas.

   If φ has no free dummy variables and no [], <>, then $(\varphi?)$ is a program. (Note that free program variables are explicitly allowed in φ.)

Parentheses are omitted at will for readibility. Program terms were defined earlier. They have all occurrences of program variables free, all occurrences of individual variables bound.

Ordinary Intuitionistic Kripke Frame Semantics

We use the Van Dalen–Troelstra intuitionistic E^+–logic Kripke frames, which we leave to the reader to look up. It simply allows function symbols for partially defined functions.

A frame $\underline{K} = ((K, \leq) , \{(D(k), =_k , \{A_{i,k} : i \in \omega\}, \{g_{j,k} : j \in \omega\}) : k \in K\})$,

consists of

(1) $(K, \leq)$, a partially ordered set of "possible worlds".

(2) D, a function assigning to each k in K a nonempty set D(k). This is the domain of the possible world k.

(3) $=_k$, an equivalence relation on D(k). This is the identity relation denoted by = of possible world k.

(4) $A_{i,k}$, a relation on D(k) of the appropriate arity. This is the relation denoted by A_i in world k.

(5) $g_{i,k}$, a partial function on D(k) of the appropriate arity. This is the partial function denoted by g_i in world k.

Further, we assume

(6) D is monotone, that is, for all k, k' in K, $k \leq k'$ implies $D(k) \subseteq D(k')$, with $C \subseteq D(k)$.

This says individuals persist in larger worlds.

(7) If $d =_k d'$ and $A_{i,k}(d)$, then $A_{i,k}(d')$. This says equality respects relations in all possible worlds.

(8) If $d =_k d'$ and $g_{i,k}(d)$ is defined, then $g_{i,k}(d')$ is defined and $g_{i,k}(d) =_k g_{i,k}(d')$. This says equality respects partial functions in all possible worlds.

(9) If $k \leq k'$ and $d =_k d'$, then $d =_{k'} d'$ This says equality persists in larger possible worlds.

(10) If $k \leq k'$, then $A_{i,k} \subseteq A_{i,k'}$ and $\mathrm{graph}(g_{i,k}) \subseteq \mathrm{graph}(g_{i,k'})$. This says partial functions persist in larger possible worlds.

(11) The terms are interpreted in each $D(k)$ as follows.

Any constant in $D(k)$ denotes the constant function with itself as value, any variable denotes a projection (identity). We use the notation $\bar{v}_k$ for the partial function denoted by term v in $D(k)$. If $v_1, \ldots, v_n$ denote $\bar{v}_1, \ldots, \bar{v}_n$, then for any function symbol g_i, $g_i(v_1, \ldots, v_n)$ denotes $g_{i,k}(\bar{v}_1, \ldots, \bar{v}_n)$ when defined. For a constant term v (no variables), by abuse of notation, $\bar{v}$ if defined is identified with its value in $D(k)$.

Let the original first order language be extended to have a name for each element of $D = \cup \{D(k) : k \in K\}$. The usual forcing relation $k \Vdash \varphi$ between possible worlds k and ordinary intuitionistic formulas (no box, no diamond) is defined as follows.

<u>Base step.</u>

1) $k \Vdash A_i(v_1, \ldots, v_n)$ iff $v_1, \ldots, v_n$ are constant terms and all $\bar{v}_i$ are defined in k and $A_{i,k}(\bar{v}_1, \ldots, \bar{v}_n)$.

2) $k \Vdash v_1 = v_2$ iff v_1 and v_2 are constant terms and $\bar{v}_1, \bar{v}_2$ are defined in k and $v_1 =_k v_2$.

3) $k \Vdash T$

<u>Induction step.</u>

4) $k \Vdash \varphi \wedge \psi$ iff $k \Vdash \varphi$ and $k \Vdash \psi$.

5) $k \Vdash \varphi \vee \psi$ iff $k \Vdash \varphi$ or $k \Vdash \psi$.

6) $k \Vdash \varphi \rightarrow \psi$ iff for all $k' \geq k$, $k' \Vdash \varphi$ implies $k' \Vdash \psi$.

7) $k \Vdash \forall y \varphi(y)$ iff for all $k' \geq k$ and for all $d \in D(k')$, $k' \Vdash \varphi(d)$.

8) $k \Vdash \exists y \varphi(y)$ iff for some $d \in D(k)$, $k \Vdash \varphi(d)$.

9) $k \Vdash \neg\varphi$ iff for all $k' \geq k$, it is not the case that $(k' \Vdash \varphi)$.

By definition $k \Vdash Es$ abbreviate $k \Vdash \exists y(y = s)$.

Kripke models for intuitionistic concurrent dynamic logic

A <u>world state</u> of a first order Kripke frame $\underline{K}$ is a pair (k, s), such that $s : X \rightarrow D(k)$, an assignment of values to program variables in a single possible world D(k). Let $S = S(\underline{K})$ be the set of all world states.

Note: The reference to k is what makes this a "world state". Even though a world state s is defined on all program variables, when we use s to evaluate terms in k we will often fail due to the partiality of functions in k. The assumption that s is defined for all x_i and instantly available corresponds to the intuition that the atomic information about the state of a register machine is simple the contents of the registers, which can be read off and used immediately.

Extend s to a corresponding (partial) map $s : T \longrightarrow D(k)$, where T is the set of all program terms, by the inductive definition.

Base step.
$s(c) = c$ for constants c.

Induction step.
If t is $g_i(t_1,..., t_n)$ then $s(t)$ is defined iff $s(t_1),..., s(t_n)$ are defined, and then $s(g_i(t_1,..., t_n)) = g_{i,k}(s(t_1),..., s(t_n))$,

Now we see where the world comes in. A program term t denotes an object in D(k) relative to $s: X \rightarrow D(k)$ iff s is defined on t, and then t denotes s(t). Only some program terms will be evaluable by s in D(k), even though all the program variables are evaluable, since the domain of s is the set X of program variables. To repeat, only some evaluations of program terms, and therefore only some assignment programs, can be carried out in world k, the ones that can be carried out grow as the world gets larger. Further, knowledge of atomic relations these terms satisfy, and therefore tests that can be performed in the world, grow as we go on to larger worlds.

Let $\underline{K}$ be a Kripke frame. Partially order the world states (k, s), (k', s') by $(k, s) \leq (k', s')$ iff s=s' and $k \leq k'$ in the Kripke frame.

The following semantics is a forcing generalization to Kripke frames of Tarski's definition of satisfaction by means of maps from the set individual variables to the domain of the relational system, often called satisfaction sequences, where the terminology of satisfaction sequences is replaced by the terminology of world states. The definition is adjusted to the many worlds of a Kripke frame. This definition requires a double induction because of the presence of < >, [ ], which themselves have inductive definitions of their accessibility relations. Each accessibility relation $R(\alpha)$ is a binary relation between world states (k, s) and sets S' of world states, with the restriction that ((k, s), S') and $(k', s') \in S'$ imply s = s'. The interpretation of this restriction is that we are using partial knowledge k of the relations and function symbols as interpreted in k to follow the execution of the program to the end, and no knowledge from larger or incomparable worlds is allowed, we have to get to the end of the execution entirely on the basis of knowledge in k.

The standard Kripke model (of intuitionistic concurrent dynamic logic) associated with first order Kripke frame $\underline{K}$ (for ordinary intuitionistic first order logic) consists of $\underline{K}$, and a concurrent transition system, with domain the set of all world states of $\underline{K}$, given by the simultaneous inductive definition below for, defining at the same time two notions.

a) forcing: $(k, s) \Vdash \varphi$ for world states (k, s) and state formulas φ,
b) accessibility: $R(p) \subseteq S \times P(S)$ for all programs $p \in P$.

Analogous definitions by double inductions will yield intuitionistic systems for deducing consequences of partial information for many other classical computer science structures, both those connected with program verification and those connected with AI. These include temporal logics, logics of knowledge and belief for many believers, dynamic logics for communicating processors, dynamic logics for shared memory machines, and sequential or concurrent PROLOGs. To get the same effect by separating the inductions, that is, in the case below, to define forcing and accesibility relations by separate single inductions, does not often work.

Base step.

1. (Atomic formulas) $(k, s) \Vdash A_i(t_1, \ldots, t_n)$ iff $s(t_1), \ldots, s(t_n)$ are defined in k, and $(s(t_1), \ldots, s(t_n)) \in A_{i,k}$.

Here we have predicate letter A_i of arity n and $t_1, \ldots, t_n$ program terms.

2. (Equality) $(s\ k) \Vdash t = t'$ iff $s(t), s(t')$ are defined in k and $s(t) =_k s(t')$

Here we have t, t' program terms.

3. (Assignment) $R(x_i \leftarrow t)$ is the set of all $((k, s_1), \{(k, s_2)\})$ such that $s_1(t)$ is defined in k and $s_2 = s_1[s_1(t)/x_i]$.

4. $(k, s) \Vdash T$

Induction step.

5. $(k, s) \Vdash \varphi \wedge \theta$ iff $(k, s) \Vdash \varphi$ and $(k, s) \Vdash \theta$.

6. $(k, s) \Vdash \varphi \vee \theta$ iff $(k, s) \Vdash \varphi$ or $(k, s) \Vdash \theta$.

7. $(k, s) \Vdash \varphi \to \theta$ iff for all $k' \geq k$, if $(k', s) \Vdash \varphi$, then $(k', s) \Vdash \theta$.

8. $(k, s) \Vdash \forall y \varphi(y)$ iff for all $k' \geq k$, and all $d \in D(k')$, we have $(k', s) \Vdash \varphi(d)$.

9. $(k, s) \Vdash \exists y \varphi(y)$ iff there is a $d \in D(k)$ such that $(k, s) \Vdash \varphi(d)$.

10. $(k, s) \Vdash \neg\varphi$ iff for all $k' \geq k$, $(k', s) \Vdash \varphi$ is false.

Every step up to now is just a copy of ordinary Kripke frame senmantics, using pairs (k, s) as forcing conditions.

11. $(k, s) \Vdash <\alpha>\varphi$ iff there is an $S' \subseteq S$ such that $((k, s), S') \in R(\alpha)$ and for all $(k, s') \in S'$, $(k, s') \Vdash \varphi$.

12. $(k, s) \Vdash [\alpha]\varphi$ iff for all $k' \geq k$, and for all $S' \subseteq S$ such that $((k, s), S') \in R(\alpha)$, $(k, s'') \Vdash \varphi$ for all $(k, s'') \in S'$.

Note. 11) says that in a check that starting with s we can terminate execution in a set of states each having a property, a check based on partial information k, we use only the information present in k to execute the computation and verify the property. If it takes more information from a larger k', we cannot execute the computation and verify the property based on k. 12) says that in a check that starting with s, we always terminate in a set of states each of which has a property, a check based on partial information k, we can use only information present in k, and similarly for all larger k'. 11, 12 are inductive clauses defining forcing from accessibility and forcing.

13. $R(\alpha; \beta)$ is the set of all $((k, s), U)$ such that there exists a V with $((k, s), V) \in R(\alpha)$ and a function defined on all the $v = (k, s')$ in V with values subsets $U_v \subseteq U$ such that for all $v \in V$, $(v, U_v) \in R(\beta)$, and $V = \cup \{U_v ; v \in V\}$.

14. $R(\alpha^*) = \cup R(\alpha^i)$.

15. $R(\varphi?) = \{((k, s), \{(k, s)\} : s \Vdash \varphi\}$, for φ without < >, [ ].
This means we don't allow tests that refer to termination or possible execution of programs, tests are ordinary intuitionistic formulas in the language of k. No other formulas. Note that 15 defines an accessibility from forcing.

16. $R(\alpha \cup \beta) = R(\alpha) \cup R(\beta)$.

17. $R(\alpha \cap \beta) = \{((k, s), U \cup V) : ((k, s), U) \in R(\alpha) \text{ and } ((k, s), V) \in R(\beta)\}$.

Proof procedures and completeness are obtained by generalizing the intuitionistic tableaux proofs and the complete systematic tableaux used in Nerode [1990]. Our treatment of intuitionistic concurrent dynamic logic uses as tableaux proofs (recursive) ω–level, ω–branching, labelled path–finite trees with a contradiction on every branch. These are generally infinite hyperarithmetic objects. These tableaux are a natural generalization of the at most binary branching, labelled, path–finite trees with a contradiction on every branch used for ordinary intuitionistic predicate logic tableaux proofs in Nerode [1990]; but the latter are of course finite by König's lemma. For ordinary intuitionistic logic the labels used on nodes in the latter were signed formulas $Tp\Vdash\varphi$, $Fp\Vdash\varphi$, where p is a finite sequence of zeroes and ones and φ is a statement of ordinary intuitionistic logic. In the generalization, φ is a statement of intuitionistic concurrent dynamic logic. The ω–branching occurs at nodes labelled $T[\alpha^*]\varphi$, $F<\alpha^*>\varphi$, where respectively all $T<\alpha^i>\varphi$, or all $F<\alpha^i>\varphi$ have to be appended at once. It is convenient to use some equivalences of concurrent dynamic logic as reduction rules within the tableaux. See Nerode–Wijesekera [1990].

---

* Research supported by NSF grant DMS–89–02797 and ARO contract DAAG 29–85–C–0018
** Thanks to Prof. Wiktor Marek and to Prof. Andre Deutz for their help in the preparation of this paper. The material in §8 has been substituted for the original exposition and is taken from Marek–Nerode–Remmel [1990].

BIBLIOGRAPHY

M. Abadi and Z. Manna, "Modal theorem proving", LCNS 230, 172–188.

K. Apt, H. Blair, A. Walker [1986], "Towards a theory of declarative knowledge", Proc. 1986 Workshop on the Foundations of Deductive Databases and Logic Programming.

L. Banachowski, A. Kreczmar, G. Mirkowska, H. Rasiowa, A. Salwicki, "An introduction to algorithmic logic", in *Mathematical Foundations of Computer Science* (A. Mazurkiewicz and Z. Pawlak, eds.), Banach Centre Publications, Polish Scientific Publishers, Warsaw.

H. Barringer [1985], *A Survey of Verification Techniques for Parallel Programs*, LCNS 191, Springer–Verlag, Berlin.

M. Ben Ari [1980], "Temporal logic proofs of concurrent programs", Tech. Rpt 80–44, Tel Aviv, Israel.

M. Ben Ari, J. Y. Halpern, and A. Pnueli [1982], "Deterministic propositional dynamic logic: finite models, complexity, and completeness", J. Comp. Syst. Sci. 25, 402–417.

M. Bozic, K. Dosen [1984], "Models for normal intuitionistic logics", Studia Logica 43, 217–245.

R. Brachman, H. Levesque, R. Reiter, eds. [1989], First Conference on Principles of Knowledge Representation and Reasoning, Toronto, Canada, May 15–18, 1989.

A. Brown [1985], "Modal propositional semantics for reason maintenance systems", Proc. 9th IJCAI, 178–184.

R. Burstall [1972], "Some techniques for proving correctness of programs which alter data structures", Machine Intelligence 7, 23–50.

R. Burstall [1974], "Program proving as hand simulation with a little induction", Information Processing 74, 308–312.

B. F. Chellas [1980], *Modal Logic: An Introduction*, Cambridge University Press.

K. L. Clark [1978], "Negation as failure", in *Logic and Databases*, Plenum Press, N. Y., 293–322.

E. Clark, E. Emerson, A. Sistla, [1986], "Automatic verification of finite state concurrent

systems using temporal logic specifications",, ACM Trans. on Programming Languages and Systems, 8, 244–263.
J. de Kleer, "An assumption–based TMS", Art. Intell. 28, 127–162.
J. de Kleer, "Extending the ATMS", Art. Intell. 28, 163–196.
J. Doyle [1979], "A truth maintenance system", Art. Intell., 231–272.
K. Dozen [1985], "Models for stonger normal intuitionistic modal logics", Studia Logica 44, 39–70.
C. Elkin [1989], "Logical characterization of nonmonotonic TMS's", Math. Found. Comp. Science 1989 (A. Kreczmar and G. Mirkowska, eds.), Springer–Verlag, Berlin, 218–224.
E. Engeler [1967], "Algorithmic properties of structures", Math. Systems Theory 1, 183–195.
E. Engeler [1968], *Languages: Automata and Structures*, Markham, Chicago.
E. Engeler [1975], "On the solvability of algorithmic problems", *Logic Colloquium '73* (H. E. Rose and J. C. Shepherdson, eds), Studies in Logic vol. 80, North–Holland, 231–251.
D. Etherington [1987], "Relating default logic and circumscription", Proc. IJCAI–87.
D. Etherington [1988], *Reasoning with Incomplete Information*, Pitman, London, 1988.
W. Ewald [1986], "Intuitionistic tense and modal logic", J.S.L. 51, 166–179.
R. Fagin and J. Y. Halpern [1988], "Belief, awareness, and limited reasoning", Art. Intell. 34, 39–76.
M. Fitting [1983], *Proof Methods for Modal and Intuitionistic Logics*, D. Reidel, Dordrecht, Holland.
M. J. Fischer and R. E. Ladner [1979], "Propositional dynamic logic of regular programs", J. Comp. Syst. Sci. 18, 194–211.
F. B. Fitch [1949], "Intuitionistic modal logic with quantifiers", Portugalia Mathematixca 7, 113–118.
R. Floyd [1967], "Assigning meaning to programs", Proc. Symp. in Applied Math., AMS 19, 19–32.
M. Font [1986], "Modality and possibility in some intuitionistic modal logics", Notre Dame Journal of Formal Logic 27, 533–546.
D. Gabbay, A. Pnueli, S. Shelah, and J. Stavi [1980], "On the temporal analysis of fairness", Proc. 7th ACM POPL, Las Vegas, 163–173.
D. Gabbay [1981], *Semantical investigations in Heyting's intuituionistic logic,* Reidel, Dordrecht.
A. Galton, ed. [1987], *Temporal Logics and their Applications*, Academic Press, N.Y.
P. Gärdenfors [1988], *Knowledge in Flux*, MIT Press, Bradford Books, Cambridge, MA.
C. Geissler and K. Konolige, "A resolution method for quantified modal logics of knowledge and Belief", in J. Y. Halpern [1986]. 309–324.
M. Gelfond, H. Przymusinska, and T. Przymusinska [1986],."The stable model semantics for logic programming", in Proc. of the 5th Logic Programming Symposium, 1070–1080 (R. Kowalski and K. Bowen, eds.), Assoc. for Logic Programming, MIT Press, Cambridge, Mass.
M. Gelfond [1987], "On stratified autoepistemic theories", Proc. AAAI–87, Amer. Assoc. for Art. Intell., Morgan–Kaufmann, Los Altos, CA, 1987.
M. Gelfond [1989], "Autoepistemic logic and the formalization of commonsense reasoning", LNCS 346, 176–186.
M. Ginsberg [1987], *Readings in Nonmonotonic Reasoning*, Morgan Kaufmann, Los Altos, Calif.
R. Goldblatt [1982], *Axiomatizing the Logic of Computer Programming*, LCNS 130, Springer–Verlag, Berlin.
R. Goldblatt [1982], "The semantics of Hoare's iteration rule", Studia Logica 41, 141–158.D. R.
R. Goldblatt [1987], *Logics of Time and Computation*, CSLI Lecture Notes 7, Center for the Study of Language and Information, Stanford, 1987.
B. Hailpern [1982], "Verifying Concurrent Processes using Temporal Logic", LCNS 68, Springer Verlag, Berlin.
J. Y. Halpern and Y. O. Moses, [1984], "Knowledge and common knowledge in a distributed environment", 3rd ACM Conference on the Principles of Distibuted Computing, 50–61. (Revised as IBM RJ 4421, 1984).
J. Y. Halpern and Y. Moses, [1984]," Towards a theory of knowledge and ignorance", Workshop on Nonmonotonic reasonong, Mohonk Mountain House, New Paltz, New York (October 17–19, 1984), available from the AAAI Office, Menlo Park, California.
J. Y. Halpern and Y. O. Moses [1985], "A guide to modal logics of knowledge and belief", in Proc. IJCAI, Los Angeles, 50–61.
J. Y. Halpern (Ed.) [1986] "Theoretical aspects of reasoning about knowledge", Proc. 1986 Conf.

S. Hanks and D. McDermott [1986], "Default reasoning, nonmonotone logics, and the frame problem", Proc. 5th Nat. Conf. on AI, Morgan–Kaufmann, 328–333.
S. Hanks and D. McDermott [1986], "Nonmonotonic logic and temporal projection", Art. Intell. 33, 379–412.
D. Harel, A. Meyer, and V. R. Pratt [1977], "Computability and completeness in logics of programs", Proc. ACM STOCS, Boulder, Colorado, 261–268.
D. Harel [1979], "First Order Dynamic Logic", LCNS 68, Springer Verlag, Berlin.
D. Harel [1983/5], "Dynamic Logic", in *Handbook of Philosophical Logic*, vol. I–III, (D. Gabbay and F. Guenthner, eds), D. Reidel, Dordrecht.
J. Hintikka [1962], *Knowledge and Belief*, Cornell University Press, Ithaca, N.Y.
C. Hoare [1969], "An axiomatic basis for computer programming", Comm. ACM 12, 576–583.
G. E. Hughes and M. J. Cresswell [1968], *Introduction to Modal Logic*, Methuen, London.
G. E. Hughes and M. J. Cresswell [1984], *A Comnpanion to Modal Logics* Methuen, London.
S. Kanger [1957], "A note on quantification and modalities", Theoria 23, 133–4.
K. Konolige [1986], *A Deductive Model of Belief*, Morgan Kaufmann, Inc., Los Altos, California.
K. Konolige [1988], "On the relation between default logic and autoepistemic logic", Artificial Intelligence 35, 343–382.
D. Kozen and R. Parikh [1982], "An elementary proof of the completeness of PDL", Theor. Comp. Sci. 14 (1981), 113–118.
D. Kozen and J. Tiuryn [1989], "Logic of programs", Cornell University Department of Computer Science Report no. 89–962.
S. Kripke [1959], "A completeness theorem in modal logic", J. Symb. Logic 24, 1–15.
S. Kripke [1963], "Semantical considerations on modal logic", Acta Phil. Fennica 16, 83–94.
S. Kripke [1963], "Semantic analysis of modal logic I: normal propositional calculi", Zeit. Math. Logik Grund. Math. 9, 67–96.
S. Kripke [1971], "Semantical considerations on modal logic", *Reference and Modality* (L. Linsky, ed.), Oxford University Press, London, 63–72.
R. E. Ladner [1977], "The computational complexity of provability in systems of modal propositional calculus", SIAM J. of Computing 6:3 467–480.
H. J. Levesque [1981], "The interaction with incomplete knowledge bases: a formal treatment", Proc. 7th IJCAI, University of British Columbia, Vancouver, B. C. (August 24–28, 1981), 240–245.
V. Lifschitz [1984], "Some results on circumscription", Proc. Non–monotonic Reasoning Workshop of AIII, New Paltz, NY., 151–164.
V. Lifschitz [1989], "The Mathematics of nonmonotonic reasoning", Proc. LICS–89, IEEE Computer Society.
V. Lifshitz [1989], "Circumscriptive theories: a logic–based framework for knowledge representation", J. Phil. Logic 17.
A. Loperic, "On the method of valuations in modal logic", Math. Logic Proc. 1st Brazilian Conf.
McDermott and J. Doyle [1980], "Non–Monotonic Logic I", Art. Intell. 13, 41–72.
J. McCarthy [1977], "Epistemological problems of artificial intelligence", Proc. 5th IJCAI, Cambridge, Mass.,1038–1044.
J. McCarthy [1980], Circumscription–a form of non–monotonic reasoning, J. Art. Intell. 13;27–39.
J. McCarthy [1986], "Applications of circumscription to formalize commonsense reasoning", Art. Intell. 28.
J. McCarthy and P. J. Hayes [1969], "Some philosophical problems from the standpoint of artificial intelligence", Machine Intelligence 4 (B. Meltzer and. D. Michie, eds.), Edinburgh University Press, 463–502.

D. McDermott [1982], "Non–monotonic logic II: Nonmonotonic modal theories", J. ACM 29, 33–57.
D. McDermott and J. Doyle [1980], "Non–monotonic logic I", Artificial Intelligence 13, (Nos. 1, 2), 41–72.
D. McDermott [1982], Non–monotonic logic II: nonmonotonic modal theories", JACM 29, (no. 1), 33–57.
Z. Manna and A. Pnueli [1981], "Verification of concurrent programs: the temporal framework", in R. S. Boyer and J. S. Moore, eds., *The Correctness Prooblem in Computer Science*, Academic Press, London, 215–273.
W. Marek [1986], "Stable theories in autoepistemic logic", Fundamenta Informaticae.

W. Marek and A. Nerode [1990], "A decision method for default logic", Mathematical Science Institute Report, Cornell University, Ithaca, N. Y.

W. Marek and A. Nerode and J. B. Remmel [1990], "Non–Monotonic Rule Systems I, II", Mathematics and Artificial Intelligence, to appear. (Abstract in IEEE LICS 90, June, 1990)

W. Marek and V. Subrahmanian [1989], "The relationship between logic program semantics and non–monotonic reasoning", Proc. 6th ICLP.

W. Marek and M. Truszczynski [1988], "Autoepistemic logic", Technical report 115–88, Computer Science Department, University of Kentucky.

W. Marek and M. Truszczynski [1989], "Stable semantics for logic programs and default theories", to appear in JACM

W. Marek and M. Truszczynski [1989], "Relating autoepistemic and default logics", in Brachmann et al. [1989].

M. Minsky [1975], "A framework for representing knowledge", *The Psychology of Computer Vision*, (P. Winston, ed.), McGraw–Hill, 211–277.

G. Mirkowska [1979], "Model existence theorem in algorithmic logic with non–deterministic programs", Fundamenta Informaticae, series iv, vol. III.2, 157–170.

G. Mirkowska [1980], "Algorithmic logic with non–deterministic programs", Fundamenta Informaticae, series IV, vol. III.1, 45–64.

R. C. Moore [1983], "Semantical considerations on nonmonotonic logic", Proc. 8th JCAI, Karlsruhe, West Germany, 272–279.

R. C. Moore [1983b], "Semantical considerations on nonmonotonic logic", SRI AI Center Tech. note 284 (June, 1983), SRI International, Menlo Park, Calif.

R. C. Moore [1984], "Possible world semantics for autoepistemic logic", in Proceedings of the nonmonotonic reasoning workshop, (October, 1984), New Paltz, N. Y. Also in *Readings on non–monotonic reasoning*, M. Ginsberg, ed., Morgan Kauffmann, 1987, 137–142.

R. C. Moore [1985], "Semantical considerations on nonmonotonic logic", Artificial Intelligence 25 (1).

R. C. Moore [1988], "Autoepistemic logic", in Smets et al.

P. H. Morris [1989], "Autoepistemic stable closure and contradiction resolution", LNCS 346, 60–73.

A. Nerode [1990], "Some Lectures on Intuitionistic Logic", Italian Summer School in Logic and Computer Science 1988, Montecatini Volume, Springer–Verlag Lecture Notes in Mathematics, to appear in 1990.

A. Nerode and D. Wijesekera [1990], "Constructive Concurrent Dynamic Logic", Mathematics and Artificial Intelligence, to appear in 1990.

I. Niemalä [1988], "Decision method for autoepistemic logic", Proc. 9th Inter. Conf. on Automated Deduction, Argonne, Ill.

I. Niemalä [1988], "Autoepistemic predicate logic", Research report A–6, Digital Systems Laboratory, Helsinki University of Technology, Espoo, Finland.

H. Ono [1977], "On some intuitionistic modal logics", Publications of the Institute of Mathematical Sciences of Kyoto University 13, 687–722.

R. Parikh [1978], "The completeness of propositional dynamic logic", Mathematical Foundations of Computer Science 1978, LCNS 64, 403–415.

R. Parikh [1978], "A decidability result for second order process logic", Proc. IEEE 19th FOCS Symposium, 177–183.

R. Parikh [1984], "Logics of knowledge, games, and dynamic logic", in *Foundations of Software Technology and Theoretical Computer Science*, LCNS 181 (M. Joseph and R. Shyamasundear, eds.), 202–222, Springer–Verlag, Berlin.

D. Peleg [1987], "Concurrent dynamic logic", JACM 34:2, 450–479.

A. Pnueli [1977], "The temporal logic of programs", Proc. 18th IEEE Symp. on Found. of Comp. Science, 46–67.

A. Pnueli [1981], "The temporal semantics of concurrent programs", Theoretical Comp. Sci 13, 45–60.

V. Pratt [1976], "Semantical considerations on Floyd–Hoare logic", 17th Annual IEEE Symp. on Found. Comp. Sci., New York, 109–121.

V. Pratt [1980], "Applications of modal logic to programming", Studia Logica 39, 257–274.

A. Prior [1967], *Past, Present, and Future*, Oxford: Clarendon Press.

M. Reinfrank, J. de Kleer, M. L. Ginsberg and E. Sandewall (Eds.) [1989] *Non-Monotonic Reasoning*, Springer-Verlag, Berlin.
R. Reiter [1980], "A logic for default reasoning", Art. Intell. 13 (1-2).
R. Reiter [1987], "Nonmonotonic reasoning", Ann. Rev. Comp. Sci. 2, 147-186.
A. Salwicki [1970], "Formalized algorithmic languages", Bull. Acad. Pol. Sci. , Ser. Sci. Math. Astron. Phy. 18, 227-232.
A. Salwicki [1977], "Algorithmic Logic", in *Logic, Foundations of Mathematics, and Computability Theory* (Butts and Hintikka, ed.), Reidel, 281-295.
Y. Shoham [1988], *Reasoning about Change*, MIT Press, Cambridge, MA
G. F. Shvarts [1990], "Autoepistemic Modal Logics", Theoretical Aspects of Reasoning About Knowledge 3, 97-109.
P. Smets, E. H. Mamdani, D. Dubois, H. Prade [1989], *Non-Standard Logics for Automated Reasoning*, Academic Press, New York.
R. Smullyan [1968], *First Order Logic*, Springer Verlag, New York.
R. Stalnaker, [1980], "A note on non-monotonic modal logic", Department of Philosophy, Cornell University, Ithaca, N.Y. (unpublished).
R. Stalnaker and G. Pearce [1981], *Ifs; Conditionals, Belief, Decision, Chance, and Time*, D. Reidel, Dordrecht; Klewer, Boston.
P. B. Thistlewhite, M. A. McRobbie, R. K. Meyer, *Automated Theorem-proving in Non-classical Logics*, Pitman, London, 1988.
A. Troelstra and D. van Dalen [1988], *Constructivism in Mathematics*, vol. 1, 2, North Holland, Amsterdam.
A. M. Turing [1949], "Checking a large routine", *Report of a Conference on High Speed Automatic Calculating Machines*, University Mathematical Laboratory, Cambridge, England, June, 1949, 67-69.
R. Turner [1984], *Logics for Artificial Intelligence*, Chichester:Ellis-Horwood.
J. van Benthem [1983], *The Logic of Time*, Reidel, Dordrecht.
J. van Benthem [1984], "Correspondence theory," in D. Gabbay and F. Guenther, eds., *Handbook of Philosophical Logic*, vol. II, Reidel, Dordrecht.
J. van Benthem [1985], *Modal Logic and Classical Logic*, Bibliopolis, Napoli.
J. van Benthem [1988], *A Manual of Intensional Logic*, 2nd ed., CSLI, Stanford.
M. Vardi [1985], "On epistemic logic and logical omniscience", in J. Y. Halpern, 293-305.
M. Vardi [1989], "On the complexity of epistemic reasoning", LICS 1989, IEEE, 243-246.
J. Von Neumann [1963], Collected Works, vol. 5, McMillan, 91-99.
M. Xiwen and G. Weide [1983], "A modal logic of knowledge", 8th IJCAI, 398-401.
L. Wallen [1990], *Automated Proof Search in Nonclassical Logics*, MIT Press.
D. Wijesekera [1990], *Constructive Modal Logics*, D. Phil. Diss., Cornell University, Ithaca, N. Y.
D. Wijesekera [1991], "Constructive Modal Logics", Ann. Pure App. Logic, to appear.

Formal Approaches to Concurrency

Wilfried Brauer

Institut für Informatik
Technische Universität München
Arcisstr. 21, D-8000 München 2, FRG

Keywords / Abstract: Petri nets, process algebras, CCS, TCSP, transition systems, bisimulation, interleaving operational semantics, step failures semantics, refinement, modular construction.

An introduction to the two main approaches to the theory of concurrent distributed systems is given: the Milner/Hoare theory of CCS and TCSP and the theory of Petri nets. We define a general abstract programming language GAP which encompasses practically all aspects of CSS, TCSP and related formalisms and give a step failures semantics for it. We introduce place/transition nets and show how GAP can be modelled by Petri nets. Moreover we report on other semantics notions for abstract programming languages and Petri nets and about techniques for modular construction and refinement of Petri nets.

0 Introduction

Formal (algebraic, combinatorial, logical) treatment of concurrent processes and of distributed systems has started rather recently only, although concurrent and distributed activities dedicated to common tasks are daily practice and have always played an important role for human societies, and reasoning about concurrent processes and distributed systems is a common and natural activity of humans, since it is vital for acting in a dynamic environment composed of a multitude of concurrently active systems. The traditional notions of computability however are all based on the concept of a single person fulfilling a task step by step. This is very explicit in Turing's work – and even if the formalism would allow for consideration of concurrency, as in the case of recursive functions defined by sets of equations, this possibility was not discussed for a long time. The reason was perhaps that until recently the computer was only considered as a tool for computing a function in a start-stop mode: Given an input it outputs, after a finite amount of time, the values of the function for certain arguments specified by the input.

As we see it now, the notion of a stand-alone sequential machine, able to compute everything which is computable, is a very strong idealization – not simply with respect to time consumption and memory capacity but also concerning the idea of a step-wise sequential computation. Even a simple Turing machine cannot be implemented without using asynchronously operating components, e.g. a unit for extending the tape if necessary (for more remarks on this see e.g. [Bra84], [Bra87], [Bra90]).

NATO ASI Series, Vol. F 79
Logic, Algebra, and Computation
Edited by F. L. Bauer

Already in the beginning of the 1960s C. A. Petri started to develop concepts and formalisms for treating systems composed of several asynchronously acting components connected by asynchronous message passing communication. Nowadays many people see the computer more as a medium for communication and an informatics system (a computer-based system) as a reactive system of concurrently operating components which often operates forever and reacts differently to the same inputs in the course of time. The fundamental concept of Petri's theory is concurrency; the formal system model is the Petri net which is based on a bipartite graph that represents distributed actions and local states, and allows also for a formal representation of the concurrent dynamic behaviour of a system. Petri nets can be seen as generalizations of traditional sequential machines to non-sequential distributed systems (see [Bra90]), and many interesting results have been obtained by combining classical theories like formal languages and automata, vector addition systems and semilinear sets or linear algebra with Petri nets (see e.g. [Bra87] and the subseries "Advances in Petri nets" of Springer's LNCS series).

By the end of the 1970s R. Milner and C. A. R. Hoare developed the second major approach to concurrency, namely a calculus for communicating systems (Milner's CCS – see [Mil80]) and a programming language for communicating sequential processes (Hoare's CSP – see [Hoa78]). Both ideas were based on the idea of systems composed of sequentially working components which operate asynchronously in general and communicate by synchronized one-way message passing. The development of CCS was based on ideas around the λ-calculus (see [Mil85] and [Tra88]) and developed into what is now called the large field of process algebras (since it deals with the now many formalisms for the description of processes by terms in an algebraic calculus). CSP was more oriented towards programming practise and became the basis of concrete programming languages (in particular the language OCCAM (see [May83], [Inm88])). Inspired by CCS, also a more abstract version of CSP, namely TCSP, was developed (in [HBR81]) and studied in many subsequent papers in the framework of process algebras. A forerunner to both CCS and CSP can be seen in the abstract programming notation COSY (see [LC75] and e.g. the paper by Best in [BRR87, part II]) which is based on the path expression notation for specifying process synchronization invented by R. H. Campbell and A. N. Habermann (see [CH74]), and which has also a close relation to Petri nets (see [LC75] and [BRR87]).

One of the reasons for the great success of the Milner-Hoare theory of concurrent systems is perhaps its rather close relationship to the traditional theory of sequential systems. The basic components of the systems are sequential systems, and the systems can be built in a modular fashion (with compositional semantics); the behaviour (the operational semantics) of the whole systems is described from the viewpoint of a single observer who can only note sequences of events such that concurrency is modelled by nondeterminism. This allows to use the large amount of knowledge from logic, automata and formal language theory and from the semantics of sequential programming languages. And the drawback of this semantics, namely that it cannot cope adequately with all important phenomena of nonsequential systems, became more widely accepted only rather recently (see e.g. [Vog91]).

In Petri's approach a clear distinction between nondeterminism and concurrency is made. Concurrency (considered as independency) is a nontransitive relation and different from (the transitive relation of) simultaneity. Its complement describes causal dependencies. A semantics respecting causality and concurrency is technically quite complicated.

Another drawback of Petri nets is that they have no inherently modular structure – this caused much research on modular composition and refinement of Petri nets (see [BC91], [BGV91], [Tau88]).

The rather recent intensification and broadening of work on concurrency is certainly due to hardware developments on the one hand and the needs coming from applications on the other hand – but, as the short remarks made above may illustrate, one has to admit that the development of theoretical informatics is very much based on its own inherent impetus, in particular on inner-informatical historical influences and on abstract (not directly hardware- or application-oriented) ideas and concepts, which often can be developed by looking at what human beings do: Turing looked at a single computing person – to imagine how nonsequential systems might look like, it is quite useful to consider a group of people (competing about scarce resources or cooperating to fulfill a complex task) and its ways of communication) in order to get ideas on systems of cooperating agents (see [Bra90]).

The two approaches we will deal with are quite different, from the formal and technical point of view as well as from the philosophical one. However, I will not concentrate on their distinctions, but treat them together from the perspective of specification and programming. Since in the case of distributed systems there is no clear distinction between specification and programming notations I shall use often the more general term "specification" to mean also programming, and in order to stress the orientation towards programming I use the term abstract programming language instead of the more abstract term process algebra.

The following five parts (corresponding to five lectures) are mainly based on work done in my research group, in particular by Dirk Taubner and by Astrid Kiehn, Volker Diekert, Robert Gold, Walter Vogler.

1 Abstract Programming Languages

1.1 A General Abstract Programming Language

Let us imagine that we should specify a distributed system composed of several agents which work rather independently but which communicate with each other (in a well organized way). To make the problem easier, we abstract from the processing of data and take as a basis simply a countably infinite alphabet

$$Alph$$

of (names of) **actions** (assuming also that the occurrence of an action in the specification of a system means that, in the realization of the system there will be an agent performing this action).

Naturally one would like to be able to describe simple systems like finite nondeterministic automata – however, we do not want to describe their structure, but their behaviour, i.e. what they should do. Therefore we use a notation similar to that of regular expressions; the main difference is that we will express the iteration (Kleene star) by **recursion**.

In addition we obviously need an operator for some sort of **parallel composition** which should include the possibility to prescribe communications or joint actions of the composed systems. There are several operators in the literature based on different ways of cooperation: the two subsystems may

- operate completely independently
- perform some actions jointly
- communicate by performing complementary actions a and $\bar{a}$ (establishing a communication link) – the joint action $(a, \bar{a})$ having no effect to the outside world (the communication is internal).

We will use an operator which encompasses all these variants.

Obviously we now need a complementary alphabet $\overline{Alph} := \{\bar{a} \mid a \in A\}$ and a notation (τ) for an action without any (visible) effect. Naturally we assume that $\bar{\bar{a}} = a$.

When we specify a system we take the point of view of an observer (or a user) who watches (or interacts with) the system and sees the effects of its actions, i.e. of actions from

$$Vis := Alph \cup \overline{Alph}.$$

According to good programming practise we would also like to be able to express hiding (abstraction) and renaming of actions. Both can be combined in the operation of applying an **action manipulation** function f to a specification (it is convenient to write this operator in postfix notation). So $af = \tau$ denotes that a is hidden. We can use this operator also to suppress (restrict) actions, if we extend its range by $\bot$, the symbol for non-action, undefinedness etc. (i.e. $af = \bot$ means action a is not allowed).

It is convenient to have a notation for unordered pairs (of jointly executed) actions:

$$EVis := \{\{a, b\} \mid a, b \in Vis\} \qquad (\{a, a\} = \{a\})$$

The set of all actions is $Act := \{\tau\} \cup Vis \cup EVis$. Let moreover $Act_\bot := Act \cup \{\bot\}$. We are now ready to define the syntax of the **general abstract programming language GAP** (i.e. the language A in [Tau88]).

The **operators** (and their intuitive meanings) are:

nil:	nullary operator	(a system which is unable to perform any action; which has stopped to work)
a:	unary operator,	used in prefix notation for each $a \in \{\tau\} \cup Vis$ (called *prefixing*; if S is a system aS is the system that can perform first a and then behaves as S)
f:	unary operator,	used in postfix notation for each $f \in Fun := \{f : Act_\bot \to Act_\bot \mid f(\bot) = \bot, f(\tau) = \tau\}$ (called *action manipulation*)
$+$:	binary operator,	used in infix notation (called *sum*; $S + S'$ behaves either like S or like S', depending on whether the first executed action belongs to S or to S')

$\ddagger$: binary operator, used in infix notation (called *general parallel composition*; $S \ddagger S'$ allows S and S' to work independently but also to perform joint actions $\{a,b\} \in EVis$ provided that S, S' can perform a, b respectively.

The notation for recursion is $rec\, r.S$, where $r \in Id$, a countably infinite set of identifiers, and S a system description in which r might occur. (This is similar to the definition of a parameterless recursive procedure r with procedure body S together with an immediate call of r.)

As usual we have the notions of free and bound identifiers, we have to use renaming of bound identifiers, we identify terms which differ only with respect to bound identifiers, and we will always assume that the Barendregt convention (see [Bar85]) is obeyed, i.e. that in each collection of terms no identifier occurring bound in one of the terms occurs also free in a term of this collection.

Now the **syntax** for GAP is given by the grammar:

$$S ::= nil \;\Big|\; r \;\Big|\; aS \;\Big|\; Sf \;\Big|\; S+S \;\Big|\; S \ddagger S \;\Big|\; rec\, r.S$$

where $r \in Id$, $a \in \{\tau\} \cup Vis$, $f \in Fun$. Let $Term_{GAP}$ be the set of all terms defined by this grammar.

1.2 Derived Operators

The above syntax is not given in order to introduce yet another abstract programming language but as a common framework for the variety of syntaxes used in the literature.

Many of the operators used in the literature can be defined with the help of the above; here are some examples. Let $S, S' \in Term_{GAP}$, $A \subseteq Vis$, then:

S **or** S' := $\tau S + \tau S'$ is the internal nondeterminism operator of TCSP (without visible effect the system decides to behave like S or like S')

$S \setminus A$:= $S\{a \mapsto \tau \mid a \in A\}$ is the hiding operator of TCSP (Here as in the following we describe a function by writing down all important argument-value pairs)

$S - A$:= $S\{a \mapsto \bot \mid a \in A\}$ is the restriction operator of CCS, however it is denoted by $S\backslash A$ there

$S \mid S'$:= $(S \ddagger S')g$, where $ag := \begin{cases} a & \text{for } a \in Vis \cup \{\tau\} \\ \tau & \text{for } a = \{b, \bar{b}\} \in EVis, \\ \bot & \text{otherwise} \end{cases}$ is the CCS parallel composition.

$S \parallel_A S'$:= $(S \ddagger S')g_A$,
where $ag_A := \begin{cases} a & \text{for } a \in \{\tau\} \cup Vis - A \\ b & \text{for } a = \{b, b\} \in EVis,\ b \in A, \\ \bot & \text{otherwise} \end{cases}$
is the TCSP parallel composition.

Milner's pure CCS, the perhaps most influential abstract programming language, developed from the middle of the 70's on, (see [Mil80], [Mil85] and [BRR87]) is basically given by the following grammar

$$S ::= nil \;\Big|\; r \;\Big|\; aS \;\Big|\; Sf \;\Big|\; S-A \;\Big|\; S+S \;\Big|\; S \mid S \;\Big|\; rec\, r.S$$

where $r \in Id$, $a \in \{\tau\} \cup Vis$,

$f \in Fun$, such that $f\lceil_{EVis} = id \wedge \forall a \in Vis : af \in Vis \wedge \overline{af} = \bar{a}f$.

$A \subseteq Vis$, such that $a \in A$ implies $\bar{a} \in A$. (where $g\lceil D$ denotes the restriction of the domain of the function g to D)

The classical operator ";" of sequential composition of two systems is not simply obtained from the prefixing operator, since we have allowed the construction of systems, which may never terminate their activities properly, i.e. may loop indefinitely or block (deadlock). We therefore introduce a particular symbol $\surd$ (called tick) which indicates successful termination. Let $Alph = Alph' \cup \{\surd, \surd_1, \surd_2\}$, where $Alph' \cap \{\surd, \surd_1, \surd_2\} = \emptyset$. Then for $S, S' \in Term_{GAP}$

$$S; S' := (Sg_1 \mid \bar{\surd}_1 S') - \{\surd_1, \bar{\surd}_1\}, \text{ where } g_1 = \{\surd \mapsto \surd_1, \bar{\surd} \to \bar{\surd}_1\}$$

Another very important abstract programming language based on Hoare's CSP ([Hoa78]) is TCSP ([BHR84], see also [BRR87]) a slightly restricted variant can be defined, using the above, by the following grammar:

$$\begin{array}{l} S ::= V \;\big|\; r \;\big|\; Sf \;\big|\; S\text{–}A \;\big|\; S \setminus A \;\big|\; S \text{ or } S \;\big|\; S; S \;\big|\; S \parallel_A S \;\big|\; rec\, r.S \\ V ::= nil \;\big|\; aS \;\big|\; V + V \end{array}$$

where $r \in Id$, $a \in Alph$, $A \subseteq Alph$, $f \in Fun \wedge f\lceil_{(Act-Alph)} = id \wedge (Alph)f \subseteq Alph \wedge \forall a \in Alph : |af^{-1}| \in \mathbb{N}$.

The main omission is the operator $[]$ of external choice, it is replaced by $+$ which can be considered as $[]$ restricted to operands which both begin by a visible action (according to the subgrammar with start symbol V). More on TCSP follows in part 2.

Examples:

(1) $rec\, r.((ar; b\surd nil) + \surd nil)$
An observer watching the system from a start action until a termination will note a sequence of actions of the form $a^n b^n \surd$, $n \geq 0$.

(2) $rec\, r.(0(rf) + \surd nil)$, where $af := \begin{cases} a+1 & \text{if } a \in \mathbb{N} \\ a & \text{otherwise} \end{cases}$, can produce each of the following action sequences:

$$\surd, 0\surd, 01\surd, 012\surd, 0123\surd, \ldots$$

(3) $rec\, r.(rf + 0\surd nil)$, where f is as above, produces only the actions $i\surd$ (where $i \in \mathbb{N}$) with increasing i (beginning with 0) if it is restarted again and again. For more details see [Tau88].

2 Semantics

2.1 Interleaving Operational Semantics

The traditional approach to the semantics of concurrent distributed systems is based on the idea of an observer (or user) watching (or interacting with) the system **without** any knowledge about its structure. This observer (user) can only operate **sequentially**, so he or she[1] will note (or cause) concurrent actions of the system in some order – thus transforming concurrency into nondeterminism.

More formally, we associate to a term of the language (i.e. in our case GAP) a **transition system** (sequential automaton) $T = \langle Z, D, z \rangle$ over Act, where
Z is the (possibly infinite) set of states,
$D \subseteq Z \times Act \times Z$ is the set of transitions
and $z \in Z$ is the start state.

Example (2) from part 1 gives the following transition system: Let S_2 denote the given term

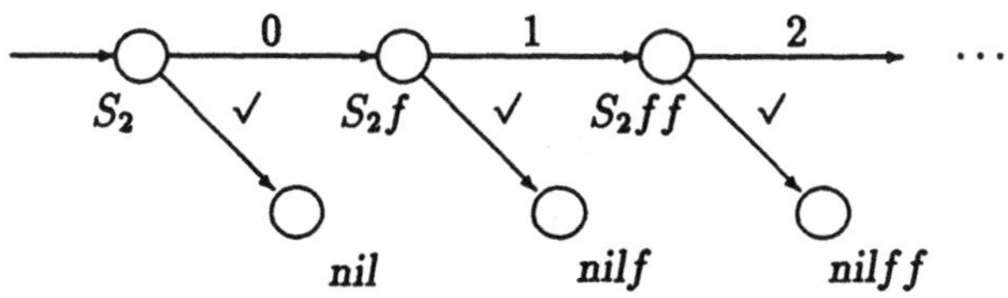

The states of the transition system for a term S are terms derived from S (where S is the start state) by the following inference rules – only the states reachable from the start state are interesting and need to be constructed in a concrete example.

$$(act) \qquad \overline{(aS, a, S)}$$

$$(fun) \qquad \frac{(S, a, S') \wedge af \neq \bot}{(Sf, af, S'f)}$$

$$(sum) \qquad \frac{(S, a, S')}{(S + R, a, S') \wedge (R + S, a, S')}$$

$$(asy) \qquad \frac{(S, a, S')}{(S \ddagger R, a, S' \ddagger R) \wedge (R \ddagger S, a, R \ddagger S')}$$

$$(syn) \qquad \frac{a, b \in Vis \wedge (S, a, S') \wedge (R, b, R')}{(S \ddagger R, \{a, b\}, S' \ddagger R')}$$

$$(rec) \qquad \frac{S \equiv \mathit{rec}\, r.R \wedge (R[S/r], a, S')}{(S, a, S')}$$

where $R[S/r]$ denotes the term obtained from R by substituting the term S for every free occurrence of the identifier r.

Example (3) from part 1 gives the transition system

[1]Instead of "he or she" I will subsequently write briefly "he" only

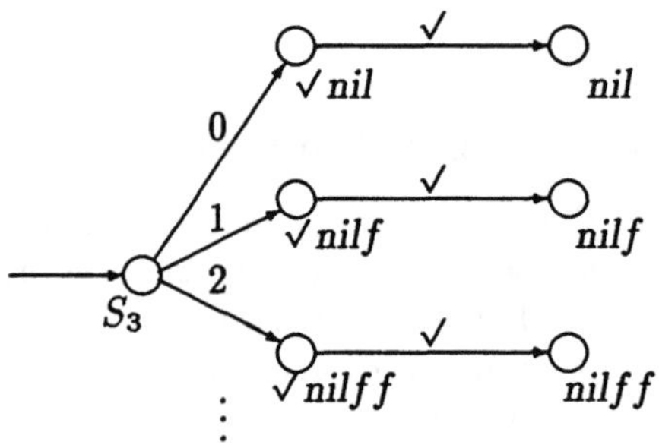

If we consider such a transition system (for a term S) as an automaton whose final states are those reached by a tick transition $(z, \surd, z')$, then the formal language accepted by it, is the set of all sequences of observations (or of actions) one can obtain from terminating runs of an implementation of the term S.

The semantics obtained is an operational one constructed according to the structured operational semantics (SOS) technique introduced by Plotkin ([Plo81]); the semantics of the parallel composition of two terms is the interleaving (or the shuffle product, if formal languages are considered,) together with the necessary synchronization of the semantics (of the sets of action sequences) of the components.

Two terms have the same meaning with respect to this semantics if the corresponding transition systems are equivalent – to compare only the sets of action sequences does not suffice, since it does not say anything about the nonterminating and the blocking behaviour. There are several equivalence notions for transition systems, we consider only the **strong bisimulation equivalence** (introduced by Milner and Park – see [Mil80], [Par81] and [Mil85]), since practically all other equivalence notions are weaker than this.

Let $T_i = \langle Z_i, D_i, z_i \rangle, i = 1,2$ be two transition systems. T_1 and T_2 are strongly bisimular (notation: $T_1 \sim T_2$) if there is $B \subseteq Z_1 \times Z_2$ such that

- $(z_1, z_2) \in B$
- $\forall (z, z') \in B, a \in Act:$

  (i) $(z, a, z_0) \in D_1 \Rightarrow \exists z'_0 : (z', a, z'_0) \in D_2 \wedge (z_0, z'_0) \in B$

  (ii) $(z', a, z'_0) \in D_2 \Rightarrow \exists z_0 : (z, a, z_0) \in D_1 \wedge (z_0, z'_0) \in B.$

Obviously two terms with bisimular transition systems produce the same sets of action sequences.

Example: $S \equiv a(rec\, r.ar) + b(rec\, q.aq)$ with $r \not\equiv q$ has the transition system

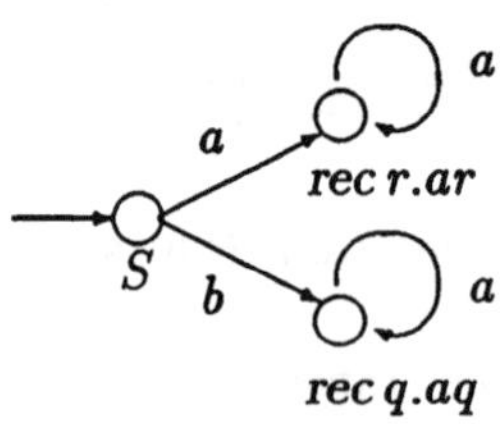

According to the Barendregt convention (see part 1) we wanted to identify *rec r.ar* and *rec q.aq*. Therefore S should be identified with

$$S' \equiv (a\ rec\, r.ar + b\ rec\, r.ar)$$

The transition system for S' is

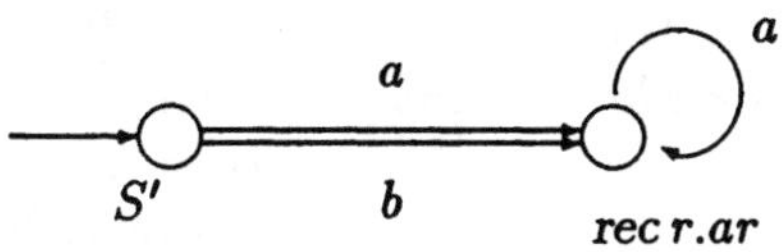

Both transition systems are strongly bisimular. For further information on this semantics see [Tau88], for other variants of the notion of bisimulation see [BGV91] and [Vog91].

2.2 A Denotational Semantics with Simultaneity

We will now refine our semantical view. If the observer (or the user) is a bit more sophisticated he may detect (or cause) some actions simultaneously. It is obvious that for the analysis of a concurrent distributed system it is useful (and sometimes necessary) to be able to describe simultaneity of actions, e.g. in order to see what effect to execution speed the increase or decrease of a number of processors would have.

For this purpose we consider a TCSP oriented variant of GAP (called GAPH, the H stands for Hoare, the author of CSP), take the (more or less) standard denotational semantics for TCSP (which however is an interleaving semantics) and equip this with the notion of step, which comes from Petri net theory.

A step is simply a finite multiset of actions which are performed simultaneously – it need not be maximal with respect to the number of simultaneous actions, since we assume that the components of the distributed system operate asynchronously. Also null steps (which do not contain any action and serve as abstractions for internal actions) are allowed – they can be interpreted as idle steps but have nothing to do with τ-actions (which are not necessary in GAPH).

Since we allow for arbitrary simultaneity and are able to argue about simultaneity in our semantics it is useful to introduce a new operator ($\#_B$) in the language which allows to restrict simultaneity of certain actions (i.e. $S \#_B$ has the effect that no step which has an element of B as a substep can be performed).

The syntax of GAPH is given by the following grammar:

$$S ::= nil \mid r \mid aS \mid Sf \mid S \#_B \mid S \,[\,]\, S \mid S \textbf{ or } S \mid S \nmid S \mid rec\, r.S$$

where $r \in Id$, $a \in Alph$, $f \in Fun$, $B \subseteq \mathbf{M}$, where $\mathbf{M}$ is the set of all finite multisets over $\Sigma := Vis \cup EVis$; $\bar{0}$ denotes the empty multiset, addition (+) and comparison ($\leq$) of multisets work componentwise.

Apart from $\#_B$ it is the operator $[\,]$ of external choice (external nondeterminism) from TCSP, which offers a choice between two systems that is resolved by the environment (and which replaces the + from GAP), that makes the difference to GAP.

The semantics is defined using the standard denotational technique: We first define a domain $\mathbb{F}$ (in our case a complete partial order) and then define for each syntactic operator *op* a corresponding continuous operator $op_{\mathbb{F}}$ on this domain.

The semantics $[\![\cdot]\!]\rho : Term_{GAPH} \to \mathbf{F}$ is then for some environment $\rho : Id \to \mathbf{F}$ given as

$$\begin{aligned}
[\![r]\!]\rho &:= \rho(r) \\
[\![op(S_1,\ldots,S_n)]\!]\rho &:= op_{\mathbb{F}}([\![S_1]\!]\rho,\ldots,[\![S_n]\!]\rho) \\
[\![rec\, r.S]\!]\rho &:= \underset{i\in\mathbb{N}}{lub}\,\varphi_i, \text{ where} \\
&\qquad \varphi_0 := \bot \\
&\qquad \varphi_{i+1} := [\![S]\!]\rho[\varphi_i/r],
\end{aligned}$$

and $\rho[X/r]$ is the same function as ρ except that X is substituted for r.

The elements of the domain are sets $F \subseteq \mathbf{M}^* \times \mathbf{P}(\mathbf{M})$ of pairs consisting of a sequence of steps (the system may perform) and of a set of steps (the system may refuse to perform after having performed the before-mentioned steps).

Example: $S \equiv a\,nil \parallel_{\emptyset} (a\,nil \,[]\, b\,nil)$, $Vis = \{a, b, \surd\}$

S may perform the step $\begin{bmatrix} a \\ b \end{bmatrix}$ (or $\begin{bmatrix} a \\ a \end{bmatrix}$) or may perform the sequence $[a][a]$ (or $[a][b]$ or $[b][a]$) of steps, and after that all steps (other than $\bar{0}$ which may occur before or after each of the steps in the sequences) can be refused; at the beginning only steps containing at least 2 simultaneous b's or at least 3 a's or one b and two a's or containing the $\surd$ can be refused; after one a or one b is performed in a single step only steps containing at least 2 simultaneous a's, one b or one $\surd$ can be refused. Therefore the step failure semantics of S (omitting $\bar{0}$ steps which are always possible) is

$$\begin{aligned}
&\{\varepsilon X \mid X \subseteq \{x \in \mathbf{M} \mid x \geq \begin{bmatrix} b \\ b \end{bmatrix} \vee x \geq \begin{bmatrix} a \\ a \\ b \end{bmatrix} \vee x \geq \begin{bmatrix} a \\ a \\ a \end{bmatrix} \vee x \geq [\surd]\}\} \cup \\
&\{[a]X, [b]X \mid X \subseteq \{x \in \mathbf{M} \mid x \geq \begin{bmatrix} a \\ a \end{bmatrix} \vee x \geq [b] \vee x \geq [\surd]\}\} \cup \\
&\{\begin{bmatrix} a \\ a \end{bmatrix} X, \begin{bmatrix} a \\ b \end{bmatrix} X, [a][a]X, [a][b]X, [b][a]X \mid X \subseteq \mathbf{M}-\{\bar{0}\}\}
\end{aligned}$$

For the description of the domain we need the concept of *stretching* of a step sequence, i.e. of replacing the step sequence by a step sequence performing the same individual actions in more (and smaller) steps; i.e. stretching means partial sequentialization plus insertion of null steps. Thus we can define the mapping $Stretch(w)$ inductively by:

$$\begin{aligned}
Stretch(\varepsilon) &:= \{\bar{0}\}^* \\
Stretch(vx) &:= Stretch(v)\{x_1x_2\ldots x_n \in \mathbf{M}^* \mid \sum_{i=1}^{n} x_i = x\}.
\end{aligned}$$

Null steps can never be refused, hence we may use $\bar{0}$ in a refusal set to indicate the possibility of divergence.

Definition: $F \subseteq \mathbf{M}^* \times \mathbf{P}(\mathbf{M})$ is an element of $\mathbb{F}$ iff

(1) $\varepsilon\emptyset \in F$

(2) $vw\emptyset \in F \Rightarrow v\emptyset \in F$

(3) $wX \in F \wedge Y \subseteq X \Rightarrow wY \in F$

(4) $wX \in F \wedge wy\emptyset \notin F \Rightarrow w(X \cup \{y\}) \in F$

(5) $(\forall Y \in p(X) : wY \in F) \Rightarrow wX \in F$

(6) $v\{\bar{0}\} \in F \Rightarrow vwX \in F$

(7) $wX \in F \wedge v \in \mathit{Stretch}(w) \Rightarrow vX \in F$

(8) $wX \in F \wedge x \in X \wedge x \leq y \Rightarrow w(X \cup \{y\}) \in F$

(9) $v\bar{0}wX \in F \Leftrightarrow vwX \in F$

(10) $(\overset{\infty}{\exists}\, a \in \Sigma : w[a]\emptyset \in F) \Rightarrow w\{\bar{0}\} \in F$

It can be shown (see [TV89]) that $(\mathbb{F}, \supseteq)$ is a cpo with bottom element $\bot = \mathbf{M}^* \times \mathbf{P}(\mathbf{M})$.

Now we can define the operators $op_{\mathcal{F}}$ which we denote by the same symbols as in the syntax.

$$
\begin{array}{lll}
nil & := & \{wX \mid w \in \{\bar{0}\}^* \wedge X \subseteq \mathbf{M}\text{--}\{\bar{0}\}\} \\[2ex]
aF & := & \{vX \mid v \in \{\bar{0}\}^* \wedge [a] \notin X \subseteq \mathbf{M}\text{--}\{\bar{0}\}\} \cup \\
 & & \{v[a]wX \mid v \in \{\bar{0}\}^* \wedge wX \in F\} \\[2ex]
F_1 \text{ or } F_2 & := & F_1 \cup F_2 \\[2ex]
F_1 \,[\,]\, F_2 & := & \{wX \mid w \in \{\bar{0}\}^* \wedge wX \in F_1 \cap F_2\} \cup \\
 & & \{wX \mid w \in \{\bar{0}\}^* \wedge \varepsilon\{\bar{0}\} \in F_1 \cup F_2 \wedge X \subseteq \mathbf{M}\} \cup \\
 & & \{wX \mid w \notin \{\bar{0}\}^* \wedge wX \in F_1 \cup F_2\} \\[2ex]
F \Vert_B & := & \{w(X \cup Y) \mid w \in (\mathbf{M}\text{--}\overline{B})^* \wedge wX \in F \wedge Y \subseteq \overline{B}\} \cup \\
 & & \{wuX \mid (w\{\bar{0}\} \in F \vee \bar{0} \in B) \wedge w \in (\mathbf{M}\text{--}\overline{B})^* \wedge \\
 & & uX \in \mathbf{M}^* \times \mathbf{P}(\mathbf{M})\}, \\
 & & \text{where } B \subseteq \mathbf{M} \text{ and } \overline{B} := \{y \mid \exists x \in B : x \leq y\}
\end{array}
$$

To define $\nmid$ we need several auxiliary definitions. For each $t : \mathit{Vis} \times \mathit{Vis} \rightarrow I\!N$ we define the following three elements of $\mathbf{M}$ (considered as mappings)

$$
\pi_1(t)(a) := \begin{cases} 0 & \text{if } a \in \mathit{EVis} \\ \sum_{b \in \mathit{Vis}} t(a, b) & \text{if } a \in \mathit{Vis} \end{cases}
$$

$$
\pi_2(t)(a) := \begin{cases} 0 & \text{if } a \in \mathit{EVis} \\ \sum_{b \in \mathit{Vis}} t(b, a) & \text{if } a \in \mathit{Vis} \end{cases}
$$

$$\varphi(t)(a) := \begin{cases} 0 & \text{if } a \in \mathit{Vis} \\ t(b,b) & \text{if } a = \{b,b\} \in \mathit{EVis} \\ t(b,c) + t(c,b) & \text{if } a = \{b,c\} \in \mathit{EVis}, \qquad b \neq c \end{cases}$$

The mapping t gives the number of communication pairs occuring in a parallel composition, the projection Π_i counts how often actions from the i-th component appear, and φ yields the number of joint actions.

Next the parallel composition of two steps yielding a set of steps is defined; here t assigns how many actions participate in communications, r_1 and r_2 represent substeps exclusive to the respective component.

Let x_1, x_2 be steps, then $x_1 \nmid x_2$ is defined by

$$\begin{aligned} x \in x_1 \nmid x_2 \quad \text{iff} \quad & \exists r_1, r_2 \in \mathbf{M}, t : \mathit{Vis} \times \mathit{Vis} \to \mathbb{N} : \\ & x_1 = \pi_1(t) + r_1 \wedge x_2 = \pi_2(t) + r_2 \wedge \\ & x = r_1 + r_2 + \varphi(t). \end{aligned}$$

Let $v = x_1 x_2 \ldots x_n$ and $w = y_1 y_2 \ldots y_n$ be step sequences of equal length, then

$$v \nmid w = \{z_1 z_2 \ldots z_n \mid z_i \in x_i \nmid y_i, \quad i = 1, \ldots, n\}$$

Now for $F_1, F_2 \in \mathbb{F}$ we have

$$\begin{aligned} F_1 \nmid F_2 := \{\, & wX \mid \exists w_1 X_1 \in F_1, w_2 X_2 \in F_2 : w \in w_1 \nmid w_2 \wedge \\ & X \subseteq \{x \in \mathbf{M} - \{\bar{0}\} \mid \forall x_1, x_2 \in \mathbf{M} : x \in x_1 \nmid x_2 \\ & \qquad \Rightarrow x_1 \in X_1 \vee x_2 \in X_2\}\} \cup \\ & \{wuX \mid \exists w_1 X_1 \in F_1, w_2 X_2 \in F_2 : w = w_1 \nmid w_2 \wedge \\ & \wedge \bar{0} \in X_1 \cup X_2 \wedge u \in \mathbf{M}^* \wedge X \subseteq \mathbf{M}\}. \end{aligned}$$

Also, to define Ff for $f \in \mathit{Fun}$ we need several auxiliary definitions:

For $x \in \mathbf{M}$ and $f \in \mathit{Fun}$ we have:

xf undefined if $\exists a \in \Sigma : x(a) > 0 \wedge af = \perp$; in all other cases xf is the step (considered as a mapping from Σ to $\mathbb{N}$) defined by

$$(xf)(a) := \sum_{b \in \Sigma \wedge bf = a} x(b).$$

This implies in particular $\bar{0}f = \bar{0}$ for each $f \in \mathit{Fun}$.

Now

$$xf^{-1} := \{y \in \mathbf{M} \mid y\lceil(\tau f^{-1} \cap \Sigma) = \bar{0} \wedge yf = x\}$$

and for $X \subseteq \mathbf{M}$

$$Xf^{-1} := \bigcup_{x \in X} xf^{-1}.$$

Let $w = x_1 x_2 \ldots x_n$ be a step sequence, then

$$wf := \begin{cases} \text{undefined} & \text{if one } w_i \text{ is undefined} \\ w_1 w_2 \ldots w_n & \text{otherwise} \end{cases}$$

where

$$w_i := \begin{cases} \varepsilon & \text{if } x_i f = \bar{0} \neq x_i \\ x_i f & \text{otherwise.} \end{cases}$$

Then for $F \in \mathbb{F}$ we get

$$
\begin{aligned}
Ff := \{wf(X \cup Y) \mid\ & w(Xf^{-1} \cup \widetilde{\tau f^{-1}}) \in F \wedge \\
& Y \subseteq \{y \in \mathbf{M} \mid yf \text{ undef.}\}\} \cup \\
\{(wf)uX \mid\ & uX \in \mathbf{M}^* \times \mathbf{P}(\mathbf{M}) \wedge \\
& (w\{\bar{0}\} \in F \vee \forall n \in I\!N : \exists v \in (\widetilde{\tau f^{-1}})^n : wv\emptyset \in F)\}
\end{aligned}
$$

where for $A \subseteq Act$ we write $\tilde{A} := \{[a] \mid a \in A - \{\tau\}\}$.

The proofs that all these operators are continuous can be found partly in [TV89] – to prove that action manipulation (application of $f \in Fun$) is continuous, one cannot simply generalize the technique used in [TV89] for the proof of the continuity of the hiding operator but has to go back to [Bro83]; however it seems to be straight-forward then.

The step failures semantics can be weakened to get new semantics by restricting the notion of steps and of refusals:

- if only null or singleton steps may be refused we get the simple step failures semantics
- if steps are restricted to be singletons only this gives the standard (linear) failures semantics.

Based on these notions of semantics we have different notions of equivalence.

Example:

$$
\begin{aligned}
S &:\equiv a\surd\, nil \parallel_{\{\surd\}} b\surd\, nil \\
T &:\equiv ab\surd\, nil \,[]\, ba\surd\, nil
\end{aligned}
$$

Then S and T have the same linear failures semantics but different (simple) step failures semantics and moreover
S, $S\,[]\,T$ and S **or** T have the same simple step failures semantics but
in the step failures semantics S and $S\,[]\,T$ have the same semantics while $S\,[]\,T$ and S **or** T have different semantics.

3 Petri Nets

3.1 Basic Ideas

Firstly we will consider distributed concurrent systems with the goal to develop a formalism that describes the structure as well as the dynamic behaviour of such systems. Using this formalism one can then specify new systems (which are to be built) also.

The basic assumptions from which we start are:

- Systems are composed of subsystems, which can communicate asynchronously with each other (and with the environment) by sending and receiving messages (or other objects).
- The subsystems can be relatively independent of each other (e.g. they can be distributed widely).

- The behaviour of a system is determined by processes which are running in subsystems and which consist of changes of the states of subsystems by actions (of communication or transport).

The formal model is developed according to the following principles.

(1) States and actions (of state change) are both explicitly represented.

(2) States (resp. actions) of the subsystems are not combined together to form global states (resp. actions) of the whole system; they are represented separately.
Consequence: We better represent these systems by (at least) two-dimensional graphics.

(3) The transport of an object (or a message) in the system can be considered as a state change. State changes can also be considered as actions of transport of objects (i.e. messages).
Consequences: For our formal description we need exactly 2 types of components and a notation for the objects and their position:

  - active components (from a set T of so-called transitions, graphically represented by □) for the representation of actions
  - passive components (from a set S of so-called places graphically represented by ○) for the representation of (local) states (of subsystems).
  - As objects we consider (in the simplest version we want to discuss here) only simple tokens (graphically represented by a dot •) which are available in a place or not (⦿ or ○).

(4) The amount of state change caused by one action is constant (i.e. always the same if the action occurs – independent of other circumstances)
Consequence: Each action component is connected to a fixed number of passive (state) components, which are subject to change by this action. Thus the system can be represented as a bipartite graph with node set $S \cup T$, where no two nodes of the same type are connected. Since state change is represented by taking away or adding tokens the graph will be directed (according to the direction of the flow of objects).

Example: Traffic light

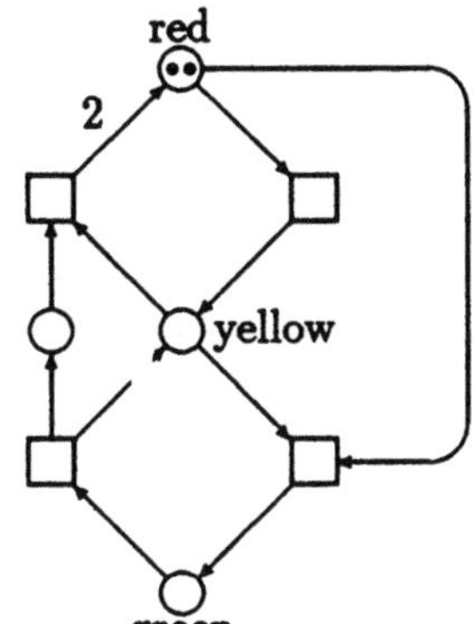

(5) An action can take place (is enabled) if the state changes to be produced by it are possible; but the action need not take place if it is enabled. Hence it may be the case that other actions will disable it again.

Remark on the example: The careful reader will notice that this traffic light may block, since both tokens may arrive at the place yellow.

3.2 Formal Definitions

There are several ways to formally describe the class of Petri nets described above. The usual definition is the graphtheoretic one, treating places and transitions equally:

(1) A **place/transition net** (P/T-net) is a triple $N = (S, T, F)$, where S and T are the disjoint sets of places and transitions and $F : S \times T \cup T \times S \rightarrow \mathbb{N}$ is the flow relation.

Often one considers labelled P/T-nets, where different transitions may be equally labelled; i.e. one adds a labelling function l from T to a set of labels. If one wants to focus attention on the transitions (labelled by actions from a set *Act* of actions) then the following is more convenient. Let $\mathbf{M}(S)$ be the set of multisets over S.

(2) A P/T-net over *Act* is a pair (S, D) where S is the set of places, $D \subseteq \mathbf{M}(S) \times Act \times \mathbf{M}(S)$ is the set of labelled transitions (together with the weighted arrows connecting the related places) (For this notation see for example [Gol88]).

For more algebraic considerations a third definition seems to be promising (see [DMM89]).

(3) Let $S^{\oplus}$ be the free commutative monoid on S (if S is finite, the elements of $S^{\oplus}$ are the multisets over S), then an mP/T-net is a quadruple $N = (S^{\oplus}, T, \sigma, \Theta)$, where T is the set of transitions and $\sigma, \Theta : T \rightarrow S^{\oplus}$ are mappings associating to every transition its pre-multiset and its post-multiset (places with the weighted arrows connecting them to the transition).

In the following we only consider P/T-nets where all weights of arrows are 1, i.e. where (in def. (1)) F maps into $\{0,1\}$, i.e. where instead of $\mathbf{M}(S)$ we can use $\mathbf{P}(S)$ in definition (2) – moreover we will use definition (2). And we briefly call these particular P/T-nets just **nets**.

A **marking** of a net is a multiset $M \in \mathbf{M}(S)$. A transition (S_1, a, S_2) is **enabled** at a marking M, if $S_1 \leq M$.

A transition d which is enabled at a marking M_1 **may occur**; if it occurs it produces the marking $M_2 := (M_1 - S_1) + S_2$; this is usually denoted by $M_1[d\rangle M_2$.

This notation is extended to arbitrary words over D by $M_1[\varepsilon\rangle M_1$ and $M_1[ud\rangle M_3$ iff $\exists M_2 : M_1[u\rangle M_2 \wedge M_2[d\rangle M_3$. Moreover we write $[M\rangle$ for the set of all markings reachable from M.

A net (S, D) with a marking M, called a marked net, is denoted as triple $(S, D; M)$.

3.3 Syntax-Driven Construction of Nets from GAP Terms

If we consider GAP as a specification language for concurrent systems and nets as formal descriptions of concurrent systems, it is natural to ask whether GAP terms can be represented as nets. Obviously the actions of GAP have to be represented by transitions, all the actions which may be performed before any other action have to be enabled by an appropriate marking; to the operators on terms should correspond operators on nets; i.e. we look for a syntax-driven modular net construction. Moreover we aim at markings with a minimal number of tokens in a place, i.e. mappings $S \to \{0,1\}$, which we represent as subsets of S. In a first step we do not consider recursion. We again denote the operators in the same way as in GAP.

$$nil \quad := \quad (\{s\}, \emptyset; \{s\})$$

Let $N = (S, D; Z)$ be a marked net, then

$$\begin{aligned} aN \quad &:= \quad (S \cup \{s\}, D \cup \{(\{s\}, a, Z)\}; \{s\}) \\ &\qquad \text{for } a \in \mathit{Vis} \cup \{\tau\}. \\ Nf \quad &:= \quad \mathit{Reach}(S, D'; Z), \text{ i.e.} \\ &\qquad \text{the subnet reachable from the initial marking of the net } (S, D'; Z), \\ &\qquad \text{where } D' = \{(M_1, af, M_2) \mid (M_1, a, M_2) \in D \wedge af \neq \bot\} \end{aligned}$$

To define the operator $+$ we introduce a restriction on the syntax: we allow only nets with a single marked place as operands (corresponding to GAP terms of the form aP) – otherwise problems would arise.

Let $N_i = (S_i, D_i; \{z_i\}), i = 1,2$ be marked nets with $S_1 \cap S_2 = \emptyset$, then:

$$\begin{aligned} N_1 + N_2 \quad &:= \quad \mathit{Reach}(S_1 \cup S_2 \cup \{z\}, D_1 \cup D_2 \cup D_+; \{z\}), \\ &\qquad \text{where } z \notin S_1 \cup S_2 \text{ and} \\ &\qquad D_+ = \{(\{z\}, a, M) \mid (\{z_1\}, a, M) \in D_1 \vee (\{z_2\}, a, M) \in D_2\} \end{aligned}$$

Let $N_i = (S_i, D_i; Z_i), i = 1,2$ be arbitrary marked nets with $S_1 \cap S_2 = \emptyset$, then:

$$\begin{aligned} N_1 \nmid N_2 \quad &:= \quad (S_1 \cup S_2, D_1 \cup D_2 \cup D_+; Z_1 \cup Z_2), \text{ where} \\ &\qquad D_+ = \{(M_1 \cup M_2, \{a_1, a_2\}, M_1' \cup M_2') \mid \\ &\qquad\qquad \forall i \in \{1,2\} : a_i \in \mathit{Vis} \wedge (M_i, a_i, M_i') \in D_i\} \end{aligned}$$

Example:

Let $P :\equiv a\,nil + b\,nil$, $Q :\equiv c\,nil$
Then $P \nmid Q$ is represented by the net

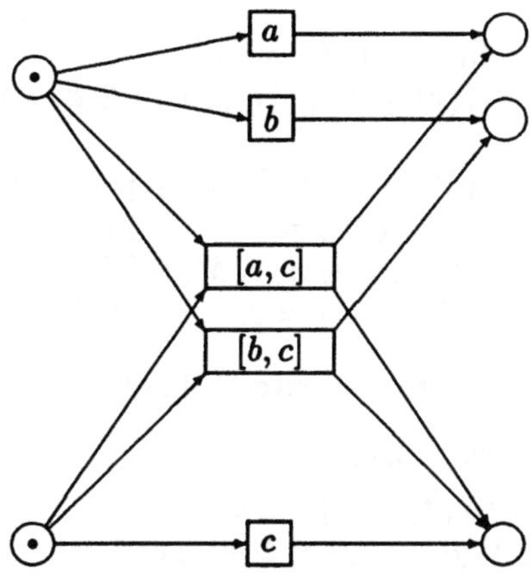

The main problem is the modelling of recursion. Therefore we make another syntactical restriction: We consider only recursion terms of the form $rec\,p.\tau Q$. It can however be shown (see [Tau88]), that semantically this is not restrictive. An important goal of the construction is to obtain finite nets in as many cases as possible. The grammar for the terms to which we associate Petri nets now is as follows

$$\begin{array}{rcl} S & ::= & nil \mid p \mid aQ \mid S + S \mid rec\,p.\tau Q \\ Q & ::= & S \mid Qf \mid Q \nmid Q. \end{array}$$

The key idea for modelling recursion is based on Milner's construction of a finite extended transition system for a CCS term (for details see [Tau88]); i.e. we enlarge the notion of a net by a means for representing identifiers and action manipulation functions: An **extended net** is a quadruple $N = (S, D, E; Z)$ where $(S, D; Z)$ is a net and

$$E \subseteq \mathbf{P}(S) \times Idf \times Fun_{\perp} \quad \text{is the set of extensions.}$$

Here $Fun_{\perp}$ is the set of action manipulation functions Fun, enlarged by the special element $\perp$, $\perp \notin Fun$.

An extension can be considered (and depicted) as a special transition with no post-set.

For $p \in Idf$ we then get the net representation

$$p := (\{z\}, \emptyset, \{(\{z\}, p, id)\}; \{z\})$$

All the other net constructions above must now be enlarged by appropriate extension sets E'

$$nil: \quad E' := \emptyset$$

$$Nf: \quad E' := \{(M,p,gf) \mid (M,p,g) \in E \wedge g \neq \bot\} \cup \{(M,p,\bot) \mid (M,p,\bot) \in E\}$$

$$N_1 + N_2: \quad E' := E_1 \cup E_2 \cup E_+ \text{ where } E_+ := \{(\{z\},p,f) \mid (\{z_1\},p,f) \in E_1 \vee (\{z_2\},p,f) \in E_2\}.$$

$$\begin{aligned} N_1 \nmid N_2: \quad E' := \{(M_1 \cup M_2, p, f) \mid\ & (M_1,p,g) \in E_1 \wedge M_2 \in [Z_2\rangle_2 \\ & \wedge (f = g \vee f = \bot \wedge \exists (U,a,U') \in D_2 \cup E_2 : U \subseteq M_2) \vee \\ & (M_2,p,g) \in E_2 \wedge M_1 \in [Z_1\rangle_1 \\ & \wedge (f = g \vee f = \bot \wedge \exists (U,a,U') \in D_1 \cup E_1 : U \subseteq M_1)\} \end{aligned}$$

To define the recursion operator for extended nets we need some more notations:

For $r \in Idf$ and $N = (S,D,E;Z)$ where
$\bot \notin E(r) := \{f \in Fun_\bot \mid (M,r,f) \in E\}$ define
$F := \{id \cdot f_1 \cdot \dots \cdot f_n \mid n \geq 0, f_1, \dots, f_n \in E(r)\}$, and
for $f \in F$ let $(S_f, D_f, E_f; Z_f) := Nf$ such that $f \neq g$ implies $S_f \cap S_g = \emptyset$.
Then

$$\text{rec}\, r.\tau N := Reach(\ \{z\} \cup \dot{\bigcup}_{f\in F} S_f,\ \ \{(\{z\},\tau,Z_{id})\} \cup D_+ \cup \bigcup_{f\in F} D_f,\ \ \bigcup_{f\in F} E_f - (\mathbf{P}(\bigcup_{f\in F} S_f) \times \{r\} \times Fun); \{z\})$$

where $z \notin S_f$ for all $f \in F$ and

$$D_+ = \{(M,\tau,Z_g) \mid (M,r,g) \in \bigcup_{f\in F} E_f\}$$

Example: (from [Tau88]) Let $P :\equiv a\,nil + b\,nil$, $Q :\equiv cp$. Then $rec\,p.\tau((P \nmid Q) - \{a,b,c\})$ gives the net

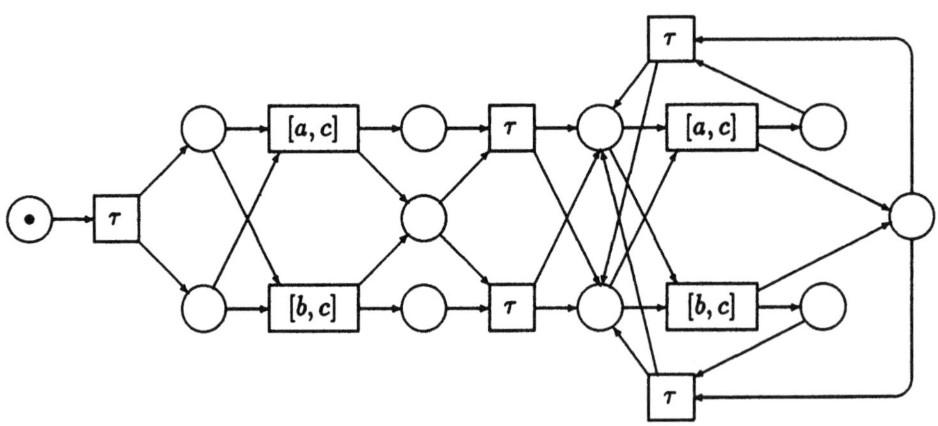

4 Semantics of Petri Nets

The simplest operational semantics of a marked net $(S, D; M)$ is the subset of all words $w \in D^*$ which denote sequences of occurrences of transitions starting from the marking M; one can refine this notion by considering an additional marking (or a set of markings) and take only those words in D^* which lead from M to this (these) marking(s). Thus a Petri net can be seen as a device to produce formal languages – and, indeed, there is a large body of interesting results on the formal languages of Petri nets; for details see the paper by M. Jantzen in [BRR87].

As Petri nets are meant to model not only relational systems (which accept input, produce output and stop) but also **reactive** systems (which are running all the time and react to interactions by users), it is useful to study also the infinite behaviour of nets – the simplest way to do so, is to study infinite sequences of transition occurrences possible in a net (see the papers by Carstensen in [CJK88] and by Valk in [BRR87]).

Since Petri nets should describe concurrent systems, it is also sensible to look after semantics that model concurrency more explicitly. Obviously the notion of steps (see part 2) can be used. A **step** of a net (S, D) is a finite multiset over D. The step x is **enabled** at a marking M if for each $s \in S$

$$\sum_{d \in D} x(d) \cdot pr_1(d)(s) \leq M(s)$$

where $pr_1(d) = \delta_1$ if $d = (\delta_1, a, \delta_2) \in \mathbf{M}(S) \times Act \times \mathbf{M}(S)$.

Now we can define finite (and infinite) step occurrence sequences in analogy to transition occurrence sequences.

Obviously we can also transfer the notion of a failures set to Petri nets (see e.g. [Vog89]).

But Petri nets offer other interesting formalisms for describing the concurrent processes going on in a distributed system: One can represent these processes again as nets.

One possibility is to unfold a net (like one can unfold a while-loop into a sequence or a transition system into a tree).

Example: (from the paper by Winskel in [BRR87])

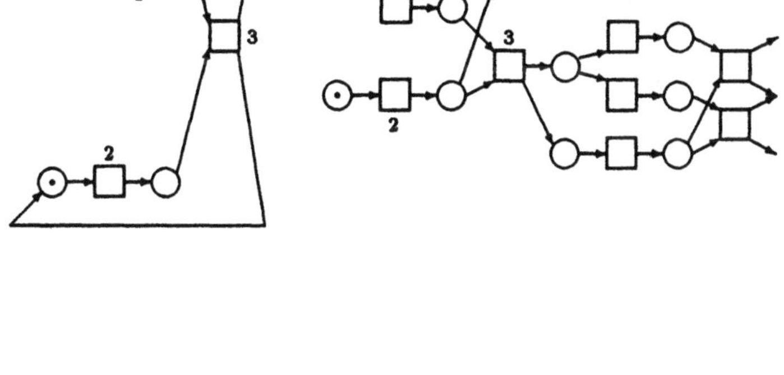

If one is only interested in the actions (the transitions) and the dependencies between the events of transition occurrences then one can erase the places in the unfolded net; one then obtains a so-called event structure – for more details on this very powerful semantical structure for describing concurrent processes see the paper by Winskel in [BRR87].

Instead of representing all possible processes in one unfolded net, one can separate them: beginning with an initial marking, only that part of the unfolding is kept where the tokens flow through during a process. Then one marked net will be represented by a possibly infinite set of possibly infinite nets in which each place is in the preset of at most one transition only. For more details on the theory of these **nonsequential processes** see the paper by Fernandez in [BRR87] and the book [BF88].

If one suppresses the places in the nonsequential process, then one gets partial orders labelled with (the names of) transitions – these labelled partial orders can be considered as generalizations of words over the set of labels (words being linear labelled orders); therefore they are often called partial words. From the language of partial words defined by a Petri net one can easily obtain the language of transition sequences as well as the language of step sequences. For more details see [Kie88] and the paper by Kiehn in [Roz88].

There is one other quite interesting idea to give a non-interleaving (i.e. concurrency and nondeterminism distinguishing) semantics for (non-labelled) Petri nets. Let us imagine an observer (as in 2.1). We now ask: what minimal global, structural information on the net do we need in order to infer from one observation sequence all other sequences possible by starting at the same marking. The answer (given by Marzurkiewicz, see his paper in [BRR87]) is: We only need to know which pairs of transitions are (always) independent such that they can be permuted in each transition sequence.

The original definition of independency of transitions has been generalized by Diekert (see [Die90]); it can however be further generalized as follows: Transitions t, t' are dependent iff $t = t'$ or $pr_3(t') \cap pr_1(t) \neq \emptyset$ or $pr_1(t') \cap pr_3(t) \neq \emptyset$. It is however not clear, what properties this generalized theory will have, since the extension of Marzurkiewicz's idea from the class of marked nets where each place can hold at most one token to general marked P/T-nets has some inherent difficulties lying in the fact that in the general case the permutability of transition occurrences in an observed sequence depends on the marking and not only on the structure of the net.

Example: (from the diploma thesis [Kie86] of A. Kiehn)

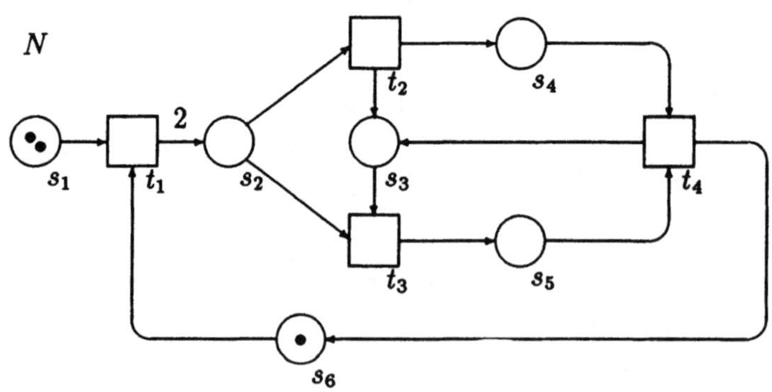

The main advantage of this approach is the following: The independency relation I on the set T of transitions can be used to define the free partially commutative monoid T^*/I of so-called traces (i.e. congruence classes of words with respect to permutation of independent transitions); the semantics of a net then becomes a subset of such a monoid; these monoids can be studied by algebraic techniques (see [Die90]).

5 Modular Construction and Refinement of Concurrent Systems

Abstract programming languages are based on the idea of modular construction; their semantics are always compositional.

Petri nets per se have no modular structure; only rather recently – based on the ideas from the field of abstract programming languages – modular construction techniques based on operators on nets have been studied (see [Tau88], [Gol88]). Also, it has been observed by Mazurkiewicz (see his paper in [Roz88]) that each unlabelled P/T-net (with arrow weights 1) can be considered as being composed of so-called atomic nets (whose sets of places contain only one element) by an operation of synchronization (i.e. composition of nets by building the disjoint union of their places but perhaps identifying some of their transitions), yielding a compositional semantics for nets based on partial orders.

Another way to build complex structures is to use refinement techniques. Abstract programming languages pose problems with respect to refinement; e.g. bisimulation is not a congruence with respect to refinement.

Example: (see [GG89]) The two terms $P :\equiv a\,nil \parallel_{\emptyset} b\,nil$ and $Q :\equiv ab\,nil + ba\,nil$ are bisimulation equivalent but if the action a is refined into the sequential composition of two actions a_1, a_2 then we obtain two systems which are not bisimular:

$$P' :\equiv (a_1 a_2\,nil) \parallel_{\emptyset} b\,nil\ ,\ Q' :\equiv a_1 a_2 b\,nil + b a_1 a_2\,nil$$

Only since recently the problem of refinement for abstract programming languages is being studied (see [GG89]).

For Petri nets refinement has been considered from the beginning on (for references see [BGV91] and [Vog90]) – there are even several methods of refinement: One can refine single transitions or single places or one may cut off a part of a net (such that the boundary along the cut consists only of transitions or only of places) and substitute a new net into the remaining net. There are two different approaches to study this: Usually refinement is studied under the aspect of preserving the behaviour of the original net; however, recently (influenced by the abstract programming languages) also the situation is studied where the behaviour is changed by the refinement, such that the same refinements in two behaviourally equivalent nets result in behaviourally equivalent refined nets (for more details see [BGV91]).

A completely different approach to refinement has been developed by A. Kiehn (see [Kie89] and [Kie90]): Instead of replacing a transition by a net, an incarnation of the refining net is called (like a subroutine). This naturally also allows for recursion. But in contrast to the perhaps too powerful possibilities of intertwining recursion and parallel composition in abstract programming languages (see [Tau88] chap. 2) in this case different incarnations cannot communicate. Using this technique a concurrent system is specified by a set of nets which may call each other recursively. For the implementation of such specifications one obviously needs some sort of stack mechanism. It is important to note that such Petri net systems in the sense of Kiehn are strictly more powerful than Petri nets. Namely the family of (the usual) Petri net languages (see chapter 4) is not closed under the operation of Kleene star (i.e. submonoid generation in the free monoid) (for this see M. Jantzen's paper in [BRR87]), while the class of languages of Kiehn's Petri net systems is closed under the Kleene star. This is due to the fact that calling a net N' from a transition t of a net N can – on the level of behaviour description – be represented simply by context-free substitution of the behaviour of N' into the behaviour of N at all positions where t occurs.

Acknowledgement: I would like to thank D. Taubner and R. Gold for many helpful suggestions; in particular D. Taubner also extended the step failure semantics to the language GAPH. Special thanks go to H. Hadwiger and D. Stein for their excellent and phantastically quick typing of this text.

References

[Bar85] H. P. Barendregt. *The Lambda-Calculus – Its Syntax and Semantics.* North-Holland Publ. Comp., Amsterdam 1985.

[BC91] L. Bernardinello, F. De Cindio. *A survey of basic net models and modular net classes.* To appear in: G. Rozenberg (ed.) Advances in Petri Nets 1990, LNCS, Springer-Verlag, Berlin, Heidelberg, New York, Tokyo, 1991.

[BF88] E. Best and C. Fernandez. *Nonsequential Processes.* EATCS Monographs on Theoretical Computer Science, vol. 13, Springer-Verlag, Berlin, Heidelberg, New York, Tokyo, 1988.

[BGV91] W. Brauer, R. Gold and W. Vogler. *Behaviour and equivalence preserving refinements of Petri nets.* To appear in: G. Rozenberg (ed.) Advances in Petri Nets 1990, LNCS, Springer-Verlag, Berlin, Heidelberg, New York, Tokyo, 1991.

[BHR84] S. D. Brookes, C. A. R. Hoare and A. W. Roscoe. *A theory of communicating sequential processes.* J. ACM **31**:560–599, 1984.

[Bra84] W. Brauer. *How to play the token game. Petri Net Newsletter*, **16**:3–13, 1984.

[Bra87] W. Brauer. *Carl Adam Petri and informatics.* In: G. Rozenberg, K. Voss and H. Genrich (eds.), *Concurrency and Nets*, pp. 13–21. Springer-Verlag, Berlin, Heidelberg, New York, Tokyo, 1987.

[Bra90] W. Brauer. *Graphs, automata, Petri nets – From sequential to distributed and concurrent systems –.* In: H. Schwärtzel, I. Mizin (eds.) *Advanced Information Processing*, pp. 15–28, Springer-Verlag, Berlin, Heidelberg, New York, Tokyo, 1990.

[Bro83] S. D. Brookes. *A Model for Communicating Sequential Processes.* Rpt. CMU-CS-83-149. Ph.D.thesis, Carnegie-Mellon Univ., 1983.

[BRR87] W. Brauer, W. Reisig, and G. Rozenberg, (eds.) *Petri Nets, Parts I and II.* LNCS vol. 254 and 255, Springer-Verlag, Berlin, Heidelberg, New York, Tokyo, 1987.

[CH74] R. H. Campbell and A. N. Habermann. *The specification of process synchronization by path expressions.* LNCS vol. 16, pp. 89–102, Springer-Verlag, Berlin, Heidelberg, New York, Tokyo, 1974.

[CJK88] M. P. Chytil, L. Janiga, and V. Koubek, (eds.) *Mathematical Foundations of Computer Science 1988.* LNCS vol. 324, Springer-Verlag, Berlin, Heidelberg, New York, Tokyo, 1988.

[Die90] V. Diekert. *Combinatorics on Traces with Applications to Petri Nets and Replacement Systems.* (Habilitationsschrift, TU München), LNCS vol. 454, Springer-Verlag, Berlin, Heidelberg, New York, Tokyo, 1990.

[DMM89] P. Degano, J. Meseguer, and U. Montanari. *Axiomatizing net computations and processes.* In: *Proc. 4th Ann. Symp. on Logic in Computer Science (LICS),* Asilomar, Ca., USA, June 5-8 1989.

[GG89] R. van Glabbeek and U. Goltz. *Partial order semantics for refinement of actions – neither necessary nor always sufficient but appropriate when used with care.* In: *EATCS Bull. No. 38,* pp. 154–163, June 1989.

[Gol88] U. Goltz. *Über die Darstellung von CCS-Programmen durch Petrinetze. GMD-Bericht,* Nr. 172, Oldenburg-Verlag, München, Wien 1988. see also [CJK88], pp. 339–350.

[HBR81] C. A. R. Hoare, S. D. Brookes and A. W. Roscoe. *A theory of communicating sequential processes.* Techn. monograph PRG–16, Oxford Univ., Progr. Research Group, 1981.

[Hoa78] C. A. R. Hoare. *Communicating sequential processes. Comm. ACM,* **21**:666–677, 1978.

[Inm88] Inmos Ltd. OCCAM 2 Reference Manual, Prentice-Hall, 1988.

[Kie86] A. Kiehn *On the Concurrent Behaviour of Petri Nets.* Techn. Report FBI–HH–B–120/86, Fachbereich Informatik, Universität Hamburg, 1986.

[Kie88] A. Kiehn. *On the interrelation between synchronized and nonsynchronized behaviour of Petri nets.* J. Inf. Proc. Cybern. *(EIK),* **24**:3–18, 1988.

[Kie89] A. Kiehn. *A structuring mechanism for Petri nets.* Report TUM-I 8902, TU München, 1989.

[Kie90] A. Kiehn. *Petri net systems and their closure properties.* In: G. Rozenberg (ed.) *Advances in Petri Nets 1989*, LNCS vol. 424, pp. 306–328, Springer-Verlag Berlin, Heidelberg, New York, Tokyo, 1990.

[LC75] P. E. Lauer and R. H. Campbell. *Formal semantics for a class of high-level primitives for coordinating concurrent processes.* Acta Informatica **5**, pp. 247–332, 1975.

[May83] OCCAM, SIGPLAN Notices **18**, 4:69–79, 1983.

[Mil80] R. Milner. *A Calculus of Communicating Systems.* LNCS vol. 92, Springer-Verlag, Berlin, Heidelberg, New York, Tokyo, 1980.

[Mil85] R. Milner. *Lectures on a calculus of communicating systems.* In: S. D. Brookes et. al., (eds.), *Seminar on Concurrency*, LNCS vol. 197, pp. 197–220. Springer-Verlag, Berlin, Heidelberg, New York, Tokyo, 1985.

[Par81] D. Park. *Concurrency and automata on infinite sequences.* In: P. Deussen (ed.), *Proc. 5th GI Conf. on Theoret. Comp. Science*, LNCS vol. 104, pp. 167–183, Springer-Verlag, Berlin, Heidelberg, New York, Tokyo, 1981.

[Plo81] G. D. Plotkin. *A structural approach to operational semantics.* Aarhus Univ. Rept. DAIMI FN–19, 1981.

[Roz88] G. Rozenberg, (ed.) *Advances in Petri Nets 1988.* LNCS vol. 340, Springer-Verlag, Berlin, Heidelberg, New York, Tokyo, 1988.

[Tau88] D. Taubner. *The Finite Representation of CCS and TCSP Programs by Automata and Petri Nets.* (Dissertation, TU München 1988) LNCS vol. 369, Springer-Verlag, Berlin, Heidelberg, New York, Tokyo, 1989.

[Tra88] B. A. Trakhtenbrot. *Comparing the Church and the Turing approaches: Two prophetical messages.* In: R. Herken (ed.) *The Universal Turing Machine, A Half-Century Survey*, pp. 603–630. Kammerer & Unverzagt, Hamburg, Berlin and Oxford University Press, Oxford, 1988.

[TV89] D. Taubner and W. Vogler. *Step failures semantics and a complete proof system.* Acta Informatica **27**:125–156, 1989.

[Vog89] W. Vogler. *Failures semantics and deadlocking of modular Petri nets.* Acta Informatica **26**:333–348, 1989.

[Vog90] W. Vogler. *Failures semantics of Petri nets and the refinement of places and transitions.* Techn. Report TUM–I 9003 Inst. f. Informatik, TU München, 1990.

[Vog91] W. Vogler. *Bisimulation and action refinement.* In: C. Choffrut, R. Jantzen (eds.) *Proc. STACS 91*, LNCS, Springer-Verlag, Berlin, Heidelberg, New York, Tokyo, 1991.

The Family of Concurrent Logic Programming Languages

Ehud Shapiro
Department of Applied Mathematics and Computer Science
The Weizmann Institute of Science
Rehovot 76100, Israel

First published in ACM Computing Surveys 21(3), 413-510 (1989).

NATO ASI Series, Vol. F 79
Logic, Algebra, and Computation
Edited by F. L. Bauer

Abstract

Concurrent logic languages are high-level programming languages for parallel and distributed systems that offer a wide range of both known and novel concurrent programming techniques. Being logic programming languages, they preserve many advantages of the abstract logic programming model, including the logical reading of programs and computations, the convenience of representing data-structures with logical terms and manipulating them using unification, and the amenability to meta-programming. Operationally, their model of computation consists of a dynamic set of concurrent processes, communicating by instantiating shared logical variables, synchronizing by waiting for variables to be instantiated, and making nondeterministic choices, possibly based on the availability of values of variables.

This paper surveys the family of concurrent logic programming languages within a uniform operational framework. It demonstrates the expressive power of even the simplest language in the family, and investigates how varying the basic synchronization and control constructs affect the expressiveness and efficiency of the resulting languages.

In addition, the paper reports on techniques for sequential and parallel implementation of languages in this family, mentions their applications to date, and relates these languages to the abstract logic programming model, to the programming language Prolog, and to other concurrent computational models and programming languages.

Categories and Subject Descriptors:

D.1.3 [Programming Techniques] Concurrent Programming.

D.3.3 [Programming Languages] Language Classification — ALPS, Concurrent Prolog, CP(↓,|), Doc, FCP, FGHC, Guarded Horn Clauses, P-Prolog, Parlog.

D.3.3 [Programming Languages] Language Constructs — Concurrent programming structures.

D.4.1 [Operating Systems] Process Management — Concurrency, Mutual Exclusion, Synchronization.

F.1.2 [Computation by Abstract Devices] Modes of Computation — Parallelism.

F.3.2 [Logics and Meanings of Programs] Semantics of Programming Languages — Operational Semantics.

F.4.1 [Mathematical Logic and Formal Languages] Mathematical Logic — Logic programming.

General Terms: Languages.

Additional Key Words and Phrases:

atomic unification, computation replay, distributed detection, incomplete messages, indeterminism, input matching, language embedding, logical variable, meta-interpretation, process structures, short-circuit, snapshots, streams.

Contents

PART I. INTRODUCTION

1. Introduction

In surveying concurrent logic programming languages, this paper:

- Introduces the computational models of logic programs, Prolog, and concurrent logic languages.
- Discusses the different role of nondeterminism in these three computational models.
- Explains the use of the logical variable as a communication channel, and the use of unification in the specification and implementation of sophisticated communication protocols.
- Demonstrates the powerful programming techniques available in concurrent logic languages, including: stream processing, the formation and manipulation of dynamic process networks, incomplete-message protocols for dialogues and network configuration, concurrent construction of shared data-structures, and short-circuit protocols for distributed termination and quiescence detection.
- Demonstrates the utility of enhanced meta-interpreters in concurrent logic programming, including their application to computation control, to the formation of live and frozen snapshots, and to computation replay and debugging.
- Exposes the spectrum of concurrent logic programming languages, ranging from the simpler and weaker ones, to the more complex and more expressive ones.
- Reports on implementation techniques for sequential and parallel computers developed for concurrent logic language, as well as on specialized architectures designed for them.

The paper does not aim at providing a historical account of the development of concurrent logic languages. Rather, it attempts to expose the core concepts of these languages, as well as the internal structure of the family and the qualities of each of its members, within a consistent operational framework. As a result, usually an idealized or a simplified version of each language is described. When applicable, the differences from the actual language, as well as relevant historical facts, are noted[1].

The paper consists of five parts. In the remainder of Part I, Section 2 surveys briefly the abstract computational model of logic programming and of (pure) Prolog, explaining the role of the logical variable, unification, and nondeterminism in this model.

Part II conveys the core concepts and techniques of concurrent logic programming. Section 3 introduces the basic concepts of concurrent logic programming, and the use of shared logical variables for communication and synchronization. Section 4 defines a simple concurrent logic language. This language is used in Section 5 to illustrate basic concurrent logic programming examples and techniques. Section 6 discusses fairness conditions for concurrent logic programs. Following that, Section 7 describes advanced concurrent logic programming techniques. Although this part uses a particular concurrent logic language, both the basic and advanced techniques shown are common to most programming languages in the family; exceptions are noted when each language is introduced.

Part III surveys the various members of the family of concurrent logic languages.

---

[1] For additional historical notes see [148,166].

Section 21 describes our method of comparing languages in the family. We compare languages for their expressiveness, simplicity, readability and efficiency. In comparing expressiveness, we explore embeddings among languages and the programming techniques provided by each language.

Section 9 discusses the semantics of concurrent logic programs. Sections 10 to 17 introduce and compare flat concurrent logic languages. A flat language is defined with respect to a given fixed set of primitive predicates (in the languages discussed these include mainly equality, inequality and arithmetic tests). In a flat language a process can perform only a simple computation, specified by a conjunction of atoms with primitive predicates, before making a committed nondeterministic choice. In non-flat languages such pre-commit computations may involve program-defined predicates, thus can be arbitrarily complex. During a computation of a non-flat language the processes form an And/Or-tree, whereas in a flat language the processes are a "flat" collection; hence their name. Non-flat concurrent logic languages are surveyed in Section 18.

Part IV describes implementations developed for concurrent logic languages, and references their applications. Implementation techniques for both sequential and parallel computers are reviewed, as well as specialized architectures designed for their efficient execution.

Part V concludes the paper by comparing the concurrent logic programming model with other approaches to programming and modeling concurrency, including Prolog, dataflow languages, functional languages, message-passing models of concurrency, object-oriented languages, and nondeterministic transition systems.

How to read the paper

The reader who wishes only to understand a single concurrent logic language can skim Part I and read Part II. There are sufficient intuitive explanations and examples so that the formal treatment of the semantics of logic programs can be skipped without loss of continuity. The reader interested in implementation techniques can read Section 19 of Part IV without reading Part III.

2. Logic Programming, Prolog, and the Power of the Logical Variable

This section introduces the logic programming computational model. It defines pure Prolog and relates it to the logic programming model. It discusses properties of the logical variable and unification and their relation to conventional data-manipulation operations.

2.1 Syntax and informal semantics of logic programs

We use the Edinburgh syntax [SS] for logical variables, terms, and predicates.

Definitions: Term, atom, clause, logic program, vocabulary.

- A *term* is a variable (e.g. X) or a function symbol of arity $n \geq 0$, applied to n terms (e.g. c and $f(a,X,g(b,Y))$).
- An *atom* is a formula of the form $p(T_1,\ldots,T_n)$, where p is a predicate of arity n and $T_1,\ldots,T_n$ are terms.
- A *definite clause* (*clause* for short) is a formula of the form:

$$A \leftarrow B_1,\ldots,B_n. \qquad n \geq 0.$$

where A is an atom $B_1,\ldots,B_n$ is a sequence of atoms. A is called the clause's *head*, and $B_1,\ldots,B_n$ its *body*. We denote the empty sequence of atoms by *true*.

- A *logic program* is a finite set of definite clauses.
- A *goal* is a sequence of atoms $A_1,A_2,\ldots,A_n$. A goal is *empty* if $n=0$, *atomic* if $n=1$, and *conjunctive* if $n>1$. Each atom in a goal is called a *goal atom*. A goal atom is often called also a *goal* for short.
- The *vocabulary* of a logic program P is the set of predicates and function symbols that occur in the clauses of P. ∎

We use the Edinburgh notation for lists (and also for streams, as discussed below). The term $[X|Xs]$ (read "X *cons* Xs") is a list whose head is X and tail is Xs, and the constant [ ] (read "*nil*") is used by convention to denote the empty list.

Informal semantics of logic programs

Logic programs can be read both declaratively and operationally. We describe these two views here informally, and make them precise in Section 2.4 below.

Declaratively, each clause in a logic program is read as a universally quantified implication. If $X_1,X_2,\ldots,X_n$ are the variables in the clause $A \leftarrow B_1,B_2,\ldots,B_k$, then the clause is read "for all $X_1,X_2,\ldots,X_n$, A is true if B_1 and B_2 and ... and B_k are true". A logic program is read as the conjunction of the universal implications corresponding to its clauses.

Operationally, logic programs can be viewed as an abstract computational model, like the Turing Machine, the Lambda Calculus, and the Random Access Machine. A computation in this model is a goal-driven deduction from the clauses of the program. Like the nondeterministic Turing machine, computations in this model are nondeterministic: from each state of the computation there may be several possible transitions. Specifically, the clauses of a logic program can be read as transition rules of a nondeterministic transition system.

The state of a computation consists of a goal (sequence of atoms) G and a substitution (assignment of values to variables) θ, and is denoted by a pair $\langle G;\theta\rangle$. A computation begins with an initial state consisting of the initial goal to be proven and the empty substitution ε, and progresses nondeterministically from state to state according to the following transition rules, Reduce and Fail. A computation can be viewed as an attempt to prove the initial goal from the program. At each state the goal represents a statement whose proof will establish the initial goal; the substitution represents the values computed so far for variables used in the computation, including the initial goal variables. A computation ends in a state whose goal is either *true* or *fail*. In the former case the computation is *successful*, and it corresponds to a successful proof of the initial goal. In the latter it is *failed*. The substitution in the terminal state, restricted to the variables in the initial goal, is called the *answer substitution* of the computation.

A successful computation has the property that its initial goal, instantiated by the answer substitution, is a logical consequence of the program.

A key step in the transitions is the unification of a goal atom with the head of a clause. Intuitively, a unifier of two terms T_1 and T_2 is a substitution θ, whose application to T_1 and T_2 yields the same term, i.e. $T_1\theta=T_2\theta$. The unification of two terms T_1 and T_2 returns their most general ("simplest") unifier θ if there is one, or *fail* if there is none. The two cases are denoted by $mgu(T_1,T_2)=\theta$ and $mgu(T_1,T_2)=fail$, respectively. For example, the most general unifier of $f(X,b)$ and $f(g(Y),Z)$ is the substitution $\{X\mapsto g(Y),Z\mapsto b\}$. Examples of other (less general) unifiers are $\{X\mapsto g(a),Z\mapsto b\}$, $\{X\mapsto g(b),Z\mapsto b\}$, and $\{X\mapsto g(g(W)),Z\mapsto b\}$.

We denote the ability to move from a state S to a state S' using a transition rule t by $S \xrightarrow{t} S'$. Substitutions can be viewed as functions from variables to values (see Section 2.4), hence we use $\theta \circ \theta'$ to denote the substitution whose application has the effect of applying θ then applying θ'. The Reduce and Fail transition rules require that the variables in the clause be consistently replaced by new variables that have not been used before in the computation. A clause to which this replacement has been applied is called ***renamed apart***. The requirement to rename a clause is inherited from the resolution rule, and ensures that clauses are "re-entrant".

There are two transition rules:

1. Reduce
   $\langle A_1,\ldots,A_i,\ldots,A_n;\theta\rangle \xrightarrow{\text{Reduce}} \langle (A_1,\ldots,B_1,\ldots,B_k,\ldots,A_n)\theta';\theta \circ \theta'\rangle$
   If $mgu(A_i,A) = \theta'$ for some renamed apart clause $A \leftarrow B_1,\ldots,B_k$ of P.
2. Fail
   $\langle A_1,\ldots,A_i,\ldots,A_n;\theta\rangle \xrightarrow{\text{Fail}} \langle fail;\theta\rangle$
   If for some i and for every renamed apart clause $A \leftarrow B_1,\ldots,B_k$ of P, $mgu(A_i,A) = fail$.

Reduce has the following property: If $\langle G,\theta\rangle \xrightarrow{\text{Reduce}} \langle G',\theta \circ \theta'\rangle$, with mgu θ', then $G\theta'$ is a logical consequence of the program and G'. This implies, by induction, that the initial goal, to which the answer substitution of a successful computation is applied, is a logical consequence of the program.

Note that there are two types of nondeterministic choices in the Reduce transition: which goal atom to reduce, and which clause to reduce it with. The first is called ***And-nondeterminism***, the second ***Or-nondeterminism***. Fail has only an And-nondeterministic choice.

A computation progresses until it reaches a ***terminal state***, which is a state to which no transition applies. By the definition of Reduce and Fail, the goal in a terminal state is either ***true*** or ***fail***.

2.2 Examples of logic programs and their computations

We show some simple logic programs and illustrate their operational behavior. The following logic program defines the predicate $sum(Xs,S)$, which holds if S is the sum of the elements of the list Xs.

```
sum(Xs,S) ←                %1
    sum'(Xs,0,S).

sum'([ ],S,S).             %2
sum'([X|Xs],P,S) ←         %3
    plus(X,P,P')
    sum'(Xs,P',S).
```

The program uses an auxiliary predicate $sum'(Xs,P,S)$, which holds if the sum of the Xs plus P is S, and the predicate $plus(X,Y,Z)$, which holds if X plus Y is Z. For the purpose of this example we assume that $plus$ is defined by a large set of facts, including:

```
plus(0,0,0).               %4
plus(0,1,1).               %5
```

```
plus(1,0,1).        %6
plus(2,0,2).        %7
plus(2,1,3).        %8
plus(2,2,4).        %9
   ⋮
```

To increase readability of the following examples of computations we annotate Reduce transition with a label (i,j) identifying the indices of goal atom and program clause that were used for reduction, Fail transitions with the index of the failing goal atom, and restrict the substitution in a state to the initial goal variables.

An example of a successful computation of the above program is:

$$\langle \text{sum}([1,2],S);\ \varepsilon\rangle \xrightarrow{\text{Reduce(1,1)}}$$
$$\langle \text{sum}'([1,2],0,S);\ \varepsilon\rangle \xrightarrow{\text{Reduce(1,3)}}$$
$$\langle \text{plus}(1,0,P),\ \text{sum}'([2],P,S);\ \varepsilon\rangle \xrightarrow{\text{Reduce(1,6)}}$$
$$\langle \text{sum}'([2],1,S);\ \varepsilon\rangle \xrightarrow{\text{Reduce(1,3)}}$$
$$\langle \text{plus}(2,1,P'),\ \text{sum}'([\ ],P',S);\ \varepsilon\rangle \xrightarrow{\text{Reduce(1,8)}}$$
$$\langle \text{sum}'([\ ],3,S);\ \varepsilon\rangle \xrightarrow{\text{Reduce(1,2)}}$$
$$\langle \textit{true};\ \{S\mapsto 3\}\rangle$$

An example of a failing computation is:

$$\langle \text{sum}([1,2],S);\ \varepsilon\rangle \xrightarrow{\text{Reduce(1,1)}}$$
$$\langle \text{sum}'([1,2],0,S);\ \varepsilon\rangle \xrightarrow{\text{Reduce(1,3)}}$$
$$\langle \text{plus}(1,0,P),\ \text{sum}'([2],P,S);\ \varepsilon\rangle \xrightarrow{\text{Reduce(2,3)}}$$
$$\langle \text{plus}(1,0,P),\ \text{plus}(2,P,P'),\ \text{sum}'([\ ],P',S);\ \varepsilon\rangle \xrightarrow{\text{Reduce(2,9)}}$$
$$\langle \text{plus}(1,0,2),\ \text{sum}'([\ ],4,S);\ \varepsilon\rangle \xrightarrow{\text{Fail(1)}}$$
$$\langle \textit{fail},\ \varepsilon\rangle$$

The failure in the last computation could have been avoided by deferring the reduction of the goal atom *plus(2,P,P')* until more information was available. The Reduce transition of Prolog, introduced in Section 2.5 below, always chooses the left-most atom in the goal for reduction. Thus a Prolog computation on an initial goal *sum(Xs,S)* whose first argument is a complete list[2] of integers and the second argument is a variable is bound to succeed. Furthermore, such a computation is deterministic, in the sense that at each step only one clause head unifies with the selected goal atom.

---

[2] A list is *complete* if every instance of it is a list [173]; $[a,b]$, $[X,b]$, and $[X,Y]$ are complete lists, and $[a,b|Xs]$, $[a|Xs]$, $[X|Xs]$ and Xs are examples of incomplete lists.

Concurrent logic languages use other mechanisms to delay the reduction of a goal atom, which do not impose such strict sequentiality.

The following logic program defines the relation *in_both(X,L1,L2)*, which holds if X is a member of both lists *L1* and *L2*. It uses the auxiliary predicate *member(X,Xs)*, which holds if X is a member of the list *Xs*.

% *in_both*$(X,L_1,L_2) \leftarrow X$ is a member of both lists L_1 and L_2.

```
in_both(X,L1,L2) ←                          %1
    member(X,L1), member(X,L2).
```

% *member*$(X,Xs) \leftarrow X$ is a member of the list Xs.

```
member(X,[X|Xs]).                           %2
member(X,[X1|Xs]) ←                         %3
    member(X,Xs).
```

Here are two possible computations from the goal *in_both(X,[a,b],[b,c])*. A failing computation, in which X is chosen to be a, and the computation of the remaining goal *member(a,[b,c])* fails:

$$\langle \text{in\_both(X,[a,b],[b,c])} \ ; \ \varepsilon\rangle \xrightarrow{\textbf{Reduce(1,1)}}$$
$$\langle \text{member(X,[a,b]), member(X,[b,c])} \ ; \ \varepsilon\rangle \xrightarrow{\textbf{Reduce(1,2)}}$$
$$\langle \text{member(a,[b,c])} \ ; \ \{\text{X}\mapsto\text{a}\}\rangle \xrightarrow{\textbf{Reduce(1,3)}}$$
$$\langle \text{member(a,[c])} \ ; \ \{\text{X}\mapsto\text{a}\}\rangle \xrightarrow{\textbf{Reduce(1,3)}}$$
$$\langle \text{member(a,[ ])} \ ; \ \{\text{X}\mapsto\text{a}\}\rangle \xrightarrow{\textbf{Fail(1)}}$$
$$\langle \textit{fail} \ ; \ \{\text{X}\mapsto\text{a}\}\rangle$$

A successful computation from the same goal, in which X is chosen to be b:

$$\langle \text{in\_both(X,[a,b],[b,c])} \ ; \ \varepsilon\rangle \xrightarrow{\textbf{Reduce(1,1)}}$$
$$\langle \text{member(X,[a,b]), member(X,[b,c])} \ ; \ \varepsilon\rangle \xrightarrow{\textbf{Reduce(1,3)}}$$
$$\langle \text{member(X,[b]), member(X,[b,c])} \ ; \ \varepsilon\rangle \xrightarrow{\textbf{Reduce(1,2)}}$$
$$\langle \text{member(b,[b,c])} \ ; \ \{\text{X}\mapsto\text{b}\}\rangle \xrightarrow{\textbf{Reduce(1,2)}}$$
$$\langle \textit{true} \ ; \ \{\text{X}\mapsto\text{b}\}\rangle$$

For this program, no ordering of goal atoms can make the computation deterministic on an initial goal atom whose first argument is a variable.

The following logic program uses the difference-list technique for efficient list concatenation, so we digress to explain it. A difference-list is a term representing a list as the difference between two (possibly incomplete) lists. By convention, the term $H\backslash T$ is used, where '\' is a binary function symbol written in infix notation; H is called the *head* and T the *tail* of the difference-list. Examples of difference-lists representing the list $[a,b,c]$ are $[a,b,c]\backslash[\,]$, $[a,b,c,d,e],\backslash[d,e]$, and $[a,b,c|Xs]\backslash Xs$. Given two difference-lists $H_1\backslash T_1$ and $H_2\backslash T_2$, if $T_1{=}H_2$ then $H_1\backslash T_2$ is their concatenation. It is easy to see that the list represented by $H_1\backslash T_2$ is the concatenation of the lists represented by $H_1\backslash T_1$ and $H_2\backslash T_2$. For example, the concatenation of $[a,b,c,d,e]\backslash[d,e]$ and $[d,e]\backslash[e]$ is $[a,b,c,d,e]\backslash[e]$. Operationally, the precondition for difference-list concatenation, i.e. $T_1{=}H_2$, is usually met by keeping T_1 a variable. For example, $[a,b,c|Xs]\backslash Xs$ can be concatenated to any difference-list. Its concatenation with $[d,e|Ys]\backslash Ys$ gives $[a,b,c,d,e|Ys]\backslash Ys$, which can be further concatenated to any list.

Difference-lists are the preferred representation of lists when concatenation is required. Programs that use difference-lists do not require explicit list concatenation using a predicate like *append*, and are thus more efficient both in time and in space. Operationally, they achieve an effect similar to that of *rplcd* in Lisp, but without destructive data-manipulation operations. However, the precondition for difference-list concatenation, i.e. $T_1{=}H_2$, cannot always be met, for example when the same list needs to be concatenated to several lists.

The third example is a recursive program for flattening a tree. It operates on trees constructed recursively from *tree(L,R)* and *leaf(X)*, where *L* and *R* are recursively trees, and *X* is the value at a leaf. The predicate *flatten(T,Xs)* holds if *Xs* is the list of values at the leaves of the tree *T*, ordered from left to right. The predicate *flatten'(T,Xs\Ys)* holds if the difference-list *Xs\Ys* represents the list thus defined.

```
flatten(T,Xs) ←                          %1
    flatten'(T,Xs\[ ]).

flatten'(leaf(X),[X|Xs]\Xs).             %2
flatten'(tree(L,R),Xs\Zs) ←              %3
    flatten'(L,Xs\Ys),
    flatten'(R,Ys\Zs).
```

The program employs several standard difference-list cliché's. The call from *flatten* to *flatten'* in Clause 1 employs the standard translation between lists and difference-lists: if $H\backslash T$ represents the list L and $T{=}[\ ]$ then $H{=}L$. *flatten'* returns a singleton difference-list in Clause 2, and implicitly concatenates the difference-lists representing the leaves of the subtrees by calling the tail of the first and the head of the second with the same name, *Ys*, in Clause 3.

The program has only deterministic and successful computations on initial goals *flatten(T,Xs)*, where *T* is a complete tree and *Xs* is a variable. For example:

⟨flatten(tree(leaf(a),tree(leaf(b),leaf(c))),Xs); ε⟩ $\xrightarrow{\text{Reduce(1,1)}}$

⟨flatten'(tree(leaf(a),tree(leaf(b),leaf(c))),Xs\[ ]); ε⟩ $\xrightarrow{\text{Reduce(1,3)}}$

⟨flatten'(leaf(a),Xs\Ys), flatten'(tree(leaf(b),leaf(c)), Ys\[ ]); ε⟩ $\xrightarrow{\text{Reduce(1,2)}}$

⟨flatten'(tree(leaf(b),leaf(c)),Ys\[ ]); {Xs↦[a|Ys]}⟩ $\xrightarrow{\text{Reduce(1,3)}}$

⟨flatten'(leaf(b),Ys\Ys'), flatten'(leaf(c),Ys'\[ ]); {Xs↦[a|Ys]}⟩ $\xrightarrow{\text{Reduce(1,2)}}$

⟨flatten'(leaf(c),Ys'\[ ]); {Xs↦[a,b|Ys']}⟩ $\xrightarrow{\text{Reduce(1,2)}}$

⟨*true*; {Xs↦[a,b,c]}⟩

The other three possible computations on the same initial goal would also be deterministic and yield the same answer substitution.

2.3 The operational view of the logical variable and unification

The main difference between logic programming and other computational models is the logical variable and its manipulation via unification.

The basic data-manipulation operation in logic programs — unification — results in a substitution. Operationally, a substitution can be thought of as a simultaneous assignment of values to variables, except that here:

- a variable can be assigned a value only once, and
- the value assigned can be itself another variable or a term containing variables.

The single-assignment property, the ability to assign one variable to another, and the ability to assign a term containing variables to a variable are all fundamental to logic programming, and are the source of many powerful logic programming techniques.

Since the basic computational step of a logic program requires the unification of a goal atom with the head of a clause, much of the effort in logic programming has been devoted to understanding both the implications of this operation and its efficient implementation. This study has led to the realization that many of the subcases of goal/clause unification correspond quite closely to basic data manipulation operations of conventional languages. For the logic programmer, this implies that these special cases can be used to achieve the effect of conventional data manipulation. For the logic programming language implementor this implies that unification can be implemented efficiently by compiling the special cases, when identifiable, into machine instructions that execute the more basic data manipulation operations.

The correspondence is illustrated in Figure 1. The left column enumerates basic data-manipulation operations of conventional languages such as Pascal or Lisp, with sample code fragments. The right column shows the corresponding special cases of unification, with the corresponding examples of goal and clause terms. In this figure $T = T'$ denotes the unification of the goal atom term T with the clause head term T'.

Note that the cases in the figure are not necessarily mutually exclusive. For example, the unification of a goal variable with a clause term both constructs the term and assigns it to the goal variable; the unification of a goal term with an incomplete clause term both tests for equality and performs data access.

2.4 Semantics of logic programs

We provide here definitions for some of the concepts used intuitively above.

Unification

A ***substitution*** is a function from variables to terms which is different from the identity function on a finite number of variables. A substitution θ is presented as the finite set of pairs $\{X_i \mapsto T_1, \ldots, X_n \mapsto T_n\}$, where $X_1, \ldots, X_n$ are the variables where θ is different from the identity function, and $T_i = \theta(X_i)$, $i = 1, \ldots, n$.

For any term T and substitution θ, $T\theta$ denotes the term obtained by replacing every variable X in T by $\theta(X)$. A term T is an ***instance*** of a term T' if $T = T'\theta$ for some substitution θ. For example, $f(X,a)$, $f(X,X)$, $f(a,a)$, $f(a,b)$, $f(g(Z),h(b))$ are all instances of $f(X,Y)$.

A substitution θ is ***more general*** than θ' if there is a substitution σ such that $\theta = \theta' \circ \sigma$, where $\circ$ denotes function composition. An equivalent condition is that

Conventional data manipulation operation	The corresponding special case of goal/clause unification
(Single-) Assignment $X := a$	Variable = Non-variable $X = a$
Equality testing $a = a?$	Term = Term $a = a$
Data access (e.g. *car* and *cdr* in Lisp, '.' in Pascal) $X := car([a,b,c])$, $Xs := cdr([a,b,c])$	Compound term = Incomplete compound term $[a,b,c] = [X \mid Xs]$
Data construction (e.g. *cons* in Lisp, *new* in Pascal) $Ys := cons(a,Xs)$	Variable = Compound term $Ys = [a \mid Xs]$
Parameter passing by value	Goal Term = Variable $f(a) = X$
Parameter passing by reference	Variable = Variable $X = Y$
No corresponding operation; similar to aliasing	Two variables = Same variable $(Y,Z) = (X,X)$

Figure 1: Basic data manipulation operations and the corresponding special cases in goal/clause unification

$T\theta'$ is an instance of $T\theta$ for any term T. For example, $\{X \mapsto Y\}$ is more general than $\{X \mapsto a, Y \mapsto a\}$ and $\{X \mapsto f(Z)\}$ is more general than $\{X \mapsto f(a)\}$.

A substitution θ is a ***unifier*** of two terms T_1 and T_2 if $T_1\theta = T_2\theta$. For example, the substitution $\{X \mapsto a, Y \mapsto f(a), Z \mapsto b\}$ is a unifier of $p(X,b)$ and $p(f(Y),Z)$, and so is the substitution $\{X \mapsto f(Y), Z \mapsto b\}$.

A substitution θ is a ***most general unifier*** (***mgu***) of T_1 and T_2 if it is a unifier of T_1 and T_2 and is more general than any other unifier of T_1 and T_2.

In the previous example the second unifier is the most general one. The most general unifier of $m(X,[X \mid Xs])$and $m(X',[a,b,c])$ is $\{X \mapsto a, X' \mapsto a, Xs \mapsto [b,c]\}$, and the most general unifier of $a([X \mid Xs], Ys, [X \mid Zs])$ and $a([a,b,c],[d,e],Zs')$ is $\{X \mapsto a, Xs \mapsto [b,c], Ys \mapsto [d,e], Zs' \mapsto [a \mid Zs]\}$.

In the previous examples there was one most general unifier. The two terms $f(X)$ and $f(Y)$ have two most general unifiers, $\{X \mapsto Y\}$ and $\{Y \mapsto X\}$.

A ***renaming*** is a substitution that permutes its domain. An example is $\{X \mapsto Y, Y \mapsto X\}$. It can be shown that all most general unifiers are equivalent up to renaming, i.e. if θ and θ' are two most general unifiers of some terms than there is a renaming ρ such that $\theta = \theta' \circ \rho$. In addition, it can be shown that if two terms have a most general unifier, then they have an idempotent most general unifier, i.e. an mgu θ for which $\theta = \theta \circ \theta$.

We define a function ***mgu***, which takes two terms and returns their set of idempotent most general unifiers, if there are any, and ***fail*** if there are none. Usually we do not care which mgu is employed; in such cases we write $mgu(T_1, T_2) = \theta$ instead of $\theta \in mgu(T_1, T_2)$.

For a detailed analysis of unification see [109]. The operational intuitions behind unification were elaborated in Section 2.3 above.

A transition system for logic programs

Transitions systems will be employed throughout this paper. We specify a transition system for logic programs, as well as general notions that will be used in subsequent transition systems for concurrent logic programs. The general style of the transition system is that of Pnueli [144]; the details are adapted from Gerth *et al.* [65].

Definition: Transition system for a logic program P.
We associate with every logic program P a *transition system* which consists of:

- A set of states.
  A *state* is a pair $\langle G;\theta \rangle$, where G (the *goal*) is either a sequence of atoms or *fail*, and θ is a substitution.
- A set of transitions.
  A *transition* is a function from states to sets of states. For states S, S' and transition t, we denote that $S' \in t(S)$ by $S \xrightarrow{t} S'$. The set includes the Reduce and Fail transitions defined in Section 2.1 above. ∎

Definition: Enabled transition, terminal state, success state, failure state.

- A transition t is *enabled* on a state S if $t(S)$ is non-empty.
- A state on which no transition is enabled is called a *terminal state*. A terminal state of the form $\langle true;\theta \rangle$ is called a *success state*, and $\langle fail;\theta \rangle$ a *failure state*. ∎

Definition: Computation
A *computation of a program P on a goal G* is a (finite or infinite) sequence of states

$$c = S_1, S_2, \ldots$$

satisfying:

- Initiation: $S_1 = \langle G;\ \varepsilon \rangle$, where ε is the empty substitution.
- Consecution: For each k, $S_{k+1} \in t(S_k)$ for some transition t.
- Termination: c is finite and of length k only if S_k is terminal. ∎

Definition: Partial computation, partial answer substitution.
Any prefix of a computation is called a *partial computation*. The *partial answer substitution* of the partial computation $\langle G,\varepsilon \rangle, \ldots, \langle G',\theta \rangle$ is θ restricted to variables of G. ∎

Soundness and completeness of the transition system

A rule that governs the And-nondeterministic choices, i.e. the choice which goal atom to reduce next, is called a *computation rule*. Formally, it is a function from a goal to one of its constituent atoms. A computation obeys a computation rule if the goal atom selected at each transition is the one specified by the rule.

Theorem: Independence of the computation rule [19,122].
Let P be a program and R a computation rule. If P has a successful computation on a goal G with answer substitution θ, then it has a successful computation on G with answer substitution θ that obeys R.

The transition system for logic programs realizes, in effect, a proof procedure for logic programs. Each Reduce transition is actually an application of an inference rule, called SLD-resolution [80,122], which is a special case of Robinson's resolution inference rule [149]. SLD-resolution, and hence the transition system, have soundness and completeness properties that link their operational view to the logical view of of logic programs [122].

Notation: If A is an atom or a clause with variables $X_1,X_2,\ldots,X_n$, $(\forall)A$ denotes $(\forall\ X_1,X_2,\ldots,X_n)A$. If P is a program with clauses $C_1,C_2,\ldots,C_n$ then $(\forall)P$ is the conjunction $(\forall)C_1 \wedge (\forall)C_2 \wedge \cdots \wedge (\forall)C_n$.

Theorem: Soundness and completeness of SLD-resolution [80,19,122].

Let P be a program and A an atom.

1. (Soundness): If P has a computation on the initial goal A with answer substitution θ, then $(\forall)A\theta$ is a logical consequence of $(\forall)P$.
2. (Completeness): If $(\forall)A'$ is a logical consequence of $(\forall)P$, where A' is an instance of the atom A, then there is a computation of P on the initial goal A with answer substitution θ, such that A' is an instance of $A\theta$. ∎

Note that, in particular, if $(\forall)A$ is a logical consequence of $(\forall)P$, then there is a computation of P from A with answer substitution θ such that $A\theta$ is equal to A up to renaming.

The soundness theorem relates a successful computation with a proof of a goal. Given a program P, let $S_1 \xrightarrow{*} S_2$ denote that there is partial computation of P leading from S_1 to S_2. A partial computation from a unit goal can be viewed as a proof of a clause, whose head is the initial goal, instantiated by the partial answer substitution, and its body is the remaining goal, as shown by the following lemma:

Lemma: If $\langle G;\varepsilon\rangle \xrightarrow{*} \langle R;\theta\rangle$, then $(\forall)(G\theta \leftarrow R)$ is a logical consequence of $(\forall)P$. ∎

Hence every partial answer substitution can be thought of as a conditional answer to the query, whose condition is the yet-to-be-proved goal [147,206].

Program equivalence and observables

For simplicity, we assume the existence of some global vocabulary V, in which all programs and goals are written in.

A fundamental question in programming language semantics is when should two programs be considered equivalent. For example, correctness of program transformation can be studied only with respect such a notion of equivalence.

Usually, program equivalence is defined by assigning to each program a mathematical object, called its *meaning*, and defining two programs to be equivalent if they have the same meaning.

The meaning of a program is usually some abstraction of its possible computations. What is abstracted away and what is kept is, to some degree, arbitrary, and depends on what we wish to identify as the observable result of a computation. Hence the meaning of a program is sometimes referred to as its *observable behavior*, or, in case it is a set, as its *observables* for short.

In the case of logic programs there are several possible notions of equivalence. One considers successful computations. Define the *success set* of a program P to be the set of ground atoms from which P has a successful computation [41]. Two programs are *success set equivalent* if they have the same success set.

Success set equivalence does not capture differences in the answer substitutions computed by two programs. Define the *answer substitution set* of a program P to be the set of pairs $\langle G,\theta\rangle$ such that P has a successful computation from the goal G with answer substitution θ [45,110]. For a further discussion of equivalence of logic programs see [60,61]. Two program are ***answer-substitution equivalent*** iff they have the same answer substitution set.

Complexity

For a discussion of the parallel complexity of logic programs see [163,140].

2.5 Prolog

Prolog is a concrete programming language based on the abstract logic programming model. Prolog employs a procedural reading of logic programs, in which each goal atom is viewed as a procedure call, and each clause $A \leftarrow B_1, B_2, \ldots, B_n$ is viewed as a definition of a procedure, similar to:

```
procedure A
begin
        call B1,
        call B2,
          ⋮
        call Bn
end
```

Such a clause is interpreted: "To execute procedure A, call B_1 and call B_2 and ... and call B_n". Prolog uses unification to realize various aspects of procedural languages such as parameter passing by reference or by value, assignment, and data selection and construction, as was shown in Figure 1 above.

Formally, this operational behavior is achieved by employing a computation rule that selects the leftmost atom in a goal, thus eliminating And-nondeterminism. Instead of the Reduce transition of logic programs, Prolog employs the following transition rule:

- $\langle A_1, A_2, \ldots, A_n; \theta\rangle \xrightarrow{\text{Reduce}_{\text{Prolog}}} \langle (B_1, \ldots, B_k, A_2, \ldots, A_n)\theta'; \theta \circ \theta'\rangle$
  If $mgu(A_i, A) = \theta'$ for some renamed apart clause $A \leftarrow B_1, \ldots, B_k$ of P.

The resulting transition still incorporates Or-nondeterminism, which is interpreted in Prolog as implicit search for all solutions. That is, Prolog attempts to explore all computations from the initial goal, returning the answer substitutions of successful computations.

Most sequential Prolog systems compute the solutions to a goal by searching depth-first the computation tree induced by different choices of clauses. Typically, one solution is produced at a time, and additional solutions are searched for only by request. Under this behavior it is possible for a program to produce several solutions, and then diverge. The point of divergence is determined by the order of clause selection. Usually a Prolog program is defined as a sequence (rather than set) of clauses, and the order of clause selection is textual order.

The possibility of divergence in the face of both successful and infinite computations makes Prolog incomplete as a proof procedure for logic programs (see Section 2.4). However, this incompleteness is not a major problem in practice. Knowing the Prolog computation rule, Prolog programmers order bodies of clauses so that infinite computations are avoided on expected goals. In the example logic programs in Section 2.1 above, Prolog computations terminate on $sum(Xs,S)$ goals whose first argument is a complete list of numbers; on *in_both(X,L1,L2)* if both L_1 and L_2 are complete lists, and on *flatten(T,Xs)* if T is a complete tree.

Prolog is a convenient language for a large class of applications. However, to be practical it augmented the pure logic programming model with extra-logical extensions. The main purpose of these extensions is to specify input/output and to realize a shared modifiable store. See [173] for a discussion. As we shall see later, this deficiency is peculiar to Prolog, and is not inherent to the logic programming model. Indeed, concurrent logic programs can specify both input/output and shared modifiable store in a pure way, relying solely on their different computation rule and different interpretation of nondeterminism.

PART II. CORE CONCEPTS AND TECHNIQUES

3. Concurrent Logic Programming

Transformational vs. reactive languages

Prolog is a sequential programming language, designed to run efficiently on a von Neumann machine by exploiting its ability to perform efficient stack management. Sequential Prolog can be parallelized, and much research is devoted to effective ways of doing so [33,123,10,208]. Nevertheless, Prolog, whether executed sequentially or in parallel, should not be termed a concurrent programming language.

To understand why Prolog and other parallelizable sequential languages cannot be termed concurrent languages, it is useful to distinguish between two types of systems, or programs: transformational and reactive [71]. The distinction is closely related to the distinction between closed and open systems [79]. A transformational (closed) system receives an input at the beginning of its operation and yields an output at its end. On the other hand the purpose of a reactive (open) system is not necessarily to obtain a final result, but to maintain some interaction with its environment. Some reactive systems, such as operating systems, database management systems, etc., ideally never terminate, and in this sense do not yield a final result at all.

All classical sequential languages in general, and Prolog in particular, were designed with the transformational view in mind. These languages contain some basic interactive input/output capabilities, but usually these capabilities are not an integrated component of the language and sometimes, as in Prolog, are completely divorced from its basic model of computation.

It may seem that the distinction between transformational and reactive systems is not directly related to concurrent systems, and perhaps there could be concurrent transformational systems as well as concurrent reactive ones. Indeed, there are concurrent systems that exploit parallelism to achieve high performance in applications that are transformational in nature, such as the solution of large numerical problems. Following Harel [70], we call concurrent systems that are transformational as a whole *parallel systems*. However, if we investigate the components of any concurrent system — whether transformational or reactive as a whole — we find these components to be reactive; they maintain continuous interaction at least with each other and possibly also with the environment.

Hence, there seems to be a common aspect to all concurrent systems or algorithms, independently of what is their target architecture, and whether they exploit concurrency to achieve higher performance, physical distribution, or better interaction with their environment. The common aspect is that a language that describes and implements them needs to specify reactive processes — their creation, interconnection, internal behavior, communication and synchronization.

Don't-know and don't-care nondeterminism

Many abstract computational models are nondeterministic, including nondeterministic Turing machines, nondeterminitic finite automata, and logic programs. Reactive systems are also nondeterministic. However, the nature of nondeterminism in the former is very different from the one employed in the latter. Kowalski [103] adequately termed nondeterminism of the first type ***don't-know*** nondeterminism, and of the second type ***don't-care*** nondeterminism[3]. Don't-care nondeterminism is often called also ***indeterminism***, and we will use these two notions interchangeably.

The don't-know interpretation of nondeterminism implies that the programmer need not know which of the choices specified in the program is the correct one; it is the responsibility of the execution of the program to choose right when several transitions are enabled. Formally, this is achieved by specifying results of only successful computations as observable. Examples of such observables are the set of strings accepted by a nondeterministic finite automaton, or goal–answer substitutions pairs of successful computations of a logic program.

Don't-know nondeterminism is a very convenient tool for specifying transformational closed systems, as witnessed by the Prolog language. However, it seems to be incompatible with reactive open systems. The essence of don't-know nondeterminism is that failing computations "don't count", and only successful computations may produce observable results. However, it is not possible, in general, to know in advance whether a computation will succeed or fail; hence a don't-know nondeterministic computation cannot produce partial output before it completes; and hence it cannot be reactive[4].

The don't-care interpretation of nondeterminism, on the other hand, requires that results of failing computations be observable. Hence a don't-care nondeterministic computation may produce partial output (partial answer substitutions, in the case of concurrent logic programs) even if it is not known whether the computation will eventually succeed or fail.

Don't-care nondeterminism seems to be unnecessary, sometimes even a nuisance, in the specification of transformational systems, but as we shall see it is essential in the specification of concurrent reactive systems.

Although the nondeterminism of abstract computational models is commonly interpreted as don't-know nondeterminism, such models are also open to the don't-care interpretation. For example, nondeterministic finite automata can be used to specify either formal languages [88] (don't-know nondeterminism), or finite-state reactive systems (don't-care nondeterminism) [126]. The logic programming model is also open to these two interpretations. Prolog takes the don't-know interpretation, whereas concurrent logic language, being geared for specifying reactive open systems, take the don't-care interpretation.

---

[3] Manna and Pnueli [126] call the first *existential nondeterminism* and the second *universal nondeterminism*.

[4] A related argument with a similar conclusion is given by Ueda [203].

Formally, the two interpretations of nondeterminism induce different notions of equivalence on the set of programs. Assume some notion of equivalence of two (either failing and successful) computations. For example, in logic programs two computations on the same initial goal are equivalent if they have the same answer substitution and same mode of termination. Under the don't-know interpretation, two programs are equivalent if they have equivalent successful computations. Under the don't-care interpretation, two programs are equivalent if they have equivalent computations, whether successful or not.

We emphasize that concurrent logic languages are not unique in adopting the don't-care interpretation of nondeterminism. Rather, almost all models of concurrency and concurrent programming languages, including CSP [86,87], CCS [130], UNITY [16], Occam [91], Ada, and others, take this approach as well. The difference is that concurrent logic languages have as an ancestor an abstract nondeterministic computational model — namely logic programs — whose nondeterminism can be interpreted both as don't-know and as don't-care. The other concurrent models and languages do not have related models or languages which incorporate don't-know nondeterminism, hence for them the questions addressed here are usually not raised.

One active research direction in logic programming explores parallel (non reactive) languages that incorporate both don't-know and don't-care nondeterminsm [210,211,152,155,156,159,72,8,
181]. The goal of these languages it to execute logic programs more efficiently by exploiting determinism, more sophisticated control, and parallelism. This research direction is outside the scope of the survey. It is discussed further in Chapter 21.

What are concurrent logic languages?

Concurrent logic languages are logic programming languages that can specify reactive open systems, and thus can be used to implement concurrent systems and parallel algorithms. A concurrent logic program is a don't-care nondeterministic logic program augmented with synchronization. A logic program thus augmented can realize the basic notions of concurrency — processes, communication, synchronization, and indeterminism.

The process reading of logic programs [42], employed by concurrent logic programs, is different from the procedural reading employed by Prolog and mentioned in Section 2.5. In the process reading of logic programs each goal atom $p(T_1,\ldots,T_n)$ is viewed as a process, whose program state ("program counter") is the predicate p/n and data state ("process registers") is the sequence of terms $T_1,\ldots,T_n$. The goal as a whole is viewed as a network of concurrent processes, whose process interconnection pattern is specified by the logical variables shared between goal atoms. Processes communicate by instantiating shared logical variables and synchronize by waiting for logical variables to be instantiated. This view is summarized in Figure 2.

The possible behaviors of a process are specified by guarded Horn clauses, which have the form:

$$Head \leftarrow Guard \mid Body.$$

A guarded Horn clause is similar to an alternative in a guarded command [38]. The head and guard specify the conditions under which the Reduce transition can use the clause, as well as the effect of the transition on the resulting state. This is explained further below. The body specifies the state of the process after taking the transition: a

Process model	Concurrent logic programming model
Process	Goal atom
Process network	Goal (collection of atoms)
Instruction for process action	Clause (See Figure 3)
Communication channel; Shared location	Shared logical variable
Communication	Instantiation of a shared variable
Synchronization	Wait until a shared variable is sufficiently instantiated

Figure 2: The process reading of logic programs

Halt:	$A \leftarrow G \mid true.$
Change (data and/or program) state (i.e., become a different process):	$A \leftarrow G \mid B.$
Become k concurrent processes:	$A \leftarrow G \mid B_1,\ldots,B_k.$

Figure 3: Clauses as instructions for process behavior

process can halt (empty body) change state (unit body), or become several concurrent processes (a conjunctive body). This is summarized in Figure 3.

Concurrent logic languages employ the don't-care interpretation of nondeterminism. Intuitively, this means that once a transition has been taken the computation is committed to it, and cannot backtrack or explore in parallel other alternatives. Formally, this is realized by making observable partial results of the computation, as well as the final results of both successful, failing, and deadlcoked computations [65,59], as explained in Section 9 below.

The head and guard of a guarded clauses specify conditions on using the clause for reduction. A guarded clause can be used to reduce a goal atom only if the conditions specified by the head and the guard are satisfied by the atom. Concurrent logic languages differ in what can be specified by the head and the guard. A flat concurrent logic language incorporates a set of primitive predicates; in the languages surveyed these include mainly equality, inequality and arithmetic predicates. A guard in a flat language consists of a (possibly empty) sequence of atoms of these predicates. In a non-flat language, on the other hand, the guard may contain both primitive and defined predicates, and thus guard computations may be arbitrarily complex. Since guards of a non-flat language are recursively defined by guarded clauses, a computation of it forms an And/Or-tree of processes. In a flat language the processes are a "flat" collection; hence their name. Flat languages have received most of the recent attention of researchers, because it was found that their simplicity and amenability to efficient implementation come at a relatively low cost in expressiveness and convenience, when compared to non-flat languages (discussed in Section 18).

Concurrent processes communicate by instantiating shared logical variables, and synchronize by waiting for variables to be instantiated. Variable instantiation is realized in most concurrent logic language by unification. Three approaches were proposed to the specification of synchronization in concurrent logic programming: input matching (also called input unification, one-way unification, or just matching) [20,24,199], read-only unification [162], and determinacy conditions [211]. All share the same general principle: the reduction of a goal atom with a clause may be suspended until the atom's arguments are further instantiated. Once the atom is sufficiently instantiated, the reduction may become enabled or terminally disabled, depending on the conditions specified by the head and guard. Since input matching is the simplest and most useful synchronization mechanism, we pr sent it here and defer the discussion of the others till the languages that employ them are introduced.

The matching of a goal atom A with a head of a clause $A' \leftarrow G|\ B$ succeeds if A is an instance of A'; in such a case it returns the most general substitution θ such that $A = A'\theta$. It fails if the goal atom and the head are not unifiable. Otherwise it suspends. More precisely,

$$match(A,A') = \begin{cases} \theta & \theta \text{ is the most general substitution such that } A = A'\theta \\ fail & \text{if } mgu(A,A') = fail \\ suspend & \text{otherwise.} \end{cases}$$

Unlike unification, there is only one most general matching substitution. Using matching for reducing a goal with a clause delays the reduction until the goal is sufficiently instantiated, so that its unification with the clause head can be completed without instantiating goal variables. Examples are given in Figure 4.

Goal	Clause head	Result
$p(a)$	$p(X)$	$\{X \mapsto a\}$
$p(X)$	$p(a)$	*suspend*
$p(a)$	$p(b)$	*fail*
$sum([1\|In],Out)$	$sum([X\|Xs],S)$	$\{X \mapsto 1, Xs \mapsto In, S \mapsto Out\}$
$sum(In,Out)$	$sum([X\|Xs],S)$	*suspend*
$sum([\,],Out)$	$sum([X\|Xs],S)$	*fail*

Figure 4: Examples of input matching of goals with clause heads

The dataflow nature of matching is evident: an "instruction" (clause) is enabled as soon as sufficient "data" (variable instantiations) arrive. Although simple, matching is found in practice sufficiently powerful for all but the most complex synchronization tasks, as demonstrated by the programming techniques in Section 7.

Languages in the concurrent logic programming family differ mainly in the capabilities of their output mechanism. On one end of the spectrum there are languages that allow only matching prior to clause selection and perform unification past clause selection. On the other end there are languages which allow both matching and unification as tests prior to such a commitment. Test unification in its most general form subsumes powerful synchronization mechanisms used in more conventional models such as multiple simultaneous test-and-set and CSP-like output guards.

These differences and others are further elaborated upon when discussing the various languages in Part III of the paper. Until then we concentrate on the common aspects of the family.

4. FCP(|) — A Simple Concurrent Logic Programming Language

We illustrate the various aspects of concurrent logic programming discussed in the previous section using a simple concurrent logic language, FCP(|) (read "FCP-commit")[5]. FCP(|) is closely related to Flat GHC [199] and to Oc (read "Oh see!") [81,83]. We use FCP(|) as the introductory language instead of the more familiar language Flat GHC since its definition is simpler, and since it can more easily express some of the programming techniques related to distributed termination detection, discussed in Section 7. However all programs shown in Sections 5 and 7 are legal Flat GHC programs as well, and, except for the termination detection programs, the difference between the behavior of these programs under the operational semantics of Flat GHC and of FCP(|) is immaterial. See the discussion of Flat GHC in Section 10.

4.1 Syntax

Definition: Guard test predicates, guarded clause, FCP(|) program.

- We assume a fixed finite set of guard test predicates, including $integer(X)$, $X < Y$, $X = Y$ and $X \neq Y$. The predicates assumed in this paper are given in Section 4.2 below.
- A *guarded clause* is a formula of the form:

  $$A \leftarrow G_1,\ldots,G_m \mid B_1,\ldots,B_n. \qquad m,n \geq 0,$$

  where A, $G_1,\ldots,G_m$, $B_1,\ldots,B_n$ are atoms, the predicate of each G_i, $i = 1,\ldots,m$ is a guard test predicate and the variables of G_i occur in A. If the quard is empty ($m = 0$) then the commit operator '|' is omitted. An empty body ($n = 0$) is denoted by *true*.
- An FCP(|) *program* is a finite sequence of guarded clauses, which contains the unit clause
  $X = X$ as the only clause with head predicate '='. ∎

Note: '=' is a primitive predicates in FCP(|) that cannot be redefined by a program. The reason for a program being a sequence of clauses, rather than a set, will become apparent when we discuss the *otherwise* predicate in Section 7.

4.2 Operational semantics

Modelling concurrency by interleaving atomic actions

We specify the behavior of concurrent logic programs in general, and FCP(|) programs in particular, using a transition system very similar to that of logic programs. In this standard approach [144,16], concurrency is modelled by the nondeterministic interleaving of the atomic actions of the processes participating in the computation. The approach requires, therefore, a precise specification of what is an atomic step of

[5] The nomenclature we use to describe concurrent logic languages is influenced by the one used by Saraswat [152], but is different from it.

execution, as differences in the grain of atomic actions may lead to radically different computational models. As we shall see, one of the major differences between the various concurrent logic languages is indeed the grain of their atomic actions.

Our transition system is not reactive: it does not model input from an outside environment. This is not a major drawback, since if we wish to investigate a reactive computation of a program P from a goal G, we can model the environment as another process G', whose behavior is specified by a program, say E, with predicates disjoint from P, and investigate computations of the program $P \cup E$ from the conjunctive goal (G, G') [59]. An alternative is to add an explicit input transition [120].

Modeling concurrency by interleaving is a common approach, which has the advantage of being simple and well understood. Its disadvantage is that concurrency is not explicit, and hence an interleaving model sometimes gives rise to artificial fairness problems, which are not present if the concurrency is explicit in the model. We defer the discussion of fairness to Section 6.

Guard test predicates and guard checking

The meaning of the guard test predicates is given via a fixed set of ground atoms T over these predicates. The predicates used in this paper and their meanings are:

$X{=}X$	for every ground term X.
$X{\neq}Y$	for every two ground terms X and Y which are not equal.
$integer(X)$	for every integer X.
$X{<}Y$	for every ground arithmetic expressions X and Y such that the value of X is less than the value of Y.
$X{\leq}Y$	for every ground arithmetic expressions X and Y such that the value of X is less than or equal the value of Y.
$X{=:=}Y$	for every ground arithmetic expressions X and Y whose values are the same.
$X \backslash{=:=}\ Y$	for every ground arithmetic expressions X and Y whose values are different.

There are three guard primitives $var(X)$, $unknown(X)$, and *otherwise*, whose semantics cannot be given simply by a set of ground atoms. It is discussed when the primitives are introduced. The set of guard test predicates of "real" concurrent logic languages is not much larger than the list above. See for example [101,172].

An atom is true in T if every instance of it is in T, and false otherwise. A conjunction of atoms is true in T if all its members are true in T, and false otherwise. The check of a guard G *succeeds* if G is true in T; it *fails* if no instance of G is true in T; and it *suspends* otherwise. In other words, the check suspends if some future instantiation of the guard may result in it being true. For example:

checking $integer(2)$ succeeds
checking $integer(a)$ fails
checking $integer(X)$ suspends

checking $3 < 5$ succeeds
checking $5 < 3$ fails
checking $3 < X$ suspends
checking $a < X$ fails

The clause try function

The only difference between the transition system of logic programs and of FCP(|) is that the Reduce and Fail transitions employ matching and guard checking instead of unification. The operation of matching and guard checking is captured by the clause try function, *try*.

The *mgu* function unifies the goal atom with the clause head, and returns a substitution or *fail*. The *try* function does the same if the goal atom is an equality and the clause is the equality clause $X{=}X$. Otherwise it matches the goal atom and clause head, and, if successful, checks the guard, instantiated by the matching substitution. It may return *suspend* if the matching or the guard check suspends. *try* is defined for equality goals as follows:

$$try(T_1{=}T_2, X{=}X) = mgu(T_1, T_2).$$

And for clauses whose head predicate is different from the equality predicate:

$$try(A,(A' \leftarrow G|B)) = \begin{cases} \theta & \text{if } match(A,A') = \theta \wedge \text{checking } G\theta \text{ succeeds} \\ fail & \text{if } mgu(A,A') = \theta \wedge \text{checking } G\theta \text{ fails} \\ & \vee\ mgu(A,A') = fail \\ suspend & \text{otherwise} \end{cases}$$

The transition system

The state of a computation of an FCP(|) program is, as in a logic program, a pair $\langle G;\theta\rangle$, where G is a goal and θ a substitution. A computation begins from an initial goal $\langle G;\epsilon\rangle$ and progresses using Reduce and Fail transitions, similar to computations of logic programs. The difference is that instead of unifying the goal atom with the clause head, the Reduce and Fail transitions use the clause try function, *try*, instead of the *mgu* function:

1. Reduce
   $\langle A_1,\ldots,A_i,\ldots,A_n;\theta\rangle \xrightarrow{\text{Reduce}} \langle (A_1,\ldots,B_1,\ldots,B_k,\ldots,A_n)\theta';\theta \circ \theta'\rangle$
   If $try(A_i,C) = \theta'$ for some renamed apart clause $C = A \leftarrow G \mid B_1,\ldots,B_k$ of P.
2. Fail
   $\langle A_1,\ldots,A_i,\ldots,A_n;\theta\rangle \xrightarrow{\text{Fail}} \langle fail;\theta\rangle$
   If for some i and for every renamed apart clause C of P, $try(A_i,C) = fail$.

Note that the *suspend* result of the try function is not used in the Reduce and Fail transitions. Its effect, therefore, is to prevent a goal atom from reducing with a clause and from failing. Specifically, if A is a goal atom for which $try(A,C){=}suspend$ for some clause C in P, and $try(A,C'){=}suspend$ or *fail* for every other clause C' in P, then A can participate neither in a Reduce transition nor in a Fail transition. Such a goal atom is called ***suspended***. A state consisting of a goal in which all atoms are suspended is terminal, as no transition applies to it. Such a state it is called a ***deadlock state***, and a computation ending in a deadlock state is called a ***deadlocked computation***[6].

The following lemma relates successful computations of an FCP(|) program to computations of the corresponding logic program.

[6] Note, however, that this terminology is appropriate only for a closed computation. In an open computation variables may be instantiated by the environment, and hence a more refined definition of deadlock is required, which takes into account which variables are accessible to the environment.

Lemma (Soundness of FCP(|)):
Let c be a non-deadlocked computation of an FCP(|) program P. Then there is a finite subset T' of T such that c is also a computation of the logic program $P \cup T'$.
Proof: Immediate. ∎

The opposite direction of this lemma is of course not true. In particular, the logic program can proceed with a Reduce transition on states which the corresponding FCP(|) program deadlocks.

Observables of concurrent logic programs

As explained in Section 3, the observables of a concurrent logic program reflect both successful and failing computations.

Definition: The ***observable behavior*** of a finite computation $c = \langle G_1,\varepsilon\rangle, \ldots, \langle G_n,\theta_n\rangle$ of a concurrent logic program P is the triple $\langle G_1,\theta,x\rangle$ where θ is the answer substitution of the computation (i.e. θ_n restricted to variables of G_1), and $x{=}G_n$ if $G_n{=}\mathit{true}$ or $G_n{=}\mathit{fail}$, and $x{=}\mathit{deadlock}$ otherwise. The ***observables*** of a concurrent logic program $P, [\![P]\!]$, are the set of observable behaviors of every finite computation c of P. ∎

4.3 Examples of concurrent logic programs

We show several simple examples of concurrent logic programs, written in FCP(|), that correspond to the logic programs shown in Section 2.2, and describe their behavior.

The following concurrent logic program defines the process *sum(Xs,S)*, which unifies S with the sum of the elements of the input stream *Xs*.

```
sum(Xs,S) ←                 %1
    sum'(Xs,0,S).

sum'([ ],P,S) ← P=S.        %2
sum'([X|Xs],P,S) ←          %3
    plus(X,P,P')
    sum'(Xs,P',S).
```

There are two differences between this program and the logic program for *sum* shown in Section 2.2. The first is that the base clause of *sum'* unifies the partial sum P with the answer S explicitly in the body. This is necessary since in FCP(|) the goal atom is matched with the clause head, not unified with it. The second is that the definition of *plus* has to be modified to reflect the direction of the computation. *plus* behaves as if defined by clauses of the form:

```
plus(0,0,X) ← X=0.
plus(0,1,X) ← X=1.
plus(2,2,X) ← X=4.
  ⋮
```

Note also that each clause has an implicit commit operator "|", which is omitted by our syntactic conventions since the guard is empty. In contrast to the logic program for *sum*, the concurrent logic program has only successful computations on an initial goal *sum(Xs,S)*, where *Xs* is a list of integers and S is a variable.

Consider the following partial computation:

$$\langle \text{sum([1,2],S)};\ \varepsilon\rangle \xrightarrow{\text{Reduce(1,1)}}$$

⟨sum′([1,2],0,S); ε⟩ $\xrightarrow{\text{Reduce(1,3)}}$

⟨plus(1,0,P), sum′([2],P,S); ε⟩ $\xrightarrow{\text{Reduce(2,3)}}$

⟨plus(1,0,P), plus(2,P,P′), sum′([ ],P′,S); ε⟩

At this state the logic program could reduce any of the three atoms. However, the concurrent logic program cannot reduce the second *plus* process: it is suspended until its second argument *P* is instantiated. A possible continuation of this partial computation is:

⟨plus(1,0,P), plus(2,P,P′), sum([ ],P′,S); ε⟩ $\xrightarrow{\text{Reduce(1,plus)}}$

⟨P=1, plus(2,P,P′), sum([ ],P′,S); ε⟩ $\xrightarrow{\text{Reduce(3,2)}}$

⟨P=1, plus(2,P,P′), P′=S; ε⟩ $\xrightarrow{\text{Reduce(3,=)}}$

⟨P=1,plus(2,P,S); ε⟩ $\xrightarrow{\text{Reduce(1,=)}}$

⟨plus(2,1,S); ε⟩ $\xrightarrow{\text{Reduce(1,plus)}}$

⟨S=3; ε⟩ $\xrightarrow{\text{Reduce(1,=)}}$

⟨*true*; {S↦3}⟩

The logic program for flattening a tree can also be easily turned into a concurrent logic program. The only syntactic change required is specifying the construction of the output difference-list explicitly, by an equality goal:

```
flatten(T,Xs) ←                    %4
    flatten′(T,Xs\[ ]).

flatten′(leaf(X),Xs\Ys) ←          %5
    Xs=[X|Ys].
flatten′(tree(L,R),Xs\Zs) ←        %6
    flatten′(L,Xs\Ys),
    flatten′(R,Ys\Zs).
```

The *flatten* process can operate on a tree that is provided incrementally, by a concurrent process, as the two clauses of *flatten′* suspend until it is known whether their first argument is *leaf*(_) or *tree*(_,_). It can also be connected to the *sum* process, which incrementally sums the list produced by *flatten*, as shown by the following computation:

⟨flatten(tree(leaf(17),leaf(19)),Xs), sum(Xs,S); ε⟩ $\xrightarrow{\text{Reduce(1,4)}}$

⟨flatten′(tree(leaf(17),leaf(19)),Xs\[ ]), sum(Xs,S); ε⟩ $\xrightarrow{\text{Reduce(1,6)}}$

⟨flatten′(leaf(17),Xs\Ys), flatten′(leaf(19),Ys\[ ]), sum(Xs,S); ε⟩ $\xrightarrow{\text{Reduce(1,5)}}$

⟨flatten′(leaf(19),Ys\[ ]), sum([17|Ys],S); {Xs↦[17|Ys]}⟩ $\xrightarrow{\text{Reduce(2,1)}}$

$\langle$flatten'(leaf(19),Ys\[ ]), sum([17|Ys],0,S); {Xs↦[17|Ys]}$\rangle \xrightarrow{\text{Reduce(2,3)}}$

$\langle$flatten'(leaf(19),Ys\[ ]), plus(17,0,P'), sum(Ys,P',S); {Xs↦[17|Ys]}$\rangle \xrightarrow{\text{Reduce(2,plus)}}$

$\langle$flatten'(leaf(19),Ys\[ ]), P'=17, sum(Ys,P',S); {Xs↦[17|Ys]}$\rangle \xrightarrow{\text{Reduce(2,=)}}$

$\langle$flatten'(leaf(19),Ys\[ ]), sum(Ys,17,S); {Xs↦[17|Ys]}$\rangle \xrightarrow{\text{Reduce(1,5)}}$

$\langle$sum([19],17,S); {Xs↦[17,19]}$\rangle \xrightarrow{\text{Reduce(1,3)}}$

$\langle$plus(19,17,P''), sum([ ],P'',S); {Xs↦[17,19]}$\rangle \xrightarrow{\text{Reduce(1,plus)}}$

$\langle$P''=36, sum([ ],P'',S); {Xs↦[17,19]}$\rangle \xrightarrow{\text{Reduce(2,2)}}$

$\langle$P''=36, P''=S; {Xs↦[17,19]}$\rangle \xrightarrow{\text{Reduce(1,=)}}$

$\langle$36=S; {Xs↦[17,19]}$\rangle \xrightarrow{\text{Reduce(1,=)}}$

$\langle$*true*; {Xs↦[17,19],S↦36}

Summing the elements of a tree can be done more efficiently by combining the two procedures, *flatten* and *sum*, into a single procedure *tree_sum*:

% *tree_sum(T,S)* ←
  S is the sum of values of the leaves of the tree *T*.

```
tree_sum(T,S) ←
    tree_sum'(T,0,S).

tree_sum'(tree(L,R),P,S) ←
    tree_sum'(L,P,P'),
    tree_sum'(R,P',S).
tree_sum'(leaf(X),P,S) ←
    plus(X,P,S).
```

This program spawns a network of linearly connected *plus* processes, which sum the leaf elements sequentially from left to right. A possible computation from the initial goal *tree_sum(tree(tree(leaf(17),leaf(19)),leaf(23)),S)* may contain the intermediate goal *plus(0,17,P)*, *plus(17,P',P'')*, *plus(23,P'',S)*, which is then reduced from left to right. (Note however that the leftmost *plus* process may be reduced even if the spawning of the *plus* processes to its right has not been completed yet.) The program demonstrates that in a recursively constructed process tree, leaf processes (*plus* processes in our case) can communicate directly even if they are not directly related.

Turning the *in_both* logic program into a concurrent logic program is more difficult, since it employs don't-know nondeterminism in an essential way: it guesses a member of one list, and verifies that it is also a member of the second. The following logic program does not need to guess which clause to use, if the two lists are given in the initial goal, as in the goal *in_both(Xs,[1,2,3,4],[3,4,5,6])*. In such a case all computations of the following program are deterministic: each goal atom created during the computation of the program unifies with exactly one clause head. The program employs the difference-list technique.

% *in_both*(Xs,L_1,L_2) ←
The list Xs contains the members of both lists L_1 and L_2.

```
in_both(Xs,[X|L1],L2) ←
    member(X,L2,Xs\Xs'),
    in_both(Xs',L1,L2).
in_both([ ],[ ],_).
```

% *member*$(X,L,Xs\backslash Xs')$ ←
X is a member of L and $Xs = [X|Xs']$ or X is not a member of L and $Xs = Xs'$.

```
member(X,[X|L],[X|Xs]\Xs).
member(X,[Y|L],Xs\Xs') ←
    X≠Y, member(X,L,Xs\Xs').
member(X,[ ],Xs\Xs).
```

In contrast to the previous program, this program returns the (possibly empty) list of elements common to the two input lists, rather than nondeterministically selecting a common single element if one exists, or failing if there is none. Here *in_both*(Xs,L_1,L_2) holds if Xs is the list of all elements common to L_1 and L_2. The multiplicity of elements in Xs is the same as in L_2. This program employs a difference-list to construct the output. *member*$(X,L_1,Xs\backslash Xs')$ holds if both X is in L_1 and X is the difference between Xs and Xs' (i.e. $Xs = [X|Xs']$) or if X is not in L_1 and the difference between Xs and Xs' is empty (i.e. $Xs = Xs'$).

Since the program is deterministic on the desired set of goals, it can be turned into a concurrent logic program quite easily. The following is such an FCP(|) program:

```
in_both(Xs,[X|L1],L2) ←
    member(X,L2,Xs\Xs'),
    in_both(Xs',L1,L2).
in_both(Xs,[ ],_) ← Xs=[ ].

member(X,[X|L],Xs\Xs') ← Xs=[X|Xs'].
member(X,[Y|L],Xs\Xs') ←
    X≠Y | member(X,L,Xs\Xs').
member(X,[ ],Xs\Xs') ← Xs=Xs'.
```

Intuitively, the program operates as follows. On a call *in_both*$(Xs,L1,L2)$ it spawns parallel *member*$(X,L2,Xs\backslash Xs')$ processes, one for each element of $L1$, using the recursive clause of *in_both*. Each of these processes searches down the list $L2$ for an element equal to its X. If it finds one, it returns X in the difference-list $Xs\backslash Xs'$, by unifying Xs with $[X|Xs']$. Otherwise it returns the empty difference-list by unifying Xs with Xs'. The difference-lists are implicitly concatenated into the output list by the recursive clause of *in_both*. The output list is closed by the base clause of *in_both*.

The program operates correctly even if the two input lists $L1$ and $L2$ are given incrementally, by some concurrent process. This is achieved since the program inspects them using matching (specified by clause heads), which suspends if the input list is still unavailable. The output, in contrast, is constructed using unification, specified explicitly by equalities in the body of the clauses.

The guard of the second clause of *member* ensures that the clause is selected only after it is determined that X is different from Y. In the other clauses the guard is empty and the commit operator is implicit.

4.4 The power of the logical variable in concurrent programming

The standard uses of the logical variable and unification were mentioned in Section 2.3. Concurrent logic programming extends its use also to process communication. By following specific conventions and protocols, a wide range of concurrent programming techniques can be a realized using shared logical variables, unification, and matching.

In this section we provide a glossary of the major uses of the logical variable in concurrent programming. These will be demonstrated by concrete examples throughout the paper. When applicable, references to the relevant sections are provided.

It is useful to view variables shared between processes and their instantiation in two complementary ways: as communication channels, which transmit message streams, and as shared locations, which are instantiated, possibly incrementally and cooperatively, to compound data structures. Both are explained below.

Shared logical variables as communication channels, and communication stream protocols

Given the single-assignment nature of logical variables, it may seem that a communication channel implemented by a shared logical variable might carry at most one message. In some sense this is true. A better way to understand the situation, however, is to view a shared logical variable as a Genie, who will grant you a single wish[7]. A good strategy to follow when encountering such a Genie is, of course, to request to have two wishes. This is realized by instantiating the logical variable to a list (*cons*) cell, whose head and tail are logical variables. The head may be used to send the current message. Its tail is a new variable, shared by the processes sharing the original variable, and can be used for subsequent communications *ad infinitum*, as in

$$Xs = [m_1|Xs1],\ Xs1 = [m_2|Xs2],\ Xs2 = [m_3|Xs3],\ \ldots$$

In this way multiple "wishes" are achieved by the multiplicity of elements in a single list.

Hence, when serving as a communication channel, a shared logical variable is typically instantiated to a stream of messages. Several protocols can be followed in constructing such a message stream. They differ in the number of processes sharing the variable, and whether they share the writing and/or reading of the stream. Useful stream communication protocols include:

- point-to-point communication (single-writer single-reader), e.g. *tree_sum* in Section 4.3 above.
- broadcast communication, (single-writer multiple-reader) (Section 7.7)
- duplex communication (two writers/readers, who use the stream both for bidirectional communication and for tight synchronization) (Section 14)
- many-to-one communication (multiple-writer, single-reader), (Section 15) and
- blackboard communication (multiple writer/reader, cooperatively reading and writing the stream) (Section 15).

These stream protocols require progressively stronger synchronization mechanisms. Only the first two can be implemented by all concurrent logic languages. The language properties required to realize the duplex protocol and multiple-producer protocols are described when the protocols are introduced.

[7] This analogy is due to Bill Silverman.

A single stream is not always the preferred data structure for communication. For high-volume many-to-one communication, nondeterministic merging of multiple single-writer streams is usually preferred over having multiple writers cooperatively produce a single stream, since it eliminates contention on the stream's tail. Stream merging is discussed in Sections 6 and 7. In addition, when the set of writer and the set of readers in a multiple-writer multiple-reader stream are disjoint, it is often the case that the total ordering of a stream imposes on its elements unnecessarily serialized sending and receiving of messages. In such a case, a more general data-structure, called *Channel*, may be appropriate [195]. A Channel is a partially ordered set of messages, which can be produced in parallel without contention, and can be read with the same degree of parallelism with which it was produced.

Incomplete-message protocols

Messages sent on a stream need not be ground (i.e. variable-free) terms. A message containing variables is called an *incomplete message*. The ability to send incomplete messages is perhaps the single most important reason for the added flexibility and expressiveness of concurrent logic programming languages over other concurrent languages. D.H.D. Warren once characterized Prolog as "pointers made easy". Adapted to concurrent logic programming, the slogan may read "communication channels made easy". Indeed, incomplete messages may be viewed as a structured and high-level way of dynamically allocating and distributing communication channels.

There are several useful protocols employing incomplete messages:

- *Back-communication protocol* (the "remote procedure call" effect) (Section 7.3). A sender sends a message with a newly allocated reply variable R, and waits for R to be instantiated. The receiver responds to the message by instantiating R to the reply. This protocol achieves essentially the effect of a remote-procedure-call mechanism [134], without adding any special constructs to the language.
  It is easy to program servers using many-to-one message streams and the back-communication protocol. There is no need for the server to know the identity of the client who sent the request, nor for specially programmed mechanisms to route replies back to the clients.
  If a server can serve multiple requests in parallel, then a many-to-many channel may be preferred to a stream.
- *Dialogue* (Section 7.4).
  The back-communication protocol constitutes only one round in a possibly longer dialogue. The reply to which the variable R is instantiated may contain another reply variable R', with which the original sender can continue the dialogue, repeating the back-communication protocol as long as desired by both parties.
- *Network formation protocol* (Section 7.2).
  Assume that two processes p and c sharing a variable X want each to become n processes $p_1,\ldots,p_n$ and $c_1,\ldots,c_n$, and establish a communication stream X_i between p_i and c_i, $i = 1,\ldots,n$. This can be achieved as follows. p sends to c the message $[X_1,\ldots,X_n]$, which contains n variables, and then creates n processes, providing the ith process with X_i. Upon receipt of the message c creates n processes and similarly provides its ith process with X_i.
  In a variant of this protocol, the stream $[X_1,\ldots,X_n]$ is constructed incrementally by p. The process c need not know the stream's length in advance, and can create the next process c_i when X_i is available. This technique is heavily used in the formation of recursive process networks, as described in Section 7.2.

- *Network reconfiguration protocol.*
  Assume a process p that shares a variable Q with a process q and a variable R with a process r. If p wants to establish direct communication between q and r it sends to q on the stream Q an incomplete message containing R. Once q receives that message, it can use R to communicate with r directly. Alternatively, p can send to both q and r a message containing a new variable X.
  This technique can be employed to form an arbitrary communication graph in a network, independently of how the network was created to begin with. In particular, in a recursively constructed network the communication graph need not follow the path of the recursion, and two "leaf" processes may communicate directly no matter how high up their common ancestor is (Section 4.3).
- *Bounded-buffer communication protocol* (Section 7.3).
  The basic stream communication protocol is asynchronous. However, using incomplete messages one can implement synchronized communication. For example, a k-bounded-buffer protocol, for $k \geq 1$, can be implemented using a single-writer single-reader stream of incomplete messages as follows [179]: Each message contains an acknowledgement variable. The reader acknowledges the message by instantiating the variable to some constant upon receipt. The writer does not send the $n{+}k^{th}$ message before receiving an acknowledgement for the nth message.

Shared logical variables as shared locations

A shared logical variable can also be viewed as a shared location that can be assigned a data structure, i.e. a logical term. The term may be compound, and its construction may proceed incrementally and in cooperation between the processes initially sharing the variable.

In the special case that the term is a stream we obtain the stream communication protocols described above. However, the fact that stream communication results in a data structure, and is not just a sequence of events that occur in time, has several ramifications. Two of them are:

- *Communication history can be kept and later examined.*
  A shared variable used as a communication channel is incrementally instantiated to a stream data structure, which contains the messages sent on it and their replies. Typically, a stream writer or reader iterates with the tail of the stream once its head was written or read, and eventually the head becomes inaccessible. Memory occupied by inaccessible data-structures is eventually reclaimed by garbage-collection. Alternatively the initial stream variable can be kept, either by the process communicating via the stream itself, or by a concurrent observer who shares the initial stream variable. The data structure kept by the observer can be used later for various purposes, such as debugging, logging, and recovery. The stream data structure reflects only the order in which messages were sent and their content, but does not record the order in which message subterms, including replies, were constructed. In addition, if a process communicates via several independent streams, their content cannot be used to determine the temporal relations between messages on different streams.
  For some applications this abstract form of communication history, represented by a stream term, is sufficient. If a more precise history of the computation is required, e.g., to diagnose transient timing bugs, a different technique for recording information about a computation, which is sufficient for its accurate reconstruction, can be used [119]. This is further discussed in Section 14.

- ***A message stream can be inspected and transformed.***
  A process may examine or transform its incoming stream before processing its messages, if it so desires. A simple illustration of this is the ability of a process to "send to self", a useful object-oriented programming paradigm. To do so a process prepends a message to its input stream and proceeds with the resulting stream.

The constructed term need not be a stream, however. For example, in the distributed database system of Reches *et al.* [146], a transaction is a tree-structured term that is constructed cooperatively by the user program and the database system. The user program constructs the term, possibly concurrently, out of terms corresponding to sub-transactions, leaving in it variables for the database system replies. The database system consumes the term, executing, possibly concurrently, the subtransactions, and instantiates the reply variables to the answers.

Another example is the type-checker of Yardeni [212]. In it, multiple processes cooperate in constructing a term representing a finite automaton that defines the type of the program being checked. The programming technique used is similar to the multiple-writer stream described in Section 15.

5. Basic Programming Examples and Techniques

This section examines the operational behavior of FCP(|) programs via examples, and illustrates basic concurrent logic programming techniques.

Writers and readers

A writer process $p(X)$, that unifies X with a and halts, can be defined using the single clause program:

$$p(X) \leftarrow X = a. \qquad \%1$$

A reader $c(X)$ that waits till X is a and then halts is defined by:

$$c(a). \qquad \%2$$

A computation starting from a writer and a reader connected by the variable X, progresses as follows:

$$\begin{array}{l}\langle p(X),\ c(X);\ \varepsilon\rangle \xrightarrow{\text{Reduce(1,1)}} \\ \langle X{=}a, c(X);\ \varepsilon\rangle \xrightarrow{\text{Reduce(1,=)}} \\ \langle c(a);\ \{X \mapsto a\}\rangle \xrightarrow{\text{Reduce(2,2)}} \\ \langle true;\ \{X \mapsto a\}\rangle\end{array}$$

The final state is a success state, and this is the only possible computation from that initial state.

Many reader may read the same value, giving the effect of a broadcast. The computation starting from the initial state:

$$\langle p(X), c(X), c(X), \ldots, c(X);\ \varepsilon\rangle$$

reduces $p(X)$, unifies $X = a$, and then reduces all the $c(a)$ goal atoms one by one in some arbitrary order.

Nondeterminism in writers and readers

A process $p_2(X)$ that nondeterministically chooses to unify X with a or with b is defined by:

$$p_2(X) \leftarrow X{=}a. \quad \%1$$
$$p_2(X) \leftarrow X{=}b. \quad \%2$$

There are two possible computations if we use the nondeterministic writer p_2 instead of p in the example above. A successful one, essentially identical to the one above, and a failing one:

$$\langle p_2(X),\ c(X);\ \varepsilon\rangle \xrightarrow{\textbf{Reduce(1,2)}}$$
$$\langle X{=}b,\ c(X);\ \varepsilon\rangle \xrightarrow{\textbf{Reduce(1,=)}}$$
$$\langle c(b);\ \{X{\mapsto}b\}\rangle \xrightarrow{\textbf{Fail(1)}}$$
$$\langle \mathit{fail};\{X{\mapsto}b\}\rangle$$

If instead of $c(X)$ we use a nondeterministic reader $c_2(X)$, which accepts either a or b as values for X:

$$c_2(a).$$
$$c_2(b).$$

then instead of failing this latter computation would proceed and terminate successfully.

The process $c_2(X)$ has two alternatives. Which one is taken is completely determined by its environment. $p_2(X)$ has also two alternatives. However, the environment has no effect on its choice. An intermediate example is the process $cp(X,Y,Z)$, which behaves as follows. If $X = a$ it unifies Z with a. If $Y = b$ it unifies Z with b. If both $X = a$ and $Y = b$ it nondeterministically chooses one of the two.

$$cp(a,Y,Z) \leftarrow Z{=}a.$$
$$cp(X,b,Z) \leftarrow Z{=}b.$$

Starting from the initial state:

$$\langle p_2(X),p_2(Y),cp(X,Y,Z);\varepsilon\rangle$$

there are several possible computations, depending on the choices of goal atoms and clauses. To focus on clause choices, assume that goal atoms are reduced from left to right. Then there are five possible computations. Three of the four choices of the two p_2 processes uniquely determine the behavior of $cp(X,Y,Z)$. For example, if $p_2(X)$ unifies $X = a$ and $p_2(Y)$ unifies $Y = a$ then cp must unify $Z = a$. If $X = b$ and $Y = a$, then cp fails and the computation fails. However, if $p_2(X)$ chooses $X = a$ and $p_2(Y)$ chooses $Y = b$, then cp has a choice: it can either reduce with the first clause, and unify $Z = a$, or with the second, and unify $Z = b$. Both computations are possible.

Streams: producers, consumers, transducers, distributors, and mergers

As mentioned in Section 3, a stream is a list constructed incrementally.

Stream producers

Assume a process $X := E$, which evaluates the arithmetic expression E when it becomes ground, and unifies X which its value (it can be defined in FCP(|) using more primitive arithmetic processes, such as *plus* shown in Section 4.3). A process *integers(From,To,Xs)*, which, given integers *From* and *To*, produces the stream [*From,From +1,...,To*], can be defined by:

% *integers(From,To,Ns)* ← *Ns* is the list of integers from *From* to *To*[8].

```
integers(From,To,Ns) ← From > To | Ns=[ ].
integers(From,To,Ns) ← From ≤ To | Ns=[From |Ns'],
    From' := From +1,
    integers(From',To,Ns').
```

A more interesting stream producer is *fib(N,Ns)*, which produces the elements of the Fibbonacci series less than or equal to *N*.

% *fib(N,Ns)* ← *Ns* is the Fibbonacci series less than or equal to *N*.

```
fib(N,Ns) ←
    fib'(N,0,1,Ns).

fib'(N,N1,N2,Ns) ← N < N1 | Ns=[ ].
fib'(N,N1,N2,Ns) ← N ≥ N1 | Ns=[N1|Ns'],
    N3 := N1+N2,
    fib'(N,N2,N3,Ns').
```

Stream consumers

A process which sums the elements of its input stream, was defined in Section 4.3.

The following process reads two vectors of equal length, represented by streams of numbers, and computes their inner product. It will form the building block of a matrix-multiplication program, shown in Section 7.2.

% *ip(Xs,Ys,S)* ← *S* is the inner product of *Xs* and *Ys*.

```
ip(Xs,Ys,S) ←
    ip1(Xs,Ys,0,S).

ip1([ ],[ ],P,S) ← P=S.
ip1([X|Xs],[Y|Ys],P,S) ←
    P' := P + X*Y,
    ip1(Xs,Ys,P',S).
```

Stream transducers

The following process multiplies its input stream by some integer, to produce an output stream:

% *multiply(In,N,Out)* ←
  Out is the stream resulting from multiplying each element of the stream *In* by *N*.

```
multiply([ ],N,Out) ← Out=[ ].
multiply([X|In],N,Out) ← Out = [Y|Out'],
    Y := X * N,
    multiply(In,N,Out').
```

The following transducer filters all multiples of an integer from a stream. It is a building block of the parallel Sieve of Eratosthenes, shown in Section 7.2.

8 The comments associated with concurrent logic programs explain only their logical reading; the reactive aspects are usually explained in the text.

% filter(In,P,Out) ←
Out is the stream resulting from deleting all multiples of *P* from the stream *In*.

```
filter([X|In],P,Out) ← 0 =/= X mod P | Out=[X|Out'],
    filter(In,P,Out').
filter([X|In],P,Out) ← 0 =:= X mod P |
    filter(In,P,Out).
filter([ ],P,Out) ← Out=[ ].
```

A stream distributor

The following stream distributor has one input stream and two output streams. If an input message *send(1,X)* is received *X* is sent on the first output stream, and if *send(2,X)* is received *X* is sent on the second output stream. A variant of this program is used in the *msg* message sending system shown in Section 7.2.

% distribute(In, Out$_1$*, Out*$_2$*)* ←
In is a stream of elements of the form *send(1,_)* and *send(2,_)*. *Out*$_1$ is the stream of *X*'s such that *send(1,X)* is in *In*, and *Out*$_2$ is the stream of *X*'s such that *send(2,X)* is in *In*.

```
distribute([send(1,X)|In],Out1,Out2) ← Out1=[X|Out1'],
    distribute(In,Out1',Out2).
distribute([send(2,X)|In],Out1,Out2) ← Out2=[X|Out2'],
    distribute(In,Out1,Out2').
distribute([ ],Out1,Out2) ← Out1=[ ], Out2=[ ].
```

A deterministic stream merger

The following process receives two ordered streams of integers, and produces an ordered merge of them. A variant of it is used in the solution to Hamming's problem and in mergesort in Section 7.

% omerge(In$_1$*,In*$_2$*, Out)* ←
If *In*$_1$ and *In*$_2$ are ordered streams of numbers, then *Out* is an ordered merge of *In*$_1$ and *In*$_2$.

```
omerge([X|In1],[Y|In2],Out) ← X≤Y | Out=[X|Out'],
    omerge(In1,[Y|In2],Out').
omerge([X|In1],[Y|In2],Out) ← X>Y | Out=[Y|Out'],
    omerge([X|In1],In2,Out').
omerge([ ],In2,Out) ← In2=Out.
omerge(In1,[ ],Out) ← In1=Out.
```

6. Fairness

A nondeterministic stream merger

The following is a nondeterministic stream merger. Its output stream is some order-preserving interleaving of its two input streams.

% merge(In$_1$*,In*$_2$*,Out)* ← The stream *Out* is an interleaving of the streams *In*$_1$ and *In*$_2$.

```
merge([X|In1],In2,Out) ← Out=[X|Out'],
```

```
        merge(In1,In2,Out').
    merge(In1,[X|In2],Out) ← Out=[X|Out'],
        merge(In1,In2,Out').
    merge([ ],In2,Out) ← In2=Out.
    merge(In1,[ ],Out) ← In1=Out.
```

The nondeterministic *merge* process thus defined guarantees nothing about the rate in which it serves its two input streams. We say that a stream merger is *fair* if every message on its input stream is eventually written to its output stream. According to the semantics of FCP(|) defined in Section 4 above, the merger thus defined is fair. Furthermore, if the merger (or any other process) is executed in parallel with a nonterminating process, e.g. $p \leftarrow p$, then there is no guarantee that it will reduce at all.

A fairness requirement states conditions under which an event that may happen must eventually happen. The purpose of incorporating fairness requirements into the definition of a language is to provide the programmer with confidence that even in the presence of nondeterminism and unbounded computations certain program steps will eventually occur [50].

In concurrent logic programming it is useful to distinguish two types of fairness: And-fairness and Or-fairness. An *And-fairness* requirement states conditions under which a certain process would eventually be reduced; thus it constrains And-nondeterminism. An *Or-fairness* requirement states conditions under which a certain clause would eventually (not) be taken, thus constraining Or-nondeterminism.

An And-fairness requirement should guarantee, for example, that even in the presence of diverging processes, deterministic stream consumers will eventually read all their stream elements; similarly for producers. However, And-fairness cannot provide such a guarantee for nondeterministic consumers such as stream mergers or interrupt handlers. This is the purpose of Or-fairness requirements. Or-fairness requirements should allow, for example, specifying a fair stream merger and an interrupt handler in the language. Together, And-fairness and Or-fairness requirements should allow one to compose a controlling process and an interruptible process (e.g. in the style of the computation controller and the interrupt-handling meta-interpreter shown in Section 7) and guarantee that the controller process can interrupt the controlled process even if the latter is nonterminating. The following fairness requirements achieve this. Their definition can be skipped without loss of continuity.

<u>And-fairness</u>
For simplicity we restrict the discussion to computations whose initial state has a unit goal. Let P be a program and $c = \langle G;\varepsilon\rangle, \ldots$ be a computation from the unit goal G. Let b be the maximal number of atoms in the body of any clause in P. We label goal atoms in the computation with strings in $\{1,\ldots,b\}^*$. The initial goal is labeled with the empty string. Let A be a process labelled s, which is reduced using the clause $A' \leftarrow B_1,\ldots,B_k$. Then each of the new atoms B_i in the new goal is labeled with the string $s\hat{}i$.

We extend the Reduce transition label to contain p, the label of the reduced process, and θ, the try substitution restricted to variables of the process p, as in Reduce(p,θ). We extend the Fail transition label to contain the label of the failing process p, as in Fail(p).

Definition: And-fairness.

A computation c is ***And-fair*** if there is no Reduce(p,θ) transition or Fail(p) transition which is almost always enabled on the states of c.[9] ▮

Notes:

1) Since the state of a process changes in a monotonic way (it can only be instantiated further), if Reduce(p,θ) or Fail(p) are infinitely often enabled, they are also almost always enabled, hence there is no distinction in this case between weak fairness (also called justice) and strong fairness [50,144].
2) We have defined the fairness condition by ruling out certain computations allowed by the transition system. An alternative approach is to define a transition system which generates only fair computations to begin with. The approach of Costa and Stirling [32] to weak fairness in CCS can be applied here as well.

Or-fairness

In programs implementing stream merging and interrupt handling, complete freedom in clause selection would result in undesirable behaviors: a merger can ignore one of its input streams indefinitely; a process may ignore a message on its interrupt stream for indefinitely time. Several approaches to constraining clause selection in such programs were suggested; none of them seems completely satisfactory.

A global fairness requirement, which states that all clauses in a program satisfying certain conditions should be used eventually seems unreasonable, because of the dynamic nature of processes, and the fact that multiple processes may share the same set of clauses. Therefore approaches which specify conditions on the selection of a clause by a process were pursued.

One approach is to impose preference on clauses, and require the Reduce transition to select the most preferred clause among the applicable ones [162]. Assuming such a preference, specified by textual order, the following program implements a fair merger. It achieves fairness by switching the two input streams when an element from the preferred stream is read:

```
merge([X|In1],In2,Out) ← Out=[X|Out'],
    merge(In2,In1,Out').
merge(In1,[X|In2],Out) ← Out=[X|Out'],
    merge(In1,In2,Out').
merge([ ],In,Out) ← In=Out.
merge(In,[ ],Out) ← In=Out.
```

A process $p(\ldots,Is)$ that responds to interrupts on the stream Is can be defined by placing the clause testing for an interrupt first:

```
p(...,[I|Is]) ← interrupt_handler(...,[I|Is]).
... other clauses for p...
```

Although simple to define operationally, strict preferences are problematic. From a methodological point of view, they destroy the clause-wise modularity of programs. This may suggest awkward programming techniques (*à la* red cuts in Prolog [173]), and make the life of program analyzers, transformers and compilers more difficult. From an implementation point of view, preferences require strict synchronization, since a second clause can be selected only if nothing happens that may enable the first one. In fact, a correct distributed implementation of strict preferences may require locking all

---

9 i.e. enabled on all but a finite number of the states of c.

the variables involved in the reduction of the first clause while attempting to reduce with the second clause. A weaker notion of preference seems more desirable, but it is not clear how it should be defined.

Another approach is to use explicit conditions on clauses. Assume a guard primitive *var*(X), which succeeds if X is a variable, and fails otherwise. Using *var*, a fair merger can be written as follows, with base cases as above:

```
merge([X|In1],In2,Out) ← Out=[X|Out'],
    merge(In2,In1,Out').
merge(In1,[X|In2],Out) ← var(In1) | Out=[X|Out'],
    merge(In1,In2,Out').
```

A process $p(\ldots,Is)$ which is sensitive to interrupts on Is can be defined by adding to all its clauses which do not serve the interrupt the test *var*(Is). The use of *var* to achieve Or-fairness has been first proposed by Kusalik [105]. The *var* test approach is better than preferences since it does not destroy clause-wise modularity, and has no effect when not used. Its drawback is that *var* seems too strong a tool for this purpose. From a methodological point of view, it offers opportunities for abuse. From an implementation point of view, *var*, like preferences, implies tight synchronization. To implement correctly an interrupt-sensitive process thus defined, the Is variable has to be locked whenever a reduction of the processes using a clause with a *var*(Is) test is attempted. If there are hundreds or thousands of such processes, all sharing the same interrupt stream Is, this would be prohibitively expensive.

For that reason a weaker primitive, called *unknown*(X), was defined [192]. Intuitively, *unknown* is similar to *var*, except that its definition allows *unknown*(X) to 'ignore' for some finite, but bounded, amount of time the fact that X was instantiated. For example, if X is instantiated to a in processor P_1, *unknown*(X) can succeed after that time in another processor P_2. However, the value of X should eventually reach P_2, preventing *unknown*(X) tests from succeeding thereafter.

In our interleaving based transition system, this intuitive definition is formalized as follows: If X is instantiated to a value in the nth state of a computation, then there exists $n' \geq n$ such that *unknown*(X) succeeds in all states up to n and fails in states following u'. The behavior of *unknown*(X) in states between n and n' is indeterminate. Using *unknown* instead of *var* in the above programs would achieve the desired effect: the merger would be fair, and the process p would eventually respond to an interrupt. This is achieved without heavy synchronization costs, and without giving the programmer too powerful a tool, since *unknown* (X) succeeding *does not* imply that X is presently not instantiated. In a language without atomic variables (see the discussion of Flat GHC in Section 10), the difference between *var* and *unknown* is immaterial.

Note that unlike the other guard test predicates introduced, the success of the *var* and *unknown* primitives is defined operationally, without reference to the notion of truth.

7. Advanced Concurrent Logic Programming Techniques

The power of concurrent logic programming languages comes from the wide range of concurrent programming techniques they support. To convey this, we have assembled a range of FCP(|) programs which demonstrate these techniques.

7.1 Static process networks

Processes operating on streams can be composed into networks. This section shows two examples of a static process network.

A static network of stream transducers: a solution to Hamming's problem

The following program [42] solves the so-called Hamming's problem [39]: generate an ordered stream of all numbers of the form $2^i 3^j 5^k$ without repetition.

% *hamming(Xs)* ← *Xs* is the ordered stream of all numbers of the form $2^i 3^j 5^k$.

```
hamming(Xs) ←
    multiply([1|Xs],2,X2),
    multiply([1|Xs],3,X3),
    multiply([1|Xs],5,X5),
    omerge'(X2,X3,X23),
    omerge'(X5,X23,Xs).
```

where *omerge'* is a variant of *omerge* shown in Section 5, which removes duplicates from its input stream.

% *omerge'*(In_1,In_2,*Out*) ←
If In_1 and In_2 are ordered streams of numbers, then *Out* is an ordered merge of In_1 and In_2 with duplicates removed.

```
omerge'([X|In1],[X|In2],Out) ← Out = [X|Out'],
    omerge'(In1,In2,Out').
omerge'([X|In1],[Y|In2],Out) ← X<Y | Out=[X|Out'],
    omerge'(In1,[Y|In2],Out').
omerge'([X|In1],[Y|In2],Out) ← X>Y | Out=[Y|Out'],
    omerge'([X|In1],In2,Out').
omerge'([ ],In2,Out) ← In2=Out.
omerge'(In1,[ ],Out) ← In1=Out.
```

multiply was defined in Section 5.

A static network: the MSG message sending system

The *msg* process network is a simple message sending system for two computer terminals. Input from each of the keyboards K_1 and K_2 is a stream of messages, including messages of the form *message*(X). Every message M on K_1 is echoed on S_1 as *1*:M. In addition, a message of the form *message*(X) on K_1 is also echoed on S_2 as *1*:*message*(X). Similarly for messages on K_2 [93]. The program uses the *merge* process and a variant of the *distribute* process defined in Section 5.

% *msg(K1,S1,K2,S2)* ←
S_1 is an interleaving of *1*:X such that X is in K_1 and *2*:*message*(X) such that *message*(X) is in K_2. Similarly S_2 is an interleaving of *2*:X such that X is in K_2 and *1*:*message*(X) such that *message*(X) is in K_1.

```
msg(K1,S1,K2,S2) ←
    distribute(1,K1,K11,K12),
    distribute(2,K2,K22,K21),
    merge(K11,K21,S1),
    merge(K22,K12,S2).
```

```
distribute(Id,[message(X)|In],Out1,Out2) ←
    Out1=[Id:message(X)|Out1'],
    Out2=[Id:message(X)|Out2'],
    distribute(Id,In,Out1',Out2').
distribute(Id,[X|In],Out1,Out2) ←
    X≠message(_) |
    Out1=[Id:X|Out1'],
    distribute(Id,In,Out1',Out2).
distribute(Id,[ ],Out1,Out2) ← Out1=[ ], Out2=[ ].
```

7.2 Dynamic process networks

We show examples of dynamic process networks of various topologies. They are dynamic since their size depends on their input. Dynamic process networks which solve algorithmic problems typically exhibit two phases of operation. A spawning phase, in which the process network is spawned, and a "systolic" phase [104], in which the processes in the network perform both local computations and communication. It is interesting to note that many of the concurrent logic programs shown below, which implement "systolic"-like parallel algorithms, are almost identical, as logic programs, to Prolog programs which implement the corresponding sequential algorithms. More on the relation between systolic algorithms and concurrent logic programming can be found in [164].

Process pipes: linear process networks

The following program is a parallel implementation of the Sieve of Eratosthenes [165]. It consists of a process generating all integers in the desired range, and a set of filter processes, one per prime number found, which erase multiples of their prime from the remaining stream. This program overlaps the dynamic construction of the process network with the computation of the result. Its network consists of a dynamic linear pipeline of transducers. It uses the *integers* and *filter* processes defined in Section 5.

% *primes(N,Ps)* ← *Ps* are all the primes up to *N*.

```
primes(N,Ps) ←
    integers(2,N,Ns), sift(Ns,Ps).
```

% *sift(Ns,Ps)* ← *Ps* are the numbers in *Ns* which are prime relative to their predecessors.

```
sift([P|Ns],Ps) ← Ps=[P|Ps'],
    filter(Ns,P,Ns'), sift(Ns1,Ps').
sift([ ],Ps) ← Ps=[ ].
```

Mergesort

The following program implements a parallel mergesort algorithm. Given a list of length N of (possibly singleton) sorted lists, it forms a pipeline of length $\log_2 N$ of *msort* processes. Each stage in the pipeline performs a pairwise order-preserving merge of the sublists, using the *omerge* procedure defined in Section 5. Each stage doubles the length of the sublists and divides their number by approximately 2. Using $\log_2 N$ processors, this program can sort an N elements list in time $O(N)$. See [190] for further discussion of the complexity of concurrent logic programs.

% *mergesort(In,Out)* ←
If *In* is a list of ordered lists of numbers then *Out* is a sorted list of these numbers.

```
mergesort([ ],Ys) ← Ys=[ ].
mergesort([Xs],Ys) ← Xs=Ys.
mergesort([X1,X2|Xs],Ys) ←
    msort([X1,X2|Xs],Zs),
    mergesort(Zs,Ys).

msort([X1,X2|Xs],Ys) ← Ys=[Y|Ys'],
    omerge(X1,X2,Y),
    msort(Xs,Ys').
msort([X],Ys) ← Ys=[X].
msort([ ],Ys) ← Ys=[ ].
```

Vector-matrix multiplication
A linear array of *ip* processes can multiply a matrix, represented by a list of vectors, by a vector. It uses the *ip* process defined in Section 5.

% *vm(Xv,Ym,Zv)* ← multiplying the vector *Xv* by the matrix *Ym* gives the vector *Zv*.

```
vm(_,[ ],Zv) ← Zv=[ ].
vm(Xv,[Yv|Ym],Zv) ← Zv=[Z|Zv'],
    ip(Xv,Yv,Z), vm(Xv,Ym,Zv').
```

Process trees
The following program merges a list of streams into one stream, by creating a balanced tree of binary merge processes. It uses the merge process defined in Section 6.

% *merger(In,Out)* ← *Out* is the merge of the list of streams *In*.

```
merger([Xs1,Xs2|In],Out) ←
    merge_layer([Xs1,Xs2|In],Out'),
    merger(Out',Out).
merger([Xs],Out) ← Xs=Out.

merge_layer([Xs1,Xs2|In],Out) ← Out=[Ys|Out'],
    merge(Xs1,Xs2,Ys),
    merge_layer(In,Out').
merge_layer([Xs],Out) ← Out=[Xs].
merge_layer([ ],Out) ← Out=[ ].
```

Note that this program operates correctly only if the complete list of streams is given, since it emits elements from the root of the tree only after the construction of the tree is completed. If the list is given incrementally, i.e. it is actually a stream of streams, then a different approach is needed. One naive solution is to create an unbalanced tree incrementally, using the following program.

```
merger1([Xs|In],Out) ←
    merge(Xs,Out',Out),
    merger1(In,Out').
merger1([ ],Out) ← Out=[ ].
```

The program builds a right linear tree top down. At each point in its construction, elements of the input streams already connected to *merge* processes can reach the root of the tree.

More sophisticated balanced merge trees can be constructed, which support the dynamic addition and deletion of merged streams, using the concept of two-three trees, as shown in this section below.

Process arrays

Two matrices can be multiplied by an array of *ip* processes, each computing the inner product of the appropriate row and a column of the matrices. We assume that the second matrix is already transposed.

% *mm(Xm, Ym, Zm)* ←
Zm is the result of multiplying the matrix *Xm* with the transposed matrix *Ym*.

```
mm([ ],_,Zm) ← Zm=[ ].
mm([Xv|Xm],Ym,Zm) ← Zm=[Zv|Zm']
    vm(Xv,Ym,Zv), mm(Xm,Ym,Zm').
```

The program uses the *vm* process defined above.

The behavior of the *mm* program is reminiscent of the well-known systolic algorithm for multiplying two matrices [104]. However, there are two differences. First, the process network is created dynamically, to fit the size of the input matrices. This suggested the name[10] "soft-systolic" to these kinds of software-oriented systolic algorithms, in contrast with the classical hardware-oriented "hard-systolic" algorithms. Another difference is that the program, as specified, does not pipeline the matrices along the connections between the *ip* processes, but rather "broadcasts" each row and column to all processes requiring it. The program, however, can be easily modified to do the pipelining. See [165,190] for further discussions of the subject.

To achieve pipelining, the *mm* program has to be modified to form direct connections between adjacent *ip* processes. A similar goal is achieved by the following *torus* program. Given a matrix *Array*, represented as a list of list of values, it spawns a torus cell processes, each with one value and with communication links to adjacent *cell* processes. The array is augmented with end-round connections, to form a torus process network. This program schema (cliché) has several applications, including array relaxation and the like.

% *torus(Array)* ←
Spawns a torus of *cell* processes, one for each element of *Array*.

```
torus(Array,...) ←
    torus'(Array,Bottoms,Tops,...),
    Bottoms=Tops.

torus'([Row|Array],Bottoms,Tops,...) ←
    row(Row,Left,Right,Bottoms,Middles,...),
    Left=Right,
    torus'(Array,Middles,Tops,...).
torus'([ ],Bottoms,Tops) ←
    Bottoms=Tops.
```

[10] Due to Vijay A. Saraswat.

```
row([Element|Row],Left,Right,Bs,Ts,...) ←
    Bs=[Bottom|Bs'], Ts=[Tops|Ts'],
    cell(Element,Left,Middle,Bottom,Top,...),
    row(Row,Middle,Right,Bs,Ts,...).
row([ ],Left,Right,Bs,Ts,...) ←
    Left=Right, Bs=[ ], Ts=[ ].

cell(Element,Left,Right,Bottom,Top,...) ←
    An application specific program.
```

The layered stream method

Search problems that are amenable to depth-first search have elegant and efficient solutions in Prolog. Assume that the solution to the search problem is in the form of a list of elements, which satisfy some consistency criterion. The incremental construction of a solution in Prolog often relies on its backtracking mechanism, where forward computation consists of extending some prefix of a solution, and backtracking occurs when it is discovered that the prefix cannot be extended further, either because of inconsistencies or because it is a complete solution and additional solutions are required.

The layered stream data structure, proposed by Okumura and Matsumoto [141], allows an incremental and parallel construction of solutions to search problems without relying on a Prolog-like backtracking mechanism. A layered stream represents a set of lists sharing a common head as the cross-product of the head and the list of tails. The list of tails in turn can be represented by a layered stream. For example the lists *[1,2,4]*, *[1,2,8]*, *[1,3,9]* are represented by the layered stream *1* * [*[2,4]*, *[2,8]*, *[3,9]*], and also by the layered stream *1* * [*2**[[*4*],[*8*]], *[3,9]*]. The function symbol '*' is used for mnemonic purposes, but of course any other function symbol would do. The product $X*[\,]$ represents the empty set of lists, and the product $X*\mathit{true}$ represents X.

The following programming methodology is associated with the layered stream. Suppose the problem is to find values for N variables ranging over some finite domain of values V. $N*|V|$ processes are initially created, one for each possible value of each variable. Denote the process associated with value v of variable n by $p_{n,v}$. Each $p_{n,v}$ process receives an input stream of partial solutions which consist of an assignment to variables $1,\ldots,n-1$, and produces an output stream of partial solutions obtained by extending each input assignment with the assignment of v to variable n, provided the resulting assignment is consistent. That is, given the input partial solution $v_1,\ldots,v_{n-1}$, if the extended partial solution $v_1,\ldots,v_{n-1},v$ is consistent, it is written to the output stream. Otherwise no output is written.

The layered stream data structure allows all processes $p_{n,v}$, $v \in V$, to share the same input partial solutions, thus saving space and hence also time. It allows the pipelining of partial solutions, hence increases the available parallelism. An example of a search program using a layered stream is the four queens program shown below [141]. The program easily generalizes to N queens by replacing the explicit construction of the filter processes by iterative procedures to do so.

% ***four_queens(Qs)*** ←
  Qs is a layered stream of all legal assignments of four queens on a 4×4 board.

```
four_queens(Qs) ←
    queen(true,Qs1), queen(Qs1,Qs2), queen(Qs2,Qs3), queen(Qs3,Qs).
```

% ***queen(In,Out)*** ←

If *In* represents the set of legal assignments of *N* queens on a 4×4 board, *Out* represents the set of legal assignments of *N+1* queens on a 4×4 board.

```
queen(In,Out) ←
    filter(In,1,1,Out1), filter(In,2,1,Out2), filter(In,3,1,Out3), filter(In,4,1,Out4),
    Out=[1*Out1,2*Out2,3*Out3,4*Out4].
```

% ***filter(In,I,D,Out)*** ←
If *In* is a set of assignments of *N* queens to consecutive columns, *Out* is the set of assignments of *N+1* queens obtained as follows: Extend each assignment in *In* with a queen on the next column and on row *I*. If the added queen does not interfere with the previous queens, incorporate the extended assignment in *Out*.

```
filter(true,_,_,Out) ← Out=true.
filter([ ],_,_,Out) ← Out=[ ].
filter([I*_|In],I,D,Out) ← filter(In,I,D,Out).                     % Same row
filter([J*_|In],I,D,Out) ← D=:=abs(I−J) | filter(In,I,D,Out).     % Same diagonal
filter([J*In'|In],I,D,Out) ← I≠J, D\=:=abs(I–J) |                 % No interference
    D':=D+1,
    filter(In',I,D',Out'),
    filter(In,I,D,Out''),
    Out=[J*Out'|Out''].
```

The answer obtained from the goal *four_queens(Qs)* is the following layered stream:

```
Qs = [1 * [3 * [ ], 4 * [2 * [ ]]],
      2 * [4 * [1 * [3 * true]]],
      3 * [1 * [4 * [2 * true]]],
      4 * [1 * [3 * [ ]], 2 * [ ]]]
```

which represents the list of lists:

```
Qs = [ [2,4,1,3], [3,1,4,2] ].
```

A comparison of the sequential and parallel performance of this program with other concurrent logic programs and Prolog programs for the *N*-queens problem is given by Tick [194].

7.3 Incomplete message protocols

Incomplete message protocols were reviewed in Section 4.4. Here we show several examples of their application. The first application is monitors. A ***monitor*** (or ***server***) is a process that maintains some local state, and serves requests to inspect and/or modify the state. It is called so since its function is similar to Hoare's original concept of a monitor [85]. Clients of a monitor are typically connected to it via a merge network, and communicate with it using incomplete messages.

Perhaps the simplest monitor is the ***counter***, which maintains a local counter, and reponds to the messages ***clear***, ***add***, and ***read***(X), the last of which is an incomplete message.

% ***counter(In)*** ←
In is a stream of ***clear***, ***add***, and ***read***(X) such that X is the number of ***add***'s since the most recent ***clear***.

```
counter(In) ← counter'(In,0).
```

```
counter'([clear|In],C) ← counter'(In,0).
counter'([add|In],C) ← C' := C+1, counter'(In,C').
counter'([read(X)|In],C) ← X=C, counter'(In,C).
counter'([ ],C).
```

The client of a counter who wishes the know its value sends it the message *read*(X), and waits for X to be instantiated.

Shared queues

A more sophisticated monitor is the following queue process. It serves requests of the form *enqueue*(X) and *dequeue*(X), by unifying the arguments of corresponding enqueue and dequeue requests, and maintaining arguments of superfluous requests. While a list is a natural data structure for representing a stack, a difference-list is most convenient for representing a queue. The arguments of superfluous requests are maintained in a difference-list data-structure, which is positive if the queue received more enqueue than dequeue requests, empty if the number of requests received of each type were equal, and negative otherwise. A difference-list, explained in Section 2.2, is a common data-structure both in Prolog and in concurrent logic programming languages [27,162,173].

% *queue*(*In*) ←
  In is a stream of *enqueue*(X) and *dequeue*(X), for which the list of X's such that *enqueue*(X) is in *In* is identical to the list of X's such that *dequeue*(X) is in *In*.

```
queue(In) ←
    queue'(In,Q\Q).
queue'([dequeue(X)|In],H\T) ← H=[X|H'],
    queue'(In,H'\T).
queue'([enqueue(X)|In],H\T) ← T=[X|T'],
    queue'(In,H\T').
queue'([ ],H\T).
```

Another useful monitor is a priority queue. A priority queue has two input streams. One of enqueue requests of the form *enqueue*(X,P), and one of dequeue requests of the form *dequeue*(X). It maintains an internal priority queue, which is a list of the elements enqueued but not dequeued, sorted by their priority. It serves a dequeue request only if the queue is non-empty.

% *pqueue*(*Es*,*Ds*) ←
  Es is a list of *enqueue*(X,P) and *Ds* is a list of *dequeue*(Y) for the which the corresponding multisets of X's and Y's are the same, and if *dequeue*(X) precedes *dequeue*(Y) in *Ds* then either *enqueue*(X,PX) precedes *enqueue*(Y,PY) in *Es* or *enqueue*(Y,PY) precedes *enqueue*(X,PX) in *Es* and $PY<PX$.

```
pqueue(Es,Ds) ←
    pqueue'(Es,Ds,[ ]).
pqueue'([enqueue(X,P)|Es],Ds,Q) ←
    insert(X,P,Q,Q'),
    pqueue'(Es,Ds,Q').
pqueue'(Es,[dequeue(X)|Ds],[(Y,P)|Q]) ← X=Y,
    pqueue'(Es,Ds,Q).
```

```
pqueue'([ ],[ ],Q).

insert(X,P,[ ],Q) ← Q=[(X,P)].
insert(X,P,[(X',P')|Q],Q') ←
    P < P' | Q'=[(X,P),(X',P')|Q].
insert(X,P,[(X',P')|Q],Q') ←
    P ≥ P' | Q'=[(X',P')|Q''],
    insert(X,P,Q,Q'').
```

Merge trees

Another application of incomplete message protocols is network reconfiguration. We show here a simple example of a dynamic merge tree that grows according to the two-three tree insertion algorithm [2]. A process *p(Xs,...)* with a stream *Xs* to the tree can create a new process *q(Ys,...)* with a stream *Ys*, and join *Ys* to the tree by sending down *Xs* the message *merge(Ys)*. For example:

```
p(Xs,...) ←
    Xs=[merge(Ys)|Xs'],
    p(Xs',...),
    q(Ys,...).
```

A balanced tree of merge processes capable of handling such messages can be composed of binary and ternary mergers, defined below. We show the clauses which handle messages on the first input stream only. Handling messages on the other streams is done by similar clauses of the same procedure. Note how a *merge2* process that receives a *merge(X)* message turns into a *merge3*, and a *merge3* process that receives such a message turns into two *merge2*'s, and sends up another *merge* message.

```
merge2([X|Xs],Ys,Zs) ←
    X≠merge(_) | Zs=[X|Zs'],
    merge2(Xs,Ys,Zs').
merge2([merge(Ws)|Xs],Ys,Zs) ←
    merge3(Ws,Xs,Ys,Zs).
merge2([ ],[ ],Zs) ← Zs=[ ].

merge3([W|Ws],Xs,Ys,Zs) ←
    W≠merge(_) | Zs=[W|Zs'],
    merge3(Ws,Xs,Ys,Zs').
merge3([merge(Ws1)|Ws],Xs,Ys,Zs) ←
    Zs=[merge(Zs1)|Zs'],
    merge2(Ws1,Ws,Zs1),
    merge2(Xs,Ys,Zs').
merge3([ ],[ ],[ ],Zs) ← Zs=[ ].
```

The merge tree described grows dynamically, but does not shrink in a balanced way. An extension based on a distributed variant of the two-three tree deletion algorithm is described in [169].

The bounded-buffer protocol

In several situations it is desirable to allow the reader of a stream some control over its writer. Examples are when the reader is much slower than the writer, and when only some prefix of the produced stream is required, but its size can only be determined by the reader at runtime. The bounded-buffer protocol [179] employs difference lists and incomplete messages to realize this kind of control.

The idea of the bounded-buffer protocol is simple: the controlling reader process maintains a difference-list $H\backslash T$ of incomplete messages, say of the form *message*(X), where X is a variable. The difference-list represents the "buffer". After the buffer is initialized to a list of n incomplete messages, the reader operates as follows: when it is ready to process the next message, it waits until the first element in the buffer is known, i.e. H=[*message*(X)|H'], where X is known, dequeues it, and enqueues an incomplete message, *message*(X'), at the tail of the difference list. When it does not desire to receive any further messages it unifies the tail with nil. What to do in such a case with the messages pending in the buffer is application dependent.

The producer is given initially the head of the difference list as its input stream. It then operates as follows. It waits until its input stream has the message *message*(X), produces the next element, unifying it with X, and iterates with the tail of its input stream. It terminates when its input stream is nil.

Schematic programs for the producer and the reader are shown below.

```
bounded_buffer_network(...) ←
    buffer(n,H\T),
    read(H\T,...),
    produce(H,...).

% buffer(N,H\T) ←
    H\T is a difference list of message(_) of size N.

buffer(0,H\T) ← H=T.
buffer(N,H\T) ←
    N>0 |
    N':=N-1, T=[message(_)|T'],
    buffer(N',H\T').

read([message(X)|H]\T,...) ←
    known(X),
    ...I want more X's... |
    T=[message(_)|T'],
    ... process X...,
    read(H\T',...).
read(H\T,...) ←
    ...I don't want more X's... |
    T=[ ],
    ...process remaining messages in H....

produce([message(X)|In],...) ←
    ...produce X...,
    produce(In,...).
produce([ ],...).
```

Several variations on this protocol are possible. For example, it is not necessary for the reader to maintain a fixed size buffer: it can increase or decrease the size of the buffer if it so desires. It is not necessary to synchronize on every message: a more efficient protocol might be to produce k stream elements per incomplete message, or to provide a parameter in the incomplete message, specifying how many more elements to produce. Finally, it is possible for the incomplete message to be simply a variable, rather than a term containing a variable.

7.4 Mutual exclusion protocols

Mutual exclusion can be achieved in FCP(|) using the following mechanism. The set of processes participating in the mutual exclusion protocol are connected via a merge network into a *mutex* process. A single round mutual exclusion protocol is as follows: all processes competing for lock send a *lock(Reply)* incomplete message to *mutex*. *mutex* grants the first *lock* request received by unifying *Reply* = *granted*, and denies the other requests by unifying *Reply* = *denied*. It is defined as follows:

% *mutex(In)* ← *In* is a list containing one *lock(granted)* followed by zero or more *lock(denied)*.

```
mutex([lock(Reply)|In]) ← Reply=granted,
    mutex'(In).

mutex'([lock(Reply)|In]) ← Reply=denied,
    mutex'(In).
mutex'([ ]).
```

The single-round mutual exclusion protocol can be used to simulate CSP with input guards [87]. A simulation of CSP with both input and output guards is discussed in Section 14.

A multiple round mutual exclusion protocol is only slightly more complex. Instead of simple back-communication, it uses a three stage dialogue: the process requests the lock, then *mutex* grants it, then the process releases the lock, and *mutex* serves the next lock request. Processes competing for permission send *lock(Reply)* as before. *mutex* answers the first by *Reply* = *granted(Done)*, and waits for *Done* = *done*. When the process to which the lock was granted ends it critical operation, it releases the lock by unifying *Done* = *done*. *mutex* then grants the next lock, and so on. If the merge network is fair, and every process that is granted a lock eventually releases it, then every lock request will eventually be granted.

The definition of the multiple-round *mutex* process is as follows. Its trivial logical reading indicates that its interest lies in its reactive aspects only.

% *mutex(In)* ← *In* is a stream of *lock(granted(done))*.

```
mutex(In) ←
    mutex'(In,done).

mutex'([lock(Reply)|In],done) ← Reply=granted(Done),
    mutex'(In,Done).
mutex'([ ],_).
```

A program schema for a perpetual process *p* participating in a multiple-round mutual exclusion protocol is shown below. We assume that initially its first argument is a stream merged to *mutex*; other arguments are application specific.

```
p(ToMutex,...) ← p_request(done,ToMutex,...).

p_request(done,ToMutex,...) ← ToMutex=[lock(Reply)|ToMutex'],
    p_wait(Reply,ToMutex',...).
```

```
p_wait(granted(Done),ToMutex',...) ←
    ... do critical operation; when done, unify Done=done...
    p_request(Done,ToMutex,...).
```

7.5 Short-circuit protocols for distributed termination, quiescence detection, and distributed event-driven simulation

The problems of distributed termination detection and quiescence detection have received considerable attention [15,16,40,51,106,127]. In concurrent logic programming, these problems have very elegant solutions, using the short-circuit protocol. The protocol is originally due to Takeuchi [177], and was later extended by Weinbaum and Shapiro [209] and Saraswat *et al.* [160]; we largely follow [160] in the following discussion. The underlying behavior of implementations of this protocol are closely related to that of distributed termination and quiescence detection algorithms based on distributed counters [106,127]. We do not know of algorithms for distributed event-driven simulation corresponding to the one based on the short-circuit shown above.

Distributed termination detection

The idea of the short circuit for termination detection is as follows. Call the computation whose termination should be detected the underlying computation, and the program it executes the underlying program. Augment each process participating in the underlying computation with two additional arguments, called *Left* and *Right*. For readability, these arguments are typically packed in one term using the '–' infix function symbol, as in *Left–Right*. The pair is called a *switch*. It is *closed* if *Left*=*Right*, *open* otherwise.

Initially, connect all processes in a chain, by unifying *Right* of the ith process with *Left* of the ith+1 process. The *Left* of the first process and the *Right* of the last process are called the ends of the short-circuit. For n processes, the chain contains n open switches.

Each process in a computation operates as follows. If it halts it unifies its *Left* and *Right* variables. If it iterates it leaves them unchanged. If it creates n new processes, it extends the short-circuit by n–1 intermediate links. This behavior is achieved by transforming the clauses of the underlying program (except the equality clause) along the following schema, where '...' denotes underlying program arguments.

```
p(...).        ⇒ p(...,L-R) ← L=R.

p(...) ←       ⇒ p(...,L-R) ←
    p'(...).         p'(...,L-R).

p(...) ←       ⇒ p(...,L-R) ←
    p1(...),         p1(...,L-X1),
    p2(...),         p2(...,X1-X2),
    :                :,
    pn(...).         pn(...,Xn-1-R).
```

In FCP(|), a correct use of the short-circuit requires threading it to the equality goal atoms in a special way. If the underlying program has a body atom *T1*=*T2*, the transformed program should have the atom (*Left*,*T1*)=(*Right*,*T2*) for the appropriate switch variables *Left* and *Right*, so that the switch would not close before the underlying unification completes.

The invariant of the short circuit under this behavior is that the number of open switches is identical to the number of processes in the computation. In particular, all switches are closed, which implies that the two ends of the initial chain are identical, if and only if all processes in the computation have terminated, which is a stable property.

Any process wishing to detect that the computation has terminated is given the initial ends of the short circuit. Assume the termination detecting process is called ***halted(Left–Right,...)***. It can be implemented in FCP(|) in two ways:

```
halted(X–X,...) ← ... report termination ...
```

or:

```
halted(Left–Right,...) ←
    Left=done, wait_for_done(Right,...)
wait_for_done(done,...) ← ... report termination ...
```

Distributed phased termination detection

Some computations consist of phases, where a process is allowed to begin computations of the next phase only if all processes have completed the previous phase [131,209]. The short circuit can be generalized to achieve phased termination detection as well. Instead of having one short circuit, a stream of short circuits is threaded through the underlying computation. Each process is augmented with a ***Left–Right*** switch as before, and with the original left and right ends of the circuit, ***LeftEnd–RightEnd***. However, instead of unifying ***Left*** and ***Right*** upon termination, it treats ***Left*** and ***Right*** as streams. At the termination of a phase it unifies the head of ***Left*** with the head of ***Right***. Following that, it waits for the heads of ***LeftEnd*** and ***RightEnd*** to be identical before it proceeds with the next phase. This is achieved by the following iterative schema:

```
p(Left–Right,LeftEnd–RightEnd,...) ←
    ... do computation of this phase, when done, do the following ...,
    Left=[X|Left'],
    Right=[X|Right'],
    p_wait(Left'–Right',LeftEnd–RightEnd,...).
p_wait(Left–Right,[X|LeftEnd]–[X|RightEnd],...) ←
    p(Left–Right,LeftEnd–RightEnd,...).
```

Process creation and termination is handled as before.

Note that the solution is completely symmetric. There is no centralized process that detects the termination of a phase; rather, the ends of the circuit are distributed to all processes, and each of them detects the end of phase independently.

Quiescence detection

Consider a network of processes participating in some underlying computation by exchanging messages. The computation begins by a designated process, which sends one or more messages to other processes. Each process that receives a message sends out zero or more messages in response. No process spontaneously initiates new messages. The computation ends when all messages sent have been received, and no new response messages need to be generated. Normally, this results in a deadlock of the underlying computation. We would like to augment the underlying computation, so that instead of deadlocking it would report quiescence [40,15,16].

This can be achieved by another variant of the short circuit protocol. In this variant, switches are embedded in messages, rather than in processes. The initial set of messages are threaded with a short circuit, as was the initial set of processes above. A process wishing to detect quiescence holds the ends of the circuit and waits for them to become identical. Each message in the underlying computation is augmented with a switch, and each process in the underlying computation is augmented to obey the following protocol. When it absorbs a message, i.e. receives a message without generating any additional messages in response, it closes the switch in the message. When it sends one message in response to a message, it includes in the outgoing message the switch of the incoming message, intact. When it generates n response messages, $n>1$, it extends the switch into n switches, and embeds the new switches in the outgoing messages.

For simplicity, assume that each process has one input stream and one output stream of messages. Mergers and distributers can be attached to these streams if necessary. The schema of an augmented process is:

```
p([m(Left-Right,...)|In],Out,...) ←          % Absorb a message
    Left=Right,
    ...,
    p(In,Out,...).
p([m(Left-Right,...)|In],Out,...) ←          % Send one message
    ...,
    Out=[m'(Left-Right,...)|Out'],
    p(In,Out',...).
p([m(Left-Right,...)|In],Out,...) ←          % Send many messages
    ...,
    Out=[m1(Left-Middle1,...),
         m2(Middle1-Middle2,...),
         ...,
         mn(Middlen-1-Right,...)|Out'],
    p(In,Out,...).
```

The invariant of this protocol is that the number of open switches is the number of messages sent (or to be sent) but not yet received. When this number reaches 0, the short circuit is closed, and quiescence can be reported. Note that this protocol requires that each message has at most one receiver. To achieve broadcasting the underlying program must be augmented with explicit distributors, which follow the same protocol.

<u>Distributed event-driven simulation</u>

One interesting application of the above techniques is distributed event-driven simulation. In event-driven simulation, in contrast to clock-driven simulation, only changes are communicated between the components participating in the simulation. This is especially important in hardware simulation, where very often only a small percentage of the simulated device is active at any given time.

An event-driven simulation is phased, since changes which occur in the next phase can be reliably communicated only when all changes related to the previous phase have been received. The method for phased termination detection, using the stream of short circuits described above could be used, except that it requires every process participating in the simulation to be activated in every cycle in order to close its segment of the short-circuit, contrary to our goal. Our solution is a combination of the quiescence detection and phased termination detection techniques.

Each message is augmented with a stream of switches and the ends of the short circuit; these are the same data structures each process is augmented with in phased termination detection. In addition, each process is augmented to behave as follows. In each phase, the process treats the first message it receives as follows: It closes the head of its switch, and keeps the tail of the switch and the circuit's ends. It then waits either for the head of the circuit's ends to close, or for additional messages. (Note that only one of them can occur, since the head of the circuit's ends close only after all messages sent in this phase have been received.) If an additional message is received, it closes the message's entire switch, after verifyng that the message-circuit's ends are identical to the ones it maintains (this is necessary to ensure that the message belongs to the current phase; otherwise it is possible that this message was sent by a process that has already detected the end of the current phase and sent a message belonging to the next phase). If the head of the circuit's ends close, it sends zero or more messages, as required by the underlying computation, each with a segment of the tail of the switch, and with the tail of the circuit's ends.

A schema of such a process follows. For simplicity a process which sends out one message per phase is shown.

```
p_dormant([m(Left–Right,LeftEnd–RightEnd,...)|In],Out,...) ←
    Left=[X|Left'],        % received first message
    Right=[X|Right'],        % acknowledge receipt
    ...,        % process and store message
    p_passive(In,Out,Left'–Right',LeftEnd–RightEnd,...).
p_passive([m(Left1–Right1,LeftEnd–RightEnd,...)|In],
        Out,Left–Right,LeftEnd–RightEnd,...) ←
    Left1=Right1,        % received additional message of current phase
    ...,        % acknowledge receipt
                % process and store message
    p_passive(In,Out,Left–Right,LeftEnd–RightEnd,...).
p_passive(In,Out,Left–Right,[X|LeftEnd]–[X|RightEnd],...) ←
    ...,        % detect end of phase
                % compute outgoing message
    Out=[m(Left–Right,LeftEnd–RightEnd,...)|Out'],
    p_dormant(In,Out',...).
```

The reason for embedding the circuit's ends in messages is efficiency. If the ends were distributed to all processes in the network initially, a process receiving a message after being dormant for some time would have to search for the tail of the end's streams. In the current scheme it receives the updated tails in the message.

More details on this subject can be found in [160,209].

7.6 Object-oriented programming, delegation, and *otherwise*

Concurrent logic programming languages naturally give rise to an object-oriented programming style, where the objects are processes communicating via message streams. Much research was devoted to understanding the relation between classical object-oriented concepts and techniques and the object-oriented style offered by concurrent logic programming [96,171,95]. For a further discussion of object-oriented programming see Section 21.

One common object-oriented technique is delegation. A process that does not understand a certain message delegates it to another process, who may be better equipped to handle it. Consider a process *p(In,...,Out)*, which receives messages on *In*. Some messages it handles by itself; others are delegated to the *Out* stream. If the set of messages it recognizes is simple, say *a* and *b*, then *p* can be coded easily:

```
p([a|In],...,Out) ←
    ..., p(In,...,Out).
p([b|In],...,Out) ←
    ..., p(In,...,Out).
p([X|In],...,Out) ←
    X ≠ a, X ≠ b | Out=[X|Out'],
    p(In,...,Out').
```

However, if the messages are complex, and have arguments which should have specific combinations of values, then the explicit specification of conditions under which the message should be delegated becomes harder. To that effect a new guard primitive, called *otherwise*, is introduced. The operational semantics of *otherwise* is given assuming an ordering on clauses (say textual order). Given a goal atom G, an *otherwise* guard in a clause C succeeds if $try(G, C') = fail$ for every clause C' preceding C.

Using *otherwise*, defaults can be handled easily:

```
p([X|In],...,Out) ←
    otherwise | Out=[X|Out'],
    p(In,...,Out').
```

Otherwise destroys clause-wise modularity, and the explicit formulations of the conditions under which it succeeds is often cumbersome[11]. This is the source of its power, but also an indication that it should not be used excessively. *Otherwise* is best thought of as a primitive exception handling mechanism, which should be used only to handle exceptions, and not in normal programming practice.

7.7 Enhanced meta-interpreters

A meta-interpreter for a language L is an interpreter for L written in L. If a language has simple meta-interpreters, then one of the most convenient ways to enhance a language, or implement sublanguages, is by starting from a meta-interpreter and enhancing it [62,151,173,180,184]. There can be several meta-interpreters for a language, which differ in what aspects of the execution model they *reify*, i.e. execute themselves, and what aspects they *absorb*, i.e. default to the underlying languages. The most useful type of meta-interpreter in logic programming is the one that reifies goal reduction and absorbs unification.

---

[11] However, K. Kahn (personal communication, 1989) notes that there is a sense in which *otherwise* enables clause modularity. If a procedure needs to specify a default case, as in this example, which applies when all other clauses don't apply, then without *otherwise* it must encode explicitly the negation of the other guards, and should be updated if the other clauses change. However, by encoding the default with *otherwise* there is no textual dependency between the default clause and the other clauses of a procedure.

Another distinction is how the meta-interpreter is composed with the program to be interpreted. One method is to pass a data-structure representing the program as a parameter to the interpreter. This approach is the most flexible, but usually imposes unacceptable runtime overhead. On the other extreme, the meta-interpreter and the program to be interpreted can be bound together at compile time. This may give the most efficient result, especially if source to source transformation techniques, such as partial evaluation, are applied to the combined program (see below). This approach, however, is very inflexible.

The most common approach in logic programming, which is also taken here, is an intermediate one in terms of efficiency and flexibility. The program to be interpreted is compiled in a special way, and an interface to the meta-interpreter is provided. The interface determines which aspects of the computation are absorbed, and hence compiled efficiently, and which are to be reified by the meta-interpreter.

A plain FCP(|) meta-interpreter

We demonstrate the approach for FCP(|). Each clause

$$A \leftarrow G \mid B$$

of the FCP(|) program to be interpreted is transformed into the unit clause

$$clause(A,X) \leftarrow G \mid X = B'.$$

where B' is the conjunction obtained by replacing every goal atom G in B whose predicate is neither *true* nor '=' by the term *goal*(G).

For example, the *omerge* program is represented by clauses like the following:

```
clause(omerge([X|In1],[Y|In2],Out), B) ← X ≤ Y |
     B=(Out=[X|Out'],goal(omerge(In1,[Y|In2],Out'))).
clause(omerge([ ],In2,Out), B) ← B=(In2=Out).
```

Given such a representation, an FCP(|) meta-interpreter can be written as follows:

% *reduce*(*Goal*) ← *Goal* is reducible using the program defined by the *clause* predicate.

```
reduce(true).                                   %1
reduce(X=Y) ← X=Y.                              %2
reduce((A,B)) ← reduce(A), reduce(B).           %3
reduce(goal(A)) ← clause(A,B), reduce(B).       %4
```

The meta-interpreter reifies process termination (clause 1) spawning (clause 3) and reduction (clause 4). Note that the meta-interpreter interprets the parallel processes (*A*,*B*) in parallel, by forking into the two processes *reduce*(*A*) and *reduce*(*B*). It absorbs unification (clause 2) by calling FCP(|)'s primitive unification predicate when interpreting a unification goal. It also absorbs goal/clause matching and guard evaluation, since these are carried by the *clause/2* predicate.

A termination detecting meta-interpreter

The meta-interpreter described is not so interesting on its own right. However, it may be enhanced in several ways, to provide useful functionalities. One example is the following meta-interpreter, employing the short-circuit technique to detect the termination of the interpreted program. On the call *reduce*(*A*,*Done*), *Done* is unified with *done* when the computation of *A* successfully terminates.

% *reduce*(*Goal*,*Done*) ← *Goal* is reducible and *Done*=*true*.

```
reduce(A,Done) ←
    reduce'(A,done–Done).

reduce'(true,L–R) ← L=R.
reduce'(X=Y,L–R) ← (X,L)=(Y,R).
reduce'((A,B),L–R) ← reduce'(A,L–M), reduce'(B,M–R).
reduce'(goal(A),L–R) ← clause(A,B), reduce'(B,L–R).
```

One of the main weaknesses of FCP(|) is that, although it can reflect on termination, it cannot reflect on failure, without reifying unification. In other words, it is not possible in FCP(|) to enclose a computation within a meta-interpreter in the style shown above, which reports failure when the computation it interprets fails, without failing itself.

This problem is alleviated in more powerful languages such as FCP(:), as discussed in Section 14.

An alternative solution is to replace FCP(|)'s unification primitive with a three-argument predicate, which returns an indication whether unification succeeded or failed. This approach is taken by Fleng [135], and is critically discussed in Section 21.

Interrupt handling

Processes in FCP(|) are anonymous. Their number and rate of creation and termination renders any conventional operating system approach to process management infeasible. Therefore the implementation of standard operating system capabilities, such as the ability to suspend, resume, and abort processes requires novel solutions. The natural unit of control in concurrent logic programming is not a process, but a (reactive) computation.[12]

In the Logix system [172] several, possibly interacting, computations can proceed concurrently. We show below a meta-interpreter that can control an interpreted reactive computation by responding to control signals.

% *reduce(Goal,Is)* ←

Is is a stream of *suspend*, *resume* and *abort* messages. *Goal* is reducible or *Is* contains *abort*.

```
reduce(true,Is).
reduce(X=Y,Is) ← X=Y.
reduce((A,B),Is) ← reduce(A,Is), reduce(B,Is).
reduce(goal(A),Is) ← clause(A,B,Is), reduce(B,Is).
reduce(A,[I|Is]) ← serve_interrupt([I|Is],A).

serve_interrupt([abort|Is],A).
serve_interrupt([suspend|Is],A) ← serve_interrupt(Is,A).
serve_interrupt([resume|Is],A) ← reduce(A,Is).
```

The plain meta-interpreter is enhanced with an interrupt stream *Is*. Whenever an interrupt is sensed, an interrupt-handling routine is called. The interrupt handler can serve the messages *suspend*, *resume*, and *abort*. To ensure that an interrupt will eventually be served, even if the interpreted computation is non-terminating, the *unknown(Is)* guard should be added to all but the last clause of *reduce*. To ensure

[12] The notion of computation employed here is related to the one used in the semantic definitions, but is different from it in being reactive. See [120].

that even a suspended process responds to an interrupt, an additional clause is added to the representation of the interpreted programs:

clause(A,B,[I|Is]) ← A=B.

Its purpose is to return the interpreted process intact when an interrupt is sensed: If an interrupt is sensed, the *clause* process terminates and returns in the body argument the goal atom it was called with. This ensures that suspended goal atoms of the interpreted computation are halted rather than being left suspended. Once the computation is resumed, the process is retried. This feature is used for another purpose by the following snapshot meta-interpreters.

Repeated live snapshots

The problem of obtaining a snapshot of the state of a distributed computation has been investigated is various models [14,15]. The meta-interpreter shown above can be enhanced to obtain repeated snapshots of the interpreted computation, by treating the short-circuit as a (possibly empty) stream of snapshot requests. To obtain a snapshot, a message *state*([ ]) is sent down the left end of the short-circuit. A process P that senses a message *state*(S) on its left-end of the switch sends the message *state*($[P|S]$) on the right end of the switch. This is achieved by augmenting the termination detection meta-interpreter shown above with the clause [151] :

reduce(A,[state(S)|L]–R) ← R=[state([A|S])|R′],
reduce(A,L,R′).

When the message *state*(S) arrives at the right end of the circuit, it contains a list of processes.

There are several delicate points to note. First, as specified, the message is guaranteed to arrive eventually only if the interpreted computation terminates or deadlocks. To improve upon this the guard *unknown*(L) can be added to the other clauses of the meta-interpreter. This ensures that if the number of processes created in the computation is bounded (i.e. the number of times a clause with more than one atom in the body is used is finite), then the message would eventually arrive, even if the computation is nonterminating. To obtain a snapshot in a computation with unbounded process creation, the frozen snapshot technique, discussed below, must be used.

Second, the distributed fashion in which the live snapshot was obtained implies that the list of processes obtained is not necessarily a possible state that actually occured in a computation [160]. For example, process A could have been added to the snapshot, then reduced, performed a unification that enabled some other reduction, which created a process B, which was then added to the snapshot. So the live snapshot may contain two processes which are causally related, and therefore could never exist simultaneously. Furthermore, processes in the snapshot could appear more instantiated than they were when added to it, due to other processes reducing before the snapshot was completed. Nevertheless, under certain circumstances[13], a live-snapshot is restartable, in the following sense. If G has a successful computation, and G' is a live snapshot of this computation, then G' also has a successful computation (but may also have failing and deadlocked ones). In spite of these limitations live snapshots are

[13] Specifically, in the case of FCP(|), that neither *var* nor *unknown* are used in the interpreted program.

useful for various purposes, including the detection of stable properties of networks. This subject is further discussed in [160].

Combining the concepts: interrupt handling, termination detection, and the computation of live and frozen snapshots

We show a meta-interpreter which combines the various features discussed. It has both an interrupt stream and a short circuit, and it uses the clause form of the interrupt-handling meta-interpreter.

```
reduce(true,Is,L-R) ← L=R.
reduce(X=Y,Is,L-R) ← (X,L)=(Y,R).
reduce((A,B),Is,L-R) ← reduce(A,Is,L-M), reduce(B,Is,M-R).
reduce(goal(A),Is,L-R) ←
    clause(A,B,Is), reduce(B,Is,L-R).
reduce(A,[I|Is],L-R) ←
    serve_interrupt([I|Is],A,L-R).

serve_interrupt([halt|Is],A,L-R) ← L=R.
serve_interrupt([suspend|Is],A,L-R) ←
    L=[Done|L'], R=[Done|R'],
    serve_interrupt(Is,A,L'-R').
serve_interrupt([resume|Is],A,L-R) ←
    reduce(A,Is,L-R).
serve_interrupt([snapshot|Is],A,L-R) ←
    L=[state(S)|L'], R=[state([A|S])|R'],
    serve_interrupt(Is,A,L'-R').
```

The meta interpreter, called with the goal *reduce(G,Is,L-R)*, can be used to obtain a live snapshot *S*, even in the presence of unbounded process creation, by providing it with the following input:

Do in parallel:
Is=[snapshot,resume|Is'], L=[state([ ])|L'], R=[state(S)|R'].

which cause each process to suspend, add its state to the snapshot, and resume immediately.

A frozen snapshot is obtained by suspending the computation, and only then collecting the state of its processes. The following sequence of unifications can be used to get a frozen snapshot and then resume a computation.

Suspend the computation:
Is=[suspend|Is'], L=[done|L'],
wait till R=[done|R'],
Take a snapshot:
Is'=[snapshot|Is''], L'=[state([ ])|L''],
wait till R'=[state(S)|R'']
Resume:
Is''=[resume|Is'''].

Specialization of meta-interpreters

We have shown that an enhanced meta-interpreter is a very convenient tool for specifying functions of computation control. However, a naive implementation of these functions via enhanced meta-interpreters could be quite costly. It is quite common that a program interpreted under an enhanced meta-interpreter runs an order of magnitude slower compared with its direct execution.

One approach to the problem employs the concept of partial evaluation [58,43], first explored in this context by Gallagher [62,63], and refined by others [112,56,64,150,151,180,102]. It is to specialize at compile-time the meta-interpreter for the execution of a given program.

For example, consider the following (inefficient) FCP(|) program for reversing a list:

```
rev([X|Xs],Ys) ← rev(Xs,Zs), append(Zs,[X],Ys).
rev([ ],Ys) ← Ys=[ ].

append([X|Xs],Ys,Zs) ← Zs=[X|Zs'], append(Xs,Ys,Zs').
append([ ],Ys,Zs) ← Ys=Zs.
```

The plain meta-interpreter, specialized to execute this program, is the program itself (although *append* can be specialized further, see [151]). In [151,152] a partial evaluator for Flat Concurrent Prolog, capable of partially evaluating meta-interpreters, was developed. As there is no partial evaluator for FCP(|), we show here examples of manual specializations of meta-interpreters. Using partial evaluation techniques similar to those of [151,150], the termination-detection meta-interpreter can be specialized to execute this list reversal program, resulting in the program:

```
rev([X|Xs],Ys,L-R) ← rev(Xs,Zs,L-M), append(Zs,[X],Ys,M-R).
rev([ ],Ys,L-R) ← (Ys,L)=([ ],R).

append([X|Xs],Ys,Zs,L-R) ← (Zs,L)=([X|Zs'],M), append(Xs,Ys,Zs',M-R).
append([ ],Ys,Zs,L-R) ← (Ys,L)=(Zs,R).
```

And the interrupt-handling meta-interpreter can be specialized to execute this program, resulting in:

```
rev([X|Xs],Ys,Is) ← rev(Xs,Zs,Is), append(Zs,[X],Ys,Is).
rev([ ],Ys,Is) ← Ys=[ ].
rev(Xs,Ys,[I|Is]) ← serve_interrupt([I|Is],rev(Xs,Ys)).

append([X|Xs],Ys,Zs,Is) ← Zs=[X|Zs'], append(Xs,Ys,Zs',Is).
append([ ],Ys,Zs,Is) ← Ys=Zs.
append(Xs,Ys,Zs,[I|Is]) ← serve_interrupt([I|Is], append(Xs,Ys,Zs)).

serve_interrupt([abort|Is],A).
serve_interrupt([suspend|Is],A) ← serve_interrupt(Is,A).
serve_interrupt([resume|Is],rev(Xs,Ys)) ← rev(Xs,Ys,Is).
serve_interrupt([resume|Is],append(Xs,Ys,Zs)) ← append(Xs,Ys,Zs,Is).
```

Note how the state of the interrupted process is passed to the *serve_interrupt* routine, and that this routine has two clauses, one for resuming *rev* and one for resuming *append*.

Such specializations eliminate the overhead of interpretation, while preserving the functionality of the enhanced meta-interpreter. The transformed programs are usually only 10% to 50% slower than the original programs, depending on the added functionality, compared to the order of magnitude slowdown of naive execution of the interpreter [84].

Techniques for proving the correctness of transformations of concurrent logic programs are not, as yet, well established. One question under debate is whether a transformation should preserve the meaning of a program, including all possible nondeterministic choices, an approach taken by [57,205], or whether a transformation could fix some choices at "compile time" thus change the meaning of a program; this approach views the source program as a specification, which may have several, nonequivalent but correct, implementations.

PART III. CONCURRENT LOGIC PROGRAMMING LANGUAGES

8. Language Comparison

In a trivial sense all reasonable programming languages are equivalent, since they are Turing-complete (i.e. can simulate a Turing machine, which is a universal computational model). However, if the differences between languages were not material, we would not have invented so many of them.

Concurrent logic languages are similar enough to allow a more precise comparison than is usual among programming languages. They all share the same abstract computational model, share the same principles, and employ very similar syntax. Therefore it it easier to focus on their differences. In comparing languages in this family, we consider mostly expressiveness, simplicity, readability, and efficiency.

In comparing languages for expressiveness, we use two methods: the first is to embed one language in another; the second is to show programming techniques available in one but not in another. We conclude that one language is more expressive, or "stronger" than another if the latter can be "naturally" embedded in the former, but not vice versa, and/or if all programming techniques of the latter are available in the former, but not vice versa.

We first define the notion of language embedding, which can be used to compare any two languages, and then discuss the finer notion of natural embedding, which is tailored for the comparison of logic programming languages. Related notions of language embedding and their application to the comparison of concurrent logic languages were studied by Saraswat [157] and Levy [115].

Definition: Language embedding
Let L_1 and L_2 be two languages, c a function from L_1 programs to L_2 programs, and v a function from observables of L_2 to observables of L_1. We say that $\langle c,v \rangle$ is an ***embedding of*** L_1 ***in*** L_2 if $v([\![c(P)]\!]_{L_2}) = [\![P]\!]_{L_1}$ for every L_1-program P. In such a case c is called the ***compiler*** of the embedding and v its ***viewer***.

We say that L_1 can be embedded in L_2 if there are effective functions c and v such that $\langle c,v \rangle$ is an embedding of L_1 in L_2. ∎

In other words, a compiler c and a viewer v form an embedding of L_1 in L_2 if the observable behavior of every L_1 program P is the same as the observable behavior of the L_2 program obtained by compiling P using c and viewing its behavior through v.

This notion of embedding is rather weak. Because of the Turing-completeness of the languages under consideration, any language L_1 can be embedded in any other language L_2, by writing in L_2 an interpreter of L_1, and "compiling" an L_1 program P to the L_2 program consisting of the interpreter augmented with a representation of P.

The real issues with embeddings from L_1 to L_2 are what is the complexity of the compilation (e.g. how complex is the L_1 interpreter written in L_2), what is the runtime overhead of compiled programs, how much of the parallelism of L_1 is preserved in the compilation, etc. This is usually related to how much of the execution mechanism of L_1 needs to be *reified* in the translation, and how much of it can be *absorbed* in the execution mechanism of L_2.

The basic execution mechanism in logic programming is unification. Therefore we are interested in embeddings from L_1 to L_2 which absorb unification, i.e. use the unification mechanism of L_2 to implement the unification mechanism of L_1. In such an embedding, logical variables of L_2 represent logical variables of L_1 and hence non-ground goals of L_2 can be used to represent similar goals of L_1. We call an embedding that maps logical variables in one language to logical variables in another *natural*. We formalize this notion by requiring that the viewer be the identity function on observables containing goals with predicates of the source program (although it may hide auxiliary predicates introduced in the target program, if any). This precludes embeddings in which the compiler encodes variables in the source language by constants of the target language and the viewer decodes the answer substitution given in terms of these constants[14].

Definition: An embedding $\langle c,v\rangle$ between two concurrent logic programming languages is *natural* if v is defined by:

$$v([\![c(P)]\!]) = \{\langle G,\theta,x\rangle \in [\![c(P)]\!] \mid G \text{ is a predicate of } P\}. \blacksquare$$

The observables of a concurrent logic program $P, [\![P]\!]$, were defined in Section 4.2. In the following discussions of embeddings we assume that the viewer is defined as above and hence discuss only the compiler.

In the following we show natural embeddings among concurrent logic programming languages, and argue (although not prove) the lack of opposite natural embeddings. Our findings are summarized in Figure 5. An arrow in the figure indicates the existence of a natural embedding.

Most of the embeddings we show have additional pleasant properties. For example, being defined clause-wise, and preserving not only the observables but also the behavior in context (the so-called compositional semantics). We do not address these aspects further here.

A second dimension of comparison is simplicity of the syntax and semantics. A simpler language is preferred since it is easier both to grasp and to be used by humans, and is more amenable to automatic program analysis and transformation. Usually a weaker language is also simpler, but this is not always so, especially when the difference lies in the granularity of the atomic operations. Usually a language with coarser granularity, i.e. with larger atomic operations, is also stronger. Sometimes, in addition, its transition system is also simpler to define. For example, the languages FCP(|), FGHC_{av}, and FGHC_{nav} discussed below, have, progressively, finer granularity and more complicated transition systems[15].

---

14 We note, however, that if the concurrent logic language has sufficiently powerful extra-logical constructs, which enable it to examine and compare logical variables, then it can construct its internal "variable-value" dictionaries. Using this dictionary encoding and decoding can be done internally by the target program. The languages discussed in this survey do not have this capability.

15 The phenomenon is true for the interleaving semantics used in this paper, as well as for the

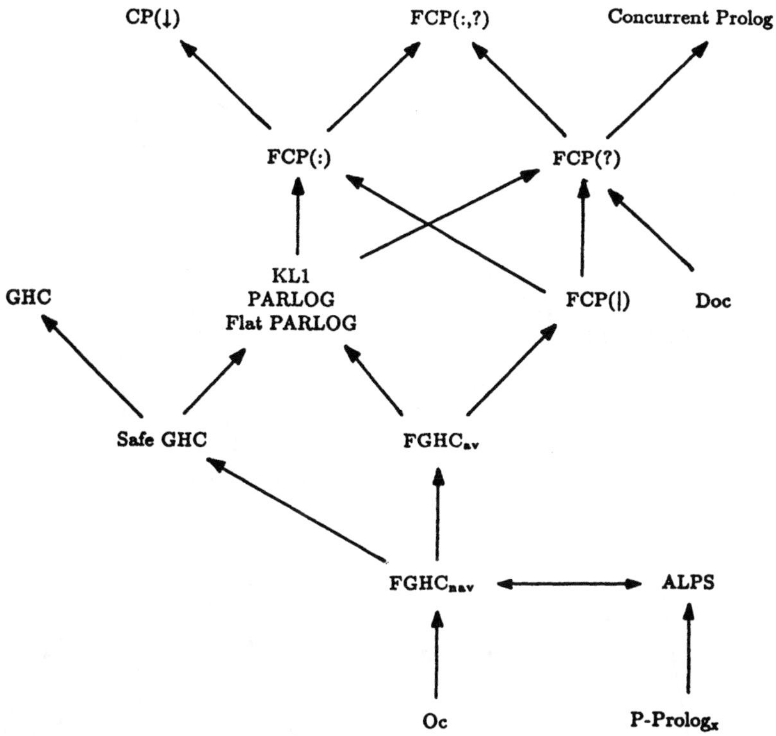

Figure 5: Natural embeddings among concurrent logic programming languages

A third dimension of comparison is readability. All languages described in this survey use guarded Horn clauses, first employed in the Relational Language [20]. Most of them follow the syntactic conventions of GHC [199], that matching is used in the head, and unification is specified explicitly in the body. Exceptions are FCP(?), P-Prolog, ALPS, and Doc; the impact of their different syntactic conventions on readability is discussed when the languages are introduced.

A fourth dimension of comparison is ease of implementation. In general, the weaker the language the easier it is to implement. In particular, the finer the granularity of the language's atomic operations, the simpler the synchronization mechanisms required by its parallel implementation. We defer the comparative discussion of implementation to Section 20.

Each of the dimensions mentioned — expressiveness, simplicity, readability, and efficiency — is only one dimension in a multidimensional design space, which usually involves design tradeoffs. For example, a more expressive language may have a more complicated semantics, and be more difficult to implement. A weaker language may need extra-lingual facilities to compensate for its lack of expressiveness. Presently there is no consensus which language in this design space is optimal as a general purpose programming language for parallel and distributed computers, and several languages are being pursued actively as candidates for this role. Notable efforts which

---

and extended by Saraswat [158].

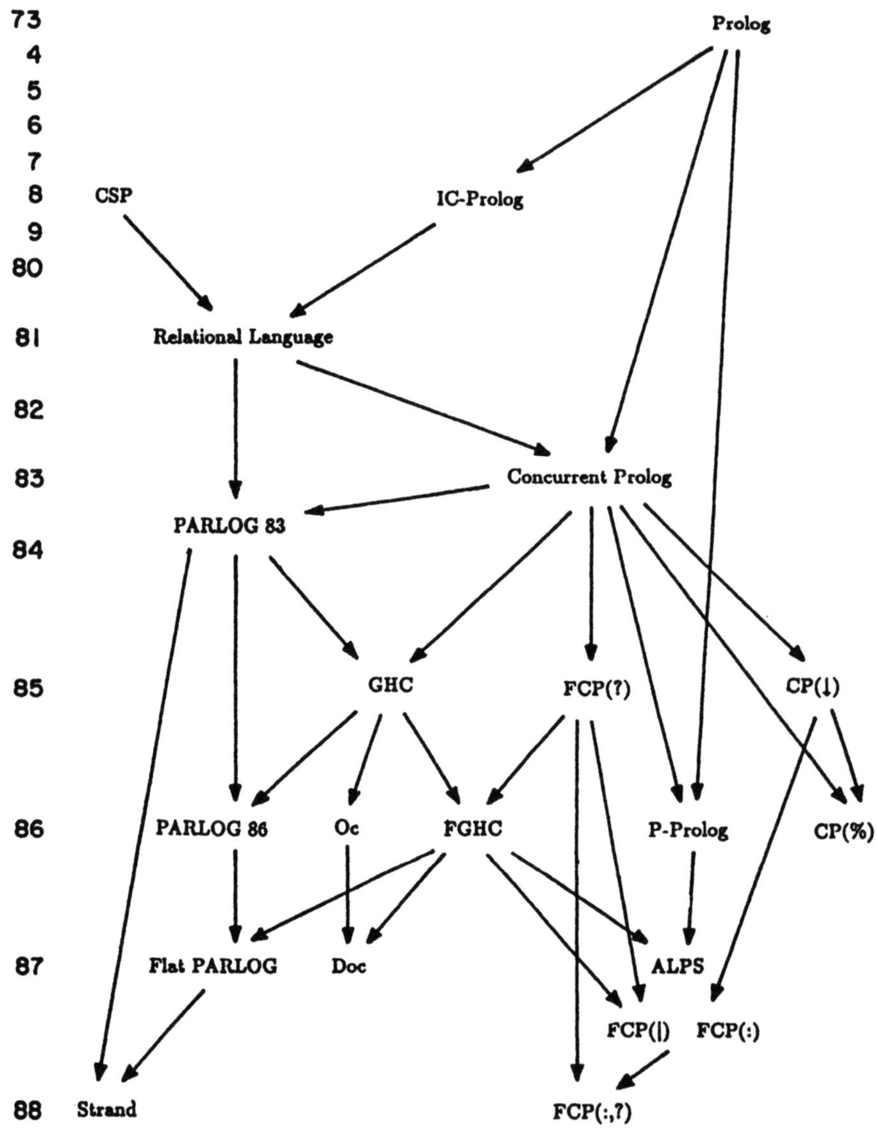

Figure 6(a): A historical chart of concurrent logic languages

comprise of both language design, system design, and sequential and parallel implementations include KL1 (Flat GHC + control meta-call) [55] and PIMOS at ICOT [18], PARLOG and its flat variants [164,66,49], and a PARLOG system [48] at the Imperial College of Science and Technology, and Flat Concurrent Prolog [164] and its variants [100], and the Logix system [84,172] at the Weizmann Institute of Science.

For completeness, we provide a historical chart of concurrent logic languages in Figure 6. It is an extension of an earlier chart by Ringwood [148]. In the chart the vertical axis denotes the time in which the language design was published, and an arrow indicate some kind of intellectual influence.

Year	Language	Section / Reference
73	Prolog	1.5 [173] Sterling and Shapiro, 1986
78	CSP	[86] Hoare, 1978; [87] 1985a
	IC-Prolog	[26] Clark et al., 1982
81	Relational Language	[20] Clark and Gregory, 1981
83	Concurrent Prolog	3.11.2 [162] Shapiro, 1983b
	PARLOG 83	[21] Clark and Gregory, 1983
85	Flat Concurrent Prolog (FCP(?))	3.8 [129] Mirowsky et al., 1985
	GHC	3.11.1 [201] Ueda, 1986a; [199] 1986b; [200] 1986c
	CP($\downarrow$, |)	3.11.2 [152] Saraswat, 1985
86	Flat GHC	3.3 [205] Ueda and Furukawa, 1988
	PARLOG 86	3.11.1 [24] Clark and Gregory, 1986; [66] Gregory, 1987
	Oc	3.10 [83] Hirata, 1987
	P-Prolog	3.9 [210] Yang, 1986; [211] Yang and Aiso, 1986
	CP(%)	[145] Ramakrishnan and Silverschatz, 1986
87	Flat PARLOG	3.4 [49] Foster and Taylor, 1987; [66] Gregory, 1987; [107] Lam and Gregory, 1987
	ALPS	3.6 [125] Maher, 1987b
	Doc	3.10 [82] Hirata, 1986
88	FCP(|)	2.2 [this paper]
	FCP(:)	3.7 [158] Saraswat, 1988a; [101] Kliger et al., 1988
	FCP(:,?)	3.9 [101] Kliger et al., 1988
	Strand	3.9 [189] Taylor and Foster, 1988

Figure 6(b): A historical chart of concurrent logic languages

9. Semantics of Concurrent Logic Programming Languages

In the following sections we investigate several concurrent logic languages. All the flat languages are defined similarly to FCP(|), and assume the same set of guard test predicate. Although small, this set turns out in practice to be sufficient for most practical purposes[16]. Their state of computation, as well as transitions, are identical to the ones of FCP(|) defined in Section 4.2. The differences between most of the flat languages are captured simply by varying the definition of the clause try functions. Although different, all try functions employed satisfy the following two properties:

a) *Suspension is not stable*:

If $try(Goal, Clause) = suspend$
then there is a substitution θ such that:
$try(Goal\theta, Clause) \neq suspend.$

b) *Failure is stable*:

If $try(Goal, Clause) = fail$

16 See [172] for a description of the guard predictaes and other primitives in a practical system.

then for every substitution θ
$try(Goal\theta, Clause) = fail.$

Property *a* implies that a suspended clause try may succeed or fail in the future, if the goal atom is further instantiated (e.g. by reducing other atoms in the goal). Property *b* implies that a failed clause try need not be tried again.

We say that a language is ***success stable*** [159] if it satisfies the following property *c*:

c) *Success is stable*:

If $try(Goal, Clause) = \theta$
then $try(Goal\theta\sigma, Clause) \notin \{suspend, fail\}$ for every substitution σ.

Most languages discussed in this survey, including FCP(|), are success stable if the guard primitives *unknown* and *var* are excluded. Exceptions will be noted when introduced.

The non-flat languages are described only informally. Transition system for non-flat languages were defined by Saraswat [156,159] and Levy [115].

The notion of language embedding as described in the previous section presupposed a definition of the observables of the source and target language. As discussed in Section 3, since concurrent logic languages employ don't-care nondeterminism, their observables record the results of failing and deadlocked computations, in addition to the results of successful ones. The observables of a concurrent logic program, in any of the languages surveyed, record the initial state and an abstraction of the final state of every computation.

Compositional semantics for concurrent logic programs that are fully abstract with respect to these observables were investigated by [59,65]. Other investigations of the semantics of concurrent logic languages include [9,44,30,111,132,205]. However, since the work on the semantics of concurrent logic languages is in a state of flux we do not survey it here.

10. Flat GHC: A Language With Non-Atomic Unification

Flat GHC is the flat subset of the language Guarded Horn Clauses [199,200] (see Section 18). Flat GHC, augmented with a control meta-call primitive discussed in Section 10.3 below, is the basis of Kernel Language 1 [55], the core language of the parallel computer system developed at ICOT as part of the Fifth Generation Project [196].

We consider two variants of Flat GHC. One called $FGHC_{av}$, for Flat GHC with atomic variables, and the other called $FGHC_{nav}$, for Flat GHC with non-atomic variables.

$FGHC_{av}$ is derived from the original definition of GHC [199], and it is quite similar to FCP(|). The difference is that in $FGHC_{av}$ a unification specified by the goal $T_1 = T_2$ need not be carried out atomically. Saying it differently, a program in $FGHC_{av}$ cannot specify that a compound unification is to be carried out as an atomic operation. We have found only one implication of this difference in terms of expressiveness: $FGHC_{av}$ requires slightly more elaborate code than FCP(|) to implement the short circuit technique.

$FGHC_{nav}$ has an even finer notion of atomic actions. Intuitively, in $FGHC_{nav}$ even the instantiation of a variable to a value need not be done atomically, and several occurrences of the same variable can be instantiated to different (conflicting) values simultaneously. If such a conflict occurs it is eventually detected and results in failure. However, in $FGHC_{nav}$ there are intermediate states of the computation in which the same variable may have different values, whereas in $FGHC_{av}$ (and in FCP(|)) this cannot happen.

This property of $FGHC_{nav}$ is a consequence of the principle of ***anti-substitutability*** [200, 198], also called ***logical referential transparency***[17]. The principle states, informally, that an occurrence of a variable X can be replaced in any context by a new variable X', provided the equality $X = X'$ is "added" to the context. The principle is motivated by semantic elegance, and it justifies a wide range of program transformations [205]. Operationally, it allows "decoupling" different occurrences of the same variable, and instantiating them to different values. In such a case the inconsistency between the instantiations is detected eventually, and failure results.

The main difference in terms of expressiveness between $FGHC_{av}$ and $FGHC_{nav}$ is that in the latter the short-circuit technique cannot be used to detect the successful termination of a computation. The reason is that the closing of the short circuit, both in the original version described for FCP(|), and the variant described for $FGHC_{av}$ below, cannot guarantee that only consistent instantiations have been made. It is still possible that two occurrences of some variable in the computation were instantiated to inconsistent values, which would result in failure past the closing of the circuit.

It seems that without additional facilities, such as the control meta-call discussed below, detection of successful termination of a computation cannot be specified in $FGHC_{nav}$. Saying it differently, $FGHC_{nav}$ cannot reflect on successful termination, unlike FCP(|) and $FGHC_{av}$.

The initial informal description of GHC [199] takes the atomic variables approach, as well as the treatment of GHC in [159]. Subsequent theoretical work on GHC embraced the anti-substitutability principle [199], and thus imply non-atomic variables. The practical work at ICOT, however, still adheres to atomic variables: the KL1 language designed and implemented at ICOT is essentially $FGHC_{av}$ augmented with a control meta-call [90].

Although the impelementation work on GHC and KL1 adopted the notion of "flatness" employed in this paper, Flat GHC was defined formally only recently [205] under the name "Theoretical Flat GHC". The notion of flatness used there is different from ours.

In the following we relate FCP(|), $FGHC_{av}$ and $FGHC_{nav}$ with regard to the short circuit technique, and show simple embeddings of $FGHC_{av}$ in FCP(|), and of $FGHC_{nav}$ in $FGHC_{av}$. The syntax and the try function are the same for $FGHC_{av}$ and $FGHC_{nav}$. The difference is captured in an additional ***Anti-substitute*** transition for $FGHC_{nav}$ described below.

10.1 The language $FGHC_{av}$

Syntax

Definition: An $FGHC_{av}$ *program* is a finite sequence of guarded clauses that includes the unit clause:

[17] By Graem Ringwood.

$$X = X.$$

and the clauses:

$$f(X_1,X_2,\ldots,X_n) = f(Y_1,Y_2,\ldots,Y_n) \leftarrow X_1{=}Y_1,\ X_2{=}Y_2,\ldots,X_n{=}Y_n$$

for every function symbol f/n, $n > 0$, occurring in some clause whose head predicate is different from '='[18]. ∎

Semantics

The fact that unification need not be done atomically is captured by the equality clauses, which allow a compound unification to be performed piecemeal.

The $FGHC_{av}$ try function is defined as follows. Let $C = (X_1{=}X_2 \leftarrow \ldots)$ be a renaming of a unification clause in P.

$$try_{FGHC}(T_1{=}T_2, C) = mgu((T_1,T_2),(X_1,X_2)).$$

The try function for the other clauses of P is defined as for FCP(|).

Notes:

1) The semantics of the unit clause $X = X$ in $FGHC_{av}$ is identical to that of FCP(|), since $mgu((T_1,T_2),(X,X)) = mgu(T_1,T_2)$, if X does not occur in T_1 and T_2.
2) The atomic operation in $FGHC_{av}$ is assigning a variable to a variable, or assigning a term whose arguments are distinct variables to a variable. The first action is allowed by the unit unification clause; the second by the other clauses. We do not prevent larger atomic actions; we simply do not require them, by permitting smaller ones.

Comparison of $FGHC_{av}$ with FCP(|)

The main implication of the lack of atomic unification in FGHC in terms of expressiveness is that $FGHC_{av}$ cannot use the short-circuit technique as specified to detect the termination of a computation. In FCP(|) one can perform the unification of the underlying computation $X = Y$ and close the short circuit $L{-}R$ atomically, within the same compound unification $(X,L) = (Y,R)$. In $FGHC_{av}$ one needs first to perform the unification $X = Y$, wait for it to complete using matching, and only then close the short-circuit[19]. This can be achieved by the procedure *unify_and_close_sc*$(X,Y,L{-}R)$, defined using the auxiliary procedure *match_and_close_sc* as follows:

```
unify_and_close_sc(X,Y,L-R) ←
    X=Y, match_and_close_sc(X,Y,L-R).

match_and_close_sc(X,X,L-R) ←
    L=R.
```

Note that to detect the termination of an underlying computation, *unify_and_close_sc* must be used instead of '=' throughout the underlying program.

The FCP(|) termination detecting meta-interpreter shown in Section 7.7 above is also an FGHC meta-interpreter. However, the unification performed in the body of the clause:

---

18 We assume here that an initial goal does not contain function symbols which do not occur in the program. Alternatively, an equality clause has to be added for every function symbol allowed in a goal, or a general recursive definition of unification, using the Prolog-like predicates *functor* and *arg* should be used.

19 This method is due to E.D. Tribble

reduce(X=Y,L–R) ← (X,L)=(Y,R).

behaves differently in FCP(|) and $FGHC_{av}$. The modified version of the short circuit can be used in an $FGHC_{av}$ termination detecting meta-interpreter, by replacing the above clause with the clause:

reduce(X=Y,L–R) ← unify_and_close_sc(X,Y,L–R).

The difference between FCP(|) and $FGHC_{av}$ can be observed by composing a program that does the compound unification $f(X, Y) = f(a,b)$ with a program that matches either $X = a$ or $Y = b$, then unifies the other variable with c. Such a program is the same in FCP(|) and FGHC:

test(a,Y) ← Y=c.
test(X,b) ← X=c.

If this FCP(|) program were to execute using the goal:

test(X,Y), f(a,b)=f(X,Y)

the terminal state would never contain a substitution in which $X \mapsto c$ or $Y \mapsto c$. As an $FGHC_{av}$ program, some executions will have $X \mapsto c$ and some $Y \mapsto c$, since $FGHC_{av}$ cannot specify that the two unifications $X = a$ and $Y = b$ be carried out atomically.

An attempt to establish the difference in power between languages with and without atomic unification was made by Saraswat [157].

An embedding of $FGHC_{av}$ in FCP(|)

$FGHC_{av}$ can be naturally embedded in FCP(|) using the following compiler. The compiler translates an $FGHC_{av}$ program P to FCP(|) clause-wise. In every clause, it replaces every body goal $X = Y$ with the goal *unify*(X, Y), where *unify*/2 is a predicate not occurring in P. It adds to the resulting program the clause:

unify(X,Y) ← X=Y.

and for every function symbol f/n occurring in P, the clause:

unify(f(X_1,X_2,...,X_n),f(Y_1,Y_2,...,Y_n)) ←
    unify(X_1,Y_2), unify(X_2,Y_2), ..., unify(X_n,Y_n).

This completes the description of the compiler.

Like in the definition of the semantics of $FGHC_{av}$, the effect of the definition of *unify*/2 is that a compound unification can be carried out either atomically, using the first clause, or non-atomically, using the other clauses.

10.2 The language $FGHC_{nav}$

Syntax

The syntax of $FGHC_{nav}$ is the same as that of $FGHC_{av}$.

Semantics

The difference between $FGHC_{av}$ and $FGHC_{nav}$, namely the anti-substitutability principles, can be modelled in our framework by extending the transition system of $FGHC_{av}$ with the following transition:

- Anti-substitute:

$$\langle G;\theta\rangle \xrightarrow{\text{Anti-substitute}} \langle G',X{=}Y;\theta\rangle$$

where Y is a variable that does not occur in G, and G' is obtained by replacing one occurrence of X by Y in G.

This transition directly models the principle of anti-substitutability. However, as stated, it allows almost any FGHC_{nav} program to diverge, by alternating the introduction and elimination of the equality goals using Anti-substitute and Reduce. To prevent this, additional complicated fairness conditions need to be incorporated.

Note that the Anti-substitute transition may cause a conditional answer substitution to contain inconsistent assignments. We have defined substitutions to be functions, i.e. have a single value for a variable. This has to be modified in order to specify observables for FGHC_{nav}.

The difficulty in modelling the semantics of FGHC_{nav} can be attributed to the need to accommodate inconsistent constraints on the values of variables. The method proposed by Maher [125] and further developed by Saraswat [158,159], suggest another method for modelling this. The method is to separate the goal atoms into "pools", each containing its own binding environment, and add explicit transitions which communicate equality constraints between pools. Failure occurs as soon as one of the pools detects inconsistency.

Comparison of FGHC_{av} with FGHC_{nav}.

FGHC_{av} could use the short-circuit technique to detect successful termination of a computation, albeit with some additional effort. The technique is not applicable in FGHC_{nav}. It is possible that the unifications executed by the monitored computation are inconsistent, without this inconsistency detected prior to the closing of the short circuit. Thus, unlike in FCP(|) or FGHC_{av}, the closing of the short-circuit is not a reliable indication that the computation has not failed. Technically, the Anti-substitute transition incorporates in the underlying computation unifications which are not threaded via the short-circuit. Even if the short-circuit closes, the new unifications introduced by the Anti-substitute transitions can subsequently fail.

An embedding of FGHC_{nav} in FGHC_{av}

An embedding of FGHC_{nav} in FGHC_{av} consists of a clause-wise compiler and the identity viewer. For each clause, the compiler iteratively performs anti-substitution to any variable that occurs more than once in clause body atoms other than equality, until no such variables are left. By doing so, the compiler "decouples" every variable that may be used for communication, and adds equalities to clause bodies. These equalities will eventually unify all decoupled occurrences of each variable.

10.3 The meta-call construct

The ability to reflect on the termination and failure of a computation is essential to a systems programming language, but FGHC_{av} (and FCP(|)) cannot do the latter, and FGHC_{nav} can do neither, without reifying unification. The problem can be solved in two different ways. One is to strengthen the basic mechanisms of the language. Atomic variables are sufficient to reflect on termination. To reflect on failure, atomic test unification is needed, as incorporated in the stronger variants of FCP(|) shown in Sections 14, 15, 16 below.

Another solution, which was taken by the developers of both GHC and PARLOG, is to add to the language a meta-level construct, which has "built-in" reflection and control capabilities. There are several variations on the construct, originally proposed by Clark and Gregory [22]. One variant, which is referred to in the following as the *control meta-call*, has the form *call(Goal,Signals,Events)*, where *Signals* is a stream of {*suspend, resume, abort*}, and *Events* is a stream of {*suspended, resumed, failed(Goal), halted, aborted*}, the last three being terminal events.

The intuitive semantics of the control meta-call is as follows. A computation of a goal G is started under the control meta-call using the goal *call(G,In,Out)*. If some goal atom G' in the computation fails, the message *failed*(G') appears on the *Out* stream. If the computation terminates, the message *halted* appears on *Out*. To suspend the computation, the message *suspend* is sent to the *In* stream, and when suspension occurs the acknowledgement message *suspended* appears on *Out*. Similarly, to resume or abort the computation the message *resume* or *abort* is sent on *In*, and the corresponding acknowledgement message *resumed* or *aborted* appears on *Out*. Using the control meta-call, a process in the language can start a computation and monitor it.

We refer to the language FGHC_{av} augmented with the control meta-call as KL1 [55]. The actual meta-call implemented as part of PIMOS [18], the KL1 operating system, also includes resource management facilities: a computation is allocated some CPU time and some memory, and when either of these is consumed it announces resource overflow and suspends. It can be resumed by providing it with additional resources.

The control meta-call eliminates much of the freedom of non-atomic variables. For example, it can be used to detect the successful termination of unification, a capability not present in FGHC_{nav}. Hence its implementation restricts the kind of algorithms that can be used in a distributed implementation of the language; in particular, the algorithms must incorporate some form of distributed termination detection.

In comparison with the meta-interpreters of FCP(|) shown in Section 7.7, the meta-call construct reflects on failure, whereas an FCP(|) meta-interpreter cannot. On the other hand, an FCP(|) meta-interpreter can produce snapshots, whereas the standard meta-call constructs cannot (although Gregory *et al.* [67] have proposed an enhanced control meta-call that does). We will come back to the meta-call when we discuss FCP(:) in Section 14.

Yet a third approach is to construct a meta-interpreter that reifies unification, and extend it in various ways. A first step in this direction was taken by Tanaka [184].

11. Flat PARLOG: FGHC Extended With Sequential-Or and Sequential-And

The PARLOG language [21], described in Section 18, preceded GHC, but went through several evolutions that made it closer to GHC [24,66,148]. In the earlier definition [21], referred to as PARLOG83 by [148], the output mechanism was assignment, rather than unification. In the latter definition [66], refered to as PARLOG86 by [148], the output mechanism was (non-atomic) unification, as employed by GHC. PARLOG consists of two sublanguages: the single-solution subset, and the all-solutions subset. The latter is essentially Or-parallel Prolog. Here we concentrate on the flat subset of the former. The non-flat language is discussed in Section 18. Presently, the main difference between the computational models of the single-solution subset of PARLOG and GHC is the sequential-Or and sequential-And constructs of PARLOG. In addition, PARLOG offers a surface syntax which contains mode declarations. For example, using modes, the PARLOG *append* program could be specified as follows:

```
mode append(?,?,↑).
append([X|Xs],Ys,[X|Zs]) ← append(Xs,Ys,Zs).
append([ ],Ys,Ys).
```

This program is then translated to PARLOG standard form:

```
append(Xs,Ys,Zs) ←
    Xs ⇐ [X|Xs'] | Zs=[X|Zs'], append(Xs',Ys,Zs').
append(Xs,Ys,Zs) ←
    Xs ⇐ [ ] | Ys=Zs.
```

where ⇐ is PARLOG's input matching primitive. This program is operationally identical to the Flat GHC program:

```
append([X|Xs],Ys,Zs) ← Zs=[X|Zs'], append(Xs,Ys,Zs').
append([ ],Ys,Zs) ← Ys=Zs.
```

Several proposals were made for a "flat" subset of PARLOG [66,107,49]. The Flat PARLOG of Foster and Taylor [49] is essentially Flat GHC with mode declarations as surface syntax. Recently, a language called Strand [189] was derived by Foster and Taylor from their Flat PARLOG language by restricting the output mechanism to be assignment, rather than unification. Strand is essentially a flat version of PARLOG83, with sequential-And and sequential-Or eliminated. PARLOG83 and Strand are not success stable.

Our definition of Flat PARLOG is based on the $KP_{AND\ Tree}$ language of Gregory [66]. The language is Flat GHC augmented with sequential-Or and sequential-And. We investigate each of the two extensions to Flat GHC separately, denoting the resultant languages FP(;) and FP(&). For the sake of uniformity we use the Flat GHC syntax instead of the PARLOG standard form syntax. The translation from the Flat PARLOG syntax to the Flat GHC syntax is straightforward.

In the following we refer to the language combining FP(;), FP(&), and the control meta-call as Flat PARLOG. Although the subject is not discussed explicitly in the PARLOG papers, we assume that PARLOG has atomic variables, and hence consider Flat PARLOG to be an extension of $FGHC_{av}$ rather than of $FGHC_{nav}$.

11.1 The language FP(;)

The language FP(;) allows the specification of sequential-Or clauses $C_1;C_2;\ldots;C_n$, where each disjunct C_i is an ordinary guarded clause. The idea of a sequential-Or clause is that the guarded clause C_i can be selected only if the clause tries of the clauses $C_1,\ldots,C_{i-1}$ fail. The connective ';' is called ***sequential-Or***. The similarity of sequential-Or clauses to if-then-else constructs in procedural languages and to conditionals in Lisp is apparent.

Syntax

Definition: Sequential-Or clause, program.

- A ***sequential-Or clause*** is a guarded clause or has the form $C_1\ ;\ C_2$ where C_1 is a guarded clause and C_2 is a sequential-Or clause.

- An FP(;) *program* is a set of sequential-Or clauses, augmented with unification clauses as in FGHC. ∎

Semantics

The $try_{FP(;)}$ function is defined as follows. For a conditional clause $C_1;C_2$

$$try_{FP(;)}(A,C_1;C_2) = \begin{cases} try_{FGHC}(A,C_1) & \text{if } try_{FGHC}(A,C_1) \neq fail \\ try_{FP(;)}(A,C_2) & \text{if } try_{FGHC}(A,C_1) = fail. \end{cases}$$

For a guarded clause C

$$try_{FP(;)}(A,C) = try_{FGHC}(A,C).$$

An embedding of FP(;) in FGHC$_{av}$ with *otherwise*

The embedding consists of a clause-wise compiler. Its method of compiling sequential-Or into *otherwise* (introduced in Section 7.6) is similar to the one used by Codish and Shapiro [29] to translate a non-flat language into a flat one.

The general idea is to reify clause selection, by explicitly programming the commitment operation. Each sequential-Or clause $C_1;C_2;\ldots;C_m$ is translated into a different procedure consisting of m guarded clauses by adding *otherwise* to the clauses $C_2,\ldots,C_m$. This ensures that a disjunct can succeed only if all previous disjuncts fail. The head predicates of clauses resulting from each disjunctive clause are renamed to form the new procedure.

A call to the original procedure is translated into a conjunctive call, one goal for each of the new procedures. The single-round mutual exclusion protocol shown in Section 7.4is used to ensure that at most one of these goals would "commit", i.e. proceed to execute the body of the selected disjunct of that sequential-Or clause. The other goals terminate quietly without causing any effect.

More specifically, an FP(;) procedure of the predicate p/n with k sequential-Or clauses is translated into $2k$ FGHC$_{av}$ procedures as follows. The ith conditional clause $C_1;C_2;\ldots;C_m$ is translated into the two FGHC$_{av}$ procedures $test\_p_i/k{+}1$ and $commit\_p_i/k{+}1$. Each disjunct $C_j = (p(T_1,T_2,\ldots,T_k) \leftarrow G \mid B)$ is translated to the clauses:

```
test_pi(T1,T2,...,Tk,Commit) ←
    otherwise, G |
    Commit=lock(Reply),
    commit_pi(Reply,T1,T2,...,Tn).

commit_pi(granted,T1,T2,...,Tn) ← B.
commit_pi(refused,_,_,...,_).
```

And the call to p/n is translated to calls to the $test\_p_i$ procedures using the clause:

```
p(X1,X2,...,Xn) ←
    test_p1(X1,X2,...,Xn,Commit1),
    test_p2(X1,X2,...,Xn,Commit2),
    ⋮
    test_pm(X1,X2,...,Xn,Commitm),
    mutex(Commit1,Commit2,...,Commitm).
```

where the ith clause of $mutex/n$ is:

```
mutex(Commit1,...,lock(Reply),...,Commitm) ←
```

Reply=granted,
Commit_1=lock(refused),
Commit_2=lock(refused),
⋮
...(excluding Commit_i)...
⋮
Commit_m=lock(refused).

As the translation shows, there is a close relationship between sequential-Or and *otherwise*, and it can be said that they were both designed to solve the same problem. Which construct to prefer is largely a matter of taste. Both destroy clause-wise modularity and are easily open to abuse, and therefore should be used sparingly. Sequential-Or is more appealing in being general and uniform. *Otherwise* is more restricted (it can be viewed as a special case of sequential-Or [156,158]), as perhaps appropriate for an exceptional construct, and the cases in which it is less convenient than sequential-Or for its purpose are rare.

11.2 FP(&)

The language FP(&) is FGHC_{av} augmented with sequential-And. Adding sequential-And to a language that supports dynamic creation of processes complicates both the definition and implementation of the language. In defining the operational semantics, the state of the computation cannot be represented by a sequence of goals. A tree of alternating sequential-And and parallel-And nodes, whose leaves contain the goals, is required. The definition of a transition is also complicated by the constraint that a goal can be selected only if it can be reached from the root by selecting the left-most branch in every sequential-And node.

Because of this complication, FP(&) does not fit the semantic framework we described. Instead, we define the syntax of FP(&), and provide it with semantics by embedding it in FGHC_{av}, using the short-circuit technique.

The compiler of the embedding translates each FP(&) program P into an FP(&) interpreter written in FGHC_{av}, augmented with the standard clausal representation of P. Since FGHC_{av} can be embedded directly in FP(&), using the identity compiler and viewer, this shows that the two languages and practically identical from an expressiveness point of view.

Syntax

Definition: FP(&) clause and program.

- An FP(&) *clause* is a formula of the form

  $$A \leftarrow G_1,\ldots,G_m \mid B_1,\ldots,B_n. \qquad n,m \geq 0$$

  where the A and G_i's are as before, and each B_i has the form:

  $$A_1 \,\&\, \ldots \,\&\, A_k \qquad (k > 0)$$

  where each A_i is an atom.
- An FP(&) *program* is a finite sequence of FP(&) clauses. ∎

Semantics

Let P be an FP(&) program. Translate each clause:

$$A \leftarrow G \mid B.$$

of P into the $FGHC_{av}$ clauses:

clause(A,X) ← G | B′=X.

where each unit goal G in B with predicate other than '=' is replaced by $goal(G)$ in B' and

A ← reduce(A).

where *clause* and *reduce* are predicates not occuring in P. Call the resulting program P'.

An interpreter I of FP(&) in $FGHC_{av}$, which assumes this representation, is defined as follows:

```
reduce(A) ←
    reduce′(A,done–Done).

reduce′(true,L–R) ← L=R.
reduce′(X=Y,L–R) ← unify_and_close_sc(X,Y,L,R).
reduce′((A,B),L–R) ← reduce′(A,L–M), reduce′(B,M–R).
reduce′(A&B,L–R) ← reduce′(A,done–Done), wait(Done,B,L–R).
reduce′(goal(A),L–R) ← clause(A,B), reduce′(B,L–R).

wait(done,A,L–R) ← reduce′(A,L–R).

unify_and_close_sc(X,Y,L,R) ← See definition in Section 10.
```

The interpreter implements A & B by executing A and suspending the execution of B until A terminates. Recursively nested sequential and parallel And's, which may be created by recursive procedures, are handled correctly, by starting a new short circuit for every sequential component.

The compiler c is defined to map P to $P' \cup I$. The viewer v is the identity function on the predicates of P, and hides the predicates *clause* and *reduce*. The observables of an FP(&) program P are then defined to be $v([[c(P)]])$.

Not only the direct definition of sequential-And is quite complex, but also its direct implementation. First, without complex data-structures it may take an unbounded amount of time to find the next process to execute — the amount is determined by the depth of nesting of sequential and parallel And's. Second, in a parallel implementation of the language, executing correctly the conjunct A & B requires performing distributed termination detection on A.

The interpreter of FP(&) in FCP(|) solves the two problems by delegating them to the underlying implementation of $FGHC_{av}$: the process suspension and activation mechanism of $FGHC_{av}$ wakes up the *wait* process when its first argument is instantiated to *done*. The short circuit technique combined with the implementation of unification with atomic variables essentially realizes a well-known distributed termination detection algorithm based on distributed counters [160] (see discussion in Section 7.5).

12. P-$Prolog_x$ — Synchronizing Deterministic Logic Programs

In the languages presented so far synchronization was achieved with matching, specified by clause heads: a clause try suspends if its matching with the clause head, or checking the guard, suspend.

An alternative approach to synchronization in concurrent logic programming was proposed by Yang and Aiso [210,211], and incorporated in the language P-Prolog. Although P-Prolog incorporates also an all-solutions Or-parallel component, we do not discuss it here. We focus on its other component, which employs a novel synchronization mechanism called *exclusive guarded Horn clauses*. We refer to this language subset as P-Prolog$_\mathbf{x}$.

P-Prolog$_\mathbf{x}$ uses goal/clause unification, rather than matching, and employs the following synchronization principle instead of matching: the reduction of a goal with a clause is enabled when it can be determined that the reduction with all alternative clauses is failed. In other words, a process is suspended as long as it has more than one clause to reduce with. It reduces if it has exactly one clause to reduce with; it fails when it has none. A process never makes an Or-nondeterministic choice.

The appeal of this synchronization principle is in the following lemma, a variant of which is due to Maher [125]. The lemma implies that the And-nondeterminism of P-Prolog$_\mathbf{x}$ does not affect the result of computations.

Lemma: Equivalence of P-Prolog$_\mathbf{x}$ computations.
If a P-Prolog$_\mathbf{x}$ program P has a successful computation from a goal G then every computation of P from G is successful and the answer substitutions of all such computations are the same (up to renaming). ∎

Syntax
The syntax of P-Prolog$_\mathbf{x}$ is the same as that of FCP(|).

Semantics
We define the P-Prolog$_\mathbf{x}$ try function, try_{PP}, using the auxiliary function try'_{PP}. Note that try'_{PP} is essentially try_{LP} augmented with guard evaluation. The program P is an additional parameter of the functions.

$$try'_{PP}(A',(A \leftarrow G|B),P) = \begin{cases} \theta & \text{if } mgu(A,A') = \theta \wedge \text{checking } G\theta \text{ succeeds} \\ fail & \text{if } mgu(A,A') = \theta \wedge \text{checking } G\theta \text{ fails} \\ & \quad \vee\; mgu(A,A') = fail \\ suspend & \text{otherwise} \end{cases}$$

$$try_{PP}(A,C,P) = \begin{cases} \theta & \text{if } try'_{PP}(A,C,P) = \theta \wedge try'_{PP}(A,C',,P) = fail \\ & \text{for every } C' \in P,\ C' \neq C \\ fail & \text{if } try'_{PP}(A,C,P) = fail \\ suspend & \text{otherwise} \end{cases}$$

Discussion
The advantage of P-Prolog$_\mathbf{x}$ is that the order of execution of processes is immaterial, since if a goal has a successful computation, then all of its computations are successful and produce the same answer substitution.

The determinism of P-Prolog$_\mathbf{x}$ limits it to algorithmic applications, since it cannot implement system programs such as a stream merger and an interrupt handler [20]. Most algorithmic concurrent logic programs can be written in P-Prolog$_\mathbf{x}$ quite easily, without the need to distinguish between matching and unification. This implies that some P-Prolog$_\mathbf{x}$ programs can be used in more than one 'mode'. Consider, for example, the P-Prolog$_\mathbf{x}$ *append* program:

---

20 Although an *ad hoc* extension to allow this was proposed [211]. Another extension, ALPS, is discussed in the next section.

```
append([X|Xs],Ys,[X|Zs]) ← append(Xs,Ys,Zs).
append([ ],Ys,Ys).
```

This program can be used to append two lists, as usual. However, it can also be used to compute the difference between a list and its prefix, using, e.g., the call:

```
append([1,2,3],Ys,[1,2,3,4,5,6]).
```

This is possible since at most one clause head unifies with the initial goal, as well as with subsequent goals, and hence goal reduction can proceed.

The practical advantage of this 'multiple-mode' ability is questionable. In practice, few logic programs are used in more than one mode. When they do, the two common modes are output generation and testing, which can be employed by all other concurrent logic languages mentioned, rather then inverting the roles of input and output within a single clause, which is unique to P-Prolog$_x$ and its superset ALPS (and is available in a more restricted sense also in FCP(?) and FCP(:,?) introduced below).

Furthermore, P-Prolog$_x$ uses unification in the head. As mentioned in the discussion of FCP(?) in Section 15 below, this generality seems to impede program readability and maintainability, since often the intended mode of use is known and fixed, but is not communicated by the code.

Embedding P-Prolog$_x$ in FCP(|)

The implementation of P-Prolog$_x$ is not trivial. A naive implementation would be to try all clauses whenever a process reduction is attempted; return to the successful clause if only one exists, or suspend on all variables instantiated during clause tries if there were more than one successful clause try. The overhead of this scheme seems unacceptable. An efficient implementation of P-Prolog$_x$ seems to require a complete analysis of all possible call patterns, which is also quite complex.

To establish the relation between P-Prolog$_x$ and other languages in the family, we show here an embedding of P-Prolog$_x$ in FCP(|). The idea of the embedding is as follows. For each goal atom in the source program we create a controlling process, and for each source clause potentially unifiable with this atom we create a reduction process simulating the attempt to reduce the goal atom with the clause. The reduction process operates as follows. If it detects that it cannot perform the simulated goal/clause reduction, it informs the controller. If it receives a permission from the controller to reduce, it simulates the reduction.

The controlling process counts the number of clause try failures, and when all but one clause have failed, it permits the remaining one to try and reduce. This behavior is achieved by the following translation[21].

Each P-Prolog$_x$ clause $A \leftarrow G|B$ is translated into an FCP(|) procedure with three clauses. The purpose of the first clause is to fail as soon as it is determined that the goal atom does not unify with the head of the source clause or the guard fails. It can never succeed. The second clause informs the controller if the first clause has failed. The third clause reduces if permission is given from the controller.

Specifically, let $C_1, C_2, \ldots, C_k$ be the clauses of the P-Prolog$_x$ procedure p/n. It is translated into k+1 FCP(|) procedures, p/n, $p_1/n+2$, $p_2/n+2$, ..., $p_k/n+2$, which

[21] This translation was developed in collaboration with M. Maher, and benefited from comments by V.A. Saraswat.

use two auxiliary procedures, as follows. The ith clause $p(T_1,T_2,\ldots,T_n) \leftarrow G|B$ of the P-Prolog$_x$ procedure p/n is translated into the FCP(|) procedure:

$$p_i(T_1,T_2,\ldots,T_n,\_,foo) \leftarrow G \mid true.$$
$$p_i(\_,\_,\ldots,\_,Failed,\_) \leftarrow otherwise \mid Failed{=}failed.$$
$$p_i(X_1,X_2,\ldots,X_n,go,\_) \leftarrow G \mid (X_1,X_2,\ldots,X_n){=}(T_1,T_2,\ldots,T_n),\ B.$$

The FCP(|) procedure p/n is defined as follows:

$$
\begin{aligned}
&p(X_1,X_2,\ldots,X_n) \leftarrow\\
&\quad p_1(X_1,X_2,\ldots,X_n,S_1,\_),\\
&\quad p_2(X_1,X_2,\ldots,X_n,S_2,\_),\\
&\quad \vdots\\
&\quad p_k(X_1,X_2,\ldots,X_n,S_k,\_),\\
&\quad xor_k(S_1,S_1,\ldots,S_k),
\end{aligned}
$$

where xor_k is defined as follows:

$$
\begin{aligned}
&xor_k(Go,failed,failed,\ldots,failed) \leftarrow Go{=}go.\\
&xor_k(failed,Go,failed,\ldots,failed) \leftarrow Go{=}go.\\
&\vdots\\
&xor_k(failed,failed,\ldots,failed,Go) \leftarrow Go{=}go.
\end{aligned}
$$

with *Go* on the diagonal, and *failed* anywhere else.

The translated program operates as follows. The procedure p/n spawns k parallel clause processes p_i, one for each of the original p/n clauses, plus a xor_k process. If the ith clause process fails it unifies the S_i variable with *failed*. The xor_k process counts $k-1$ failures, and unifies *go* with remaining variable, which enables the remaining clause process to reduce if it has not failed yet. Note that the FCP(|) program fails whenever the source P-Prolog$_x$ program fails.

The translation assumes that the unification implied by P-Prolog$_x$'s Reduce transition is atomic. If it is non-atomic, then the exact same embedding can be used with FGHC$_{av}$ or FGHC$_{nav}$ as the target language, depending on the kind of nonatomicity allowed.

13. ALPS — An Integration of P-Prolog$_x$ and FGHC

ALPS was proposed by Maher [125] as an algorithmic concurrent logic programming language. ALPS goal reduction rule states that a goal can be reduced with a clause if either this is the only candidate clause left (the P-Prolog$_x$ rule), or the reduction does not instantiate variables of the goal (the FGHC and FCP(|) rule).

In particular, the FGHC unification primitive is definable in ALPS using the single unit clause:

$$X = X.$$

The reduction of the goal $T_1 = T_2$ with this clause is enabled if T_1 and T_2 are unifiable, using the P-Prolog rule, since this is the only candidate clause. Unlike FCP(|), and like FGHC, the unification specified by such a goal need not be carried out atomically. In particular, the transition system of ALPS defined by Maher realizes non-atomic variables, as in FGHC$_{nav}$.

ALPS was defined in the general setting of constraint logic programming [92]; we address this aspect of the language in Section 21.

Embedding $FGHC_{nav}$ and P-$Prolog_x$ in ALPS

$FGHC_{nav}$ can be embedded in ALPS using a compiler that duplicates each clause, and the identity viewer. Clause duplication prevents the resulting ALPS program from "eagerly" reducing using the determinacy rule, since no goal is ever determinate[22]. P-$Prolog_x$ can be embedded using the embedding into FCP(|), shown in the previous section, assuming unification need not be carried out atomically. ALPS can be embedded in FGHC much the same way that P-$Prolog_x$ was embedded in FCP(|).

Discussion

The transition rules of ALPS are more 'eager' than those of FGHC. This means that some programs which deadlock as FGHC programs may proceed as ALPS programs. The practical implications of this difference are yet to be determined. The benefits in terms of added expressiveness are unclear, and the comments on P-$Prolog_x$ apply here as well. In addition, the difficulties in efficient implementation of the ALPS language, compared with FGHC, seem substantial.

14. FCP(:) — FCP(|) Extended With Atomic Test Unification

In FCP(|), FGHC and Flat PARLOG, a program can perform only matching prior to clause selection. In the next set of languages shown, FCP(:), FCP(?), and FCP(:,?)[23], a program can perform unification as part of the test for clause selection, prior to commitment. If the unification fails, it should leave no trace of its attempted execution; in other words, the unification attempt should be atomic. We call unification which is tried before commit *atomic test unification*. In FCP(|), atomic unification is a special predicate. In FCP(:) and FCP(:,?) it is definable, and in this sense these languages are natural generalizations of FCP(|).

The first flat language to combine input matching and atomic test unification is Saraswat's FCP(↓,|) [152,156]. This idea was generalized by Saraswat in the Ask-and-Tell framework [158], which gave rise to the concurrent constraint languages cc(↓,|) [158,159] and the language FCP(:) [101] described below.

14.1 The language FCP(:)

Syntax

Definition: FCP(:) clause and program.

- An FCP(:) *clause* has the form:

  $$A \leftarrow Ask : Tell \mid Body.$$

  where *Ask* and *Tell* are possibly empty conjunctions of atoms, *Ask* atoms have guard test predicates, and *Tell* contains only equality atoms. If *Tell* is empty, the colon is omitted.

---

[22] The clause duplication method is due to M. Maher.

[23] More precise but also more cumbersome names for these languages are, respectively, FCP(:,|), FCP(?,|) and FCP(:,?,|).

- An FCP(:) *program* is a sequence of FCP(:) clauses. ∎

Semantics

The effect of a clause try of a goal A with an FCP(:) clause with an empty tell part is the same as in FCP(|). If the tell part is not empty, the effect is as follows. First, the goal/head input matching and the guard checking are performed. If they fail or suspend, the clause try fails or suspends, respectively. If they succeed, then the unification specified by the tell is performed, which can either succeed or fail, but not suspend. If it succeeds, the result of the clause try is the substitution combining the ask substitution and the tell substitution. If it fails, the clause try fails.

Definition: Try function for FCP(:).

- Let $Tell = (X_1{=}Y_1,\ldots,X_n{=}Y_n)$ be a conjunction of equality atoms. We define $mgu(Tell) = mgu((X_1,\ldots,X_n),(Y_1,\ldots,Y_n))$, and the try function to be:

$$try_{\text{FCP(:)}}(A,(A' \leftarrow Ask{:}Tell|B)) =$$

$$\begin{cases} \theta \circ \theta' & \text{if } match(A,A') = \theta \wedge \text{checking } Ask\theta \text{ succeeds} \\ & \qquad \wedge\ mgu(Tell\theta) = \theta' \\ fail & \text{if } mgu(A,A') = fail \vee \\ & (mgu(A,A') = \theta \wedge \text{checking } Ask\theta \text{ fails}) \vee \\ & match(A,A') = \theta \wedge \text{checking } Ask\theta \text{ succeeds} \\ & \qquad \wedge\ mgu(Tell\theta) = fail \\ suspend & \text{otherwise} \end{cases}$$

Embedding of FCP(|) in FCP(:)

The embedding of FCP(|) in FCP(:) is trivial. All the compiler does is to replace the unification clause $X = X$ by the clause

X=Y ← true : X=Y | true.

This clause is necessary since '=' is a primitive in FCP(|) but not in FCP(:).

14.2 Programming in FCP(:)

Atomic test unification enables numerous programming techniques not available in any of the weaker languages introduced so far. These include multiple writers on shared variables, which can be used to realize sophisticated synchronization protocols and blackboard-like shared data structures; the ability to reflect on failure of unification, which enables the construction of failsafe meta-interpreters that can be used to realize the control meta-call; the ability to record the logical time in which a unification occurs, which is essential for computation replay and hence essential to concurrent algorithmic debugging; and the ability to simulate Prolog's test unification, and hence the ability to naturally embed Or-parallel Prolog and similar languages. We discuss these techniques below.

Mutual exclusion and multiple-writer streams

Using atomic test unification, single-round mutual exclusion can be achieved with less machinery than needed in FCP(|). Let $p_1,\ldots,p_n$ be the processes wishing to participate in a single-round mutual exclusion protocol, with unique identifiers $I_1,\ldots,I_n$. Add to each process variable ME as an argument. Each process p_k competing for a lock attempts nondeterministically unify its identifier I_k with ME, or to check that ME is already instantiated to some $I \neq I_k$.

A schematic description of each process is as follows. The kth process call is $p(ME,I_k,\ldots)$.

p(ME,I,...) ← true : ME=I | ... *lock granted* ...
p(ME,I,...) ← ME ≠ I | ... *lock denied* ...

This technique is not a substitute to the multiple-round mutual exclusion protocol shown in Section 7. However, in the special case that in each round the number of competing processes decreases by one, it can be generalized, as follows.

Assume a set of processes $p_1,\ldots,p_k$, where each p_i may wish to deposit a message m_i on a shared stream *Ms*. Furthermore, assume that the messages are pairwise not unifiable. One solution is to create a merge network for all these processes. However, if the number of processes actually wishing to deposit their message on the stream is much smaller than k (as is the case with exceptional message streams), this solution is very wasteful. A more efficient solution in this case is to extend the single-round mutual exclusion protocol above to streams, as follows. When wishing to deposit a message on *Ms*, the process nondeterministically attempts to do so, or to check that another message is already there. In the second case it calls itself recursively with the tail of the stream. Assume each process p_i is called with *Ms* as its first argument and m_i as its second, the code of a process is as follows:

p(Ms,M,...) ← true : Ms=[M|_] | ... *message sent; do other things...*
p(Ms,M,...) ← Ms=[_|Ms′] | p(Ms′,M,...).

Using this protocol, if the number of messages to be placed on *Ms* is finite, every process wishing to place a message on *Ms* will eventually do so (assuming And-fairness).

The dining philosophers

The seminal problem of mutual exclusion is that of the dining philosophers [37]. In this problem n philosophers are sitting at a round table, with one fork between each two philosophers. To eat, a philosopher requires two forks. Each philosopher goes through a cycle of eating and thinking. The problem is to provide the philosophers with an algorithm that guarantees that they will not deadlock, and that no philosopher will starve.

Using atomic test unification on multiple-writer streams it is easy to specify a deadlock-free behavior for philosopher:

```
phil(Id,[eating(LeftId,done)|Left],Right) ←      % Left is eating, wait till
     phil(Id,Left,Right).                         % he is done.
phil(Id,Left,[eating(RightId,done)|Right]) ←     % Right is eating wait till
     phil(Id,Left,Right).                         % he is done.
phil(Id,Left,Right) ←                             % Atomically grab both forks
     true : Left=[eating(Id,Done)|Left′],
     Right=[eating(Id,Done)|Right′] |
     ... eat, when done unify Done=done,
     then think, then become:
          phil(Id,Left′,Right′).
```

The program is independent of the number of philosophers dining. A dinner of n philosophers can be specified by the goal:

phil(1,Fork1,Fork2), phil(2,Fork2,Fork3),..., phil(n,Fork$n$,Fork1).

whose execution results in each of the ***Fork*** variables being incrementally instantiated to a stream of terms ***eating(Id,done)***, with the ***Id***'s on each ***Fork*** reflecting the order in which its two adjacent philosophers use it. For example, a partial run of this program on a dinner of 5 philosophers provided the substitution:

Fork1 = [eating(1, done), eating(5, done), eating(1, done), eating(5, _) | _]

Fork2 = [eating(1, done), eating(2, done), eating(1, done), eating(2, _) | _]

Fork3 = [eating(3, done), eating(2, done), eating(3, done), eating(2, _) | _]

Fork4 = [eating(3, done), eating(4, done), eating(4, done), eating(3, done), eating(4, done) | _]

Fork5 = [eating(4, done), eating(4, done), eating(5, done), eating(4, done), eating(5, _) | _]

The run was suspended midstream in a state in which ***Fork4*** is free and the the 2nd and 5th philosophers are eating. Up to that point each of the philosophers ate twice, except 4 which ate three times.

This program is much simpler then the Parlog86 program for the dining philosophers in [148]. The key to its simplicity is indeed the ability of FCP(:) to specify atomic test unification: a philosopher atomically tries to grab both forks, excluding other philosophers from grabbing them. The mutual exclusion is obtained by unifying the head of the ***Fork*** stream with a term containing the unique ***Id*** of the philosopher.

The deadlock-freedom of the program is guaranteed by the language semantics. The program can be further enhanced to achieve starvation freedom as well.

The duplex stream protocol

Processes placing messages on a shared stream need not be competing; they can also cooperate, and use the shared stream for both communication and tight synchronization.

For example, consider a stream producer and a stream consumer, wishing to participate in the following interaction. When the consumer reads the stream, it wants to read all the messages produced so far by the producer. The producer produces messages asynchronously, but wishes to know whenever all messages it has produced so far have been read. This can be achieved using the following duplex stream protocol [154]. The producer places a message ***M*** on the stream wrapped as ***write(M)***. The consumer, when reaching the end of the stream, places on it a ***read*** message. From the consumer's point of view, successfully placing a ***read*** on the stream indicates that it has read all messages produced so far. From the producer's point of view, failing to place a ***write(M)*** message, due to the existence of a ***read*** message, is an indication that all previous messages have been read. This is realized by the following code, where ***produce(M,Ms,Ms',Status)*** places the message ***M*** on ***Ms***, returning the remaining stream ***Ms'***, and ***Status=new*** if all messages previous to ***M*** have been already read, ***Status=old*** otherwise. ***consume(Ms,Ms',Rs)*** returns in ***Rs*** the messages ready in ***Ms***, and in ***Ms'*** the remaining stream.

```
produce(M,Ms,Ms',Status) ← true : Ms = [write(M)|Ms'] | Status=old.
produce(M,[read|Ms],Ms',Status) ← Ms=[write(M)|Ms'], Status=new.

consume([M|Ms],Ms',Rs) ← consume'([M|Ms],Ms',Rs).

consume'(Ms,Ms',Rs) ← true : Ms=[read|Ms'] | Rs=[ ].
```

consume′([write(M)|Ms],Ms′,Rs) ← Rs=[M|Rs′], consume′(Ms,Ms′,Rs′).

consume is two-staged so that it would not place a *read* message on an initially empty stream.

If the producer waits every so often for the consumer to catch up, then *consume* always terminates.

The duplex protocol gives rise to a much more efficient and more flexible bounded-buffer protocol than the FCP(|) protocol shown in Section 7.3. It is more efficient, since there is no acknowledgement for every message, only one per 'batch'. It is more flexible, since the producer can change its mind on how many messages to send without an acknowledgement, without consulting or affecting the consumer, and with no need to change 'buffer-size'.

CSP with both input and output guards

To demonstrate the power of atomic test unification, we show an FCP(:) simulation of CSP with output guards [87]. CSP with output guards is notoriously difficult to implement, and hence Occam [91], the practical realization of CSP, adopts only input guards. This restriction is significant since the natural implementation of bounded buffers in CSP requires output guards. It is interesting to note that a logic programming language with matching is sufficient to simulate CSP with input guards, but a language with both matching and atomic test unification seems to be required to simulate CSP with both input and output guards.

Consider two sets of processes $p_1,\ldots,p_n$, $c_1,\ldots,c_n$, wishing to participate in the following interaction. Some (possibly all) of the p_i's wish each to interact with exactly one of the c_i's, but they do not care which. Some (possibly all) of the c_i's wish each to interact with exactly with one of the p_i's, but they do not care which. We would like a protocol, which, if there are $i \leq n$ p's and $j \leq n$ c's willing to interact, then $min(i,j)$ pairs will do so. The protocol should be independent of i and j, and allow i and j to increase dynamically.

The protocol is as follows [150]. Each p willing to interact sends to all the c's the incomplete message *hello*(X). All messages sent by the same p have the same variable X, and the variables in messages sent by different p's are distinct. Each c willing to interact does the following: it nondeterministically and atomically selects one of its incoming *hello*(X) messages and unifies X with its unique *Id*.

The program for the case of two p's and two c's is as follows:

p(X,ToC1,ToC2) ← ToC1=hello(X), ToC2=hello(X).

c(Id,hello(X_1),_) ← true : Id=X_1 | true.
c(Id,_,hello(X_2)) ← true : Id=X_2 | true.

The initial process network is:

p(X_1,M_{11},M_{12}), p(X_2,M_{21},M_{22}), c(a,M_{11},M_{21}), c(b,M_{12},M_{22}),

which corresponds to the CSP program $P_1 \| P_2 \| C_1 \| C_2$, where P_i is $M_{i1}!$ message $\square$ $M_{i2}!$ message and C_i is $M_{1i}?X \;\square\; M_{2i}?X$.

This process network terminates, and at the end of its execution exactly one of X_1 and X_2 will be instantiated to a, and the other b.

In this example the two p's and two c's were both willing to interact. However, the definition of p and c is applicable also in the more general case, in which less are willing to interact on each side, or that processes are added dynamically.

This demonstration of the power of atomic test unification also indicates that distributed implementation is far from being trivial. It is discussed in Section 20.

Otherwise and reflection on failure

In FCP(|) it is possible to prevent failure of user-defined processes, by appending to each procedure *p* the clause:

p(...) ← otherwise | ... *report failure* ...

However, there is no way to prevent the failure of the primitive unification process '='.

In FCP(:), on the other hand, since unification is definable, it is possible also to define failsafe unification using the clauses:

X = Y ← true : X=Y | true.
X = Y ← X ≠ Y | ... *report failure of unification* ...

More generally, it is possible to define a failsafe FCP(:) meta-interpreter, which, instead of failing when the interpreted program fails, simply reports the failure. To achieve this we modify the clause representation of the interpreted program, by appending to it the clause:

clause(A,B) ← otherwise | B = failed(A).

Using this representation, a termination detecting failsafe meta-interpreter for FCP(:) is defined as follows:

```
reduce(A,Result) ←
    reduce'(A,[ ]-Result).

reduce'(true,L-R) ← L = R.
reduce'((A,B),L-R) ← reduce(A,L-M), reduce(B,M-R).
reduce'(failed(A),L-R) ← R=[failed(A)|L].
reduce'(goal(A),L-R) ← clause(A,B), reduce'(B,L-R).
```

On a call *reduce(A,Result)*, *Result* is instantiated to the (possibly empty) stream of goals failed during the computation. The stream is closed when the computation terminates.

14.3 Embedding KL1 and Flat PARLOG in FCP(:)

The inability to reflect on failure without reifying unification made all the previous languages unable to implement the control meta-call efficiently. Therefore to make them practical this construct has to be introduced as a primitive into the language as discussed in Section 10.3.

We show how the control meta-call can be implemented in FCP(:), and thus provide an embedding of KL1 in FCP(:). Combined with the techniques used to embed FP(;) and FP(&) in $FGHC_{av}$, discussed in Section 11, the implementation of the control meta-call can be enhanced to provide an embedding of Flat PARLOG in FCP(:).

An implementation of the control meta-call in FCP(:)

The meta-call implementation consists of two components: a meta-interpreter, which can produce events and is sensitive to interrupts, and a computation monitor, which provides the user interface.

The meta-interpreter requires the same clause representation of the FCP(|) interruptible meta-interpreter shown in Section 7.7, augmented with the *otherwise* clause shown above and an interrupt-sensitive clause. Each FCP(:) clause (including the unification clause)

```
A ← Ask : Tell | B.
```

is translated into:

```
clause(A,X,Is) ← Ask : Tell | X=B'.
```

where B' is B transformed as in previous meta-interpreters, and two clauses are appended:

```
clause(A,B,Is) ← otherwise | B=failed(A).
clause(A,B,[I|Is]) ← A=B.
```

The first reports failure of a reduction attempt. The second aborts the reduction attempt when sensing an interrupt. Note that the order of the last two clauses is important: if they were switched, then the meta-level process *clause(A,B,Is)* executing a failing object-level process *A* will suspend on the interrupt stream *Is* rather then reporting failure. This is another demonstration of both the subtlety and power of *otherwise*.

Using this representation, the following meta-interpreter achieves the desired behavior:

```
reduce(true,Is,Ss,L-R) ← L=R.
reduce((A,B),Is,Ss,L-R) ← reduce(A,Is,Ss,L-M), reduce(B,Is,Ss,M-R).
reduce(goal(A),Is,Ss,L-R) ←
    clause(A,B,Is), reduce(B,Is,Ss,L-R).
reduce(failed(A),Is,Ss,L-R) ←
    write(failed(A),Ss), L=R.
reduce(A,[I|Is],Ss,L-R) ←
    serve_interrupt([I|Is],A,Ss,L-R).

write(M,Ms) ← true : Ms=[M|_] | true.
write(M,[M|Ms]) ← write(M,Ms).
```

The differences between this and the snapshot meta-interpreter of FCP(|) shown in Section 7.7 above are the additional signals stream *Ss*, the clause added for handling process failure and the lack of a special clause for unification. The latter is not needed since the clause defining '=' is a normal FCP(:) clause which requires no special treatment. Failure is handled by placing an appropriate message on the signals stream, using the multiple-writer stream protocol. This is an example where creating a merger for each forked process would have had an unacceptable overhead. Assuming either low rate of process failure, or that the computation is suspended by the controller as soon as failure is detected, the multiple-writer protocol would exhibit a much better performance.

Note that if two unifiable processes fail, only one message is produced on the signals stream. This oddity can be solved either by allocating unique identifiers to the meta-interpreter processes (which is inelegant and quite expensive), or, in FCP(:,?), using the anonymous mutual exclusion protocol, discussed in Section 15.

An alternative solution which does not use a multiple-writer stream is to use the short-circuit in order to report failure, as in the failsafe FCP(:) meta-interpreter shown above. The disadvantage of this approach is that the list of failed goals will be seen only upon termination of the computation. A computation monitor, which suspends the computation as soon as a failed goal is sensed, cannot be programmed using this technique.

The definition of the computation monitor should be quite obvious now. Its top level is the same as the meta call, *call(Goal,Signals,Events)*. It invokes the meta-interpreter, keeping hold of the ends of its short circuit streams. It serves signals coming from the outside by forwarding them to the meta-interpreter, via the interrupt stream, and reports on events that happen during the computation by placing them on the *Events* stream.

The meta-interpreter given serves as a specification of the required functionality of the control meta-call. This functionality can be implemented by source to source transformation. The transformation presently employed in the Logix system [84] which achieves this functionality results in about 30% increase in runtime and 80% increase in code size. In [46], Foster reports an experimental study that quantifies the cost of direct support for metacontrol functions, and compares this with the cost of support by program transformation. The same paper describes extensions to an existing abstract machine [49] required to support these functions. This study indicates that direct support for the control meta-call need not be expensive, nor require complex implementation mechanisms.

<u>Discussion: atomic test unification vs. non-atomic unification</u>

It is a subject of ongoing debate whether it is preferable to have a stronger language which can embed meta-level functions such as the control meta-call, or to have a weaker language and provide specific meta-level functions as language extensions.

The issue seems to be a tradeoff between simplicity at the implementation level versus elegance and expressiveness at the language level. On one side of the debate are Flat GHC and Flat PARLOG, with non-atomic unification. On the other side are FCP(↓,|), FCP(:), FCP(?), and FCP(:,?), languages, with atomic test-unification.

The main arguments for a weaker language, with non-atomic unification and a built-in control meta-call, are:

- The base language is simpler to implement;
- The specialized meta-level construct can be added with less overhead than via a general-purpose language mechanism.
- The base language has simpler formal semantics, and is therefore better amenable to theoretical treatment such as verification and transformation.
- Atomicity of unification is not assumed by the theory of (pure) logic programming. Therefore, it is important to write programs without relying on atomic unification whenever possible, and a language with non-atomic unification encourages it. The resulting programs allow better declarative reading[24].

The main arguments for a stronger language, which has atomic test unification and can implement meta-level constructs via interpretation and transformation are:

- Providing semantics for any specific meta-level construct as part of the base language is both complicated and *ad hoc* (we know of no formal semantics for the control meta-call or similar constructs, other then the one implied by the semantics of FCP(:) combined with the definition of the control meta-call).

[24] These last two points were communicated by K. Ueda.

- The need for stronger meta-level constructs is continuously evolving (e.g. live and frozen snapshots, sophisticated debuggers, etc. which are not provided by the control meta-call). If these needs are met at the language definition level, rather than by interpretation and transformation, the language semantics as well as implementation have to be continuously modified.
- When atomic test unification is not employed, there is little or no runtime penalty compared to implementations of the weaker languages.
- Should the efficiency of a direct implementation of a certain meta-level function be required, it can be provided without affecting the language semantics. Such a direct implementation can be viewed as a (possibly hand-coded) specialization of a function that could be provided by the language itself.
- There are other applications in which the added strength of atomic test unification is employed, such as embedding Or-parallel Prolog [167], and debugging (see below).
- It is not obvious at present that the semantics of the weaker languages is indeed simpler.

Recently, Saraswat has proposed combining both atomic test unification and non-atomic unification in a single language [159]. Such a language inherits the complexities of both approaches, and it is not clear at present what performance gains it allows.

14.4 Computation replay and debugging

One type of bug which is most difficult to diagnose in concurrent programs are transient, or lurking, bugs. Once a bug occurs in a sequential deterministic language, it is possible to repeat the computation and analyze it with various tools. This is not always possible in a concurrent program, unless special measures are taken. Specifically, all communication and all nondeterministic (scheduler and program) choices made during a computation must be recorded, so that if an erroneous behavior is observed, the computation can be repeated.

We show an FCP(:) meta-interpreter that records scheduler and program (i.e. And- and Or-nondeterministic) choices made by the interpreted program. This information is sufficient in order to reconstruct closed (non-reactive) computations, in which all communication happens internally. The meta-interpreter computes a tree data-structure called a *trace*, which reflects the process reductions occurred in the computation. Each node in the trace contains the pair ⟨*Time,Index*⟩, with the time in which the process in that node reduced, and the identity of the clause used for reduction. Given an initial goal and a trace of its computation, the computation can be repeated by redoing the process reductions specified by the trace in the order specified by the *Time* field of each node, and for each reduction selecting the clause specified by the *Index* of its node.

To construct such a trace, we assume that the underlying machine maintains logical clocks [108], and that the language provides a new primitive, *time*(*Time*), which unifies *Time* with the present value of the local logical clock. The clause representation is modified, to provide additional information on the clause reduction: the logical time in which it took place, and the identity of the clause chosen[25].

[25] Inability to record the time in which a unification occurs is what prevents the weaker languages shown from replaying computations.

The ith clause $A \leftarrow Ask : Tell \mid B$ of the program is transformed into the clause:

$$\text{clause(A,X,Index,Time)} \leftarrow \text{Ask : Tell, time(Time)} \mid \text{X} = \text{B}', \text{Index} = i.$$

Using this representation, a meta-interpreter that constructs a trace is defined as follows:

```
reduce(true,true).
reduce((A,B),T) ← T=(T1,T2), reduce(A,T1), reduce(B,T2).
reduce(goal(A),T) ← T=trace(Index,Time,SubTrace),
    clause(A,B,Index,Time), reduce(B,SubTrace).
```

A computation reconstructor, which repeats a computation given an initial goal and a trace, can be written quite elegantly using incomplete data-structures. It first serializes the trace using the *Time* field, then executes the reductions in order, one by one. We do not show it here.

Given the ability to reconstruct a computation, algorithmic debugging techniques [161] can be applied to concurrent programs as well. See [178,119,121] for details.

14.5 An embedding of Or-parallel Prolog in FCP(:)

The question of how to provide the capabilities of Prolog in a concurrent logic programming languages has received considerable attention since the beginning.

One approach was pursued by PARLOG [24,25,66], namely to provide two sub-languages with an interface: the single-solution sublanguage, which is the counterpart of other concurrent logic programming languages, and the all-solutions sublanguage, which is essentially an all-solutions Or-parallel Prolog. A stream-like interface allows single-solution programs to invoke and control all-solution programs.

Another approach was to embed Prolog in a concurrent logic language. The first success in this direction was Kahn's Or-parallel Prolog interpreter in Concurrent Prolog, discussed in Section 18. However, this interpreter relies in an essential way on the non-flat nature of Concurrent Prolog. Initial attempts by Ueda [201, 202] and Codish and Shapiro [29], were successful in producing efficient translations when the mode of unification of the source Prolog program could be determined at compiler time. A more general, but less efficient, solution is described in [167], in the form of an Or-parallel Prolog interpreter written in FCP(?), a language introduced in Section 15. Although originally written in FCP(?), the interpreter does not exploit properties of it not available in FCP(:), and can be easily converted to this language. This implementation is not as direct as the interpreter in Concurrent Prolog, but is still quite simple. Furthermore, if the mode of subprograms can be determined, the interpreter can be gracefully interfaced to programs implemented using the transformations proposed by Ueda. The execution algorithm employed by the interpreter was proposed independently for other purposes [4,5,28]; nevertheless, its practicality is still under debate.

This embedding employs atomic test unification to implement Prolog's unification. Hence, unlike the disjoint-sublanguages approach, or the mode-based compilation, which is applicable to any concurrent logic language, the embedding approach is not applicable to languages such as FCP(|), (Flat) GHC, and (Flat) PARLOG. Should the execution algorithm employed by the embedded-language approach prove efficient in practice, its advantage over the disjoint-sublanguages approach would become apparent, especially in the presence of specialized hardware for the execution of concurrent logic languages [6,74].

A variant of the algorithm can be implemented also in Flat PARLOG or KL1, using the control meta-call. However, such an implementation would be hopelessly inefficient, since it would require a new meta-call at every choice point, and cannot prune alternatives using test unification as done in direct implementations of Prolog or in the Prolog interpreter in FCP(?).

Another approach, pursued by Saraswat [152,155,159] and Yang and Aiso [210,211] was to incorporate in concurrent logic languages don't-know nondeterminism. As the resulting languages cannot specify reactive concurrent systems, it is not an extension or a substitute for concurrent logic languages. Assuming that an underlying reactive concurrent logic language is still desired, the problem of integrating if with a parallel don't-know nondeterministic logic languages is much the same as that of integrating Prolog: it can either be implemented separately, with some all-solutions interface, as in the two sublanguages approach, or it can be compiled into a concurrent logic language, as in the embedding approach. This is discussed further in Section 21.

15. FCP(?) — Dynamic Synchronization With Read-Only Variables

The language Concurrent Prolog [162] introduced a different approach to synchronization, using read-only variables and read-only unification. The approach is preserved in its flat subset Flat Concurrent Prolog [129], also known as FCP, and called throughout the paper FCP(?) (read "FCP read-only").

FCP(?) assumes two types of variables, writable (ordinary) variables, and read-only variables, and uses read-only unification, which is an extention of ordinary unification, to unify terms containing read-only variables. The *read-only operator*, *?*, is a mapping from writable to read-only variables. When applied to a writable variable X, the read-only operator yields a *corresponding read-only variable* $X?$. The read-only operator is the identity function on terms other than writable variables.

In the absence of read-only variables read-only unification is just like ordinary unification. However, a read-only variable $X?$ cannot be unified with a value. An attempt to unify $X?$ with a term other than a writable variable suspends. When the writable variable X is instantiated to some value T (by some concurrent unification) its corresponding read-only variable $X?$ receives the value $T?$. This may release a unification suspended in an attempt to unify $X?$ with some value.

Whereas synchronization with matching is specified clause-wise and statically, synchronization with read-only unification is specified term-wise and dynamically. Read-only unification can be used to achieve various forms of dynamic synchronization, not acheivable otherwise.

15.1 The language

Syntax

The syntax of FCP(?) is the same as FCP(|), except that a clause may contain read-only variables.

Semantics

The semantics of the language is similar to FCP(|), except that goals may contain read-only variables, and the goal and the clause head are unified using read-only unification instead of matching.

Definition: Admissible substitution, read-only extension, read-only mgu.

- A substitution θ is *admissible* if $X?\theta = X?$ for every variable X.
- The *read-only extension* of an admissible substitution θ is the unique idempotent substitution $\theta_?$ satisfying $X(\theta_?) = X\theta$ and $X?\theta_? = (X\theta)?$ for every writable variable X.
- The *read-only mgu*, $mgu_?$, of two terms T_1 and T_2 is defined by:

$$mgu_?(T_1,T_2) = \begin{cases} \theta_? & \text{if } mgu(T_1,T_2) = \theta,\ \theta \text{ admissible} \\ fail & \text{if } mgu(T_1,T_2) = fail \\ suspend & \text{otherwise} \end{cases} \qquad \blacksquare$$

For example, $\{X \mapsto a\}$ is admissible but $\{X? \mapsto a\}$ and $\{X \mapsto a, X? \mapsto a\}$ are not. The read-only extension of $\{X \mapsto a, Y \mapsto Z\}$ is $\{X \mapsto a, X? \mapsto a, Y \mapsto Z, Y? \mapsto Z?\}$. $mgu_?(f(X,Y),f(a,Z)) = \{(X \mapsto a, X? \mapsto a, Y \mapsto Z, Y? \mapsto Z?\}$, and both $mgu_?(X?,a)$, $mgu_?(f(X,X?),\ f(a,a))$, and $mgu_?(f(X,X?),\ f(a,b)) = suspend$[26].

The try function of FCP(?) is the same as that of FCP(|), except that it uses $mgu_?$ instead of *match*, and it returns *suspend* if the read-only unification of the goal and the head is inadmissible due to read-only goal variable, and *fail* if it fails or is inadmissible due to a read-only clause variable only (since the latter state is stable).

15.2 FCP(?) programming techniques

Standard programming techniques

All standard programming techniques shown for FCP(|) and FCP(:) are realizable also in FCP(?). However, for most of the simple synchronization tasks, the generality and the dynamic nature of read-only unification turns out to be more of a burden than an asset. Since read-only unification is an extension of unification, using it for goal/clause unification is closer to the original model of logic programming and Prolog. Nevertheless, in concurrent logic programming, matching is used more often than unification. The default in FCP(?) encourages programmers to use unification even when matching is needed, and instead restrict the use of the procedure by placing read-only variables in the caller. For example, consider the FCP(?) procedure *append*.

```
append([X|Xs],Ys,[X|Zs]) ← append(Xs?,Ys,Zs).
append([ ],Ys,Ys).
```

The procedure is almost identical to the logic program (and Prolog program) *append*. The only difference is the read-only annotation in the recursive call. Nevertheless, this program has awkward behavior. Although its head specifies unification, the intention is that the first argument be matched. The program ensures this for recursive calls, but not for the initial call. If the initial goal is *append(Xs, Ys, Zs)* rather than *append(Xs?, Ys, Zs)*, the first (or second) clause can be chosen erroneously. Placing this responsibility on the caller is a source of non-modularity and bugs. In addition, matching can be compiled more efficiently than unification [100]. Without global analysis, which infers that the caller always places a read-only variable in the appropriate posi-

---

26 This definition of read-only unification is different from the original one [162], in that it is order-independent and disallows "self-feeding", i.e. the success of $f(X,X?)=f(a,a)$. The revision was influenced by criticism of the earlier definition [154,198], and by the language CP(%) of Ramakrishnan and Silberschatz [145].

tion [30,187], an FCP(?) program would compile less efficiently than a corresponding program in a language with input unification.

A later definition of FCP [172] allowed both a matching predicate =?= and unification in the guard. Using a matching guard, the recursive clause of append could be specified as:

$$\text{append(Xs,Ys,[X|Zs])} \leftarrow \text{Xs =?= [X|Xs']} \mid \text{append(Xs',Ys,Zs).}$$

However, since this syntax is more verbose than the default one, programmers would still use the previous style, resulting in programs which are both more error prone and less efficient. In addition, it turned out to be difficult to define cleanly the try function for guards which contained a free mix of matching and unification predicates [175].

It seems, therefore, that the approach taken by the other flat languages, namely to use matching as the default, is better. The language FCP(:,?), discussed in the next section, attempts to unify the expressiveness of FCP(?) with the more convenient and efficient programming style of the other languages.

Test-and-set

One use of read-only variables is to implement various forms of a test-and-set operation. A variable can be tested to be unbound and then set to a non-variable term T in two stages: First unify it with a new read-only variable $X?$ and if successful unify T with X:

$$\text{test\_and\_set(X?,T)} \leftarrow \text{X=T.}$$

The definition of read-only unification implies that the clause try will succeed with the goal *test_and_set*(X,T) if and only if X is a variable at the time of the try. The technique directly generalizes to simultaneous test-and-set of several variables.

The ability to implement test-and-set implies that FCP(?) is not success stable. For example, *test_and_set*(X,a) succeeds with X instantiated to a, but *test_and_set*(a,a) fails.

We note that test-and-set can be also realized in FCP(:), augmented with the *var* guard primitive, but not in any of the weaker languages.

Anonymous mutual-exclusion, multiple-writer streams and distributed queues

The ability to test-and-set can be used to implement anonymous mutual exclusion, that is, mutual exclusion without unique identifiers. For example, a multiple-writer stream, which preserves message multiplicity even in the presence of unifiable messages (in contrast to the FCP(:) program shown in Section 14 above) can be defined as follows:

$$\text{write(M,[X?|Ms],Ms)} \leftarrow \text{M=X.}$$

$$\text{write(M,[\_|Ms],Ms')} \leftarrow \text{write(M,Ms,Ms').}$$

The third argument *Ms'* can be used to place subsequent messages on the stream. It ensures that the next message is placed after the previous one, so a writer can ensure that its own messages are ordered. Even if a writer placing several messages on a stream does not care for their order, he could still use *Ms'* instead of *Ms* for subsequent messages, to increase efficiency.

Using this procedure, placing n messages by n writers on one stream requires $O(n^2)$ steps. By introducing a special abstract data type, called *mutual-reference* [170], the three argument *write* operation specified by the above program can be implemented by a destructive assignment so that the cost of sending n messages is $O(n)$. The implementation is also 'better' than the specification in another respect: assuming And-fairness it guarantees that every *write* operation will eventually complete, even in the presence of an unbounded number of writers, a property not guaranteed by the program above. Mutual-references are the standard technique for realizing efficient stream mergers. Whenever we use a multiplicity-preserving multiple-writer stream in a program we assume it is implemented efficiently and fairly using mutual-references.

Another application of anonymous mutual exclusion is a distributed queue [167]. In it, client processes are at the leaves and queue processes are at the internal nodes of a process tree. Each *enqueue(X,ME)* or *dequeue(X,ME)* request is sent up the tree from the leaf process which generated it, with X carrying the element to enqueue or dequeue and ME being a new mutual exclusion variable. If a queue process at a node can satisfy the request by matching it with a locally stored corresponding request, it does so. Otherwise it keeps a copy of the request in its local queue, and also sends a copy of it to its parent. A request is matched with a corresponding request by atomically testing the ME fields of the two requests to be variables and setting them to some value. When attempting to match the requests, the queue process also nondeterministically checks whether the ME field of any of the requests has been set by another queue process; this indicates that the request has been satisfied by some other queue, and so it is discarded.

Such a distributed queue can be used for dynamic load balancing, where workers off-load work by enqueing, and request work by dequeing [192]. It is very suitable for this application since requests are satisfied locally whenever possible, but eventually get to the most global queue (the root queue) if necessary.

Protected data-structures

Another important application of read-only variables is to protect processes communicating across trust boundaries. Consider an operating system process interacting with a possibly faulty user process via an incomplete message protocol, or by incrementally producing some data structure.

If the user process does not obey the protocol, and instead of waiting for the operating system process to instantiate some variable it instantiates this variable to some erroneous value, it may cause the operating system process to fail.

Several proposals were made to solve this problem. One is to restrict the type of communication protocols allowed between user processes and system processes, and provide user processes only with complete data-structures, with no 'holes' to mess with. This solution greatly decreases the flexibility of the interaction, and puts a heavy synchronization and termination detection burden on the operating system.

Another solution is to isolate the components of the operating system interacting with user processes, and provide them with robust failure handling mechanisms. This solution also seems infeasible, since incomplete data structures can be passed asynchronously between system components, and therefore user processes may share variables with arbitrarily 'deep' operating system components.

Another solution, adopted by the operating system designed by ICOT, is to use specialized filter processes to monitor user-system interaction. These processes forward back and forth instantiations done by the interacting processes, as long as the user processes obey the protocol which the operating system expects. When a violation by the user is detected, the filter does not pass it further to the system. Foster [48] describes three techniques for achieving robustness in operating systems implemented in languages that do not support read-only variables: at-source (by transformation of user programs), en-route (by filters) or at-destination (by making system programs fail-safe). The second technique is shown to be generally the most effective.

Read-only variables allow a simpler solution [76]. An operating system component which produces a data-structure incrementally can protect the incomplete part of the data structure from outside intervention. This is done by making it read-only to its readers, and keeping the writable access to oneself. This is achieved by the producer placing a read-only variable *X?* in every 'hole' in the data structure, and by keeping an occurrence of *X* for further writing. For example, a protected-stream producer can be defined as follows:

p([X|Xs?],...) ← p(Xs,...).

If, when *p(Xs,...)* is invoked, it has the only writable occurrence of its first argument *Xs*, this invariant will hold in all future iterations of the process, and no consumer can interfere with the stream production. If the *Message* itself is also produced incrementally, it could also be protected using the same technique.

Discussion

The advantages of read-only unification over matching is that it is a generalization of unification, rather than a special case of it: read-only unification in the absence of read-only variables is just unification. Hence read-only unification achieves both communication and synchronization with a single notion. Second, read-only unification is symmetric: unlike matching, it does not distinguish between the goal and the clause, and the read-only unification of any two terms behaves alike. Third, it is dynamic. Read-only variables can be embedded in any data-structure, hence synchronization can be associated with data, not only with procedures.

Some of the disadvantages of read-only unification come from its strenght: Not being success stable makes it harder to analyze statically FCP(?) programs, and often makes FCP(?) less readable compared to programs using input matching. Its non-monotonic nature makes it more difficult to analyze theoretically, compared to languages which use only input matching and unification. Finally, it has some points of singularity (e.g. the unification of *X* with *X?*), which do not seem to have acceptable intuition behind them.

An alternative concept, called locks, was proposed by M. Miller and E.D. Tribble and formalized by Saraswat [158]. Its motivation was to provide more reasonable semantics to the unification *X*=*X?*. In FCP(?), this unification subtracts the writing capability from *X*, making it read-only. In the alternate proposal, its effect it to make both *X* and *X?* writable. The ability of a read-only variable to become writable gives rise to both additional complications and additional programming techniques, though it has not been pursued to completion.

16. FCP(:,?) — An Integration of FCP(:) and FCP(?)

The language FCP(:,?) [101] attempts to integrate the convenience and efficiency of matching with the expressiveness of atomic test unification and read-only variables. In addition, it has the added pragmatic advantage over FCP(?) of being a superset of Flat GHC, FCP(|), and FCP(:), in the sense that every program in these languages would execute correctly as an FCP(:,?) program.

FCP(:,?) is as strong as any other language in the family, in the sense that there are natural embeddings of all languages in the family into it. It is the target language of the implementation effort at the Weizmann Institute [100].

Syntax

The syntax of FCP(:,?) is the same as that of FCP(:), except that the tell and body parts may contain read-only variables.

Semantics

The semantics of the language is also the same as FCP(:), except that in the tell part read-only unification is used instead of ordinary unification. This is reflected in the try function of FCP(:,?), which is the same as that of FCP(:), except that it uses $mgu_?$ instead of *mgu*, and returns ***suspend*** if the read-only unification in the ***tell*** part suspends on a read-only goal variable, and ***fail*** if it fails or suspends on a read-only clause variable (since the latter kind of suspension is stable).

Programming in FCP(:,?)

As mentioned above, any FGHC, FCP(|) or FCP(:) would execute correctly as an FCP(:,?) program. The FCP(?) programs shown in the previous section easily translate into FCP(:,?). For example, the multiple-writers stream is written as follows:

```
write(M,Ms,Ms') ← true : Ms=[X?|Ms'] | M=X.
write(M,[_|Ms],Ms') ← write(M,Ms,Ms').
```

and the protected stream producer as follows:

```
p(Xs,...) ← true : Xs=[Message|Xs'?] | p(Xs',...).
```

17. Doc — "$X = X$ Considered Harmful"

The language Doc (Directed Oc) by M. Hirata [82], is a successor to Oc [81,83]. Oc is essentially FGHC_{nav} with no guards. Doc is a further restriction, which follows the motto "$X = X$ considered harmful". Doc is a concurrent logic programming language in which every variable has at most one writer and at most one reader, i.e. one process which instantiates a variable, and one process that matches it. This restriction is enforced syntactically, by annotating each variable occurrence as either a writable or a read-only, and requiring that each variable may occur at most once in each mode in a clause.

The motivation for this restriction is that the cost of broadcasting information in a distributed environment may be too expensive to be supported at the language level.

Discussion

Although the removal of variable-to-variable unification from logic programming seems a rather drastic proposal, its effect is not fatal, and the resulting language is still usable. The techniques available in Doc (except for protected data structures) are a subset of those available in FGHC_{nav}. In particular, the short-circuit technique and any of the techniques relying on atomic unification are not available in Doc. Furthermore, broadcasting is not available in Doc, and should be implemented by an explicit distributor process, which receives a message and distributes it separately to each recipient. In addition, Doc's read-only annotation is a reminiscent of the read-only variable, and indeed it can employ the protected data-structures technique; actually, a Doc process *must* protect any incomplete structure it intends to produce, by the syntactic restrictions of the language. Because of the ability to specify protected data-structures, it seems that Doc cannot be embedded in a language that does not contain the equivalent of read-only variables.

An embedding of Doc in broadcast-free FCP(?)

The similarity of Doc's annotations to writable and read-only variables in FCP(?) is apparent. Indeed, it is natural to consider a subset of FCP(?), which may be called[27] *broadcast-free* FCP(?), in which every variable may occur at most once read-only and at most once writable in each clause. Doc programs can be trivially translated into broadcast-free FCP(?).

This translation is valid, in the sense that every computation of the resulting FCP(?) program corresponds to a possible computation of the source Doc program. However, the translation is not an embedding in the sense used so far. Since the read-only unification used in FCP(?) is atomic, some executions of a Doc program cannot be realized by the corresponding FCP(?) program. This can be remedied by further "decoupling" variables in the clause, as done in the embedding of $FGHC_{nav}$ in $FGHC_{av}$ in Section 10, which masks the atomicity of unification of FCP(?). For each variable X that occurs both writable and read-only in a clause, replace $X?$ by a new variable $Y?$, and add the goal *send*($X?,Y$) to the body of the clause. *send* is defined as follows. For every function symbol *f/n* in the program, $n \geq 0$, *send* has the clause:

$$\begin{aligned}&\text{send}(\text{f}(X_1,X_2,\ldots,X_n),\text{f}(Y_1?,Y_2?,\ldots,Y_n?)) \leftarrow\\&\quad \text{send}(X_1?,Y_1),\ \text{send}(X_2?,Y_2),\ldots,\ \text{send}(X_n?,Y_n).\end{aligned}$$

We note that broadcast-free FCP(?) is still stronger than Doc, since it provides a variant of the short circuit technique. In this variant a ground message is sent around the circuit in a particular direction. Its arrival at the other end indicates termination.

18. Non-Flat Concurrent Logic Programming Languages: PARLOG, GHC, Concurrent Prolog, and CP(↓,|)

A concurrent logic programming language is non-flat if the guard of a clause may contain program defined predicates. Several of the flat languages described above — Flat GHC, Flat PARLOG, FCP(|,↓), and FCP(?) — were actually derived from their non-flat ancestors simply by restricting the guard to contain predefined predicates only.

The ability to define guard predicates implies that guard computations may be unbounded and, in general, may fail to terminate. Nevertheless, as in flat languages, a clause try is an atomic operation: it succeeds, suspends, or fails, and if it suspends or fails it leaves no trace of its attempt.

Two approaches were taken to ensure atomicity of a clause try; they are also reflected in the corresponding flat languages. One approach is to forbid guard computations to assign goal variables. This way several clauses can be tried in parallel for the same goal without interference. This approach is taken by PARLOG and GHC,

[27] The name was suggested by V.A. Saraswat.

and is reflected in their flat subsets in the restriction that guards can only do matching, not unification. The second approach is to allow guard computations to assign goal variables, but to make such assignments visible only upon commitment. This is reflected in the FCP languages, which allow test unification in guards, but require the unification attempt to be atomic.

We discuss the non-flat languages informally. Transition systems for non-flat languages are given by Saraswat [156] and Levy [115].

18.1 PARLOG and GHC

PARLOG and GHC are similar in their requirement that guard computations do not instantiate goal variables, but differ in the way they realize this requirement. In PARLOG, a syntactic compile-time check, called a *safety check* is performed to ensure that the program has no computations in which guards assign goal variables [23]. Since the question whether a program is safe is undecidable in general [29], any algorithm for determining safety can only perform an approximate check, and if it correctly rejects all unsafe programs then it is bound to reject some safe programs as well. This leads to the awkward situation in which the set of legal PARLOG programs is either undecidable, or is determined by an algorithm, whose specification may be both quite complex and evolving. The practice of PARLOG programming seems to be that the safety check is not done, and the responsibility of producing safe programs is placed on the programmer's intuition.

The design of GHC [199] was influenced by an earlier design of PARLOG [24], called PARLOG83 in [148], which employed output assignment instead of tell unification, and by critical examination of Concurrent Prolog [197]. Rather than ruling out the possibility of the guard instantiating goal variables by a syntactic check, GHC ensures this with its synchronization rule. In fact, the sole synchronization rule of GHC states that a unification in the head or the guard that attempts to instantiate a variable in the goal suspends.

The implementation of this synchronization rule in full GHC requires recording for each variable to which level in the process tree it 'belongs' to, which imposes considerable complications in the runtime data-structures and algorithms [115,186]. Therefore two subsets of GHC were identified: one is the flat subset, introduced in Section 10, another is the safe subset, defined as follows. A GHC program is *safe* if it has no execution in which a body unification suspends. Note that a Flat GHC program is trivially safe. Of course whether a GHC program is safe is also undecidable.

As in their flat subsets, the main difference between Safe GHC and PARLOG is the availability of sequential-And and sequential-Or in the latter.

Although PARLOG and GHC predate their flat subsets, there are almost no examples which show that the former languages are significantly more expressive than the latter ones. Perhaps the one interesting example is that of unbounded nondeterministic choice, implemented by recursion in the guard. Consider a process $c(In,\ldots)$ which has an unbounded list (or stream) of streams In. On each iteration, c wishes to extract one element from one of the streams, if such an element is ready, and iterate with In', which contains the tail of that stream and the unmodified remaining streams. If all the streams close, the process terminates. Using non-flat guards, the program can be written in GHC as follows:

```
c(In,...) ←
```

```
        get(X,In,In') |
        ... do something with X ...
        c(In',...).
    c(In,...) ←
        halt(In) | true.

    get(X,[[X'|Xs]|In],In') ← In'=[Xs|In], X=X'.
    get(X,[Xs|In],In') ← get(X',In,In'') | X=X', In'=[Xs|In''].

    halt([[ ]|In]) ← halt(In).
    halt([ ]).
```

The intermediate variables *X'* and *In''* are needed to ensure that the recursive call of *get* does not suspend because of an attempt to assign the goal variables *X* or *In'*.

Note the difference between *get* and *halt*. Both are recursive, but *halt* iterates in the body, since it tests for a conjunctive condition (all streams have terminated), whereas *get* iterates in the guard, since it tests for a disjunctive one (there is an element on one of the streams).

The program cannot be specified directly in a flat language, since it requires nondeterminism of unbounded degree in process reduction. However, its purpose can usually be achieved using a merge network, which is specifiable in any flat language.

Embedding Safe GHC in FCP(:)

Safe GHC can be embedded in FCP(:) using a technique for compiling Or-parallelism into And-parallelism, developed by Codish and Shapiro [29]. The idea is to spawn And-parallel processes to evaluate Or-parallel guards, and thread these processes using two short-circuits: a success circuit, which reports the success of one of the guards, and a failure circuit, which reports the failure of all guards. The hierarchical And/Or tree is implemented by a hierarchy of success and failure circuits. The power of FCP(:) is needed since the method requires reflection on the failure of unification.

A mutual exclusion protocol ensures that at most one guard can commit for each goal. Although the mutual exclusion protocol used in the original embedding [29] relies on atomic unification (Section 14), the less efficient single-round mutual exclusion protocol (Section 7.4) can be used as well. The technique was later enhanced by Levy and Shapiro [117], into a compiler from Safe GHC to FCP(?).

The technique cannot be used to embed (unsafe) GHC in a flat language, since a correct implementation of GHC requires recording the guard in which a variable is allocated. This problem is further discussion in Section 19.

Embedding PARLOG in FCP(:)

The technique for compiling Or-parallelism into And-parallelism can be combined with the FCP(:) implementation of the control meta-call to form an embedding of Safe GHC + the control meta-call in FCP(:). It can be further combined with the techniques for embedding FP(&) and FP(;) in $FGHC_{av}$, to embed PARLOG in FCP(:).

18.2 Concurrent Prolog and CP(↓, |)

Concurrent Prolog [162] is the ancestor of FCP(?). Similarly, the language CP(|,↓) [156,159,153] is the ancestor of FCP(:). Unlike GHC and PARLOG, both allow guard computations to instantiate goal variables. However, to achieve atomicity of a clause try, these instantiations should not be visible outside the calling goal prior to the commitment of the clause. In order to perform Or-parallel clause evaluation in Concurrent Prolog, a 'multiple-environments' mechanism is necessary. This mechanism allows competing clauses to make temporary and hidden instantiations to goal variables, which become permanent and visible only upon commitment. Several approaches to the construction of such a mechanism were investigated [115], but none have lead to satisfactory results. The difficulty in constructing such a mechanism can be understood by examining the power of Concurrent Prolog. It can specify almost trivially an Or-Parallel Prolog interpreter, which simulates the don't-know nondeterminism of Prolog by recursion in guards.

An embedding of Or-parallel Prolog in Concurrent Prolog

The Or-parallel Prolog interpreter assumes that the Prolog program is represented by the Concurrent Prolog procedure, *clauses*/2, which returns on the call *clauses*(A,Cs) the list of clauses Cs potentially unifiable with the goal A. In principle Cs can be the entire Prolog program, but indexing on procedure names or even on goal arguments can be used to reduce the number of clauses returned. Each Prolog clause $A \leftarrow B_1,\ldots,B_k$ is translated into a term in the list Cs of the form $(A\leftarrow[B_1,\ldots B_k|Bs]\backslash Bs)$. Note that it represents the (possibly empty) body by a (possibly empty) difference-list of goals. Given this translation, an Or-Parallel Prolog interpreter can be written in Concurrent Prolog as follows[28]:

```
solve([ ]).
solve([A|As]) ←
     clauses(A,Cs), resolve(A,Cs?,As?).

resolve(A,[(A←Bs\As)|Cs],As) ←
     solve(Bs) | true.
resolve(A,[_|Cs],As) ←
     resolve(A,Cs,As) | true.
```

The interpreter as defined can return only one answer to a goal. This limitation, however, is shared also by Prolog meta-interpreters. To collect all solutions to a goal, a set abstraction is incorporated in Prolog. It is typically implemented by storing the solution found (using a side-effect) and inducing failure. The approaches of Ueda [201,202] and Shapiro [167], in comparison, naturally collect all solutions to a goal.

The simplicity of this interpreter indicates that the implementation of the multiple-environments mechanism of Concurrent Prolog is at least as difficult as the direct implementation of Or-Parallel Prolog. Presently it seems that the added complexity of Concurrent Prolog over its flat subset outweighs its added expressiveness.

PART IV. IMPLEMENTATIONS AND APPLICATIONS

19. Implementations of Concurrent Logic Programming Languages

Considerable effort was invested in efficient implementations of concurrent logic programming languages, both for sequential and parallel computers.

---

[28] This interpreter is due to Kenneth M. Kahn.

19.1 Sequential implementations

We consider in depth implementation techniques for flat languages, then mention briefly techniques for non-flat languages.

There are several implementations of flat languages [49,98,89]. All employ some variant of an abstract machine developed by a group at the Weizmann Institute, first incorporated in an interpreter for FCP [129], and later refined and integrated with techniques for compiling unification, developed by Warren [207], within a compiler/emulator based implementation [89].

The sequential abstract machine

The key concepts of the machine are as follows. The machine represents the goal by an active queue and a set of suspension lists. Each process in the goal is either in the active queue or in one or more suspension lists. Each suspension list is associated with an unbounded variable, and may consist on several processes.

The basic operation of the machine is to dequeue a process from the active queue, and try to reduce it with some clause in the program. This operation is called a ***process try***. A process try is composed of a sequence of clause tries. In each clause try the try function of the process and the clause is computed (see Section 4.2). A process try succeeds if one of its clause tries succeeds; it suspends if none succeed, but at least one suspends; it fails if all clause tries fail. When a process try succeeds the try substitution θ is computed. When a process try suspends a set of suspension variables is computed; a variable is included in the set if its being instantiated in the future may release some clause try from suspension, i.e. cause it to succeed or fail.

If a process try succeeds with a substitution θ then the goals in the body of the successful clause are added to the active queue, and θ is applied to the state of the computation. In addition, processes in suspension lists of variables in the domain of θ are moved to the active queue. If the process try suspends with a suspension set S then the process is added to the suspension lists of each of the variables in S. If the process try fails the machine halts with an error state.

Note that a process can suspend on several variables, and be activated and suspended several times before succeeding or failing. A mutual exclusion mechanism, described below, ensures that a process is activated at most once per suspended process try.

The machine is connected to one or more external input devices, realized by data streams, including a keyboard, and typically has a process consuming each stream. The machine terminates successfully when all external input streams are closed, and there are no processes left. It terminates with deadlock if all input streams are closed and only suspended processes are left.

The machine maintains all dynamic data structures in a single address space, called a ***heap***. The heap grows when terms are allocated and processes are created, and shrinks by garbage-collection. The structures in the heap are variables, terms, process records, suspension records, activation records, and programs.

A variable is represented by one memory word, which is either empty or points to a suspension list. When a variable is instantiated to a term, its memory word becomes a reference (pointer) to the term unless the term can be stored in one word (e.g. an integer), in which case it is stored in place of the variable. Other terms are represented using standard techniques. A process with a predicate p/n is represented by $n+2$ words: one for the program counter, which points at the code of the procedure p/n, n words for the process arguments, and one word for chaining the process in

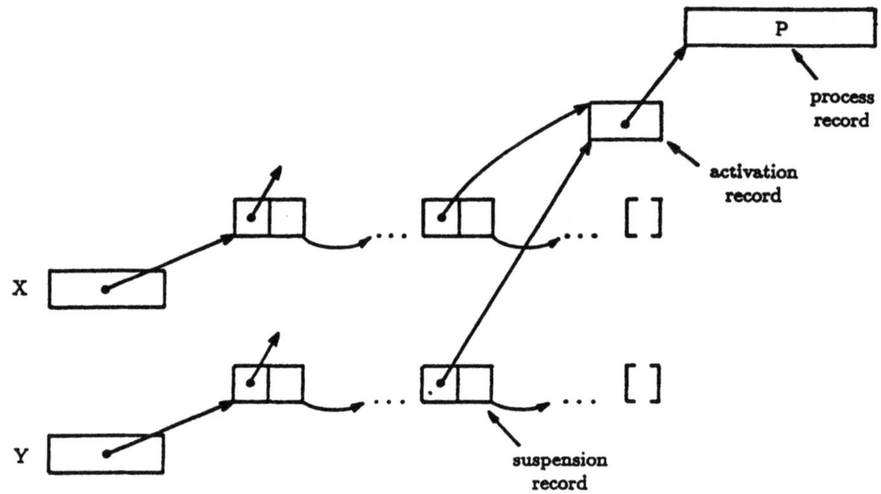

Figure 7: Suspending a process on two variables

the active queue. The active queue consists of chained processes. A suspension list consists of a list of suspension records (which could be list cells). Each suspension record points to an activation record and to the next suspension record if there is one. The activation record realizes the mutual exclusion mechanism which prevents multiple activations of the same process. It either points to a process record or is null, if the process has already been activated. If a process suspends on several variables, the suspension records in the suspension lists of these variables all point to the same process activation record, which in turn points to the process. The first variable to be assigned activates the process by enqueing it to the active queue, and sets its activation record to null. This prevents the other variables from re-activating this process. A process suspended on two variables is shown in Figure 7.

In addition to the heap, the machine has global registers for the active queue front and back, top-of-heap pointer, current process, current program counter, etc. In a language with test unification, the machine also has a trail. The trail is used to record assignments made during test unification in a clause try, so that they can be undone if the test unification subsequently suspends or fails. Unlike the standard Prolog trail, which needs to support deep backtracking, the trail in flat languages needs to support only shallow backtracking, and is reset on every clause try. As a result it can be rather small (e.g. 256 words).

The machine employs several optimizations, the most important being tail-recursion optimization[29]. Each dequeued process is given a time-slice t (e.g. t = 25). When a process A with time-slice t is reduced to the processes $B_1,\ldots,B_k$, $k \geq 1$, then one of them, say B_1, reuses A's process record (if it is large enough), inherits the time-slice t–1, and is immediately tried if $t > 1$. For the other processes $B_2,\ldots,B_k$ new process records are allocated, and they are enqueued to the back of the active

29 This name is kept for historical reasons. The optimization applies to any clause, not necessarily recursive, and not necessarily to the tail call.

queue. If $t = 1$ then B_1 is also enqueued. This scheme maintains And-fairness while decreasing process switch and memory access (assuming some process arguments are maintained in processor registers during a time-slice).

To increase the chance of a process record being reused, minimal size records are allocated (e.g. 10 words). In addition, free-lists of process records, suspension records, and activation records are maintained between garbage collections, to improve storage utilization.

Implementations of non-flat languages

One way to achieve atomicity of a clause try in a non-flat language is to try and reduce goals in some order; when reducing a goal, try each clause in some order; and for each clause guard apply this execution algorithm recursively. This is the algorithm incorporated in the first interpreter for Concurrent Prolog, written in Prolog [162]. Variants of it were implemented on top of Prolog, both for Concurrent Prolog and for GHC and CP(|,↓,&) [204, 155]. This execution algorithm, however, does not satisfy any fairness requirements. For example, an attempt to reduce a faulty process (with a nonterminating guard) may block the rest of the system forever.

Several other executions algorithms for Concurrent Prolog which do not suffer from this problem were investigated [116,128,143]. Their complexity, however, seemed unacceptable, and was partially a motivation for the development of Flat Concurrent Prolog and the simpler non-flat languages, GHC, PARLOG, and CP(|,↓). An abstract machine for PARLOG was developed by Gregory *et al.* [68], and later optimized by Crammond [34]. Its basic design differs from the FCP abstract machine [89] in that it explicitly maintains a process tree. Another abstract machine for GHC was derived from the FCP machine by Levy [114]. Although GHC is simpler than Concurrent Prolog, its implementation still required fairly heavy machinery. Therefore Safe GHC was investigated, and a compiler from Safe GHC to FCP was developed [117].

Compilation of unification

The basic data manipulation operation of logic languages is unification. Warren [207] has developed a method for compiling unification efficiently by identifying its various special cases which are specified in a clause head, and generating special instructions for them.

Warren's scheme was designed for Prolog's general unification, and is applicable both to FCP's read-only unification [89], and to the input unification employed by FGHC [98] and PARLOG. Using it, an abstract machine along the lines described above can achieve the same uniprocessor performance as the Warren abstract machine for Prolog.

However, for input matching one can do better than Warren's scheme. The input matching component of a set of clauses of the same procedure can be jointly compiled into a decision tree, which combines shared matchings and finds more efficiently the set of applicable clauses [100].

Processor architectures

Two processor architectures specialized for the execution of a concurrent logic programming language, namely FCP(?), were developed. The first architecture, Carmel [74,75], takes the RISC approach. It augments a simple processor architecture with mechanisms to support the expensive or frequent operations of FCP(?). By carefully tuning the instruction set and processor architecture, impressive performance is obtained.

The second architecture, by Alkalaj and Shapiro [6], takes the view that internal concurrency in a processor combined with a carefully designed memory hierarchy is the key to high performance. The architecture consists of several specialized processing units, each with its own memory hierarchy. The reduction and tag processors are at the root of the hierarchy. They are supported by three additional processing units: an instruction processor, a data-trail processor, and a goal-management processor. The instruction processor employs standard techniques for instruction prefetching and caching. The data-trail processor employs a data cache enhanced to support shallow backtracking, required in the implementation of atomic test unification. The goal-management processor manages the top of the process queue in a way analogous to how a RISC processor manages the top of the activation stack. The goal-management processor manages process switching, spawning, activation, and suspension, using a bank of register windows. The execution algorithms of this architecture are specified using an FCP(?) program, by hardware description techniques developed by Suzuki [174] and Weinbaum and Shapiro [209]. The specification forms a working simulator of the architecture. The performance of this architecture is yet to be evaluated. How these two processors can be integrated in a multiprocessor architecture is an open question.

The PSI-II processor was designed for the execution of Prolog, but was re-microcoded to implement KL1 [196]. It is the building block of the multi-PSI parallel machine.

19.2 Parallel implementations

We review the concepts behind two types of parallel implementations: distributed and shared-memory. The implementations include a distributed implementation of FCP [191], a distributed implementation of FGHC [90], a distributed implementation of Flat PARLOG [47], and shared-memory implementation of PARLOG [34].

The core operation in these implementations is unification.

Distributed atomic unification

In a distributed implementation each processor executes a variant of the sequential abstract machine, described above, and takes special actions when a clause try involves variables shared with other processors. These actions realize a distributed unification algorithm.

Since non-variable terms are immutable data structures, they can be replicated upon demand throughout a processor network without any special consistency maintenance mechanisms. The writing on a variable, however, needs to be coordinated. In particular, in a language with atomic unification, a unification that involves writing on several variables should either succeed in writing on all of them, or write on none. Hence, from a distributed implementation point of view, an atomic unification is best viewed as an atomic transaction, which may read from and write to several logical variables. Standard database concurrency-control techniques for realizing atomic transactions can be adapted to the particular requirements of unification.

One approach, applicable to a network of processors without shared memory, is as follows. It uses the messages *read*, *lock*, *become_value*, and *become_local*. A variable shared by several processors is represented by a directed tree, with edges pointing towards the root. Each processor sharing a variable stores a node of the tree in its local memory, which contains the address of the node it is pointing to if it is not the root. An occurrence of a variable is called *remote* if it is an internal node in the tree; *local* if it is the root of the tree.

An attempt to read a shared remote variable is called a *read-fault*. A processor executing a process which has had a read-fault sends a *read* request up the tree, and adds the faulting process to the remote variable's suspension list. When a processor storing the root of the tree receives a *read* message, it operates as follows. If the variable has been assigned a term T, a *become_value*(T) message is sent in reply. If the variable is still unbound, the *read* request is stored in the variable's suspension queue, and will be replied to when the variable is assigned.

A shared variable can be written only at its root. Write-permission is transferred between processors by changing and redirecting edges in the tree. A processor with a local shared variable (i.e. the root of a shared variable tree) may write on it when it pleases. It ensures that a unification that involves writing on several shared variables is atomic by not responding to messages, including *read* messages, while performing a clause try.

An attempt to write on a remote shared variable is called a *write-fault*. A processor executing a process which has had a write-fault sends a *lock* message up the variable's tree, and suspends the faulting process on the remote variable. The processor receiving this message replies with a *become_value*(T) if the variable has already been assigned a term T, or with a *become_local*(*Reads*) if the variable is still unbound, and changes its local variable to be a remote variable pointing at the sender's variable. *Reads* is the (possibly empty) list of suspended *read* requests on the sender's local variable suspension queue, to which a request from the sender's own variable is added in case it has local processes suspended on it. The receiver of a *become_local*(*Reads*) message changes its variable from remote to local, wakes up all processes suspended on it, and adds the *Reads* to the variable's suspension list.

The scheme as described may result in livelock, if two processors keep sending *lock* requests to each other, and none accumulates enough local variables to perform a process reduction. To prevent this, a 2-phase-locking scheme can be incorporated [191,193]. The scheme requires additional bookkeeping by a write-faulting processor, but not additional messages. We do not describe its details here.

Another question to address is how to handle variable to variable unifications. One approach is to lock (i.e. make local) the two variables when assigning one to the other. This ensures that no cycles are created, but may cause superfluous contention in applications using the short-circuit technique. A second approach is to impose some ordering on variables, and to respect this ordering when unifying two variables. Another approach is not to prevent the creation of cycles, but to break them when they are detected.

Implementation of non-atomic unification and the meta-call construct

In languages without atomic unification, such as GHC, PARLOG, and their flat subsets, simpler algorithms than the one described above apply. For example, when unifying a remote variable X with a term T it is not necessary to bring X locally before assigning it; instead, a message can be sent to X, requesting it to unify with T. If the unification fails, the machine halts with an error state (or simply notes the inconsistency and proceeds).

Since neither of these behavior is acceptable in a multi-tasking operating system, the meta-call construct, described in Section 10.3, was developed. The implementation of the meta-call construct must be integrated with the distributed unification algorithm in order to detect termination and to correctly ascribe failure. One approach, taken in the distributed implementation of FGHC [90], is to associate with every computation (invocation of a meta-call) a unique identifier, and maintain tables associating computation identifiers with the appropriate streams of the meta-call. When a unification fails, this fact is reported to the computation by placing a message on the appropriate stream. Since the short-circuit technique is not applicable, distributed termination of a computation is detected by maintaining an explicit distributed counter for each computation, at the language implementation level.

Foster [47] describes an alternative approach to the distributed implementation of the control metacall, which avoids the complexity of FGHC's distributed counters. Only uniprocessor computations are supported directly in the implementations and remote structure-to-structure unification operations are performed locally. Acknowledgement messages and message counting on individual nodes hence suffice for termination detection. Termination detection in distributed (multi-node) tasks is programmed in PARLOG using the usual techniques. An even simpler parallel execution model of systolic concurrent logic programs is described by Tamaki [182].

A shared-memory implementation

Crammond [34] describes a parallel implementation of PARLOG on a shared-memory multiprocessor. In this implementation each processor has its own data areas, although processors may access each other's areas in order to read the value of a shared variable, to assign a shared variable, or to take work (processes) from each other. A simple locking mechanism is employed, where a processor that modifies a shared object (e.g. a process queue or a shared logical variable) locks it, and a processor attempting to lock a locked object busy waits ("spins") until this object is unlocked. Since PARLOG does not have atomic unification, a processor needs at most one lock at a time, and hence this locking scheme does not result in deadlock. An extension of this implementation scheme to languages with atomic unification would require some concurrency control mechanism similar to the one discussed above for distributed atomic unification.

A simple load balancing scheme is employed in this implementation, where a processor dequeues processes from its own queue as long as it is not empty, and dequeues from some other processor with a nonempty queue if its own queue is empty. Using such a scheme, this implementation obtained a speedup of up to 15 using 20 processors. Alternative load balancing schemes can be incorporated in this implementation with little difficulty.

An analysis of a shared-memory implementation of Flat GHC is reported by Tick [194].

19.3 Process to processor mapping

The question of how to map processes to processors is not unique to concurrent logic programming, and any general approach or solution may be applicable. Approaches to the problem fall into two general categories: methods in which the program itself (or programs associated with it) specify the mapping, and dynamic mapping techniques, incorporating load-balancing algorithms. Hybrid techniques are also possible.

We show how instances of the two approaches can be realized using distributed meta-interpreters. The interpreters are shown in FCP(:), although they could be written in any flat language.

Mapping with Turtle-programs

The use of Turtle-programs for mapping processes to processors was suggested and demonstrated in [165]. Assume that the parallel machine is a (finite or infinite) two dimensional grid. View each process as a LOGO-like Turtle, which has a position on the grid, and a heading. With each process activation (body goal) *P* we can association a LOGO-like Turtle program *TP*, as in *P@TP*. The meaning of the call *P@TP* is that *P* should have the position and heading obtained by applying *TP* to the position and heading of its parent process, and execute in the processor corresponding to that position. Processes without an associated Turtle program simply inherit their parent position and heading.

Using this notation, a sequance of processes can be easily mapped on a sequence of processors. For example, consider the *vm* vector-matrix multiplication program in Section 7.2. Adding the @forward Turtle program to the recursive call to *vm*, cause the inner-product processes *ip* to be placed on a sequence of adjacent processors:

% *vm(Xv,Ym,Zv)* ← multipyling the vector *Xv* by the matrix *Ym* gives the vector *Zv*.

```
vm(_,[ ],Zv) ← Zv=[ ].
vm(Xv,[Yv|Ym],Zv) ← Zv=[Z|Zv'],
    ip(Xv,Yv,Z), vm(Xv,Ym,Zv')@forward.
```

Mapping process arrays to processor arrays is just as easy. Consider the matrix multiplication program *mm* in Section 7.2. Adding the *@forward* Turtle program to the recursive call to *mm*, and *@right* to the initial call to *vm* maps the array of *ip* processes to an isomorphic array of processors:

% *mm(Xm,Ym,Zm)* ←
Zm is the result of multiplying the matrix *Xm* with the transposed matrix *Ym*.

```
mm([ ],_,Zm) ← Zm=[ ].
mm([Xv|Xm],Ym,Zm) ← Zm=[Zv|Zm'],
    vm(Xv,Ym,Zv)@right, mm(Xm,Ym,Zm)@forward.
```

The mapping of additional process structures is discussed in [165]. An alternative mapping strategy is described in [TNM]. Here we show an enhanced distributed meta-interpreter which implements Turtle program mapping [TAS].

We assume that the underlying machine is a torus-connected mesh of processors (a virtual torus can be mapped on a two dimensional mesh by placing four virtual processors per physical one). The interpreter consists of a torus *processor* processes. We assume that these processes are mapped to the underlying processors using the *torus* program shown in Section 7.2.

Each *processor* process has four outgoing streams to its neighbors. Its four incoming streams are merged into one. An interpreted process has a heading, and possibly also a Turtle program. A headed process is represented by a pair (*Goal,Heading*) where *Heading* is one of {*north,south,east,west*}. To a headed process (*G,H*) a Turtle program *TP* may be attached, as in (*G,H*)@*TP*. We assume the process in each processor is called *processor(In,[In,ToNorth,ToSouth,ToEast,ToWest])*, where the first argument is the merge of its neighbors' outgoing streams, and its second argument is the list of its five outgoing streams, one to itself and four to its neighbors. The *processor*'s code is as follows:

```
processor([(Goal,Heading)|In],Out) ←
```

```
        reduce(Goal,Heading,In),
        processor(In,Out),
    processor([(G,H)@TP|In],Out) ←
        route(G,H,TP,Out,Out'),
        processor(In,Out').
```

It receives goals on its input stream. If a goal has a Turtle program it routes it to its appropriate output stream. Otherwise it executes it locally. Its execution may result in new goals, possibly with Turtle programs. They are merged into its input stream, and treated normally. The meta-interpreter reduces goals, maintaining their heading, and when it encounters a goal with a Turtle program it sends it to the processor for routing.

```
    reduce(true,_,_).
    reduce((A,B),H,Out) ←
        reduce(A,H,Out), reduce(B,H,Out).
    reduce(goal(A),H,Out) ←
        clause(A,B), reduce(B,H,Out).
    reduce(A@TP,H,Out) ←
        write((A,H)@TP,Out).
```

The router is specified, without showing its code:

```
    route(Goal,Heading,TP,Out,Out') ←
        Send Goal according to TP and Heading on the appropriate
        Out stream, with an updated heading and possibly with a residual
        Turtle program, and return the updated streams Out'.
```

The torus of *processor* processes can be mapped on an underlying torus using Turtle programs; but who will interpret these Turtle programs? Booting an initial process network on the processor network is necessary, and can be done using standard techniques. One solution is described in [188].

In this scheme the underlying parallel implementation of the language does not have to support remote process spawning in addition to distributed unification, since it is implemented at the language level by standard message passing between meta-interpreter (or runtime support) processes. Another mapping notation is described in [176].

Mapping with dynamic load-balancing

Dynamic load balancing requires that processors off-load work when they are too busy, and request work when they are idle. A good dynamic load balancing algorithm distributes work evenly and with little overhead. If the underlying machine has a notion of locality, i.e. communication costs between processors are not uniform, then a dynamic load balancing algorithm should prefer local distribution of work over global one, when possible.

We show here a simple implementation of dynamic load balancing using a centralized queue. The scheme can be enhanced to use a distributed queue [167], and thus reduce contention and increase locality.

Assume a network of processors, and a *next* mapping command which places the process in the next processor in some processor ordering. A distributed meta-interpreter performing dynamic load balancing can be defined as follows:

```
    processors(N,ToQ) ←
```

```
        queue(ToQ),
        processors'(N,ToQ)@next.

    processors'(0,_).
    processors'(N,ToQ) ←
        N>0 | N':=N-1,
        processor(ToQ),
        processors'(N',ToQ)@next.

    processor(ToQ) ←
        reduce(true,ToQ).

    reduce(true,ToQ) ←
        write(dequeue(A),ToQ,ToQ'),
        reduce(A,ToQ').
    reduce((A,B),ToQ) ←
        write(enqueue(B),ToQ,ToQ'), reduce(A,ToQ').
    reduce(A,ToQ) ←
        clause(A,B), reduce(B,ToQ).

    queue(In) ←
        See Section 7.3.
```

Communication can be reduced, at the expense of slightly slower distribution of work, by placing a buffer in each processor. The buffer forwards requests to the global queue only if it overflows (has too many *enqueue* requests) or underflows (cannot satisfy a *dequeue* request). For example, in experiments made on a 16 processor computer on a particular application a buffer size of about 10 was found optimal [167].

Code management

General solutions to the code management problem are also applicable to concurrent logic programming languages. One approach to the problem is described in [188].

20. Applications of Concurrent Logic Programming Languages

Since their beginning, the design of concurrent logic languages was closely coupled with the development of prototype applications, which were used as feedback to the design process. The application programs were those which testified to the little difference between flat and non-flat languages from an expressiveness point of view. The systems programs were those which stretched the synchronization capabilities of logic languages to their limits, and provided examples where the power of atomic test unification and read-only unification shows through.

A description of numerous applications, as well as further references, can be found in the Concurrent Prolog book [166]. The book reports on the implementation of parallel and distributed algorithms, systems programming, and the implementation of embedded languages, among others. Other applications of concurrent logic languages include [35,67,99,100,119,139,146,148,150,159,160,174,183,185]. Combined, these applications witness to the generality and versatility of the concurrent logic programming approach.

PART V. CONCLUSIONS

21. Relation to Other Languages and Computational Models

21.1 Prolog, parallel logic languages and concurrent constraint languages

Prolog

Concurrent logic languages, as presently defined, are not an alternative to Prolog. They are, in a sense, lower level languages, which exhibit their strength mainly in the implementation of parallel algorithms, distributed systems, reactive systems, and in systems programming. Hence the question of the integration of these languages with higher-level languages in general, and with Prolog in particular, has received considerable attention.

One of the initial goals in the design of Concurrent Prolog [162] was the definition of a language which includes Prolog as a subset. It seemed that this goal was not realized in the initial design of the language, and hence this design was termed "a subset of of Concurrent Prolog". Later, it was found out that an Or-parallel Prolog interpreter can be specified easily in that subset (the interpreter is shown in Section 18) and, as a consequence, that the original design did achieve this goal. However, the move to flat languages opened up again the question of the integration of Prolog and concurrent logic languages.

Two solutions were discussed in Section 14. One is to provide some interface between two separate languages: some form of Prolog, and some concurrent logic language [24,25]. Another is to embed some form of Prolog into a concurrent logic language [29,167,199,201]. A third solution is to provide some of the mechanism of concurrent logic language via extensions to Prolog, such as freeze [12] and wait declarations [133]. The problem with the first solution is that it really does not address the essence of the problem, namely to find an integrated solution in which the various strengths of logic programming can brought to bear. It is applicable to any two programming languages, not necessarily logic programming ones. The problem with the second solution is performance. Techniques for efficient implementation of concurrent logic languages lag one step behind those of sequential Prolog, and there are claims that the algorithms employed in the embedding of Prolog in concurrent logic programming are not feasible. The third solution is largely limited to transformational applications, since it cannot change the basic fact that Prolog is not a reactive programming language.

CP(↓,|,&), Andorra, and Pandora

The synchronization and commitment mechanisms of concurrent logic languages are useful also in non-reactive applications. This motivated a different line of research — the design of non-reactive languages that attempt to supersede Prolog in expressiveness and performance, without being rooted in its sequential execution model.

Saraswat [152,155,159] investigated a parallel logic language, called CP(↓,|,&), that incorporates both don't-care and con't-know nondeterminism, and synchronization by input matching. Although an efficient implementation on top of sequential Prolog is described [155], the language seems even more difficult to implement "for real" than the non-flat languages discussed in Section 18.

Yang and Aiso [210,211] also propose a language with don't-care and don't-know nondeterminism, called P-Prolog, but use a different synchronization mechanism — the determinacy conditions described in Section 12 on P-Prolog$_x$.

Recently, an elegant integration of the ideas of P-Prolog and of Or-parallel Prolog, called the Andorra model, was psoposed by D.H.D. Warren (personal communication), and integrated in the Andorra language [72]. The idea is as follows: reduce in parallel determinate goal atoms as long as possible (And-parallelism). When no determinate atoms remain, choose one atom for an *Or-split*. Create two or more subgoals, one for each clause unifiable with the chosen atom, and continue in parallel reducing the resulting independent goals (Or-parallelism). Under the Andorra model pure logic programs may exhibit the synchronization behavior of concurrent logic programs, yet enjoy a complete proof procedure. If in an Or-split the leftmost atom is chosen, Andorra is more efficient (in terms of the number of reductions required) than ordinary Or-parallel Prolog, since it prunes the search space better.

The ideas in the Andorra model can also be employed in an implementation of the flat subset of CP(↓,|,&). Another recent proposal along these lines is Pandora [8] — a parallel logic language incorporating PARLOG-like synchronization, and a mechanism for specifying which goal atom to choose for an Or-split.

Concurrent constraint logic programming

The framework of constraint logic programming [31,92] proved in recent years to be a powerful generalization of logic programming, both from a theoretical and from a practical point of view. Maher [125] suggested using concepts of constraint logic programming to provide a logical characterization of synchronization in concurrent logic programming. The conditions for the success of input matching and guard checking of a goal atom A with a clause $A' \leftarrow G \mid B$ are customarily defined operationally, as in this paper. Maher showed how this condition can be specified logically, as the requirement that the accumulated constraint (corresponding to the accumulated substitution in our model) entails the existential constraint $(\exists)A{=}A' \wedge G$, where the existential quantifier ranges over the variables local to the clause. Saraswat [158,159] developed these ideas further. He developed a framework of concurrent constraint logic programming in which a computation progresses by agents that communicate by placing constraints in a global store and synchronize by checking that constraints are entailed by the store. Agents correspond to goal atoms, placing constraints correspond to unification, and checking constraints correspond to matching and guard checking in concurrent logic programming. Employing the concepts of consistency and entailment between partial information (i.e. constraints), Saraswat was able to provide a logical characterization of constraint-based constructs that correspond to non-atomic unification, atomic test unification, read-only unification, test-and-set, and others. Constraint logic programming offer a logical framework for dealing with domains other than Herbrand terms, such as boolean, integer, and real arithmetic. Saraswat showed how such domains and others can be incorporated in concurrent logic languages using this framework.

The initial work on concurrent constraint logic programming is very promising, and one may expect that it will have as much theoretical and practical impact on concurrent logic programming as constraint logic programming had on logic programming.

21.2 Distant relatives — Delta Prolog and Fleng

Delta Prolog

Delta Prolog [142] is Prolog augmented with CSP-like communication primitives. Delta Prolog is different from the other languages surveyed in two respects. First, it is not a logic programming language in the sense that a successful computation corresponds to a proof of a goal statement, and a partial computation corresponds to proofs of a conditional statement. Specifically, the role of the communication primitives of Delta Prolog in the declarative reading of programs is unclear. In concurrent logic languages the synchronization primitives can be ignored in the declarative reading, since they affect only which answer substitution is found, but not the substitution itself. This is not the case in Delta Prolog. Although Delta Prolog can be given axiomatic semantics, this can be done for any programming language, not only for a logic programming one. The second difference between Delta Prolog and the other languages surveyed is that Delta Prolog is not reactive, since it may backtrack on communication.

It is not clear yet in which application area the particular features of Delta Prolog show their advantage.

Fleng

Fleng [136,137,138] is a simple concurrent programming language inspired by GHC and Kernel PARLOG [23]. Its syntax uses (guardless) Horn clauses. Like GHC, it uses goal/clause matching for synchronization, and its unification is non-atomic. Unlike GHC, unification, as well as any other primitive, reports termination.

Fleng has no notion of failure. Every primitive operation terminates and reports its termination status. For example, the unification primitive *unify(X, Y, Result)*, attempts to unify *X* and *Y*. If it succeeds it assigns *Result* the value *true*, if it fails it assigns the value *false*.

In spite of its appearance, Fleng is not a logic programming language, since not every successful computation corresponds to a proof of the goal statement. In particular, the goal *unify(a,a,true)*, terminates successfully, but apparently so does *unify(a,b,true)*.

The insistence that a successful computation should correspond to a proof is not a mere nicety, and Fleng cannot simply drop the title of being a logic programming language and live happily ever after. The concept of failure serves the fundamental role of an exception mechanism in logic programming. In its absence, some other mechanism must be developed. As is evident from other languages [73], a sound exception handling mechanism is not a trivial component of a language, and its incorporation in Fleng would certainly complicate its semantics. Specifically, if Fleng's present exception handling mechanism (namely the *Result* variable of each primitive) cannot be used to report the exception, as in the call *unify(a,b,true)*, what exception should be raised? The most natural one is to fail the computation, which brings us back to square one...

If failure is reinstated in Fleng, then it becomes similar in expressiveness to KL1, since it can be naturally embedded in KL1 and vice versa.

21.3 Dataflow languages

Concurrent logic languages share with dataflow languages [1] single-assignment (or write-once) variables and dataflow synchronization. However, this is mainly a similarity in spirit, not in implementation. The basic operation that is synchronized by dataflow in concurrent logic languages is the process try. It corresponds typically to several tens, up to several hundreds, of conventional machine instructions. In contrast, the synchronized operation in dataflow models corresponds typically to one conventional machine instruction. This difference explains why realizations of concurrent logic languages on conventional hardware have acceptable synchronization overhead, whereas dataflow language seem to necessitate a specialized architecture.

Other differences between the two models is that dataflow languages are typically deterministic, whereas concurrent logic languages are not, and that dataflow languages and architectures are typically geared for scalar operations, whereas logic languages operate mainly on compound data-structures, which may contain logical variables.

21.4 Functional languages

Much has been said on the relation between functional and logic languages [36]. In the context of concurrent programming, the major observation is that functional languages are, by design and ideology, transformational, rather than reactive. Functional programs denote time-independent functions from inputs to output, and notions of state, synchronization, communication, and nondeterminism are alien to them.

Functional programs can be parallelized, and often yield efficient parallel algorithms. However, without major extensions [7,52,53,54,69,77], which seem to undermine their original motivation and 'semantic elegance', functional programming languages cannot be used for the specification and implementation of reactive systems.

Concurrent logic languages, on the other hand, have explicit notions of processes, process state, communication, synchronization, and nondeterminism. Furthermore, processes can have several outputs, and inputs and outputs of processes can be combined into arbitrary process networks. These, combined with properties of the logical variable, seem to be the source of their power as concurrent languages; all are absent from the base model of functional languages.

In addition, it seems that there are usually simple translations from concurrent functional languages to concurrent logic languages [118]. Thus, a possible architecture of a parallel computer system, which provides both styles of programming, each for the application it suits best, is a system in which the base language is a concurrent logic programming language, which implements the underlying operating system and programming environment, and higher-level functional languages are implemented by translation to it. Such an architecture is proposed by [118].

21.5 Message-passing models of concurrency

The origins of concurrent logic programming languages can be traced back to the work of Kahn and MacQueen [94], which offered a model of concurrency based on deterministic asynchronous processes computing relations over data streams. van Emden and de Lucena [42] were intrigued by this notion, and showed how one can use logic programs to specify such processes. Clark and Gregory [20] took these ideas a crucial step further and, influenced by the notions of CSP [86,87], introduced synchronization and committed-choice nondeterminism into logic programs.

Concurrent logic languages are similar to CSP and Occam [91] in their notion of processes, nondeterminism, and synchronization via communication. They are similar to Occam, and different from CSP and Actors [78], in that processes communicate via

CCS/CSP operators	Guarded Horn clauses
action prefix	guard
parallel composition	conjunction
restriction	clause head
choice	alternative clauses
relabeling	(implicit) variable renaming
recursion	recursion

Figure 8: Analogy between CCS/CSP operators and guarded Horn clauses

'ports' (realized by logical variables) rather than by naming the destination process or object.

One difference between CSP and Occam on the one hand and concurrent logic languages on the other hand are the type of communication and synchronization they employ. In the former communication is synchronous; in the latter asynchronous. In the former a communication channel is necessarily point-to-point. In the latter it is, in the general case, many-to-many. We find the added flexibility of the communication protocols available in concurrent logic languages over those of CSP and Occam quite apparent. The additional overhead entailed by this added flexibility is yet to be determined. Presently, it is not clear for which tasks Occam-like point-to-point synchronous protocols are inherently more efficient than the general asynchronous protocols employed in concurrent logic languages, and vice versa.

Another fundamental difference is that CSP and Occam can operate on and communicate only "ground" data, whereas the ability to communicate and share incomplete data structures, i.e. data structures containing logical variables, is fundamental to concurrent logic languages, and is their main source of expressive power.

Being concrete programming languages, concurrent logic languages are not directly comparable to abstract computation models such as CCS. However, it seems that if one abstracts away the details of the data domain (i.e. terms and unification), and concentrates on the synchronization aspect of concurrent logic languages, then models which can be thought of as the asynchronous counterparts of CCS [130] emerge [65].

Although the syntax of CSP and CCS seems superficially different from that of concurrent logic languages, there is a close analogy between the basic operators of the two families, shown in Figure 8.

21.6 Concurrent object-oriented programming

The underlying operational model of concurrent logic languages resembles that of concurrent object-oriented models, such as Actors [78], in that both consist of a dynamic collection of light-weight processes, computing by performing local computations and exchanging messages. There are, however, several apparent differences.

First, Actor objects, like CSP processes, address each other by name, and not via channels. The advantage of channels over object names is modularity and abstraction;

this had led Occam's designers to depart from CSP in this respect. It is easier to connect one process network to another by assigning output channels of one to input channels of the other, than by informing one the names or mail addresses of the appropriate processes in the other. Channels are also more abstract, since knowing a channel does not imply knowing who receives or sends messages on that channel. A process can have several input channels, which provide different access modes to its local data; this feature can be the basis of a capability system. Several processes may listen on the same channel, each handling a different set of messages, or handling a different aspect of a message.

If one is able to pass channels in messages, as in logic languages, then channels have another, perhaps more fundamental, advantage over name-based addressing. Process-names in messages, like incomplete messages, can be used for network reconfiguration. However, this is only one particular application of incomplete messages. The use of incomplete messages in the back-communication protocol, in dialogues, in the bounded-buffer protocol, in the duplex-stream protocol, and in others is based on the ability to allocate communication channels on the fly, and on the fact that the channel implicitly embeds some context information, which is used in the protocol. There is no natural way to achieve these effects in name-based addressing.

The drawback of concurrent logic languages, compared to Actor-like languages, is not their underlying operational model, but rather the verbose syntax required for expressing object-oriented programs. The description of an object with one input channel and some state variables in a concurrent logic language has the typical form:

> p([Message|In],...*state variables*...) ←
> ... *handle* Message, *update state variables*...,
> p(In,...*new state variables*...).

Furthermore, when several processes share the same output channel ("talk to the same object"), then some protocol, such as the spawning of a merge network, need to be followed. This is in contrast to Actor-like languages, in which state variables are assumed not to change unless a change is stated explicitly, and explicit mergers need not be created in front of receiving objects, since they are assumed implicitly.

Another bookkeeping service provided automatically by object-oriented languages is object deallocation; when there are no more references to an object, it is deallocated, and its storage is reclaimed. In concurrent logic languages, unreferenced data-structures are reclaimed by garbage collection, but the conditions for process termination must be specified explicitly, by one or more unit clauses. Sometimes the burden of doing so manually should better be avoided.

A mechanism for detecting that a variable is referenced only by one process [17] can be used for garbage collecting processes: A process that detects that it is the only one referencing its input stream may perform some cleanup operations (e.g. close its output streams or unify its segment of a short-circuit) and terminate (K. Kahn, personal communication, 1987). Although the pragmatics of this mechanism is quite well understood, its logical semantics still needs to be worked out.

The question of the proper integration of inheritance in a concurrent object-oriented framework is still open. Delegation was suggested as a mechanism which is more suitable to a concurrent framework. As discussed in Section 7.6, objects which delegate incomprehensible messages can be specified in concurrent logic languages by augmenting the process with an additional output stream, and adding a delegating clause which uses the *otherwise* construct. This mechanism, however, is also quite verbose.

These observations have lead to the design of new object-oriented languages, such as Vulcan [96], and POLKA [35]. These languages attempt to enjoy the best of both worlds. They adopt the channel concept of concurrent logic languages, but do not require explicit repetition of state variables, explicit mergers, or explicit delegation mechanism. Another important design consideration for these languages was that their implementation be in terms of natural and efficient translations to concurrent logic programming. This allows the exploitation of implementations of such languages, as well as support integration between applications that are best described by an object-oriented language, and applications that enjoy the full power of concurrent logic languages.

Consider the standard bank account example. In the Vulcan language, a process with the same behavior is specified by the following program:

```
class(account, [Balance=0, Name="No Name Given", Errors,...]).

account :: deposit(Amount) →
    new Balance := Balance + Amount.

account :: balance(Balance).

account :: withdraw(Amount) →
    Balance ≥ Amount
    ifTrue new Balance := Balance - Amount
    ifFalse Errors : Overdrawn(Name, Balance, Amount, ...).
```

A more conservative solution in the same direction was to devise a new "surface syntax" for concurrent logic programs, rather than a completely new language. The surface syntax, called *logic programs with implicit variables* [97], allows specifying only what has changed in the process' state during a transition, rather than the entire old and new states explicitly, as required by plain logic programs. In addition, it has a special notation to support stream communication, and array operations. For example, the bank account in framed FCP(|) logic programs would be specified as follows:

```
procedure account(In)+(Balance=0,Name="No Name Given",Errors,...).

account ←
    In ? deposit(Amount) |
    Balance' := Balance + Amount,
    account.

account ←
    In ? balance(Balance) |
    account.

account ←
    In ? withdraw(Amount),
    Balance ≥ Amount |
    Balance' := Balance - Amount,
    account.

account ←
    In ? withdraw(Amount),
```

Balance < Amount |
Errors ! Overdrawn(Name,Balance,Amount,...),
account.

The variable X' specifies the new value of the process argument X. The stream notation M ? Xs is a shorthand for the input unification $Xs=[M|Xs']$, and M ! Xs is a shorthand for the same unification.

Unlike in Vulcan this notation is employed only for stream, rather then channel [195] communication. An extension of this approach to incorporate channels as an abstract data type is being investigated.

21.7 Linda

Linda [3,13] is a set of primitives that operate concurrently on a multiset of tuples, called a Tuple Space. Tuples in a Tuple Space are accessed associatively using a degenerate form of unification between tuples and tuple templates. The basic operations are *out*(T) (insert a tuple T to the Tuple Space) *in*(T) (delete a tuple matching T, instantiating variables in T; block if a matching tuple is not available) and *rd*(T) (find a tuple matching T, instantiate variables in T). A fourth primitive, *eval*, support process forking. Augmenting a conventional sequential programming language with these Linda promitives results in a concurrent programming language in which processes communicate and synchronize via the Tuple Space.

A comparison of Linda and concurrent logic programming is given in [13]. A critique of this comparison, which demonstrates an embedding of Linda's primitives in a variant of FCP(:) is given in [168].

21.8 Nondeterministic transition systems and UNITY

Nondeterministic transition systems are a natural method for specifying concurrent systems. Indeed, we have given the semantics of concurrent logic programming languages using nondeterministic transition systems. Recently, a notation was proposed for specifying concurrency called UNITY [16]. UNITY is based on unbounded iterative nondeterministic transitions.

Concurrent logic languages share with UNITY the goal of being a foundation for a general purpose concurrent programming language, the belief that the execution model of such a language should be abstract, rather then being tied with a concrete architecture, and the conviction that nondeterminism is an essential component in such a model. Another point in common between UNITY and the stronger concurrent logic languages is the size of the atomic operation: both the simultaneous assignment of UNITY and atomic unification in languages such as FCP(:) involve atomic transactions which read from and write to several variables.

One difference between UNITY and concurrent logic languages is the notion of a process. A UNITY program has one global state, and transitions operating on it, possibly concurrently; it does not have an explicit notion of a process. Concurrent logic programs have a natural notion of a process. However, this difference is only apparent. The notion of a process in concurrent logic programs is in the eyes of the beholder — it is not an inherent part of the transition system of concurrent logic programs. Similarly, one can often identify "processes" in UNITY programs, if one so desires.

Another difference between UNITY and concurrent logic languages is the notion of termination. Concurrent logic programs terminated by explicit instructions. UNITY programs terminate implicitly, by reaching a fixpoint. One implication of this decision is that there is no distinction between successful termination and deadlock. We feel that this difference is mostly a matter of definition: one can define a different model of concurrent logic programs in which termination is by fixpoint; similarly, one can define "NITY", which is like UNITY except that there are explicit termination conditions. To our opinion, explicit termination is preferable both from the programmer's and from the implementor's point of view in both models.

We find the fundamental difference between UNITY and concurrent logic languages in the notion of a variable. In UNITY, variables are mutable; therefore a transition must exclude other transitions from writing on variables it reads from, and from accessing variables it writes to. In concurrent logic languages, variables are single-assignment, therefore no mutual exclusion mechanisms are required when reading a variable. The effect of mutable unshared variables can be achieved nonetheless in concurrent logic languages, as explained in Section 7, using recurrent processes.

It seems that this fundamental difference is the source of another difference between UNITY and concurrent logic languages, namely their attitude to architectures. Although both are architecture independent, the gap between the general UNITY model and concrete architectures, such as a non-shared memory parallel computer, is sufficiently large that the authors of UNITY suggest that special sublanguages should be tailored for particular parallel architectures. In contrast, authors of concurrent logic languages believe their languages are suitable for all architectures. The burden of matching the application to the architecture resides solely with the algorithm designer and programmer. The belief, which is backed by the implementation efforts, is that concurrent logic languages are suitable for a wide range of architectures, including synchronous and asynchronous shared-memory computers, and tightly and loosely coupled non-shared memory computers. The difference between these architectures is not necessarily in the concurrent logic language suitable for them, but rather in the tradeoffs in communication and computation they offer, which determine which algorithms will better match a particular architecture.

This difference is not a coincidence. The single assignment property of logic variables means that even in a language with atomic test unification, locking of variables is very rarely necessary. Specifically, it is necessary almost only when the atomicity of unification is actually exploited to achieve some synchronization task. For example, in simple benchmarks of the parallel implementation of FCP(?) on the iPSC hypercube, more than 95% of the message traffic was associated with reading remote values (which does not require locking because of the single assignment property), and less than 5% with locking remote variables [192]. This is achieved without any special compilation or program analysis techniques. In UNITY, on the other hand, in the absence of additional information, every transition which accesses more than one variable requires locking all variables accessed. Therefore special sublanguages, which are structured to mimic the underlying architecture, have to be employed to make the model realistic.

On a methodological level, there are other differences between the approach of UNITY and that of concurrent logic languages. UNITY does not attempt to address questions of meta-programming and systems programming, or, more generally, how would a parallel computer system, whose base language is UNITY, be constructed. This question has been fundamental to concurrent logic programming from its beginning.

22. Conclusion

This survey attempted to convey the soundness, breadth, and potential of the logic programming approach to concurrency. Progress in the following can foster fully realizing this potential:

- Provide competitive implementations of concurrent logic languages for sequenial, parallel and distributed computers.
- Develop simpler semantic foundations for concurrent logic languages.
- Exploit the simplicity of these languages to provide advanced program development environments and tools.
- Exploit the simplicity of these languages to provide advanced program analysis, transformation, and optimization techniques, to aid in their efficient implementation.
- Further develop programming methodologies and techniques for these languages.
- Enhance concurrent logic programming by incorporating ideas and methods from constraint logic programming.
- Further explore techniques for embedding higher-level languages, and design higher-level languages (such as parallel constraint programming languages) especially suitable for embedding in concurrent logic languages.

Acknowledgements

Comments by Denis Dams, Andrew Davison, Moreno Falaschi, Yishai Feldman, Ian Foster, Haim Gaifman, John Gallagher, Steve Gregory, David Harel, Ken Kahn, Michael Maher, Martin Nilsson, Vijay Saraswat, Jiro Tanaka, Kazunori Ueda, Peter Wegner, Eyal Yardeni, and the anonymous referees are gratefully acknowledged.

References

[1] Ackerman, W.B., Data flow languages, *IEEE Computer* **15**(2), pp.15-25, Feb. 1982.

[2] Aho, A.V., Hopcraft, J.E., and Ullman, J.D., *The Design and Analysis of Computer Algorithms*, Addison-Wesley, 1974.

[3] Ahuja, S., Carriero, N., and Gelernter, D., Linda and friends, *IEEE Computer* **19**(8), 26–34, Aug. 1986.

[4] Ali, K.A.M., Or-parallel execution of Prolog on a multi-sequential machine, SICS Technical Report, 1986.

[5] Ali, K.A.M., Or-parallel execution of Horn clause programs based on the WAM and shared control information, SICS Technical Report, 1986.

[6] Alkalaj, L., and Shapiro, E., An architectural model for a Flat Concurrent Prolog processor, in Bowen, K. and Kowalski, R.A. (eds.), *Proc. 5th International Conference Symposium on Logic Programming*, pp. 1277–1297, MIT Press, 1988.

[7] Bage, G., and Lindstrom, G., Committed choice functional programming, *Proc. International Conference on Fifth Generation Computer Systems*, pp. 666–674, ICOT, Tokyo, 1988.

[8] Bahgat, R., and Gregory, S., Pandora: Non-deterministic parallel logic programming, To appear in *Proc. 6th International Conference on Logic Programming*, Lisbon, MIT Press, 1989.

[9] de Bakker, J.W., and Kok, J.N., Uniform abstraction, atomicity and contractions in the comparative semantics of Concurrent Prolog, *Proc. International Conference on Fifth Generation Computer Systems*, pp. 347–355, ICOT, Tokyo, 1988.

[10] Baron, U., Chassin de Kergommeaux, J., Hailperin, M., Ratcliffe, M., Robert, P., Syre, J.-C., and Westphal, H., The parallel ECRC Prolog system PEPSys: An overview and evaluation results, *Proc. International Conference on Fifth Generation Computer Systems*, pp. 841–850, ICOT, Tokyo, 1988.

[11] Bowen, D.L., Byrd, L., Pereira, L.M., Pereira, F.C.N., and Warren, D.H.D., PROLOG on the DECSystem-10 User's Manual, Technical Report, Department of Artificial Intelligence, University of Edinburgh, 1981.

[12] Carlsson, M., Freeze, indexing, and other implementation issues in the WAM, in Lassez, J.-L. (ed.), *Proc. 4th International Conference on Logic Programming*, pp. 40–58, MIT Press, 1987.

[13] Carriero, N., and Gelernter, D., Linda in context, *Comm. ACM* **32**(4), 444–458, 1988.

[14] Chandy, M., and Lamport, L., Distributed snapshots: determining global states of distributed systems, *Transactions on Computer Systems* **3**(1), pp. 63–75, 1985.

[15] Chandy, K.M., and Misra, J., A paradigm for detecting quiescent properties in distributed computations, in Apt, K.R. (ed.), *Logics and Models of Concurrent Systems*, pp. 325–342, Springer-Verlag, 1985.

[16] Chandy, K.M., and Misra, J., *Parallel Program Design*, Addison-Wesley, 1988.

[17] Chikayama, T., and Kimura, Y., Multiple reference management in Flat GHC, in Lassez, J.-L. (ed.), *Proc. 4th International Conference on Logic Programming*, pp. 276–293, MIT Press, 1987.

[18] Chikayama, T., Sato, H., and Miyazaki, T., Overview of the parallel inference

machine operating system (PIMOS), *Proc. International Conference on Fifth Generation Computer Systems*, pp. 230–251, ICOT, Tokyo, 1988.

[19] Clark, K.L., Predicate logic as a computational formalism, Research Report DOC 79/59, Department of Computing, Imperial College, London, 1979.

[20] Clark, K.L., and Gregory, S., A relational language for parallel programming, *Proc. ACM Conference on Functional Languages and Computer Architecture*, pp. 171–178, 1981. Also Chapter 1 in [166].

[21] Clark, K.L., and Gregory, S., PARLOG: A parallel logic programming language, Research Report DOC 83/5, Department of Computing, Imperial College, London, 1983.

[22] Clark, K.L., and Gregory, S., Notes on systems programming in PARLOG, *Proc. International Conference on Fifth Generation Computer Systems*, pp. 299–306, ICOT, Tokyo, 1984.

[23] Clark, K.L., and Gregory, S., Notes on the implementation of PARLOG, Research Report DOC84/16, 1984. Also in *J. Logic Programming* **2**(1), pp. 17–42, 1985.

[24] Clark, K.L., and Gregory, S., PARLOG: Parallel programming in logic, *ACM TOPLAS* **8**(1), pp. 1–49, 1986. Revised as Chapter 3 in [166].

[25] Clark, K.L., and Gregory, S., PARLOG and PROLOG united, in Lassez, J.-L. (ed.), *Proc. 4th International Conference on Logic Programming*, pp. 927–961, MIT Press, 1987.

[26] Clark, K.L., McCabe, F.G., and Gregory, S., IC-PROLOG — language features, in Clark, K.L., and Tärnlund, S.-A. (eds.), *Logic Programming*, pp. 253–266, Academic Press, London, 1982.

[27] Clark, K.L., and Tärnlund, S.-A., A first-order theory of data and programs, in Gilchrist, B. (ed.), *Information Processing* **77**, pp. 939–944, North-Holland, 1977.

[28] Clocksin, W.R., and Alshawi, H., A method for efficiently executing Horn clause programs using multiple processors, Technical Report, Department of Computer Science, Cambridge University, Cambridge, 1986.

[29] Codish, M., and Shapiro, E., Compiling Or-parallelism into And-parallelism, *New Generation Computing* **5**(1), pp. 45–61, 1987. Also Chapter 32 in [166].

[30] Codish, M., Gallagher, J., and Shapiro, E., Using safe approximations of fixed points for analysis of logic programs, *Proc. META88, Workshop on Meta-Programming in Logic Programming*, Bristol, 1988.

[31] Colmerauer, A., Opening the Prolog-III universe, *BYTE Magazine* **12**(9), August 1987.

[32] Costa, G., and Stirling, C., Weak and strong fairness in CCS, *Information and Computation* **73**, pp.207–244, 1987.

[33] Crammond, J., A comparative study of unification algorithms for Or-parallel execution of logic languages, *Proc. IEEE International Conference on Parallel Processing*, pp. 131–138, 1985.

[34] Crammond, J.A., Implementation of committed choice logic languages on shared memory multiprocessors, Ph.D. Thesis, Department of Computer Science, Heriot-Watt University, May 1988. Also Technical Report PAR 88/4, Department of Computing, Imperial College, London, 1988.

[35] Davison, A., POLKA: A PARLOG object oriented language, Department of Computing, Imperial College, London, 1988.

[36] DeGroot, D., and Lindstrom, G. (eds.), *Logic Programming — Functions, Relations and Equations*, Prentice-Hall, New Jersey, 1986.

[37] Dijkstra, E.W, Hierarchical ordering of sequential processes, *Acta Informatica* **1**, 115–138, 1971.

[38] Dijkstra, E.W., Guarded commands, nondeterminacy, and formal derivation of programs, *CACM* **18**(8), pp. 453–457, 1975.

[39] Dijkstra, E.W., *A Discipline of Programming*, Prentice-Hall, New Jersey, 1976.

[40] Dijkstra, E.W., and Scholten, C.S., Termination detection for diffusing computations, *Information Processing Letters* **11**(1), pp. 1–4, 1980.

[41] van Emden, M.H., and Kowalski, R.A., The semantics of predicate logic as a programming language, *J. ACM* **23**(4), pp. 733–742, 1976.

[42] van Emden, M.H., and de Lucena, G.J., Predicate logic as a language for parallel programming, in Clark, K.L., and Tärnlund, S.-A. (eds.), *Logic Programming*, pp. 189–198, Academic Press, London, 1982.

[43] Ershov, A.P., et al. (eds.), Special Issue: Selected papers from the Workshop on Partial Evaluation and Mixed Computation, 1987, *New Generation Computing* **6**(2,3), 1988.

[44] Falaschi, M., and Levi, G., Finite failures and partial computations in concurrent logic languages, *Proc. International Conference on Fifth Generation Computer Systems*, pp. 364–373, ICOT, Tokyo, 1988.

[45] Falaschi, M., Levi, G., Martelli, M., and Palamidessi, G., A new declarative semantics for logic languages, in K. Bowen and R.A. Kowalski (eds.), *Proc. 5th International Conference Symposium on Logic Programming*, pp. 993–1005, MIT Press, 1988.

[46] Foster, I., Efficient metacontrol in parallel logic programming, Research Report PAR 87/18, Department of Computing, Imperial College, London, 1987.

[47] Foster, I., Parallel implementation of PARLOG, *Proc. International Conference of Parallel Processing*, 1988.

[48] Foster, I., *PARLOG as a Systems Programming Language*, Ph.D. Thesis, Department of Computing, Imperial College, London, 1988.

[49] Foster, I., and Taylor, S., Flat Parlog: A basis for comparison, *International J. of Parallel Programming* **16**(2), 1987.

[50] Francez, N., *Fairness*, Springer-Verlag, 1987.

[51] Francez, N., and Rodeh, M., Achieving distributed termination without freezing, *IEEE Transactions on Software Engineering* **SE-8**(3), pp. 359–385, 1982.

[52] Friedman, D.P., and Wise, D.S., Aspects of applicative programming for parallel processing, *IEEE Trans. on Computers* **C-27**(4), pp. 289–296, 1978.

[53] Friedman, D.P., and Wise, D.S., An approach to fair applicative multiprogramming, in Kahn, G. (ed.), *Semantics of Concurrent Computation*, *LNCS* **70**, pp. 203–226, Springer-Verlag, 1979.

[54] Friedman, D.P., and Wise, D.S., An indeterminate constructor for applicative programming, *Conference Record 7th ACM Symposium on Principles of Programming Languages*, pp. 245–250, 1980.

[55] Fuchi, K., and Furukawa, K., The role of logic programming in the Fifth Generation Computer Project, *New Generation Computing* **5**(1), pp. 3–28, 1987.

[56] Furukawa, K., and Ueda, K., GHC process fusion by program transformation, *Proc. 2nd Conference on Japan Society of Software Science and Technology*, pp. 89–92, 1985.

[57] Furukawa, K., Okumura, A., and Murakami, M., Unfolding rules for GHC programs, in Bjorner, D. et al. (eds.), *Proc. Workshop on Partial and Mixed Computation*, Gl. Avernaes, 1987.

[58] Futamura, Y., Partial evaluation of computation process — an approach to a compiler-compiler, *Systems, Computers, Controls* 2(5), pp. 721–728, 1971.

[59] Gaifman, H., Maher, M.J., and Shapiro, E., Reactive behaviors for concurrent constraint logic languages, in Lusk, E. and Overbeek, R. (eds.), *Proc. 1989 North America Conference on Logic Programming*, pp. 553–572, MIT Press, 1989.

[60] Gaifman, H., and Shapiro, E., Fully abstract compositional semantics for logic programs, *Proc. ACM Symposium on Principles of Programming Languages*, pp. 134–142, 1989.

[61] Gaifman, H., and Shapiro, E., Proof Theory and Semantics of Logic Programs, *Proc. IEEE Symposium on Logic in Computer Science*, 1989.

[62] Gallagher, J., *An Approach to the Control of Logic Programs*, Ph.D. Thesis, Department of Computer Science, Trinity College, Dublin, 1983.

[63] Gallagher, J., Transforming logic programs by specialising interpreters, *Proc. 7th European Conference on Artificial Intelligence*, pp. 109–122, Brighton, 1986.

[64] Gallagher, J., Codish, M., and Shapiro, E., Specialisation of Prolog and FCP programs using abstract interpretation, *New Generation Computing* 6, pp. 159–186, 1988.

[65] Gerth R., Codish M., Lichtenstein Y., and Shapiro E., Fully abstract denotational semantics for Flat Concurrent Prolog, *Proc. IEEE Symposium on Logic in Computer Science*, pp. 320–333, 1988.

[66] Gregory, S., *Parallel Logic Programming in PARLOG*, Addison-Wesley, 1987.

[67] Gregory, S., Neely, R., and Ringwood, G.A., PARLOG for specification, verification and simulation, in Koomen, C.J., and Moto-Oka, T. (eds.), *Proc. 7th International Symposium on Computer Hardware Description Languages and their Applications*, pp. 139–148, Elsevier/North-Holland, Amsterdam, 1985.

[68] Gregory, S., Foster, I.T., Burt, A.D., and Ringwood, G.A., An abstract machine for the implementation of PARLOG on uniprocessors, *New Generation Computing* 6(4), 389–420, 1986.

[69] Halstead, R.H., MultiLisp — A language for concurrent symbolic computation, *ACM Trans. on Programming Languages and Systems* 7(4), pp. 501–538, 1985.

[70] Harel, D., *Algorithmics: The Spirit of Computing*, Addison-Wesley, 1987.

[71] Harel, D., and Pnueli, A., On the development of reactive systems, in Apt, K.R. (ed.), *Logics and Models of Concurrent Systems*, Springer Verlag, 1985.

[72] Haridi, S., and Brand, P., ANDORRA Prolog — an integration if Prolog and committed choice languages, *Proc. International Conference on Fifth Generation Computer Systems*, pp. 745–754, ICOT, Tokyo, 1988.

[73] Harper, R., MacQueen, D., and Milner, R., Standard ML, Technical Report ECS-LFCS-86-2, University of Edinburgh, 1986.

[74] Harsat, A., and Ginosar, R., CARMEL - a VLSI architecture for Flat Concurrent Prolog, EE PUB. Technical Report, Department of Computer Science, Technion, Haifa, 1987.

[75] Harsat, A., and Ginosar, R., CARMEL-2: a second generation VLSI architecture for Flat Concurrent Prolog, *Proc. International Conference on Fifth Generation Computer Systems*, pp. 962–969, ICOT, Tokyo, 1988.

[76] Hellerstein, L., and Shapiro, E., Implementing parallel algorithms in Concurrent

Prolog: The MAXFLOW experience, *J. Logic Programming* **3**(2), pp. 157–184, 1984. Also Chapter 9 in [166].

[77] Henderson, P., Purely functional operating systems, in Darlington, J., Henderson, P., and Turner, D. (eds.), ***Functional Programming and Its Applications***, Cambridge University Press, 1982.

[78] Hewitt, C., A universal, modular Actor formalism for artificial intelligence, *Proc. International Joint Conference on Artificial Intelligence*, 1973.

[79] Hewitt, C., The challenge of open systems, *Byte Magazin*, pp. 223–242, 1985.

[80] Hill, R., LUSH-resolution and its completeness, DCL Memo 78, Department of Artificial Intelligence, University of Edinburgh, 1974.

[81] Hirata, M., Letter to the editor, *SIGPLAN Notices*, pp. 16–17, May 1986.

[82] Hirata, M., Programming language Doc and its self-description, or, X=X is considered harmful, *Proc. 3rd Conference of Japan Society of Software Science and Technology*, pp. 69–72, 1986.

[83] Hirata, M., Parallel list processing language Oc and its self-description, *Computer Software* **4**(3), pp. 41–64, 1987 (in Japanese).

[84] Hirsch, M., Silverman, W., and Shapiro, E., Computation control and protection in the Logix system, Chapter 20 in [166].

[85] Hoare, C.A.R., Monitors: an operating systems structuring concept, *Comm. ACM* **17**(10), pp. 549-557, 1974.

[86] Hoare, C.A.R., Communicating sequential processes, *Comm. ACM* **21**(8), pp. 666-677, 1978.

[87] Hoare, C.A.R., *Communicating Sequential Processes*, Prentice-Hall, New Jersey, 1985.

[88] Hopcroft, J.E., and Ullman, J.D., *Introduction to Automata Theory, Languages, and Computation*, Addison-Wesley, 1979.

[89] Houri, A., and Shapiro, E., A sequential abstract machine for Flat Concurrent Prolog, Chapter 38 in [166].

[90] Ichiyoshi, N., Miyazaki, T., and Taki, K., A distributed implementation of Flat GHC on the Multi-PSI, in Lassez, J.-L. (ed.), *Proc. 4th International Conference of Logic Programming*, pp. 257–275, MIT Press, 1987.

[91] INMOS Ltd., *OCCAM Programming Manual*, Prentice-Hall, New Jersey, 1984.

[92] Jaffar, J., and Lassez, J-L., Constraint logic programming, *ACM Symposium on Principles of Programming Languages*, Munich, 1987.

[93] Johnson, S.D., Circuits and systems: Implementing communications with streams, Technical Report 116, Computer Science Department, Indiana University, 1981.

[94] Kahn, G., and MacQueen, D., Coroutines and networks of parallel processes, in Gilchrist, B. (ed.), *Information Processing* **77**, *Proc. IFIP Congress*, pp. 993–998, North-Holland, 1977.

[95] Kahn, K., Objects: A fresh look, *Proc. European Conference on Object Oriented Programming*, Nottingham, July 1989.

[96] Kahn, K., Tribble, E.D., Miller, M., and Bobrow, D.G., Vulcan: Logical concurrent objects, in Shriver, B., and Wegner, P. (eds.), *Research Directions in Object-Oriented Programming*, MIT Press, 1987. Also Chapter 30 in [166].

[97] Kahn, K., Silverman, W., and Shapiro, E., Logic programs with implicit variables, unpublished, 1988.

[98] Kimura, Y., and Chikayama, T., An abstract KL1 machine and its instruction

set, *Proc. IEEE Symposium on Logic Programming*, pp. 468–477,San Francisco, 1987.

[99] Kishimoto, M., et al., An evaluation of the FGHC via practical application programs, ICOT Technical Report TR-232, Institute for New Generation Computer Technology, Tokyo, 1987.

[100] Kliger, S. and Shapiro, E., A decision tree compilation algorithm for FCP(|,:,?), in Bowen, K. and Kowalski, R.A. (eds.), *Proc. 5th International Conference Symposium on Logic Programming*, pp. 1315–1336, MIT Press, 1988.

[101] Kliger, S., Yardeni, E., Kahn, K., and Shapiro, E., The language FCP(:,?), *Proc. International Conference on Fifth Generation Computer Systems*, pp. 763–773, ICOT, Tokyo, 1988.

[102] Kohda, Y., and Tanaka, J., Deriving a compilation method for parallel logic languages, *Logic Programming*, *LNCS* **315**, pp. 80–94, Springer-Verlag, 1988.

[103] Kowalski, R.A., *Logic for Problem Solving*, Elsevier, North-Holland, 1979.

[104] Kung, H.T., Why systolic architectures?, *IEEE Computer* **15**(1), pp. 37–46, 1982.

[105] Kusalik, A.J., Bounded-wait merge in Shapiro's Concurrent Prolog, *New Generation Computing* 1(12), pp. 157–169, 1984.

[106] Lai, T.H., Termination detection for dynamically distributed systems with non-first-in-first-out communication, *J. Parallel and Distributed Computing* **3**, pp. 577–599, 1986.

[107] Lam, M., and Gregory, S., PARLOG and ALICE: a marriage of convenience, in Lassez, J.-L. (ed.), *Proc. 4th International Conference on Logic Programming*, pp. 294–310, MIT Press, 1987.

[108] Lamport, L., Time, Clocks, and the ordering of events in a distributed system, *Communications of the ACM* **25**, pp. 558–564, July 1978.

[109] Lassez, J.-L., Maher, M.J., and Marriot, K., Unification revisited, in J. Minker (ed.), *Foundations of Deductive Databases and Logic Programming*, pp. 587–626, Morgan Kaufmann, 1987.

[110] Levi, G., Models, unfolding rules and fixpoint semantics, in Bowen, K. and Kowalski, R.A. (eds.), *Proc. 5th International Conference Symposium on Logic Programming*, pp. 1649–1665, MIT Press, 1988.

[111] Levi, G., and Palamidessi, C., An approach to the declarative semantics of synchronization in logic languages, in Lassez, J.-L. (ed.), *Proc. 4th International Conference on Logic Programming*, pp. 877–893, MIT Press, 1987.

[112] Levi, G., and Sardu, G., Partial evaluation of meta-programs in a multiple world's logic language, in Bjorner, D., Ershov, A.P. and Jones, N.D. (eds.), *Workshop on Partial Evaluation and Mixed Computation*, Gl. Avernaes, 1987.

[113] Levy, J., A unification algorithm for Concurrent Prolog, Tärnlund, S.-A. (ed.), *Proc. 2nd International Conference on Logic Programming*, pp. 333–341, Uppsala, 1984.

[114] Levy, J., A GHC abstract machine and instruction set, in Shapiro, E. (ed.), *Proc. 3rd International Conference on Logic Programming*, *LNCS* **225**, pp. 157–171, Springer-Verlag, 1986.

[115] Levy, J., *Concurrent Prolog and Related Languages*, Ph.D. Thesis, Department of Computer Science, The Weizmann Institute of Science, Rehovot, 1988.

[116] Levy, J., and Friedman, N., Concurrent Prolog implementations — two new schemes, Technical Report CS86-13, Department of Computer Science, The

Weizmann Institute of Science, Rehovot, 1986.

[117] Levy, J., and Shapiro, E., Translation of Safe GHC and Safe Concurrent Prolog to FCP, Chapter 33 in [166].

[118] Levy, J., and Shapiro, E., CFL — A concurrent functional language embedded in a concurrent logic programming environment, Chapter 35 in [166].

[119] Lichtenstein, Y., and Shapiro, E., Concurrent algorithmic debugging, *Proc. ACM Workshop on Parallel Debugging*. Also Technical Report CS87-20, Department of Computer Science, The Weizmann Institute of Science, Rehovot, 1987.

[120] Lichtenstein, Y., Codish, M., and Shapiro, E., Representation and enumeration of Flat Concurrent Prolog computations, Chapter 27 in [166].

[121] Lichtenstein, Y. and Shapiro, E., Abstract algorithmic debugging, in Bowen, K. and Kowalski, R.A. (eds.), *Proc. 5th International Conference and Symposium on Logic Programming*, pp. 1315–1336, MIT Press, 1988.

[122] Lloyd, J.W., *Foundations of Logic Programming*, Second Edition, Springer-Verlag, 1987.

[123] Lusk, E., Butler, R., Disz, T., Olson, R., Overbeek, R., Stevens, R., Warren, D.H.D, Calderwood, A., Szeredi, P., Haridi, S., Brand, P., Carlsson, M., Ciepielewski, A., and Hausman, B., The Aurora Or-Parallel Prolog system, *Proc. International Conference on Fifth Generation Computer Systems*, pp. 819–830, ICOT, Tokyo, 1988.

[124] Maher, M.J., Equivalences of logic programs, in J. Minker (ed.), *Foundations of Deductive Databases and Logic Programming*, pp. 627–658, Morgan Kaufmann Publishers, Los Altos, 1987.

[125] Maher, M.J., Logic semantics for a class of committed-choice programs, in Lassez, J.-L. (ed.), *Proc. 4th International Conference on Logic Programming*, pp. 858–876, MIT Press, 1987.

[126] Manna, Z., and Pnueli, A., Specification and verification of concurrent programs by $\forall$-automata, Report STAN-CS-88-1230, Department of Computer Science, Stanford University, Stanford, 1988.

[127] Mattern, F., Algorithms for distributed termination detection, *Distributed Computing* **2**, pp. 161–175, 1987.

[128] Miyazaki, T., Takeuchi, A., and Chikayama, T., A sequential implementation of Concurrent Prolog based on the shallow binding scheme, *IEEE Symposium on Logic Programming* pp. 110-118, 1985. Also Chapter 37 in [166].

[129] Mierowsky, C., Taylor, S., Shapiro, E., Levy, J., and Safra, S., The design and implementation of Flat Concurrent Prolog, Technical Report CS85-09, Department of Computer Science, The Weizmann Institute of Science, Rehovot, 1985.

[130] Milner, R., *A Calculus of Communicating Systems*, *LNCS* **92**, Springer-Verlag, 1980.

[131] Misra, J., Distributed discrete-event simulation, *Computing Surveys* **18**(1), pp. 39–65, 1986.

[132] Murakami, M., A declarative semantics of parallel logic programs with perpetual processes, *Proc. International Conference on Fifth Generation Computer Systems*, pp. 374–381, ICOT, Tokyo, 1988.

[133] Naish, L., *MU-Prolog 3.1db Reference Manual*, Internal Memorandum, Department of Computer Science, University of Melbourne, 1984.

[134] Nelson, B.J., Remote procedure call, Ph.D. Thesis, Report CMU-CS-81-119,

Department of Computer Science, Carnegie-Mellon University, May 1981.

[135] Nilsson, M., and Tanaka, H., Fleng Prolog — The language which turns supercomputers into Prolog machines, in Wada, E. (ed.), *Proc. Japanese Logic Programming Conference*, pp. 209–216, ICOT, Tokyo, 1986. Proceedings also published as Springer *LNCS* **264**.

[136] Nilsson, M., and Tanaka, H., The art of building a parallel logic programming system or from zero to full GHC in ten pages, in Wada, E. (ed.), *Proc. Japanese Logic Programming Conference*, pp. 155–163, ICOT, Tokyo, 1987. Proceedings also to appear as Springer *LNCS*.

[137] Nilsson, M., and Tanaka, H., Massively parallel implementation of Flat GHC on the connection machine, *Proc. International Conference on Fifth Generation Computer Systems*, pp. 1031–1040, ICOT, Tokyo, 1988.

[138] Nilsson, M., and Tanaka, H., A Flat GHC implementation for supercomputers, in Bowen, K. and Kowalski, R.A. (eds.), *Proc. 5th International Conference Symposium on Logic Programming*, pp. 1337–1350, MIT Press, 1988.

[139] Ohki, M., et al., An object-oriented programming language based on a parallel logic programming language KL1, ICOT Technical Report TR-222, Institute for New Generation Computer Technology, Tokyo, 1987.

[140] Okabe, Y., and Yajima, S., Parallel computational complexity of logic programs and alternating turing machines, *Proc. International Conference on Fifth Generation Computer Systems*, pp. 356–363, ICOT, Tokyo, 1988.

[141] Okumura, A., and Matsumoto, Y., Parallel programming with layered streams, *Proc. IEEE Symposium on Logic Programming*, pp. 224–231, San Francisco, 1987.

[142] Pereira, L.M., and Nasr, R., Delta-Prolog: a distributed logic programming language, *Proc. International Conference on Fifth Generation Computer Systems*, pp. 283–291, ICOT, Tokyo, 1984.

[143] Picca, R., Bellone, J., and Levy, J., Or-parallel And-interleaving execution of Concurrent Prolog, Technical Report CS87-07, Department of Computer Science, The Weizmann Institute of Science, Rehovot, 1987.

[144] Pnueli, A., Applications of temporal logic to the specification and verification of reactive systems: A survey of current trends, in de Bakker, J.W., de Roever, W.P., and Rozenberg, G. (eds.), *Current Trends in Concurrency, Overviews and Tutorials*, *LNCS* **224**, pp. 510–584, Springer-Verlag, 1986.

[145] Ramakrishnan, R., and Silberschatz, A., Annotations for distributed programming in logic, *Conference Record 13^{en} ACM Symposium on Principles of Programming Languages*, pp. 255–262, 1986.

[146] Reches, E., Gudes, E., and Shapiro, E., Parallel access to a distributed database and its implementation in Flat Concurrent Prolog, Technical Report CS88-11, Department of Computer Science, The Weizmann Institute of Science, Rehovot, 1988.

[147] Ringwood, G.A., Pattern-directed, Markovian, linear, guarded definite clause resolution, Department of Computing, Imperial College, London, 1987.

[148] Ringwood, G.A., PARLOG86 and the dining logicians, *CACM* **31**(1), pp. 10–25, 1988.

[149] Robinson, J.A., A machine oriented logic based on the resolution principle, *J. ACM* **12**(1), pp. 23–41, 1965.

[150] Safra, S., *Partial Evaluation of Concurrent Prolog and Its Implications*, M.Sc.

Thesis, Technical Report CS86-24, Department of Computer Science, The Weizmann Institute of Science, Rehovot, 1986.

[151] Safra, S., and Shapiro, E., Meta-interpreters for real, ***Information Processing 86***, pp. 271–278, North-Holland, 1986. Also Chapter 25 in [166].

[152] Saraswat, V.A., Partial correctness semantics for CP[↓,|,&], ***Proc. 5th Conference on Foundations of Software Technology and Theoretical Computer Science, LNCS*** **206**, pp. 347–368, New Delhi, 1985.

[153] Saraswat, V.A., Problems with Concurrent Prolog, Technical Report CS86-100, Carnegie-Mellon University, 1986.

[154] Saraswat, V.A., Merging many streams efficiently: The importance of atomic commitment, Chapter 16 in [166].

[155] Saraswat, V.A., A compiler of CP(↓,|,&) on top of Prolog, Technical Report CS-87-174, Carnegie-Mellon University, 1987.

[156] Saraswat, V.A., The concurrent logic programming language CP: Definition and operational semantics, ***Proc. ACM Symposium on Principles of Programming Languages***, pp. 49–63, 1987.

[157] Saraswat, V.A., The language GHC: Operational semantics, problems and relationship with CP[↓,|], ***Proc. IEEE Symposium on Logic Programming***, pp. 347–358,San Francisco, 1987.

[158] Saraswat, V.A., A somewhat logical formulation of CLP synchronization primitives, in Bowen, K. and Kowalski, R.A. (eds.), ***Proc. 5th International Conference Symposium on Logic Programming***, pp. 1298–1314, MIT Press, 1988.

[159] Saraswat, V.A., ***Concurrent Constraint Programming Languages***, Ph.D. Thesis, Department of Computer Science, Carnegie-Mellon University, 1989.

[160] Saraswat, V.A., Weinbaum, D., Kahn, K., and Shapiro, E., Detecting stable properties of networks in concurrent logic programming languages, ***Proc. ACM Conference on Principles of Distributed Computing***, pp. 210–222, Aug. 1988.

[161] Shapiro, E., ***Algorithmic Program Debugging***, MIT Press, 1983.

[162] Shapiro, E., A subset of Concurrent Prolog and its interpreter, ICOT Technical Report TR-003, Institute for New Generation Computer Technology, Tokyo, 1983. Also chapter 2 in [166].

[163] Shapiro, E., Alternation and the computational complexity of logic programs, ***J. Logic Programming*** **1**(1), pp. 19–33, 1984.

[164] Shapiro, E., Concurrent Prolog: A progress report, ***IEEE Computer*** **19**(8), pp. 44–58, Aug. 1984. Also Chapter 5 in [166].

[165] Shapiro, E., Systolic programming: A paradigm of parallel processing, ***Proc. International Conference on Fifth Generation Computer Systems***, pp. 458–471, 1984. Also Chapter 7 in [166].

[166] Shapiro, E. (Editor), ***Concurrent Prolog: Collected Papers***, Vols. 1 and 2, MIT Press, 1987.

[167] Shapiro, E., Or-parallel Prolog in Flat Concurrent Prolog, Chapter 34 in [166].

[168] Shapiro, E., Embedding Linda and other joys of concurrent logic programming, Technical Report CS89-07, Department of Computer Science, The Weizmann Institute of Science, Rehovot, 1989.

[169] Shapiro, E., and Mierowsky, C., Fair, biased, and self-balancing merge operators: Their specification and implementation in Concurrent Prolog, ***New Generation Computing*** **2**(3), pp. 221–240, 1984. Also Chapter 14 in [166].

[170] Shapiro, E., and Safra, S., Multiway merge with constant delay in Concurrent

Prolog, *New Generation Computing* 4(2), pp. 211–216, 1986. Also Chapter 15 in [166].

[171] Shapiro, E., and Takeuchi, A., Object-oriented programming in Concurrent Prolog, *New Generation Computing* 1(1), pp. 25–49, 1983. Also Chapter 29 in [166].

[172] Silverman, W., Hirsch, M., Houri, A., and Shapiro, E., The Logix system user manual, Version 1.21, Chapter 21 in [166].

[173] Sterling, L.S., and Shapiro, E., *The Art of Prolog*, MIT Press, 1986.

[174] Suzuki, N., Experience with specification and verification of complex computer using Concurrent Prolog, in Warren, D.H.D, and van Caneghem, M. (eds.), *Logic Programming and Its Applications*, pp. 188–209, Ablex Pub. Co., New Jersey, 1986.

[175] Szöke, D., *Distributed Flat Concurrent Prolog on a Network Architecture*, M.Sc. Thesis, Department of Computer Science, The Weizmann Institute of Science, Rehovot, 1987.

[176] Takeda, Y., Nakashima, H., Masuda, K., Chikayama, T., and Taki, K., A load balancing mechanism for large scale multiprocessor systems and its implementation, *Proc. International Conference on Fifth Generation Computer Systems*, pp. 978–986, ICOT, Tokyo, 1988.

[177] Takeuchi, A., How to solve it in Concurrent Prolog, 1983 (unpublished note).

[178] Takeuchi, A., Algorithmic debugging of GHC programs and its implementation in GHC, Chapter 26 in [166].

[179] Takeuchi, A., and Furukawa, K., Bounded-buffer communication in Concurrent Prolog, *New Generation Computing* 3(2), pp. 145–155, 1985. Also Chapter 18 in [166].

[180] Takeuchi, A., and Furukawa, K., Partial evaluation of Prolog programs and its application to meta-programming, *Information Processing 86*, pp. 415–420, North-Holland, 1986.

[181] Takeuchi, A., et al., A description language with AND/OR parallelism for concurrent systems and its stream-based realization, ICOT Technical Report TR-229, Institute for New Generation Computer Technology, Tokyo, 1987.

[182] Tamaki, H., A distributed unification scheme for systolic programs, *Proc. International Conference on Parallel Processing*, pp. 552–559, 1985.

[183] Tanaka, J., A simple programming system written in GHC and its reflective operations, *Proc. Japanese Logic Programming Conference*, pp. 143–149, ICOT, Tokyo, 1988.

[184] Tanaka, J., Meta-interpreters and reflective operations in GHC, *Proc. International Conference on Fifth Generation Computer Systems*, pp. 774–783, ICOT, Tokyo, 1988.

[185] Tanaka, J., Ueda, K., Miyazaki, T., Takeuchi, A., Matsumoto, Y., and Furukawa, K., Guarded Horn clauses and experiences with parallel logic programming, *Proc. FJCC ACM*, pp. 948–954, Dallas, 1986.

[186] Taylor, H., Localizing the GHC suspension test, in Bowen, K. and Kowalski, R.A. (eds.), *Proc. 5th International Conference Symposium on Logic Programming*, pp. 1257–1271, MIT Press, 1988.

[187] Taylor, S., *Parallel Logic Programming Techniques*, Ph.D. Thesis, Department of Computer Science, The Weizmann Institute of Science, Rehovot, 1988.

[188] Taylor, S., Av-Ron, E., and Shapiro, E., A layered method for process and code mapping, *J. New Generation Computing* 5(2), 1987. Also Chapter 22 in [166].

[189] Taylor, S., and Foster, I., *Strand Language Reference Manual*, Technical Report PAR 88/10, Department of Computing, Imperial College, London, 1988.

[190] Taylor, S., Hellerstein, L., Safra, S., and Shapiro, E., Notes on the complexity of systolic programs, *J. Parallel and Distributed Computing* 4(3), 1987. Also Chapter 8 in [166].

[191] Taylor, S., Safra, S., and Shapiro E., A parallel implementation of Flat Concurrent Prolog, *J. Parallel Programming* **15**(3), pp. 245–275, 1987. Also Chapter 39 in [166].

[192] Taylor, S., Shapiro, R., and Shapiro, E., FCP: A summary of performance results, in Fox, G. (ed.), *Proc. 3rd Conference on Hypercube Concurrent Computers and Applications*, pp. 1364–1373, ACM Press, 1988.

[193] Taylor, S., and Shapiro, E., An improved parallel algorithm for Flat Concurrent Prolog, Technical Report CS88-09, Department of Computer Science, The Weizmann Institute of Science, Rehovot, 1988.

[194] Tick, E., A performance comparison of And- and Or-parallel logic programming architectures, ICOT Technical Report TR-421, Institute for New Generation Computer Technology, Tokyo, 1988.

[195] Tribble, E.D., Miller, M.S., Kahn, K., Bobrow, D.G. and Abbott, C., Channels: A generalization of streams, in Lassez, J.-L. (ed.), *Proc. 4th International Conference of Logic Programming*, pp. 839–857, MIT Press, 1987. Also Chapter 17 in [166].

[196] Uchida, S., Taki, K., Nakashima, K., Goto, A., and Chikayama, T., Research and development of the parallel inference system in the intermediate stage of the FGCS project, *Proc. International Conference on Fifth Generation Computer Systems*, pp. 16–36, ICOT, Tokyo, 1988.

[197] Ueda, K., Concurrent Prolog re-examined, ICOT Technical Report TR-102, Institute for New Generation Computer Technology, Tokyo, 1985.

[198] Ueda, K., *Guarded Horn Clauses*, Ph.D. Thesis, Information Engineering Course, University of Tokyo, Tokyo, 1986.

[199] Ueda, K., Guarded Horn Clauses, in Wada, E. (ed.), *Logic Programming, LNCS* **221**, pp. 168–179, Springer-Verlag, 1986. Also Chapter 4 in [166].

[200] Ueda, K., Guarded Horn Clauses: A parallel logic programming language with the concept of a guard, ICOT Technical Report TR-208, Institute for New Generation Computer Technology, Tokyo, 1986 (revised in 1987). Also in Nivat, M., and Fuchi, K. (eds.), Programming of Future Generation Computers, pp. 441–456, North-Holland, 1988.

[201] Ueda, K., Making exhaustive search programs deterministic, *New Generation Computing* 5(1), pp. 29–44, 1987.

[202] Ueda, K., Making exhaustive search programs deterministic, Part II, ICOT Technical Report TR-249, Institute for New Generation Computer Technology, Tokyo, 1987.

[203] Ueda, K., Parallelism in logic programming, *Proc. IFIP Congress*, 1989.

[204] Ueda, K., and Chikayama, T., Concurrent Prolog compiler on top of Prolog, *Proc. IEEE Symposium on Logic Programming*, pp. 119–126, 1985.

[205] Ueda, K., and Furukawa, K., Transformation rules for GHC programs, *Proc. International Conference on Fifth Generation Computer Systems*, pp. 582–591, ICOT, Tokyo, 1988.

[206] Vasey, P., Qualified answers and their application to transformation, in Shapiro,

E. (ed.), *Proc. 3rd International Conference on Logic Programming*, *LNCS* **225**, pp. 425–432, Springer-Verlag, 1986.

[207] Warren, D.H.D., An abstract Prolog instruction set, Technical Report 309, Artificial Intelligence Center, SRI International, 1983.

[208] Warren, D.H.D., The SRI model for Or-parallel execution of Prolog — abstract design and implementation, *Proc. IEEE Symposium on Logic Programming*, pp. 92–102, San Francisco, 1987.

[209] Weinbaum, D., and Shapiro, E., Hardware description and simulation using Concurrent Prolog, *Proc. CHDL '87*, pp. 9–27, Elsevier Science Publishing, 1987. Also Chapter 36 in [166].

[210] Yang, R., *A Parallel Logic Programming Language and Its Implementation*, Ph.D. Thesis, Department of Computer Science, Keio University, 1986.

[211] Yang, R., and Aiso, H., P-Prolog: a parallel logic language based on exclusive relation, in Shapiro, E. (ed.), *Proc. 3rd International Conference on Logic Programming*, *LNCS* **225**, pp. 255–269, Springer-Verlag, 1986.

[212] Yardeni, E., and Shapiro, E., A type system for logic programs, Chapter 28 in [166].

NATO ASI Series F

Including Special Programmes on Sensory Systems for Robotic Control (ROB) and on Advanced Educational Technology (AET)

Vol. 1: Issues in Acoustic Signal – Image Processing and Recognition. Edited by C. H. Chen. VIII, 333 pages. 1983.

Vol. 2: Image Sequence Processing and Dynamic Scene Analysis. Edited by T. S. Huang. IX, 749 pages. 1983.

Vol. 3: Electronic Systems Effectiveness and Life Cycle Costing. Edited by J. K. Skwirzynski. XVII, 732 pages. 1983.

Vol. 4: Pictorial Data Analysis. Edited by R. M. Haralick. VIII, 468 pages. 1983.

Vol. 5: International Calibration Study of Traffic Conflict Techniques. Edited by E. Asmussen. VII, 229 pages. 1984.

Vol. 6: Information Technology and the Computer Network. Edited by K. G. Beauchamp. VIII, 271 pages. 1984.

Vol. 7: High-Speed Computation. Edited by J. S. Kowalik. IX, 441 pages. 1984.

Vol. 8: Program Transformation and Programming Environments. Report on a Workshop directed by F. L. Bauer and H. Remus. Edited by P. Pepper. XIV, 378 pages. 1984.

Vol. 9: Computer Aided Analysis and Optimization of Mechanical System Dynamics. Edited by E. J. Haug. XXII, 700 pages. 1984.

Vol. 10: Simulation and Model-Based Methodologies: An Integrative View. Edited by T. I. Ören, B. P. Zeigler, M. S. Elzas. XIII, 651 pages. 1984.

Vol. 11: Robotics and Artificial Intelligence. Edited by M. Brady, L. A. Gerhardt, H. F. Davidson. XVII, 693 pages. 1984.

Vol. 12: Combinatorial Algorithms on Words. Edited by A. Apostolico, Z. Galil. VIII, 361 pages. 1985.

Vol. 13: Logics and Models of Concurrent Systems. Edited by K. R. Apt. VIII, 498 pages. 1985.

Vol. 14: Control Flow and Data Flow: Concepts of Distributed Programming. Edited by M. Broy. VIII, 525 pages. 1985.

Vol. 15: Computational Mathematical Programming. Edited by K. Schittkowski. VIII, 451 pages. 1985.

Vol. 16: New Systems and Architectures for Automatic Speech Recognition and Synthesis. Edited by R. De Mori, C.Y. Suen. XIII, 630 pages. 1985.

Vol. 17: Fundamental Algorithms for Computer Graphics. Edited by R.A. Earnshaw. XVI, 1042 pages. 1985.

Vol. 18: Computer Architectures for Spatially Distributed Data. Edited by H. Freeman and G. G. Pieroni. VIII, 391 pages. 1985.

Vol. 19: Pictorial Information Systems in Medicine. Edited by K. H. Höhne. XII, 525 pages. 1986.

Vol. 20: Disordered Systems and Biological Organization. Edited by E. Bienenstock, F. Fogelman Soulié, G. Weisbuch. XXI, 405 pages.1986.

Vol. 21: Intelligent Decision Support in Process Environments. Edited by E. Hollnagel, G. Mancini, D.D. Woods. XV, 524 pages. 1986.

NATO ASI Series F

Vol. 22: Software System Design Methods. The Challenge of Advanced Computing Technology. Edited by J.K. Skwirzynski. XIII, 747 pages. 1986.

Vol. 23: Designing Computer-Based Learning Materials. Edited by H. Weinstock and A. Bork. IX, 285 pages. 1986.

Vol. 24: Database Machines. Modern Trends and Applications. Edited by A.K. Sood and A.H. Qureshi. VIII, 570 pages. 1986.

Vol. 25: Pyramidal Systems for Computer Vision. Edited by V. Cantoni and S. Levialdi. VIII, 392 pages. 1986. *(ROB)*

Vol. 26: Modelling and Analysis in Arms Control. Edited by R. Avenhaus, R.K. Huber and J.D. Kettelle. VIII, 488 pages. 1986.

Vol. 27: Computer Aided Optimal Design: Structural and Mechanical Systems. Edited by C.A. Mota Soares. XIII, 1029 pages. 1987.

Vol. 28: Distributed Operating Systems. Theory und Practice. Edited by Y. Paker, J.-P. Banatre and M. Bozyiğit. X, 379 pages. 1987.

Vol. 29: Languages for Sensor-Based Control in Robotics. Edited by U. Rembold and K. Hörmann. IX, 625 pages. 1987. *(ROB)*

Vol. 30: Pattern Recognition Theory and Applications. Edited by P.A. Devijver and J. Kittler. XI, 543 pages. 1987.

Vol. 31: Decision Support Systems: Theory and Application. Edited by C.W. Holsapple and A.B. Whinston. X, 500 pages. 1987.

Vol. 32: Information Systems: Failure Analysis. Edited by J.A. Wise and A. Debons. XV, 338 pages. 1987.

Vol. 33: Machine Intelligence and Knowledge Engineering for Robotic Applications. Edited by A.K.C. Wong and A. Pugh. XIV, 486 pages. 1987. *(ROB)*

Vol. 34: Modelling, Robustness and Sensitivity Reduction in Control Systems. Edited by R.F. Curtain. IX, 492 pages. 1987.

Vol. 35: Expert Judgment and Expert Systems. Edited by J.L. Mumpower, L.D. Phillips, O. Renn and V.R.R. Uppuluri. VIII, 361 pages. 1987.

Vol. 36: Logic of Programming and Calculi of Discrete Design. Edited by M. Broy. VII, 415 pages. 1987.

Vol. 37: Dynamics of Infinite Dimensional Systems. Edited by S.-N. Chow and J.K. Hale. IX, 514 pages. 1987.

Vol. 38: Flow Control of Congested Networks. Edited by A.R. Odoni, L. Bianco and G. Szegö. XII, 355 pages. 1987.

Vol. 39: Mathematics and Computer Science in Medical Imaging. Edited by M.A. Viergever and A. Todd-Pokropek. VIII, 546 pages. 1988.

Vol. 40: Theoretical Foundations of Computer Graphics and CAD. Edited by R.A. Earnshaw. XX, 1246 pages. 1988.

Vol. 41: Neural Computers. Edited by R. Eckmiller and Ch. v. d. Malsburg. XIII, 566 pages. 1988.

NATO ASI Series F

Vol. 42: Real-Time Object Measurement and Classification. Edited by A. K. Jain. VIII, 407 pages. 1988. *(ROB)*

Vol. 43: Sensors and Sensory Systems for Advanced Robots. Edited by P. Dario. XI, 597 pages. 1988. *(ROB)*

Vol. 44: Signal Processing and Pattern Recognition in Nondestructive Evaluation of Materials. Edited by C.H. Chen. VIII, 344 pages. 1988. *(ROB)*

Vol. 45: Syntactic and Structural Pattern Recognition. Edited by G. Ferraté, T. Pavlidis, A. Sanfeliu and H. Bunke. XVI, 467 pages. 1988. *(ROB)*

Vol. 46: Recent Advances in Speech Understanding and Dialog Systems. Edited by H. Niemann, M. Lang and G. Sagerer. X, 521 pages. 1988.

Vol. 47: Advanced Computing Concepts and Techniques in Control Engineering. Edited by M.J. Denham and A.J. Laub. XI, 518 pages. 1988.

Vol. 48: Mathematical Models for Decision Support. Edited by G. Mitra. IX, 762 pages. 1988.

Vol. 49: Computer Integrated Manufacturing. Edited by I. B. Turksen. VIII, 568 pages. 1988.

Vol. 50: CAD Based Programming for Sensory Robots. Edited by B. Ravani. IX, 565 pages. 1988. *(ROB)*

Vol. 51: Algorithms and Model Formulations in Mathematical Programming. Edited by S.W. Wallace. IX, 190 pages. 1989.

Vol. 52: Sensor Devices and Systems for Robotics. Edited by A. Casals. IX, 362 pages. 1989. *(ROB)*

Vol. 53: Advanced Information Technologies for Industrial Material Flow Systems. Edited by S.Y. Nof and C.L. Moodie. IX, 710 pages. 1989.

Vol. 54: A Reappraisal of the Efficiency of Financial Markets. Edited by R.M.C. Guimarães, B.G. Kingsman and S.J. Taylor. X, 804 pages. 1989.

Vol. 55: Constructive Methods in Computing Science. Edited by M. Broy. VII, 478 pages. 1989.

Vol. 56: Multiple Criteria Decision Making and Risk Analysis Using Microcomputers. Edited by B. Karpak and S. Zionts. VII, 399 pages. 1989.

Vol. 57: Kinematics and Dynamic Issues in Sensor Based Control. Edited by G. E. Taylor. XI, 456 pages. 1990. *(ROB)*

Vol. 58: Highly Redundant Sensing in Robotic Systems. Edited by J.T. Tou and J.G. Balchen. X, 322 pages. 1990. *(ROB)*

Vol. 59: Superconducting Electronics. Edited by H. Weinstock and M. Nisenoff. X, 441 pages. 1989.

Vol. 60: 3D Imaging in Medicine. Algorithms, Systems, Applications. Edited by K.H. Höhne, H. Fuchs and S.M. Pizer. IX, 460 pages. 1990.

Vol. 61: Knowledge, Data and Computer-Assisted Decisions. Edited by M. Schader and W. Gaul. VIII, 421 pages. 1990.

Vol. 62: Supercomputing. Edited by J.S. Kowalik. X, 425 pages. 1990.

Vol. 63: Traditional and Non-Traditional Robotic Sensors. Edited by T.C. Henderson. VIII, 468 pages. 1990. *(ROB)*

Vol. 64: Sensory Robotics for the Handling of Limp Materials. Edited by P.M. Taylor. IX, 343 pages. 1990. *(ROB)*

Vol. 65: Mapping and Spatial Modelling for Navigation. Edited by L.F. Pau. VIII, 357 pages. 1990. *(ROB)*

NATO ASI Series F

Vol. 66: Sensor-Based Robots: Algorithms and Architectures. Edited by C.S.G. Lee. X, 285 pages. 1990. *(ROB)*

Vol. 67: Designing Hypermedia for Learning. Edited by D.H. Jonassen and H. Mandl. XXV, 457 pages. 1990. *(AET)*

Vol. 68: Neurocomputing. Algorithms, Architectures and Applications. Edited by F. Fogelman Soulié and J. Hérault. XI, 455 pages. 1990.

Vol. 69: Real-Time Integration Methods for Mechanical System Simulation. Edited by E.J. Haug and R.C. Deyo. VIII, 352 pages. 1990.

Vol. 70: Numerical Linear Algebra. Digital Signal Processing and Parallel Algorithms. Edited by G. Golub and P. Van Dooren. XIII, 729 pages. 1991.

Vol. 71: Expert Systems and Robotics. Edited by T. Jordanides and B. Torby. XII, 744 pages. 1991.

Vol. 72: High-Capacity Local and Metropolitan Area Networks. Architecture and Performance Issues. Edited by G. Pujolle. X, 536 pages. 1991.

Vol. 73: Automation and Systems Issues in Air Traffic Control. Edited by J.A. Wise, V.D. Hopkin and M.L. Smith. XIX, 589 pages. 1991.

Vol. 74: Picture Archiving and Communication Systems (PACS) in Medicine. Edited by H.K. Huang, O. Ratib, A.R. Bakker and G. Witte. XII, 445 pages. 1991.

Vol. 75: Speech Recognition and Understanding. Recent Advances, Trends and Applications. Edited by P. Laface and R. De Mori. 1991.

Vol. 76: Multisensor Fusion for Computer Vision. Edited by J.K. Aggarwal. 1991. *(ROB)*

Vol. 77: Computer Algorithms for Solving Linear Algebraic Equations: The State of the Art. Edited by E. Spedicato. VIII, 352 pages. 1991.

Vol. 78: Integrating Advanced Technology into Technology Education. Edited by M. Hacker, A. Gordon, M. de Vries. VIII, 185 pages. 1991. *(AET)*

Vol. 79: Logic, Algebra, and Computation. Edited by F.L. Bauer. VII, 485 pages. 1991.

MIX
Papier aus verantwortungsvollen Quellen
Paper from responsible sources
FSC® C105338
FSC
www.fsc.org

If you have any concerns about our products,
you can contact us on
ProductSafety@springernature.com

In case Publisher is established outside the EU,
the EU authorized representative is:
Springer Nature Customer Service Center GmbH
Europaplatz 3, 69115 Heidelberg, Germany

Printed by Libri Plureos GmbH
in Hamburg, Germany